W9-CER-343

THERMAL RADIATIVE TRANSFER AND PROPERTIES

THERMAL RADIATIVE TRANSFER AND PROPERTIES

M. QUINN BREWSTER
Department of Mechanical and Industrial Engineering
University of Illinois,
Urbana, Illinois

A Wiley-Interscience Publication
JOHN WILEY & SONS, INC.
New York / Chichester / Brisbane / Toronto / Singapore

In recognition of the importance of preserving what has been written, it is a policy of John Wiley & Sons, Inc., to have books of enduring value published in the United States printed on acid-free paper, and we exert our best efforts to that end.

Library of Congress Cataloging in Publication Data:

Brewster, M. Quinn.

Thermal radiative transfer and properties/M. Quinn Brewster.

p. cm.

"A Wiley-Interscience publication."

Includes bibliographical references.

1. Heat—Radiation and absorption. I. Title.

TJ260.B74 1991

621.402′2—dc 20 91-12888

ISBN 0-471-53982-1 (alk. paper) CIP

Printed and bound in the United States of America by Braun-Brumfield, Inc.

10 9 8 7 6 5 4 3 2 1

To Linda
and
to Kenji, Jonathan, and Kylee

"In the beginning God created the heaven and the earth. And the earth was without form, and void; and darkness was upon the face of the deep. And the Spirit of God moved upon the face of the waters."

"And God said, Let there be light: and there was light."

Genesis 1: 1 – 3

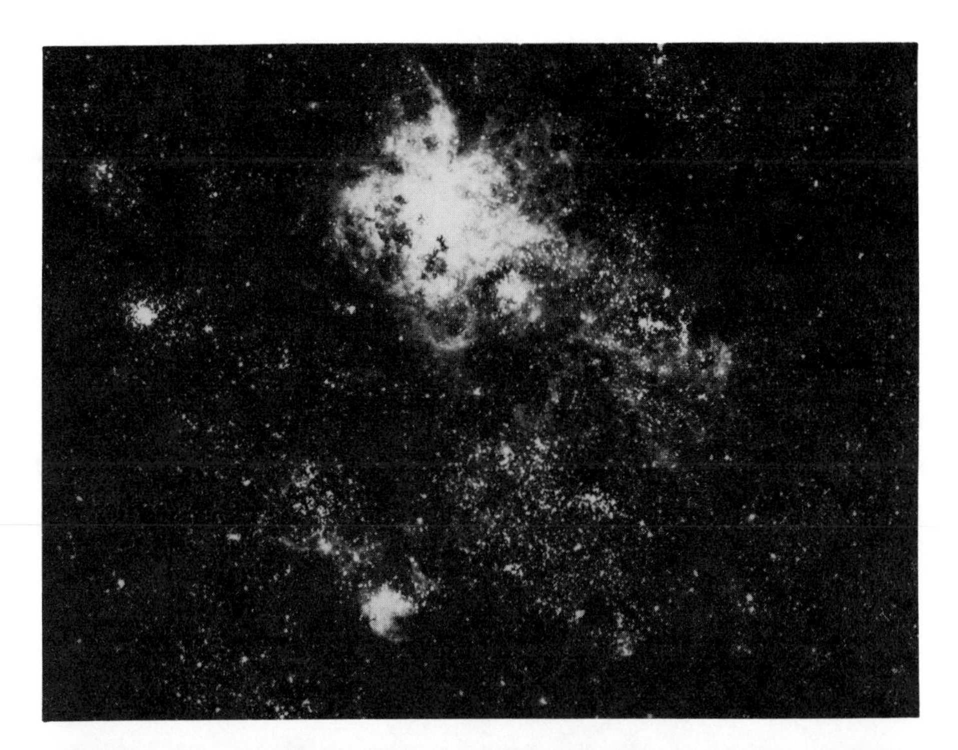

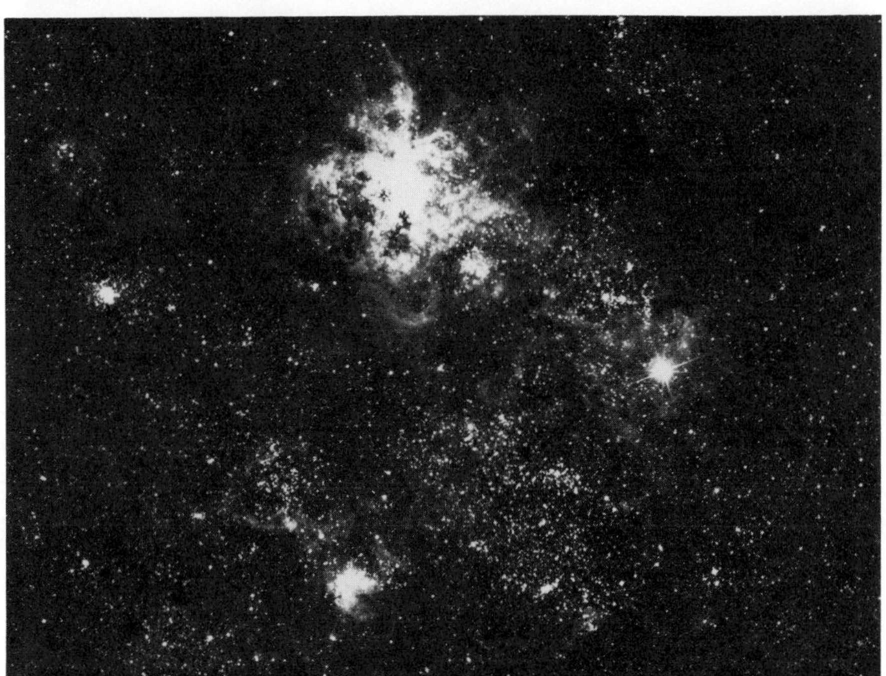

Before and after images of Supernova 1987A in the Large Magellanic Cloud. The top image was taken in 1969 by V. Blanco and the bottom image was taken February 26, 1987 by W. Roberts. (Used by permission of National Optical Astronomy Observatories.)

PREFACE

Radiative transfer is one of the most fundamental and pervasive processes in the universe. According to both the Bible (see Genesis 1:1–3) and scientific theory (the "big bang"), a tremendous production of light and radiative energy was associated with the beginning of the universe as we know it.

In this day and age, radiative transfer is still one of the most fundamental and pervasive processes on our planet. To a greater or lesser degree, radiative transfer plays a role in every natural and man-made system on earth. From global warming to optical computing, almost every emerging technological challenge will require scientists and engineers to have a good working knowledge of radiative transfer and properties.

This book is an attempt to help students gain confidence in their ability to include radiation as part of their thermal design and analysis. It is also an attempt to help students gain a greater appreciation for radiative transfer processes in engineering systems as well as in nature.

There are two important aspects of thermal radiation. One is the transfer aspect and the other is the properties aspect. One aspect is not complete without the other. This is the underlying philosophy of this book; hence the title, *Thermal Radiative Transfer and Properties*. Five chapters (2, 4, 5, 8, and 9) emphasize radiative properties of surfaces (2, 4, and 5) and participating media (8 and 9). The other chapters emphasize transfer between surfaces (3 and 6) and in participating media (10 through 14).

This book was originally designed as a text for a one-semester graduate course on radiation heat transfer. It can also be used at the undergraduate level. Chapters 1, 2, and 3 were written so that they stand alone and can be used as part of an undergraduate course covering introductory radiative heat transfer (surface transfer).

There are two unique features of this book that distinguish it from other books currently in the marketplace. One is the presentation of dispersion theory (Chapter 5). Dispersion theory is the fundamental basis for the radiative behavior of all materials. The discussion of classical dispersion theory in Chapter 5 lays a foundation for the discussion of radiative properties that appears throughout the rest of the book. The other unique feature of this book is the treatment of particle radiative properties (Chapter 9). The importance of particle radiative properties has been increasingly recognized in the field of radiative heat transfer. Therefore a detailed description of particle radiative properties has been included in Chapter 9, including real particle size distribution effects.

A conscious effort has been made to include illustrative examples that are both realistic and instructive. Similarly, an effort has been made to develop homework problems that are also realistic, instructive, and of current interest. Homework exercises and illustrative examples include problems in satellite thermal control, planetary heat transfer, atmospheric radiation, and radiation in industrial and propulsion combustion systems.

Many people have contributed to the development of this book. The assistance of many students over the years is greatly appreciated as are the insightful comments of Professors E. M. Sparrow, D. K. Edwards, and M. I. Flik, who reviewed the manuscript. I am also indebted to my Ph.D. advisor Professor C-L. Tien for his guidance and enthusiasm and to Professors T. Makino and R. O. Buckius for many enlightening discussions on thermal radiative transfer and properties.

QUINN BREWSTER

CONTENTS

APPENDIXES

SYMBOLS

Numbers in parentheses refer to equation numbers.

A	surface area, effective bandwidth
a	soot total optical depth parameter in (10-55)
a_n	Mie coefficients (9-42)
$\mathbf{B}$	magnetic induction vector
B	semi-isotropic back-scattering fraction, molecular rotation constant
b	self-broadening parameter in (8-69)
b_n	Mie coefficients (9-43)
C	optically thick mean beam length correction factor, flow constant (13-73)
$C_{e,s,a}$	extinction, scattering, or absorption cross section
C_p	specific heat
$C_{1,2}$	blackbody radiation constants
C_3	Wien's displacement law constant
c	speed of light in nonvacuum, interparticle clearance
c_0	speed of light in vacuum
$\mathbf{D}$	electric displacement vector
D	sphere diameter
d	particle diameter, line spacing
$\mathbf{E}$	electric vector
E	energy
E_n	exponential integral function of order n
e	electron charge
e_b	blackbody hemispherical flux
$\mathbf{e}$	unit vector
F	force, nondimensional stream function (13-61)
F_{ij}	view factor

f	blackbody fractional function, oscillator strength, function of optical constants defined by (10-52)
f_i	internal fractional function
f_v	particle volume fraction
$f_{1,2}$	functions of optical constants defined in (9-12) and (9-13)
G	particle projected area
g	multiple scattering asymmetry parameter, statistical degeneracy, gravitational constant
$\mathbf{H}$	magnetic vector
h	Planck's constant, heat transfer coefficient, film thickness, enthalpy
I	intensity, molecular moment of inertia
i	electric current
J	source function, Bessel function
$\mathbf{j}$	free charge current density vector
j	rotational quantum number
$K_{e,s,a}$	extinction, scattering, or absorption coefficient
k	absorption index, thermal conductivity
k_B	Boltzmann's constant
L	slab thickness, characteristic length scale for intensity change
L_m	mean beam length
L_{m_0}	geometric mean beam length
LTE	local thermodynamic equilibrium
$\mathbf{M}$	magnetization vector
m	mass of particle, wedge angle parameter
MFP	mean free photon path length
$\mathbf{N}$	normal vector
N	particle number density, number of bundles emitted per unit time, conduction-radiation parameter
Nu	Nusselt number
n	refractive index, parameter (8-69)
$\mathbf{P}$	electric polarization vector
P	probability function, pressure, perimeter
P_n	Legendre polynomial
P_n^1	associated Legendre function
Pr	Prandtl number
Pr_m	radiation-modified Prandtl number
p	scattering phase function
$\langle p \rangle$	single-scattering asymmetry parameter
p^*	multiple scattering asymmetry parameter for radiation diffusion
Q	heat transfer rate
$Q_{e,s,a}$	extinction, scattering, or absorption efficiency
q	heat flux
q'''	volumetric energy source/sink

R	cumulative probability function, random number, sphere radius
R_i, R_{ij}	network resistance factor
Ra	Rayleigh number
Re	Reynolds number
$\Re$	Fresnel reflectivity of optically smooth, plane interface or layer
$\mathbf{r}$	position vector
r	particle or charge separation distance, Fresnel reflectance coefficient, particle radius
r_e	electrical resistivity or effective separation distance
$\mathbf{S}$	Poynting vector
S	integrated line intensity, number of bundles absorbed per unit time
$S_{1,2}$	Mie intensity coefficients (9-50) and (9-51)
s	path length, standard deviation
T	temperature, internal transmittance
$\mathscr{T}$	Fresnel transmissivity of optically smooth, plane interface or layer
t	optical depth, time, Fresnel transmittance coefficient
u	gas band optical depth parameter (optical depth at band head or band center), radiative energy density, parameter defined in (8-73)
u, v	velocity components, complex variable components defined in (4-66)
U	bulk velocity
$\mathbf{V}$	velocity vector
V	volume, eigenvector
v	vibrational quantum number
w	energy per bundle, Gaussian quadrature weight factors
X	mass path length
x	particle size parameter
x, y, z	spatial coordinates
α	absorptivity, thermal diffusivity, integrated band intensity
β	line overlap parameter, phase shift parameter (4-80), parameter defined in (9-37), volumetric thermal expansion coefficient, wedge angle over π, local surface polar angle in Chap. 9
χ	electric susceptibility, matrix defined in (3-69)
Δ	line half-width, small increment
Δq	partial (directional) flux
δ	incremental vibrational quantum number, Dirac delta, particle center-to-center spacing, hydrodynamic boundary layer length scale

δ_T	thermal boundary layer length scale
δ_{ij}	Kronecker delta
ε	emissivity, permittivity
ε_r	relative permittivity (dielectric constant)
ε_m	turbulent momentum diffusivity
ε_H	turbulent thermal (heat) diffusivity
Φ	dissipation function, matrix defined in (3-66), statistical mechanical function in (8-76)
ϕ	azimuthal angle
Γ	gamma function
γ	damping frequency, line width, line broading parameter
η	wavenumber, temperature ratio parameter, similarity variable (13-60)
κ	mass absorption coefficient
Λ	matrix defined in (3-67)
λ	wavelength, eigenvalue
μ	magnetic permeability, cos θ, discrete ordinates
ν	frequency, kinematic viscosity
π_n	Mie angular coefficients (9-52)
θ	polar angle
ρ	reflectivity, density
ρ_e	electric charge density
σ	Stefan-Boltzmann constant or surface roughness
σ_e	electrical conductivity
τ	transmissivity
τ_n	Mie angular coefficients (9-53)
Ω	solid angle or direction
ω	circular frequency, bandwidth parameter
ω_0	single-scattering albedo
ξ	Ricatti-Bessel function
Ψ	statistical mechanical function in (8-75)
ψ	slip coefficient, matrix defined in (3-68), Riccatti-Bessel function
ζ	size distribution width parameter in (9-85), parameter in (12-70), stream function

Superscripts

$'$	directional quantity, real component of complex quantity, dummy variable of integration, upper energy state, single derivative, or fluctuating component
$''$	bi-directional quantity, imaginary component of complex quantity, dummy variable of integration, lower energy state, double derivative
$'''$	volumetric quantity, or triple derivative

$\bar{}$	average quantity over wavelength, direction, particle size, distance, or time
$\tilde{}$	complex or nondimensional quantity
$+$	forward direction
$-$	backward direction
$*$	dummy variable of integration, known view factor, nondimensional quantity, complex conjugate

Subscripts

a	absorbed, absorbing species
avg	average
B	Brewster angle, Boltzmann
b	blackbody
c	critical or cutoff, conduction, convection, collimated
D	Debye
d	diffuse component
dc	direct current (zero frequency)
e	emitted or extinction, environment, electric or electronic, effective, equivalent
eff	effective
em	emission
g	gas
h	homogeneous solution, hydraulic (diameter)
I	isotropic
i, j, k	species, line, band, or element
L	lower limit
m	maximum (12-102) and (12-103)
mp	most probable
n	normal direction
nb	narrow band
0	vacuum or reference condition, center value
P	Planck mean value
p	particle geometry, plasma, particular solution
R	Rosseland mean value
r	reflected, radiation, reference, relative
r, rot	rotational
rms	root-mean-square
s	specular or scattered component, slab geometry, surface
T	thermal
t	transmitted
U	upper limit
v, vib	vibrational
WL	wall layer
w	wall

wb	wide band
∞	freestream condition
$\perp$	perpendicular polarization or direction
$\parallel$	parallel polarization or direction
λ	spectrally (wavelength) dependent
η	spectrally (wavenumber) dependent
32	32 moment (Sauter mean), (9-64)
63	63 moment, (9-80)

THERMAL RADIATIVE TRANSFER AND PROPERTIES

CHAPTER 1

THE NATURE OF THERMAL RADIATION

Thermal radiation is electromagnetic radiation emitted by particles of matter (molecules, atoms, ions, and electrons) as they undergo internal energy state transitions. The radiative energy produced by these transitions is usually in the ultraviolet, visible, and infrared portions of the electromagnetic spectrum. Like all forms of electromagnetic radiation, thermal radiation travels at the speed of light.

Thermal radiation is also a form of heat. *Heat*, defined as thermal energy transfer from one body of matter to another due to a temperature difference, appears in two fundamental forms: conduction and radiation.* The fundamental mechanism of energy transport in conduction is direct exchange of kinetic energy between particles of matter. In radiation, the fundamental mechanism of energy transport is by electromagnetic waves (or photons) that are emitted and absorbed by the particles of matter as they undergo energy state transitions.

Like conduction, thermal radiation is in harmony with the second law of thermodynamics such that, in the absence of work, thermal energy is radiated spontaneously from higher temperature matter to lower temperature matter. Unlike conduction, however, which requires a material path, radiative transfer can occur between two spatially remote bodies of matter at different temperatures even when the intervening space is a vacuum. The most prominent example of thermal radiation on the earth is solar radiation, which is transported across the vacuum of space to the earth.

*Convection or advection is thermal energy transport associated with bulk fluid motion and as such is not a form of heat.

Thermal radiation plays a critical role in the operation of many natural and man-made systems. Thermal radiation in the earth's atmosphere determines the temperature on the surface of our planet and ensures the possibility of life here. Thermal radiation is also important in satellite and other space systems because of the absence of convective heat transfer with a surrounding atmosphere. Space systems are subject to severe radiant heating and cooling conditions as they are exposed to both solar radiation and the near-zero kelvin blackbody of deep space. In terrestrial energy systems thermal radiation is also important. For example, thermal radiation is a significant heat transfer mechanism in fossil-fuel-fired utility boilers and in both thermal and photovoltaic solar energy systems. In new materials processing methods, such as laser materials processing, as well as in conventional methods, thermal radiation is also of fundamental importance.

INTENSITY

Electromagnetic energy propagation can be described at various levels of complexity, depending on the features of the radiation field to be described. For example, if polarization is important, the Stokes parameters [Born and Wolf, 1970; Bohren and Huffman, 1983] are used to describe the intensity field. If wave effects are important (e.g., if the characteristic length scale for transport, such as spacing between surfaces, is on the order of the wavelength), the vector wave quantities from electromagnetic theory such as electric vector, magnetic vector, and Poynting vector must be used. For calculating heat transfer in most situations, however, this level of description is overly complicated.

For most heat transfer applications thermal radiation can be treated as unpolarized* and incoherent.† Furthermore, the length scale for transport is usually much larger than the wavelength of radiation. Therefore, the limiting description of geometric optics ($\lambda \rightarrow 0$) can be applied wherein the waves are described as bundles of rays carrying energy in a small volume associated with the solid angle in the direction of the rays. The fundamental quantity used for describing unpolarized radiant energy transport in the limit of $\lambda \rightarrow 0$ is the *spectral intensity* I_λ, defined as the radiant power per unit area

*Unpolarized radiation can give rise to a polarized field upon reflection or scattering. This effect is discussed more in Chapters 3 and 9. For calculating heat transfer, however, multiple reflection and scattering usually nullify polarization effects and an unpolarized description is adequate.

†Coherence refers to waves or photons being in phase. Radiation produced by stimulated emission (e.g., laser radiation) is coherent since the emitted photon has the same frequency and phase as the stimulating photon. Radiation produced by spontaneous emission is, by definition, incoherent, since no stimulating photon is involved in the process. For heat transfer purposes stimulated emission can be treated as negative absorption, and thus coherence effects can be ignored in the description of intensity.

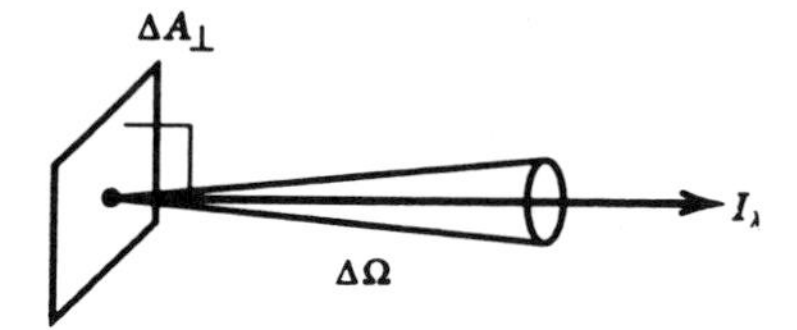

Figure 1-1 Definition of radiant intensity

perpendicular to the direction of travel per unit solid angle and per unit wavelength interval.

$$I_\lambda = \frac{\text{power}}{\Delta A_\perp \, \Delta\Omega \, \Delta\lambda} = \frac{\text{W}}{\text{m}^2 \text{ sr } \mu\text{m}} \tag{1-1}$$

Defined in this way, the intensity is a conserved quantity for radiation propagating in a nonattenuating medium. That is, intensity is constant with distance (s) along any given direction of travel in a nonscattering, nonabsorbing medium.

$$\frac{dI_\lambda}{ds} = 0 \qquad I_\lambda = \text{constant} \tag{1-2}$$

The radiant intensity passing through a particular point in space (x, y, z) in a particular direction (θ, ϕ) at a particular wavelength (λ) is depicted in Fig. 1-1.

Consider the pupil of your eye to be a differential area perpendicular to the direction of travel of visible radiation $\Delta A_\perp$. As you hold your head in one position (x, y, z) and scan a distant scene, the variation of intensity with direction (θ, ϕ) is readily evident, especially as a bright source is contrasted to a dark surface, such as an outside window compared to a blackboard, or the white background of this page compared with the black printing. Spectral variations are evident in visual observations by the variety of colors, which are indicative of different wavelengths of light.

The solid angle Ω appearing in the definition of intensity is defined in a manner similar to the plane angle θ. *Plane angle* is defined as the arc length along the circumference of a circle of radius R divided by R and has the dimensionless unit of radians (rad). Similarly, *solid angle* is defined as the area on the surface of a sphere of radius R divided by R^2 and has the dimensionless unit of steradians (sr).

The total solid angle in a sphere is 4π and the solid angle in a hemisphere is 2π. For a pencil of rays confined to a differential solid angle $\Delta\Omega$ (as in the definition of intensity), a small solid angle approximation can be used that replaces the true spherical surface with the nearly equivalent planar surface (Fig. 1-2).

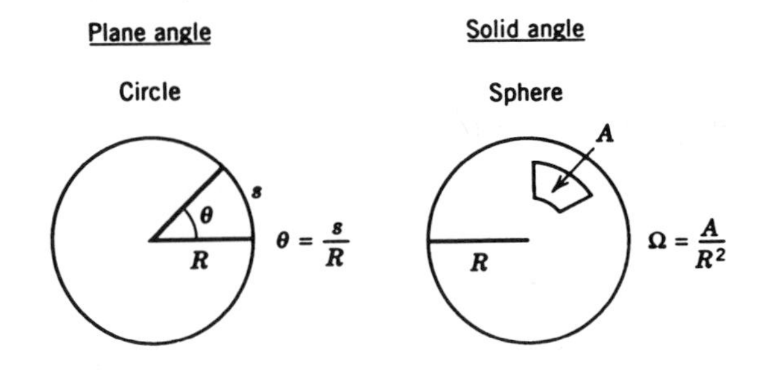

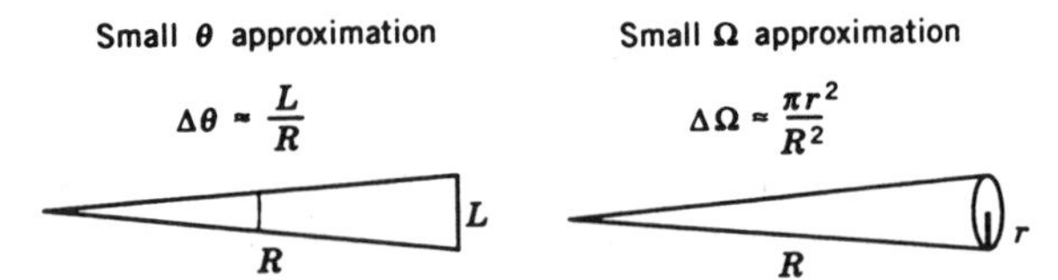

Figure 1-2 Definition of solid angle.

The direction of travel of intensity is established relative to a reference direction. The reference direction is the normal to a surface of interest ΔA as shown in Fig. 1-3. The reference area ΔA often corresponds to a material surface, as in the case of radiative absorption, emission, and reflection by a solid surface. However, ΔA may also correspond to a nonmaterial surface in space, as in the case of radiative transfer in a medium (either participating or nonparticipating). The coordinate system most widely used to describe the

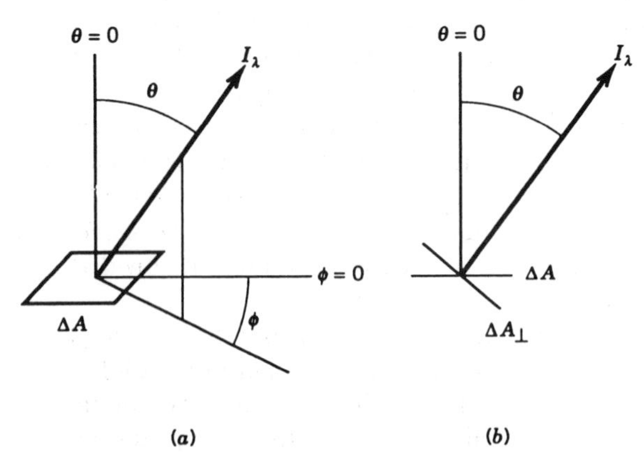

Figure 1-3 (a) Polar coordinate system and (b) ΔA vs. $\Delta A_\perp$.

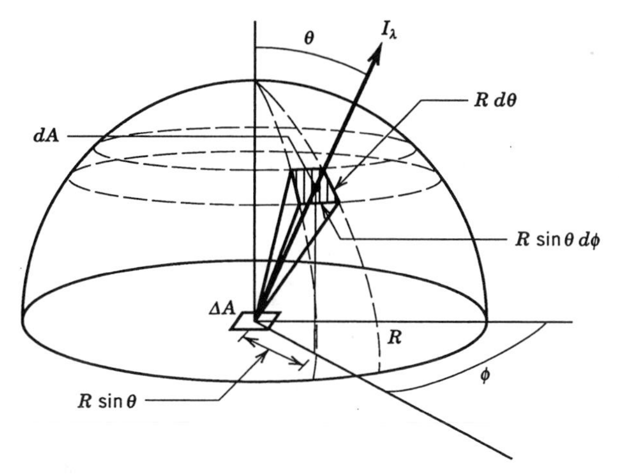

Figure 1-4 Solid angle in polar coordinates.

direction of travel of intensity is the spherical polar coordinate system. This coordinate system uses a polar angle θ and an azimuthal angle ϕ as shown in Fig. 1-3*a*. The reference area ΔA and the area appearing in the definition of intensity $\Delta A_\perp$ are related to each other by a factor of $\cos\theta$ as shown in Fig. 1-3*b* and Eq. (1-3).

$$\Delta A_\perp = \Delta A \cos\theta \tag{1-3}$$

The differential solid angle associated with the intensity in a given direction can be expressed in terms of the polar and azimuthal angles θ and ϕ as follows. With reference to Fig. 1-4, the area of a differential element dA on the surface of a hemisphere above ΔA is given by Eq. (1-4).

$$dA = R^2 \sin\theta \, d\theta \, d\phi \tag{1-4}$$

From the definition of solid angle (Fig. 1-2) the differential solid angle $d\Omega$ subtended by the differential area dA is

$$d\Omega = \sin\theta \, d\theta \, d\phi \tag{1-5}$$

Most of the perceived mathematical complexity usually associated with radiative transfer analysis arises from the directional (θ, ϕ) and spectral (λ) variations of intensity. Mathematically, the spectral intensity can be expressed as a function of spatial position (x, y, z) direction (θ, ϕ) and wavelength (λ).

$$I_\lambda = I_\lambda(x, y, z, \theta, \phi, \lambda) \tag{1-6}$$

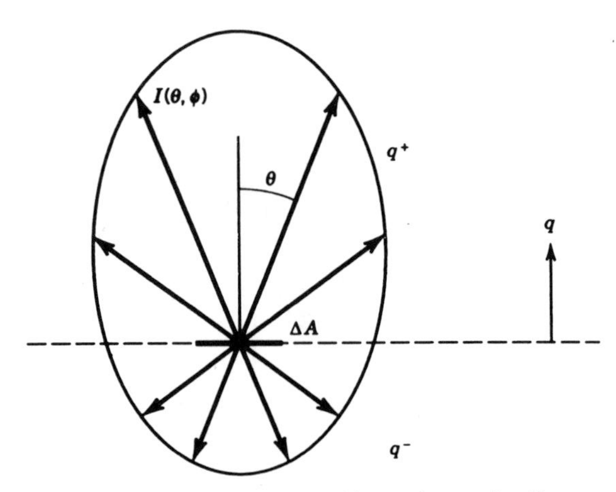

Figure 1-5 Integration of intensity to give flux.

The goal of radiative transfer theory is to determine I_λ as a function of these independent variables. With the spectral intensity known, the *total intensity* can be determined by integrating with respect to wavelength.

$$I(x, y, z, \theta, \phi) = \int_0^\infty I_\lambda(x, y, z, \theta, \phi, \lambda)\, d\lambda \tag{1-7}$$

Once the total intensity is known, the flux can be determined by integrating with respect to direction, as discussed in the next section.

FLUX

In order to incorporate radiant energy into a thermal energy balance, it is necessary to include the contribution from intensity over all directions. Consider Fig. 1-5, which shows a two-dimensional profile of an arbitrary intensity distribution at a point in space. The reference area ΔA is an elemental area on a surface of interest (such as a heat transfer surface on the outside of a tube). The component of the *net flux* through ΔA is the radiant power per unit reference area that crosses ΔA.

$$q_\lambda = \frac{\text{power}}{\Delta A\, \Delta\lambda} \quad \left(\frac{\text{W}}{\text{m}^2\, \mu\text{m}}\right) \quad \text{(spectral net flux)} \tag{1-8}$$

$$q = \int_0^\infty q_\lambda\, d\lambda = \frac{\text{power}}{\Delta A} \quad \left(\frac{\text{W}}{\text{m}^2}\right) \quad \text{(total net flux)} \tag{1-9}$$

The *spectral net flux* q_λ is based on a unit wavelength interval and *total net flux* q is integrated over all wavelengths.

The component of the net flux through ΔA is obtained by integrating the intensity field over all directions (i.e., over 4π steradians) with a weight factor of $\cos\theta$ included to convert from a unit area basis of $\Delta A_\perp$ (which appears in the definition of intensity) to a common unit area basis of ΔA. The integration can be performed over all 4π steradians at once

$$q = \int_{4\pi} I \cos\theta \, d\Omega \tag{1-10}$$

$$q = \int_0^{2\pi} \int_0^{\pi} I(\theta, \phi) \cos\theta \sin\theta \, d\theta \, d\phi \tag{1-11}$$

or broken up into forward and backward hemispheres as follows:

$$q^+ = \int_0^{2\pi} \int_0^{\pi/2} I(\theta, \phi) \cos\theta \sin\theta \, d\theta \, d\phi \tag{1-12}$$

$$q^- = \int_0^{2\pi} \int_{\pi}^{\pi/2} I(\theta, \phi) \cos\theta \sin\theta \, d\theta \, d\phi \tag{1-13}$$

$$q = q^+ - q^- \tag{1-14}$$

The quantities q^+ and q^- are referred to as *hemispherical fluxes* since only the rays from a hemisphere of 2π steradians have been included. In this manner the intensity in all directions is normalized to a common unit area ΔA and added up over all directions (or solid angles) to give the component of the net flux in the direction perpendicular to ΔA. When the area ΔA coincides with an actual material surface (as in the case of surface interchange analysis), by convention, the $\theta = 0$ direction is taken to be normal away from the surface. In that case the hemispherical flux leaving the surface q^+ is referred to as the *radiosity* and the hemispherical flux incident upon the surface q^- is referred to as the *irradiation*.

The foregoing equations are also valid on a spectral basis. Since wavelength and direction are independent variables, the order of integration with respect to wavelength and solid angle is interchangeable.

The component of the net flux associated with intensity traveling in a small solid angle $\Delta\Omega$ is referred to as the *partial flux* Δq (Fig. 1-6). The partial flux is obtained by integrating intensity over a narrow solid angle $\Delta\Omega$. If the intensity and $\cos\theta$ are constant over $\Delta\Omega$, this integration reduces to

$$\Delta q = \int_{\Delta\Omega} I \cos\theta \, d\Omega = I \cos\theta \, \Delta\Omega \quad (I, \cos\theta \text{ constant over } \Delta\Omega) \tag{1-15}$$

A special case of significant practical interest is the case of an *isotropic* or *diffuse* intensity field (Fig. 1-7). This is the case when the intensity is

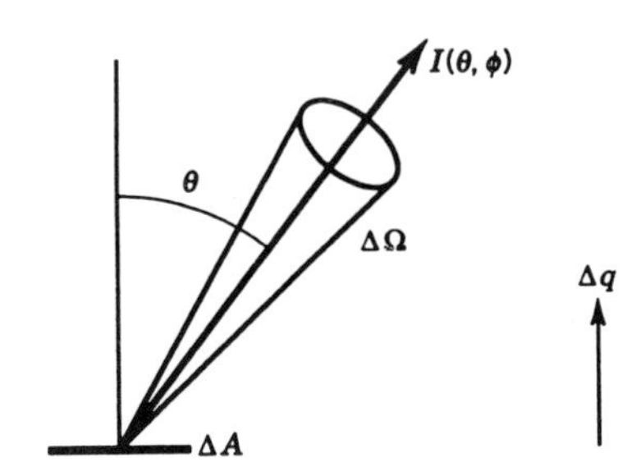

Figure 1-6 Partial flux.

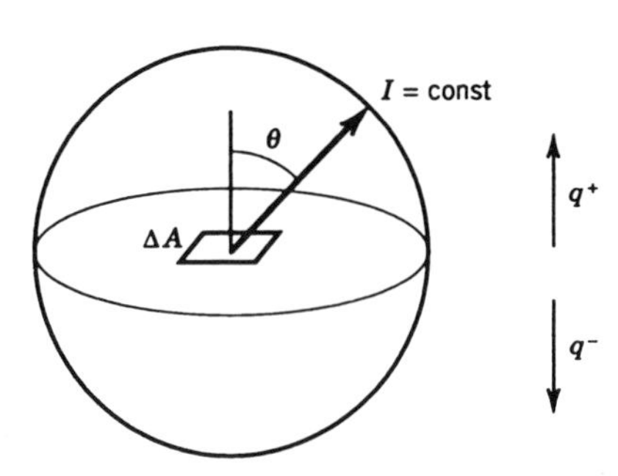

Figure 1-7 Isotropic intensity distribution.

independent of direction, $I_\lambda \neq I_\lambda(\theta, \phi)$. In this case the integration of intensity over direction can be readily performed. The result is that the hemispherical fluxes q^+ and q^- are related to the magnitude of the (constant) intensity by a simple factor of π (see Problem 1-10).

$$q^+ = q^- = \pi I \quad \text{(isotropic intensity)} \tag{1-16}$$

Since the forward and backward hemispherical fluxes are equal, the net flux is zero ($q = 0$) at a point in space where the intensity is isotropic. Often, a radiation field is assumed to be semi-isotropic. A semi-isotropic radiation field has a constant intensity distribution in each hemisphere,

$$q^+ = \pi I^+ \quad \text{(semi-isotropic intensity)} \tag{1-17}$$

$$q^- = \pi I^- \quad \text{(semi-isotropic intensity)} \tag{1-18}$$

but the net flux $\pi(I^+ - I^-)$ is not necessarily zero. This assumption is commonly made in evaluating the net radiative heat flux at a surface that is exchanging thermal radiative energy with its surroundings.

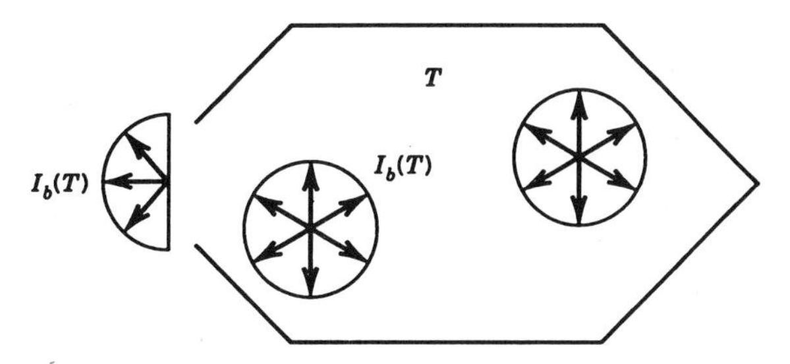

Figure 1-8 Hohlraum or blackbody (isothermal enclosure).

BLACKBODY RADIATION

Blackbody radiation is a special type of thermal radiation. Blackbody radiation is thermal radiation that exists inside an isothermal enclosure. By definition, a blackbody is a hohlraum or an isothermal enclosure (Fig. 1-8).

Blackbody radiation is isotropic. At any point inside an isothermal enclosure the magnitude of the intensity is equal in all directions. Furthermore, the magnitude of the intensity is the same at every point in space inside the enclosure. Thus the hemispherical intensity distribution emerging from a small hole in the enclosure is semi-isotropic.

The magnitude and spectral distribution of blackbody radiation are given by the Planck function. In 1901 Max Planck showed [Planck, 1959] from thermodynamics and the newly developed quantum theory that the *spectral blackbody intensity* I_{b_λ} is a function of only wavelength and temperature as given by the following relation*:

$$I_{b_\lambda}(\lambda, T) = \frac{C_1/\pi}{\lambda^5(\exp[C_2/(\lambda T)] - 1)} \tag{1-19}$$

$$C_1 = 2\pi h c_0^2 = 37{,}413 \ \frac{\mathrm{W}\,\mu\mathrm{m}^4}{\mathrm{cm}^2} \tag{1-20}$$

$$C_2 = \frac{h c_0}{k_B} = 14{,}388 \ \mu\mathrm{m\ K} \tag{1-21}$$

Since blackbody radiation is isotropic, the *spectral hemispherical blackbody flux* e_{b_λ} in any direction is just a factor of π times the intensity.

$$e_{b_\lambda} = \pi I_{b_\lambda} = \frac{C_1}{\lambda^5(\exp[C_2/(\lambda T)] - 1)} \tag{1-22}$$

*Here it is assumed that the refractive index of the medium enclosed in and surrounding the isothermal cavity is one, which is true exactly for a vacuum and is a good approximation for gases.

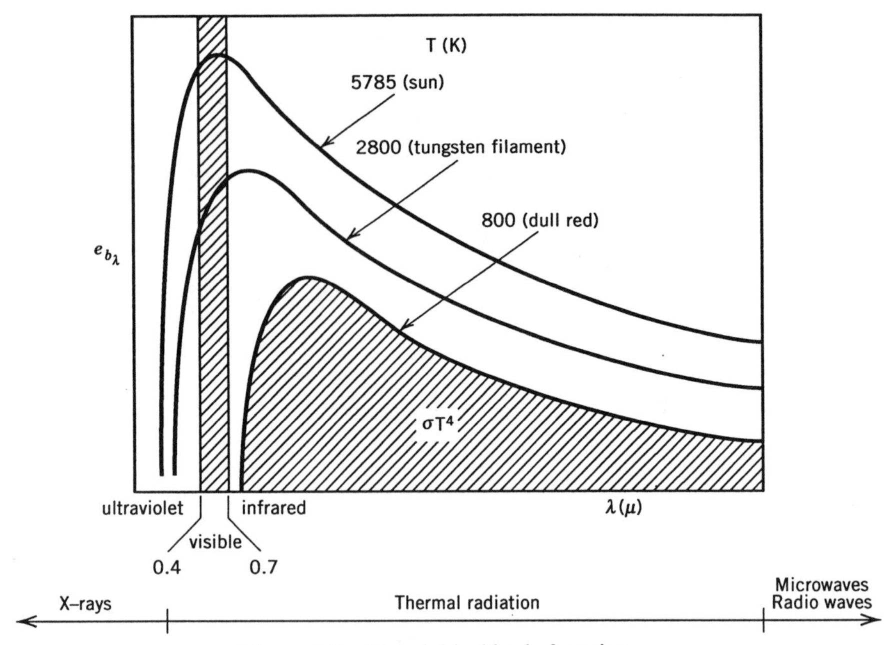

Figure 1-9 Planck blackbody function.

A brief table of e_{b_λ} is given in Appendix A. A qualitative plot of the Planck function versus wavelength for different temperatures is given in Fig. 1-9.

Figure 1-9 shows that at small wavelengths the magnitude of e_{b_λ} is relatively small and as wavelength increases, e_{b_λ} increases exponentially. This is because, for a given temperature, at small enough wavelengths, the $\exp[C_2/(\lambda T)]$ term in the denominator of the Planck function dominates the expression. At intermediate wavelengths the λ^5 term becomes comparable in magnitude to the $\exp[C_2/(\lambda T)]$ term, and as a result Fig. 1-9 shows that e_{b_λ} goes through a maximum. At long wavelengths the λ^5 term begins to dominate, and e_{b_λ} decreases as λ^{-4}.

Figure 1-9 also shows that the location of the maximum in the spectral energy distribution depends on temperature. At low temperatures most of the energy lies in the infrared region. As temperature increases the spectral energy distribution shifts more to the visible and ultraviolet regions. For example, at 800 K the Planck function is just beginning to enter the visible region. This temperature also corresponds to the point at which a heated object just begins to glow with a dull reddish color. As the temperature is increased to 2800 K (the temperature of a tungsten filament in an incandescent lamp), the proportion of visible energy has increased significantly, although most of the energy is still in the infrared region. At 5785 K (the effective blackbody temperature of the surface region of the sun*) the maximum spectral blackbody emission occurs in the middle of the visible region, with a significant portion in the ultraviolet region as well.

The location of the maximum spectral blackbody radiation can be calculated by differentiating the Planck function with respect to wavelength and setting the result equal to zero. This procedure gives a transcendental equation that can be solved for the wavelength temperature product.

$$(\lambda T)_{\max} = 2898\ \mu\text{m K} \tag{1-23}$$

This relation is called Wien's displacement law† and can be used to calculate the wavelength of maximum blackbody radiation for a given temperature. For example, the wavelength of maximum e_{b_λ} for a flame at 2000 K is 1.5 μm while that for a cryogenic insulation at 50 K is 58 μm. Figure 1-9 also shows that the region of the electromagnetic spectrum occupied by thermal radiation extends from the ultraviolet through the infrared region.

The *total hemispherical blackbody flux* e_b can be obtained by integrating the spectral flux over all wavelengths. Graphically, e_b corresponds to the area under the curve in Fig. 1-9.

$$e_b(T) = \int_0^\infty e_{b_\lambda}(\lambda, T)\, d\lambda = \sigma T^4 \tag{1-24}$$

The constant σ is the Stefan-Boltzmann constant. The numerical value can be obtained by substituting the Planck function and carrying out the integration.

$$\sigma = \int_0^\infty \frac{e_{b_\lambda}}{T^5}\, d(\lambda T) \tag{1-25}$$

$$\frac{e_{b_\lambda}}{T^5} = \frac{C_1}{(\lambda T)^5(\exp[C_2/(\lambda T)] - 1)} \tag{1-26}$$

$$\sigma = 5.67 \times 10^{-12}\ \frac{\text{W}}{\text{cm}^2\ \text{K}^4} \tag{1-27}$$

The *total blackbody intensity* I_b is obtained by integrating the spectral intensity or by dividing the hemispherical flux by π [recall Eq. (1-16) and the

*The solar flux incident on the earth outside the atmosphere is 1390 W/m^2, which corresponds to an effective blackbody temperature of 5785 K. At the surface of the earth the flux is somewhat less due to atmospheric absorption and scattering. On a clear day and for moderate (near normal) slant paths through the atmosphere, the actual incident flux at the ground is closer to 1190 W/m^2, which corresponds to an effective blackbody temperature of 5600 K.

†It should be noted that the constant in Wien's displacement law depends on the spectral variable chosen. A different value results if wavenumber or some other spectral variable is chosen as a basis for spectral intensity (see Problem 1-8).

fact that blackbody radiation is isotropic].

$$I_b(T) = \int_0^{\infty} I_{b_\lambda}(\lambda, T)\, d\lambda = \frac{\sigma T^4}{\pi} \qquad (1\text{-}28)$$

There are two limiting expressions for the Planck function that are worth mentioning for historical reasons alone, if not for the practical significance. In the limit of small wavelengths and/or low temperatures the term $\exp[C_2/(\lambda T)]$ in the Planck function is much larger than one and the following expression results, called the Wien's limit.

$$\frac{C_2}{\lambda T} \gg 1: \quad \frac{e_{b_\lambda}}{T^5} = \frac{C_1}{(\lambda T)^5 \exp[C_2/(\lambda T)]} \qquad (1\text{-}29)$$

In the opposite limit of long wavelengths and/or high temperatures the term $\exp[C_2/(\lambda T)]$ can be expanded in a Taylor series.

$$e^x = 1 + x + \cdots \qquad (1\text{-}30)$$

The resulting expression is known as the Rayleigh-Jeans limit.

$$\frac{C_2}{\lambda T} \ll 1: \quad \frac{e_{b_\lambda}}{T^5} = \frac{C_1}{C_2(\lambda T)^4} \qquad (1\text{-}31)$$

These two limiting expressions for spectral hemispherical blackbody flux e_{b_λ} are plotted in Fig. 1-10 along with the exact Planck function.

Historically, the two limiting expressions were developed by Willy Wien, Lord Rayleigh, and Sir James Jeans in the late 1800s using thermodynamic arguments, before Planck had formulated the quantum theory. Existing experimental data tended to fall along the line shown by the Planck function in Fig. 1-10, indicating that a discrepancy probably existed in the Wien formula at large values of λT, and in the Rayleigh-Jeans formula at small values of λT. The missing link was provided by Planck who first determined the mathematical formula necessary to fit the experimental data over the full range of λT and then formulated the hypotheses leading to the development of quantum theory, with which he was able to predict exactly the correct functional form of the blackbody function. This development turned out to be one of the crowning achievements of scientific progress in modern times. For more discussion of the historical development of radiation theory see Siegel and Howell [1988].

Blackbody Behavior of an Isothermal Cavity

A practical blackbody source can be built by constructing an enclosure that is as nearly isothermal as possible with a small hole in it for radiation to escape,

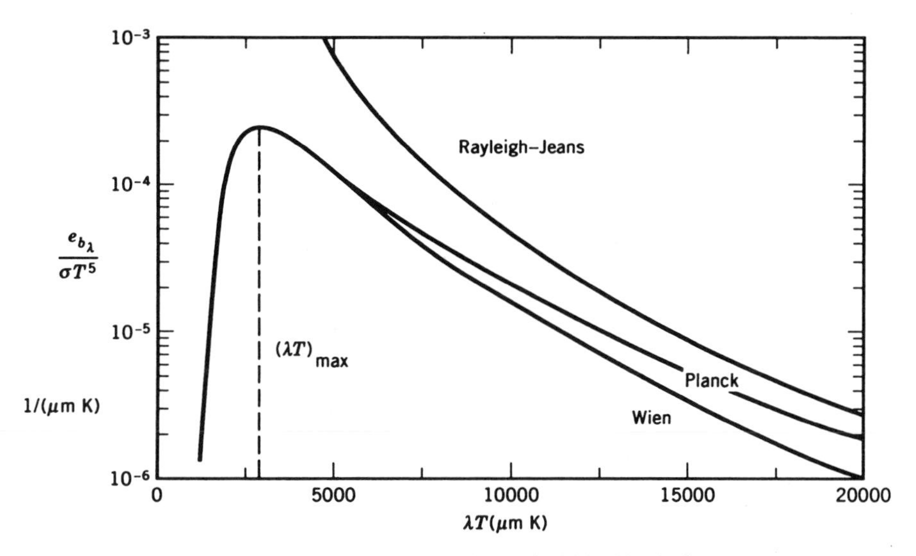

Figure 1-10 Spectral hemispherical blackbody flux.

as pictured in Fig. 1-8. The area of the hole should be small relative to the inside surface area of the enclosure.

One of the most critical requirements of a good blackbody is the isothermal condition. To approach isothermal performance requires careful thermal design of the enclosure, the heat source distribution, and the surrounding insulation. Ideally, the inside surface should be constructed of a material with a high thermal conductivity, such as a metal. Most metals with high thermal conductivity, however, have relatively low melting points. Higher temperature operation can be achieved with refractory materials (graphite, ceramics, etc.) at the expense of lower thermal conductivity. A low temperature blackbody (< 1300 K) with a metal cavity can achieve 0.1% of perfect blackbody performance (apparent emissivity of 0.999), if it is properly designed.

An interesting characteristic of blackbodies is that is it not necessary for the inside surface of the enclosure to be a black surface. Blackbody radiation is generated inside an isothermal enclosure, regardless of the emissivity of the surface, as long as the enclosure is isothermal. The reason the emissivity of the surface is unimportant is that reflection compensates for less than ideal emission. If a surface inside the enclosure emits less than the blackbody intensity due to its material properties, the difference is made up by reflection of radiation that is incident on the surface. In practice, however, it should be noted that since a small hole must be introduced into the enclosure to allow radiation to escape, the performance of the blackbody is always better, the closer the enclosure surface itself is to behaving like a black surface.

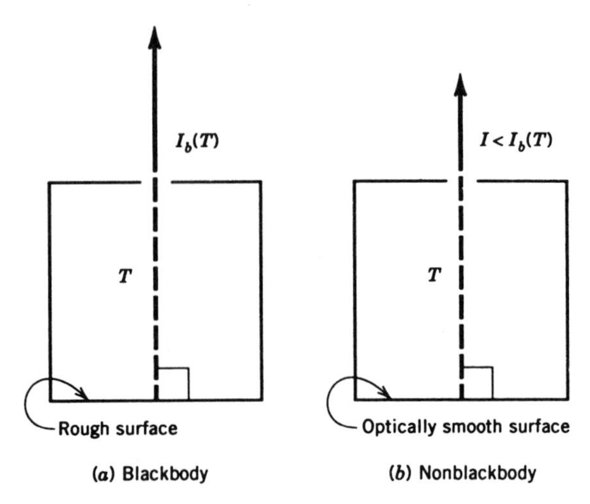

Figure 1-11 Two isothermal enclosures (*a*) blackbody and (*b*) nonblackbody.

Another interesting feature of an isothermal enclosure is that blackbody radiation exists inside the cavity regardless of the surface finish of the inside of the enclosure. An optically smooth, mirror-like surface finish will still produce isotropic radiation in the enclosure.

There is one practical problem, however, that arises when there is a small hole in the enclosure. This problem is illustrated by Fig. 1-11, which shows two enclosures that are both isothermal. One enclosure (a) is a blackbody and the other (b) is not. The difference is that enclosure (a) has a rough surface that reflects radiation diffusely and enclosure (b) has a smooth surface that reflects radiation in a specular or mirror-like fashion. Furthermore, the smooth surface in enclosure (b) can be viewed from outside the enclosure through the hole at an angle normal to the surface. The problem is that the intensity that emerges from enclosure (b) along the line of sight shown is only comprised of radiation emitted directly from the optically smooth surface, while the intensity emerging from enclosure (a) is made up of both emitted and reflected radiation. It is impossible for blackbody radiation streaming around inside enclosure (b) to be reflected from the smooth surface into the line of sight shown. Thus the intensity leaving enclosure (b) in the direction shown is less than the blackbody intensity. This problem can be avoided by making sure that the surface viewed through the hole is either not optically smooth (as in Fig. 1-11*a*) or not normal to the line of sight (as in Fig. 1-8), or both.

A practical result of the foregoing considerations is that it is very difficult to measure the emissivity of a diffuse surface inside a heated cavity, and impossible if the surface of interest and all the surrounding surfaces are at the same temperature. On the other hand, it is possible to determine the emissivity of a surface inside an isothermal cavity if the surface is a specular reflector and the line of sight is normal to the surface. In fact, the emissivity of molten metals is measured this way, with an inert atmosphere maintained inside the cavity to prevent the formation of a diffuse oxide layer.

Example 1-1

A small detector with active area ΔA_2 is exposed to radiation from a small hole in a furnace wall ΔA_1 as well as radiation from the surroundings at T_e (Fig. 1-12). The furnace is at $T_1 \gg T_e$. The detector is located at a distance (s) from the hole, such that $\Delta A_{1,2} \ll s^2$ angle between the normal to the area ΔA_1 and the line connecting ΔA_1 and ΔA_2 is θ_1 and similarly for θ_2. Assuming that the air between ΔA_1 and ΔA_2 is nonparticipating radiatively show that the total power incident on the detector is $\sigma T_e^4 \, \Delta A_2 + C\sigma T_1^4$ where $C = \cos\theta_1 \cos\theta_2 \, \Delta A_1 \, \Delta A_2/(\pi s^2)$.

The spectral, hemispherical flux incident at the detector is

$$q_{\lambda_2}^- = \int_{2\pi} I_{\lambda_2}^- \cos\theta_2' \, d\Omega_2' \tag{1-32}$$

where θ_2' and Ω_2' denote dummy variables of integration. The spectral intensity incident on the detector from the surroundings is $I_{b_\lambda}(T_e)$. The spectral intensity incident from ΔA_1 is $I_{\lambda_1}^+ = I_{b_\lambda}(T_1)$. Hence, the incident intensity distribution is

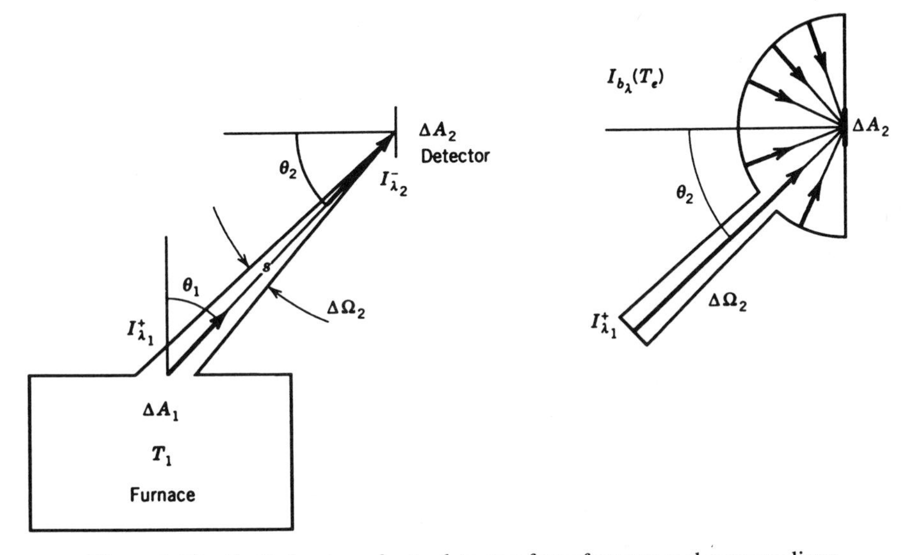

Figure 1-12 Radiative transfer to detector from furnace and surroundings.

$$I_{\lambda_2}^- = \begin{cases} I_{b_\lambda}(T_1) & \text{over } \Delta\Omega_2 \\ I_{b_\lambda}(T_e) & \text{all other directions} \end{cases} \tag{1-33}$$

and

$$q_{\lambda_2}^- = \int_{2\pi - \Delta\Omega_2} I_{b_\lambda}(T_e)\cos\theta_2'\, d\Omega_2' + \int_{\Delta\Omega_2} I_{b_\lambda}(T_1)\cos\theta_2'\, d\Omega_2' \tag{1-34}$$

which can be rearranged and also written as

$$q_{\lambda_2}^- = \int_{2\pi} I_{b_\lambda}(T_e)\cos\theta_2'\, d\Omega_2' + \int_{\Delta\Omega_2} \left[I_{b_\lambda}(T_1) - I_{b_\lambda}(T_e)\right]\cos\theta_2'\, d\Omega_2' \tag{1-35}$$

The first integral is simply $e_{b_\lambda}(T_e)$ because I_{b_λ} is independent of direction. Furthermore, since

$$\Delta\Omega_2 = \frac{\Delta A_1 \cos\theta_1}{s^2} \ll 1 \tag{1-36}$$

it is reasonable to assume that θ_2' is constant over $\Delta\Omega_2$ at the value of θ_2 in the second integral, giving

$$q_{\lambda_2}^- = e_{b_\lambda}(T_e) + \left[I_{b_\lambda}(T_1) - I_{b_\lambda}(T_e)\right]\cos\theta_2\, \Delta\Omega_2 \tag{1-37}$$

Integrating over all wavelengths gives the total, hemispherical incident flux as

$$q_2^- = \sigma T_e^4 + \frac{\sigma}{\pi}\left[T_1^4 - T_e^4\right]\cos\theta_2\, \Delta\Omega_2 \tag{1-38}$$

Ignoring T_e^4 relative to T_1^4 gives the total power incident on the detector as

$$Q_2^- = \sigma T_e^4\, \Delta A_2 + C\sigma T_1^4 \tag{1-39}$$

where

$$C = \frac{\cos\theta_1 \cos\theta_2\, \Delta A_1\, \Delta A_2}{\pi s^2} \tag{1-40}$$

Point Source of Radiation

An important concept in radiative transfer theory is that of a point source. A point source of radiation is any source viewed from a distance (s) that is large relative to the dimensions of the source. In Example 1-1, ΔA_1 acts like a point source of radiation when viewed from a distance $s \gg \sqrt{\Delta A_1}$. Example 1-1 also demonstrates that the flux emanating from a point source decays like $1/s^2$.

$$\Delta q_2^- = I_1^+ \cos\theta_2 \frac{\Delta A_1 \cos\theta_1}{s^2} \tag{1-41}$$

This dependence ($\Delta q^{-} \sim 1/s^2$) is consistent with the behavior of *spherical waves* whose power density or flux decays as $1/s^2$. However, *for a small change* in distance ($\Delta s \ll s$) the small solid angle subtended by the source actually changes very little

$$\Delta\Omega_2 = \frac{\Delta A_1 \cos\theta_1}{(s + \Delta s)^2} \sim \frac{\Delta A_1 \cos\theta_1}{s^2} \ll 1 \tag{1-42}$$

and thus the flux Δq_2^- remains essentially constant on a localized basis. This *local* dependence ($\Delta q^{-} \sim s^0$) fits the description of *plane waves*, which propagate rectilinearly in a single direction with a small, nearly constant divergence ($\Delta\Omega \ll 1$) and hence constant flux. Thus the radiation field from a point source behaves locally (i.e., over relatively short distances) like a collimated flux or a series of plane waves.

Fractional Function

One of the most important blackbody functions is the *fractional function* $f(\lambda T)$. This is defined as the fraction of hemispherical blackbody flux (or intensity) between zero and an arbitrary wavelength λ.

$$f(\lambda T) = \frac{\int_0^\lambda e_{b_\lambda}(\lambda, T)\, d\lambda}{\int_0^\infty e_{b_\lambda}(\lambda, T)\, d\lambda} = \int_0^{\lambda T} \frac{e_{b_\lambda}\, d(\lambda T)}{\sigma T^5} \tag{1-43}$$

By definition, the limiting values of the fractional function are zero and one; $f(0) = 0$ and $f(\infty) = 1$. Tabulated values of $f(\lambda T)$ are given in Appendix A. A graphical interpretation of $f(\lambda T)$ is given in Fig. 1-13.

The differential form of the fractional function is given by the following expression:

$$df(\lambda T) = \frac{e_{b_\lambda}(T)}{\sigma T^4}\, d\lambda \tag{1-44}$$

In this form the fractional function can be thought of as a variable transformation for wavelength to evaluate integrals of spectral quantities that are weighted by e_{b_λ} or I_{b_λ}. This procedure is illustrated in Chapter 2.

The fractional function can also be used to calculate the fraction of blackbody energy in a spectral region between two wavelengths, λ_1 and λ_2. This fraction is given by the difference, $f(\lambda_2 T) - f(\lambda_1 T)$, where it is assumed that $\lambda_1 < \lambda_2$.

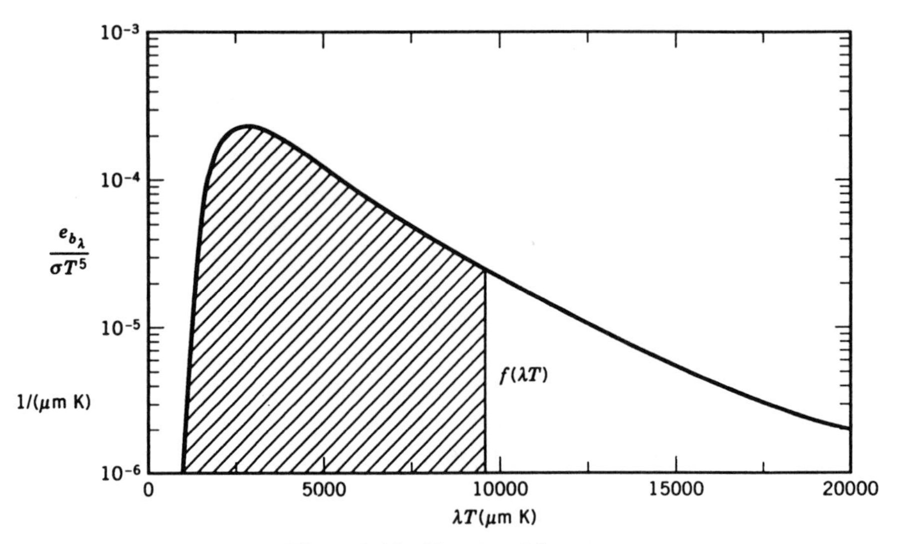

Figure 1-13 Fractional function.

Example 1-2

What is the fraction of solar emission in the visible region assuming the sun emits like a blackbody at 5785 K and the visible region is from 0.4 to 0.7 μm wavelength?

Based on a temperature of $T = 5785$ K, the respective wavelength-temperature products at the lower and upper limits of the visible region are

$$\lambda_1 T = (0.4\ \mu\text{m})(5785\ \text{K}) = 2314\ \mu\text{m K} \tag{1-45}$$

$$\lambda_2 T = (0.7\ \mu\text{m})(5785\ \text{K}) = 4050\ \mu\text{m K} \tag{1-46}$$

The fraction of blackbody energy between these limits can be obtained using the fractional function from Appendix A as

$$f(\lambda_2 T) - f(\lambda_1 T) = 0.49 - 0.12 = 0.37 \tag{1-47}$$

or 37% of the solar emission lies in the visible region.

SPECTRAL VARIABLES

In the foregoing discussion wavelength has been used as the variable to describe the spectral quality of radiation. Other equally valid spectral variables can be employed that are based either on the laws of wave propagation or the photon description of electromagnetic energy.

One of the most commonly used spectral variables is frequency. *Frequency* is defined as the ratio of wave speed (speed of light) to wavelength.

$$\nu = \frac{c_0}{\lambda_0} = \frac{c}{\lambda} \quad \left(\frac{1}{\text{sec}}\right) \tag{1-48}$$

c_0 = speed of light in a vacuum
c = speed of light in medium of interest
λ_0 = wavelength in vacuum
λ = wavelength in medium of interest

One of the advantages of using frequency as a spectral variable is that it remains constant as radiation propagates through various media, whereas wavelength does not.

The ratio of the wavelength (or speed of light) in vacuum to that in another medium defines the refractive index of the medium, n.

$$\frac{\lambda_0}{\lambda} = \frac{c_0}{c} = n \tag{1-49}$$

When radiation passes from one medium into another (with different optical properties), the speed of propagation and wavelength change as well as the direction of propagation. The change in wavelength and speed is dictated by the magnitude of the refractive index and is such that the frequency remains constant.

Consider the following familiar example of visible light passing into water. As visible light passes from air ($n = 1$) into water ($n = 1.33$), the speed of propagation decreases by a factor of 1.33. To accommodate this change in speed and still maintain wave continuity through the interface, the direction of propagation bends toward the normal to the interface and the wavelength decreases by a factor of 1.33, as indicated in Fig. 1-14. The relationship between a small increment in frequency and a small increment in wavelength can be obtained by differentiating Eq. (1-48) as follows:

$$d\nu = \frac{-c}{\lambda^2}\, d\lambda = \frac{-\nu^2}{c}\, d\lambda \tag{1-50}$$

Circular frequency ω is another common spectral variable. Circular frequency is simply a multiple of frequency.

$$\omega = 2\pi\nu \quad \left(\frac{\text{rad}}{\text{sec}}\right) \tag{1-51}$$

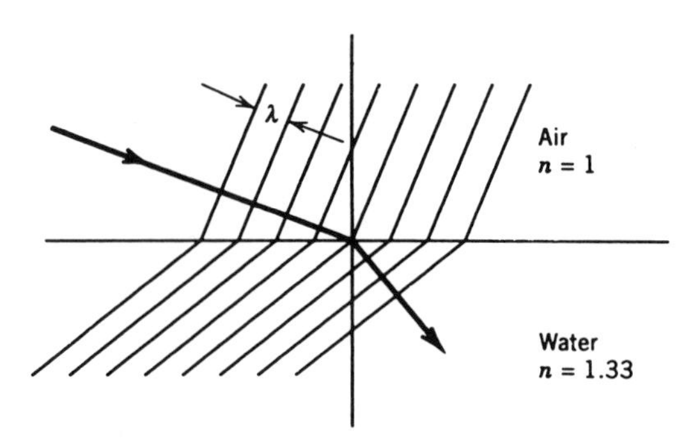

Figure 1-14 Change in wavelength and speed.

Wavenumber η is a spectral variable commonly used in spectroscopy. It is defined as the inverse of wavelength and is usually given in units of inverse centimeters.

$$\eta = \frac{1}{\lambda} \quad \left(\frac{1}{\text{cm}}\right) \tag{1-52}$$

Differentiation gives the relationship between a small increment of wavenumber and a small increment of wavelength.

$$d\eta = \frac{-1}{\lambda^2}\, d\lambda = -\eta^2\, d\lambda \tag{1-53}$$

The definition of spectral intensity depends on the spectral variable selected. The relationship between spectral intensity based on wavenumber and that based on wavelength can be obtained from the relation between $\Delta\eta$ and $\Delta\lambda$ [Eq. (1-53)] and the definition of spectral intensity as follows.

$$I_\eta = \frac{\text{power}}{\Delta A_\perp\, \Delta\Omega(-\Delta\eta)} = \frac{\text{power}}{\Delta A_\perp\, \Delta\Omega\, \Delta\lambda} \cdot \lambda^2 = I_\lambda \cdot \lambda^2 \tag{1-54}$$

Photon energy E is also often used as a variable to describe the spectral quality of electromagnetic radiation.

$$E = h\nu = \frac{hc}{\lambda} = \frac{1.24}{\lambda\ (\mu\text{m})} \quad (\text{eV}) \tag{1-55}$$

Radiation Energy Density and Pressure

There is an instantaneous energy density associated with photons analogous to the internal energy from thermodynamics that is associated with particles of matter. The radiative energy density at a point in space can be calculated by integrating the intensity at that point over all directions and dividing by the speed of light.

$$u_\lambda = \frac{1}{c}\int_{4\pi} I_\lambda \, d\Omega \quad \left(\frac{\text{kJ}}{\text{m}^3}\right) \tag{1-56}$$

For blackbody radiation (isotropic) the energy density over all wavelengths is readily obtained as follows:

$$u_b = \frac{4\sigma T^4}{c_0} \tag{1-57}$$

There is also a radiation pressure associated with the momentum of photons, which can be calculated from the energy density.

$$P = \frac{u}{3} \quad \text{(perfect reflecting surface)} \tag{1-58}$$

$$P = \frac{u}{6} \quad \text{(black absorbing surface)} \tag{1-59}$$

As would be expected, more momentum is imparted to a surface if the photons are reflected than if they are absorbed.

Radiation pressure is not usually an important force in the momentum considerations of thermal engineering systems, as can be verified by substituting characteristic temperatures into the relations above. However, radiation pressure can be a significant force in plasmas and is one of the dominant forces acting in the dynamic equilibrium of stars.

SPONTANEOUS VERSUS STIMULATED EMISSION

There are two types of emission, spontaneous and stimulated emission. Spontaneous emission occurs when a particle of matter (atom, ion, electron, or molecule) undergoes a spontaneous transition from an upper (relatively excited) energy state* to a lower state (perhaps the ground state). Stimulated emission occurs when a particle of matter undergoes such a transition due to the influence of a stimulating photon. In the case of stimulated emission the emitted photon has the same frequency and phase as the stimulating photon. Since the two photons are identical, the process of stimulated emission can be thought of as having the opposite effect on the net radiative transport as the absorption process. In fact, stimulated emission is usually treated as negative absorption.

The relative importance of stimulated emission can be seen from Fig. 1-10. The Planck function shown in Fig. 1-10 represents the emission leaving an

*The energy state includes any one or all of the following modes that have an associated oscillating dipole or accelerating charge: electronic, vibration, and rotation.

isothermal cavity including both spontaneous and stimulated emission. In contrast, it can be shown that the Wien distribution represents the emission that would leave an isothermal cavity if only spontaneous emission were included. Thus it can be seen that the contribution of stimulated emission to thermal radiation is relatively small on a total energy basis. On a spectral basis, stimulated emission takes on more importance at longer wavelengths for a given temperature. However, the bulk of thermal radiative emission is due to spontaneous energy state transitions.

Since spontaneous emission accounts for most thermally emitted energy, consider the frequency distribution of the radiative energy emitted by spontaneous transitions. The frequency of radiation emitted during an energy state transition is determined by the energy difference between the two states. Using E to also represent the *internal energy of the particle*,* the change in energy is obtained from Eq. (1-55) as

$$\Delta E = h\nu = \frac{hc}{\lambda} \tag{1-60}$$

The spectral distribution of the radiant energy is determined by the energy state population distribution, which is determined by the thermodynamic state of the system (i.e., temperature, pressure, and mole numbers).

At low temperatures, most of the particles occupy the lower energy states (i.e., the ground state and energy states just above the ground state). Thus, when a spontaneous radiative transition occurs, it generally has low energy E, low frequency ν, and long wavelength λ. Therefore the radiant energy is distributed at long wavelengths and the magnitude of the energy density (in terms of either flux or intensity) is relatively low. This condition corresponds to the curves and transitions labeled "low T" in Fig. 1-15.

At high temperatures, more particles occupy the upper excited states than at low temperatures. Thus it is more likely for "high energy" transitions to occur such as those labeled "high T" in Fig. 1-15. As a result, two things happen to the radiant energy distribution as temperature increases; the magnitude increases (at all wavelengths) and the distribution shifts to shorter wavelengths (see curves labeled "high T" in Fig. 1-15). These arguments account for the spectral distribution of the Planck function that describes blackbody radiation.

Figure 1-15 also denotes a fundamental difference between the behavior of condensed systems (liquids and solids) and gases. Particles of condensed matter have many degrees of freedom and therefore many possible energy states. In effect, there is a continuum of allowable transitions that condensed phase particles can undergo. Therefore condensed matter generally emits in a continuum similar to blackbody radiation. On the other hand, gaseous particles have relatively few degrees of freedom and relatively few allowable energy transitions. As a result gases emit energy in narrow lines or bands.

*Dual usage of the symbol E for both photon energy and particle internal energy is quite common.

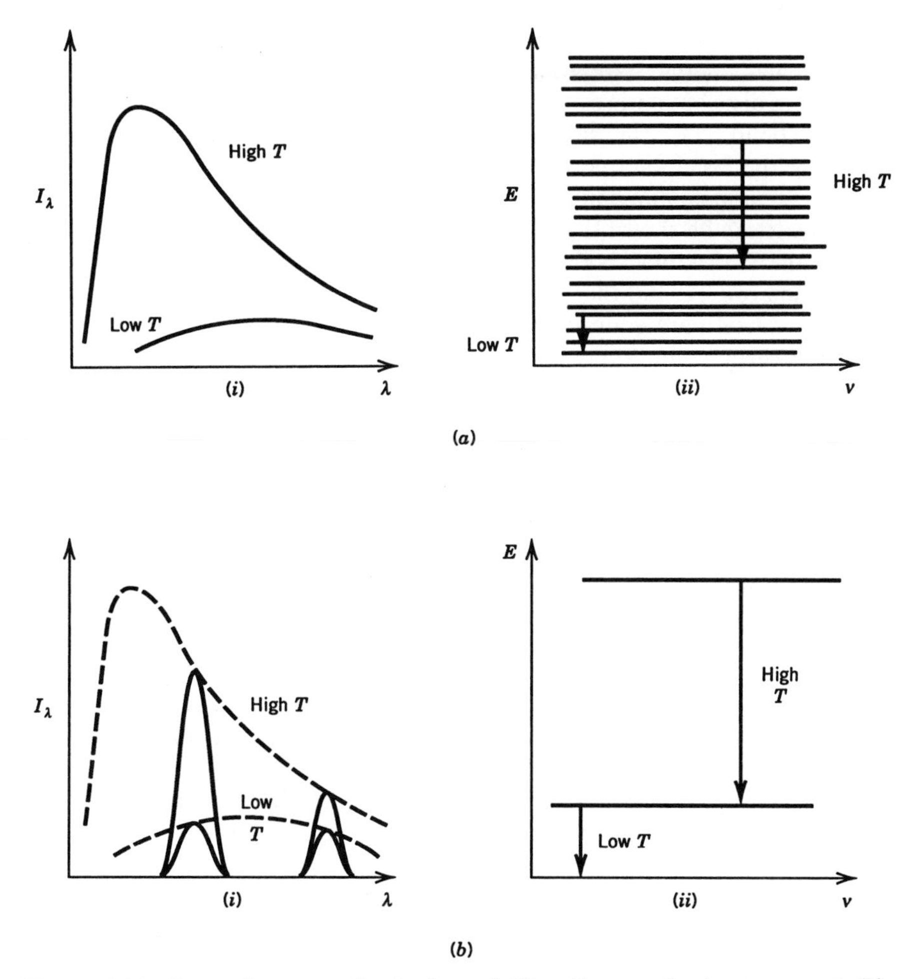

Figure 1-15 Spectral energy distribution of (i) radiant emitted energy and (ii) particle internal energy for (a) solids and liquids and (b) gases.

Nevertheless, the radiation emitted by both condensed and gaseous matter generally follows the Planck distribution in those spectral regions that correspond to allowable energy state transitions of the system.

THERMAL VERSUS NONTHERMAL RADIATION

Radiation is usually categorized as either thermal or nonthermal. Thermal radiation is radiation emitted by a physical system in local thermodynamic equilibrium (LTE). The concept of LTE arises from statistical thermodynamics and a detailed discussion is beyond the scope of this text (see [Goody and Yung, 1989]). Briefly, the attainment of LTE depends on the rate of transition between energy states through collisional interactions between molecules. If collisional energy state transitions occur rapidly enough, a Boltzmann population distribution is established among the energy states and the system is said to be in LTE. The radiation emitted by a system in LTE is determined solely by the local thermodynamic state of the system (i.e., temperature, pressure, and mole numbers). Most systems of engineering interest fall in this category. It can be shown that the source function for emission by a system in LTE is the Planck blackbody function [Goody and Yung, 1989]. That is, the assumption of LTE and that of the Planck function as the source function for thermal emission are equivalent assumptions.

Under certain conditions, collisional transitions are unable to maintain a Boltzmann distribution among the internal energy states and LTE is not achieved. Such conditions include low density and low pressure, high incident photon flux, and rapid state change. For example, in the upper atmosphere, low densities and pressures result in relatively infrequent collisional transitions, such that radiative transitions (i.e., spontaneous and induced emission, and absorption) affect the energy state populations. During rapid expansion and compression processes, such as flow in rocket nozzles and shock waves, the thermodynamic state of the flowing gas changes so rapidly that collisional transitions are often unable to maintain LTE. Laser radiation is also nonthermal radiation that is produced by an energy state population inversion and selectively amplified in a resonator cavity. Under nonequilibrium conditions, such as the examples just cited, the Planck function is not the appropriate source function for thermal emission. In fact, temperature is not defined* for a nonequilibrium system, and one would not know what temperature to use in Eq. (1-19) to predict the thermal emission. Instead the radiative energy emitted by a nonequilibrium system is determined by the relative rate processes associated with collisional and radiative relaxation.

SUMMARY

Thermal radiation is radiant energy emitted by particles of matter in LTE undergoing internal energy state transitions. The basic measure of the rate of propagation of radiant energy in a particular direction at a particular point in space is the intensity. It is defined as the radiant power per unit area perpendicular to the direction of travel, per unit solid angle. Integration of the intensity over solid angle gives the component of the radiant flux in a particular direction. This quantity, the component of the net flux in a given direction, is the primary quantity of interest in thermal design and heat transfer analysis.

Blackbody radiation is a special, ideal type of equilibrium thermal radiation produced inside an opaque enclosure that is in macroscopic thermody-

*More precisely, a single temperature for the system is not defined. For example, rotational states may be in equilibrium with each other but not with vibrational states, thus defining different rotational and vibrational temperatures.

namic equilibrium. Blackbody intensity is independent of direction (isotropic) and its magnitude as a function of wavelength is given by the Planck function.

REFERENCES

Bohren, C. F. and Huffman, D. R. (1983), *Absorption and Scattering of Light by Small Particles*, John Wiley & Sons, New York.

Born, M. and Wolf, E. (1970), *Principles of Optics*, Pergamon Press, Oxford.

Goody, R. M. and Yung, Y. L. (1989), *Atmospheric Radiation Theoretical Basis*, 2nd ed., Oxford University Press, New York.

Planck, M. (1959), *The Theory of Heat Radiation*, Dover Publications, New York.

Siegel, R. and Howell, J. R. (1988), *Thermal Radiation Heat Transfer*, 2nd ed., Taylor, Francis, Hemisphere, New York.

REFERENCES FOR FURTHER READING

Edwards, D. K., *Radiation Heat Transfer Notes*, Hemisphere, New York, 1981.

Hottel, H. C., "Radiant Heat Transmission," in *Heat Transmission*, 3rd ed., W. H. McAdams, Ed., Chap. 4, McGraw-Hill, New York, 1954.

Ozisik, M. N., *Radiative Transfer and Interactions with Conduction and Convection*, Werbel and Peck, New York, 1985.

Sparrow, E. M. and R. D. Cess, *Radiation Heat Transfer*, Augmented Edition, McGraw-Hill and Hemisphere, New York, 1978.

PROBLEMS

1. Consider a lamp filament with area ΔA_s and temperature T_s to be a blackbody source surrounded by a nonattenuating medium that can be taken to be a blackbody at 0 K.

(a) What is the total intensity leaving the filament in the direction normal to ΔA_s?

(b) What is the total intensity in the direction normal to and away from ΔA_s at a distance s_1?

(c) What is the total hemispherical flux in the direction normal to and away from ΔA_s at a distance s_1? (Assume $\sqrt{\Delta A_s} \ll s_1$ such that $\cos \theta_s \approx 1$ for all rays between the point at s_1 and points on ΔA_s and thereby the small solid angle approximation holds.)

2. A steam tube of length l and diameter d_t is subject to radiative heating inside a blackbody enclosure at temperature T_e.

(a) What is the total flux incident on the tube?

(b) What is the total power incident on the tube?

(c) Repeat parts (a) and (b) for the case of incident radiation that is collimated with intensity I_0 and divergence $\Delta\Omega$.

3. Consider the sun to be a blackbody emitter at 5785 K.

(a) What is the total intensity (W/m^2 sr) in any direction away from the sun at its surface? [2.02×10^7 W/(m^2 sr)]

(b) What is the total hemispherical flux (W/m^2) in a direction normal to and away from the sun at its surface? (6.35×10^7 W/m^2)

(c) What is the total power (W) emitted from the sun (diameter = 1.39×10^9 m)? (3.85×10^{26} W)

(d) What is the wavelength of maximum emitted spectral intensity? (0.50 μm)

(e) What is the spectral intensity I_λ^+ at λ_{max} in any direction away from the sun at its surface? [2.65×10^7 W/(m^2 μm sr)]

(f) What is the spectral flux q_λ^+ at λ_{max} in a direction normal to and away from the sun at its surface? [8.34×10^7 W/(m^2 μm)]

(g) What is the spectral intensity in terms of wavenumber I_η^+ (W/m^2 cm^{-1} sr) at λ_{max}? [663 W/(m^2 cm^{-1} sr)]

(h) What is the fraction of solar emission in the ultraviolet region (0.1–0.4 μm)? (12%)

4. The sun's diameter is 1.39×10^9 m, the earth's diameter is 1.27×10^7 m, and the radius of the earth's orbit is 1.50×10^{11} m. Assume that the sun is a blackbody emitter at 5785 K.

(a) What is the total flux (W/m^2) incident on a surface parallel to the earth's surface at 23° north latitude, at noon on the summer solstice (June 21)? (1360 W/m^2)

(b) What is the total flux (W/m^2) incident on a surface parallel to the earth's surface at the north pole on the summer solstice? (531 W/m^2)

(c) Integrate the incident flux over the surface of the earth to obtain the total power (W) incident on the earth from the sun? (1.72×10^{17} W)

5. Consider radiative transfer from a small hole $\Delta A_1 = 100$ mm^2 in a furnace at $T_1 = 1000$ K to a small detector with area $\Delta A_2 = 10$ mm^2 located at a distance $s = 0.1$ m from the hole (see Example 1-1). The angle between the normal to the area ΔA_1 and the line connecting ΔA_1 and ΔA_2 is $\theta_1 = 30°$ and similarly, $\theta_2 = 60°$. Assuming that the medium between the areas is nonparticipating radiatively and that the surroundings (other than the hole) are isothermal at $T_e = 300$ K, what is the total incident power at the detector? (5.36 mW)

6. On a clear day the intensity incident on the surface of the earth can be approximated as a two step function. Over the small solid angle subtending the sun's disk ($\Delta\Omega_0 = 6.74 \times 10^{-5}$ sr) the total incident intensity is constant at $I_0 = 1.52 \times 10^7$ W/(m^2 sr) due to direct solar emission and over the rest of the hemisphere it is constant at $I_s = 75$ W/(m^2 sr) due to scattering and re-emission by the earth's atmosphere. Determine the total incident flux at a particular location on the surface of the earth where the angle between the normal to the surface and the direction to the sun is $\theta_0 = 45°$. (960 W/m^2)

7. At a point in space the total intensity distribution is $I = I_0 + I_1 \cos\theta$ where I_0 and I_1 are constants.

(a) Are there any mathematical constraints on the allowable values of I_0 and I_1?

(b) Determine the *net flux* at that point in the $\theta = 0$ direction.

8. Show that the maximum of e_{b_η} occurs at a different wavelength than that for e_{b_λ} for a given temperature by determining the value of λT at which e_{b_η} is a maximum. Calculate for both solar ($T = 5785$ K) and terrestrial radiation ($T = 250$ K) the wavelength and wavenumber at which both e_{b_η} and e_{b_λ} are at a maximum.

9. Show that intensity is constant along a nonattenuating path. Hint: Consider isotropic emission by a point source. Evaluate the intensity at a distance r_1 from the source and again at a distance r_2.

10. For a semi-isotropic intensity distribution carry out the integration indicated in Eq. (1-12) and thereby verify Eq. (1-17).

11. The toy radiometer has four vanes spaced at 90°, which are painted black on one side and white on the other. The vane assembly is balanced on a needle point and enclosed inside a glass envelope under a near vacuum. When placed near a source of visible or near infrared radiation, the radiometer rotates in the direction shown in the figure below. Discuss the possible reasons for the rotation. Could radiation pressure be responsible for this motion?

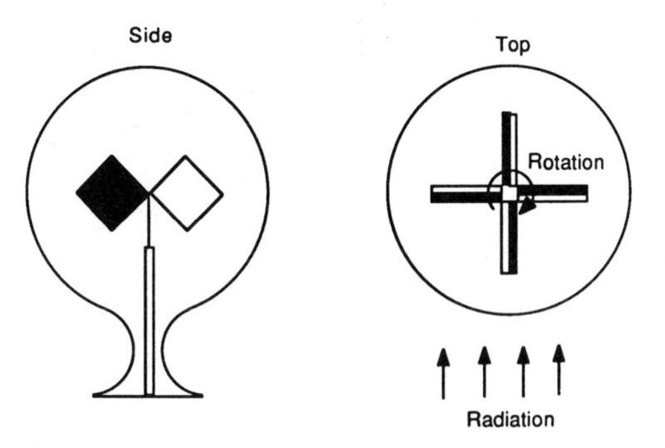

Figure P11

CHAPTER 2

RADIATIVE PROPERTIES AND SIMPLE TRANSFER

Thermal radiative heat transfer analysis consists of two basic parts, evaluation of radiative properties and solution of the transfer problem. The radiative properties are a measure of the tendency of a given surface or participating medium to emit, absorb, and reflect (or scatter) radiation. Once the properties have been evaluated, the transfer problem can be solved to determine the net transfer rate of radiant energy to or from a given surface or volume element of mass.

In this chapter the groundwork for solving general radiative transfer problems is laid by introducing the definitions of the basic radiative properties. These properties pertain to both surfaces and participating media. However, the main emphasis of this chapter is surface properties. Additional properties used in connection with participating media are discussed in Chapter 7. Two common examples of simple radiative transfer are also presented in this chapter: (1) radiation from a point source and (2) radiation from a large, isothermal surroundings.

ENGINEERING RADIATIVE PROPERTY DEFINITIONS

The set of engineering radiative properties consists of emissivity, absorptivity, reflectivity, and transmissivity.

$$\text{Engineering radiative properties} = \begin{cases} \text{emissivity} & \varepsilon \\ \text{absorptivity} & \alpha \\ \text{reflectivity} & \rho \\ \text{transmissivity} & \tau \end{cases}$$

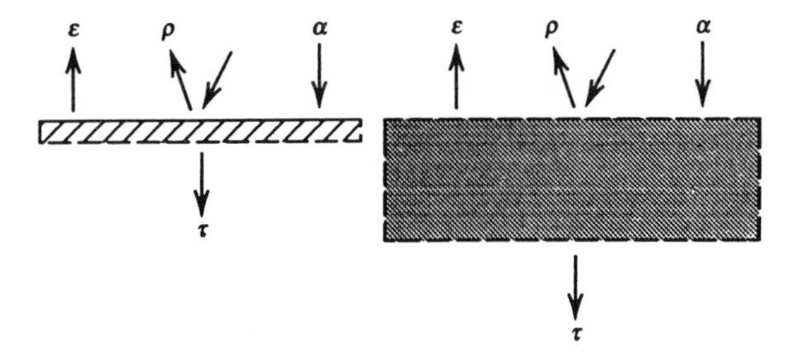

Figure 2-1 Engineering radiative properties.

While other physical properties, such as refractive index, may also be considered to be radiative properties, the set listed above is the set commonly referred to as engineering radiative properties. The definitions of these properties that follow apply to both surfaces and participating media. In the case of a participating medium the edge or boundary of the medium can be considered as the effective surface to which these properties apply* (Fig. 2-1).

All of the engineering radiative properties are defined as the ratio of a sample flux to a reference flux.

$$\varepsilon, \alpha, \rho, \tau = \frac{\text{sample flux}}{\text{reference flux}}$$

The sample flux is the flux of interest, i.e., the actual flux emitted from, absorbed at, reflected from, or transmitted through the surface. The reference flux is the theoretical flux available for emission, absorption, reflection, or transmission. In the case of emissivity the reference flux is blackbody flux because this is the maximum theoretical emitted flux possible (for a system in thermodynamic equilibrium), and for absorption, reflection, and transmission the reference flux is the incident flux, because this is the maximum flux available to be absorbed, reflected, or transmitted.

$$\varepsilon = \frac{\text{emitted flux}}{\text{blackbody flux}} = \frac{q_e}{e_b}$$

$$\alpha, \rho, \tau = \frac{\text{absorbed, reflected, transmitted flux}}{\text{incident flux}} = \frac{q_{a,r,t}}{q^-}$$

*The distinction between surfaces and participating media is mostly academic. Opaque surfaces are merely a limiting case of participating media.

	Directional	Hemispherical
Spectral	Δq_λ	q_λ
Total	Δq	q

Figure 2-2 Sample or reference flux notation.

Each flux referred to above can also be classified according to its directional and spectral characteristics. There are four combinations possible, depending on whether a narrow pencil of rays in one direction (partial or directional flux) or all directions (hemispherical flux) are considered and whether a narrow band of wavelengths (spectral flux) or all wavelengths (total flux) are included. The matrix of possible combinations is illustrated in Fig. 2-2.

EMISSIVITY

The concept of a blackbody or isothermal enclosure introduced in Chapter 1 gives rise to the notion of a black surface. A black surface is any surface that is a perfect absorber and a perfect emitter; it absorbs any and all radiation incident upon it and reflects none, like a small hole in an isothermal enclosure. A black surface also emits blackbody radiation. Of course the concept of a black surface is an idealization. No real surface exhibits perfectly the properties of an ideal black surface. The emissivity provides a means of describing the nonideal emission of actual material surfaces.

Consider a surface with temperature T or the boundary of a participating medium that is characterized by a maximum temperature of T. The directional distribution of intensity emitted from the surface at a given wavelength λ is pictured in Fig. 2-3 along with the corresponding blackbody intensity. The actual emitted intensity I_{λ_e} in any direction is less in magnitude than the blackbody intensity and in general varies in magnitude with changes in direction.

The *spectral directional emissivity* ε'_λ is defined as the ratio of spectral directional (or partial) emitted flux in a given direction (θ, ϕ) at a given wavelength (λ) to the spectral directional blackbody flux. Using the definitions of flux and intensity introduced in Chapter 1, the definition of ε'_λ can be expressed as shown in Eq. (2-1).

$$\varepsilon'_\lambda = \frac{\Delta q_{\lambda_e}}{\Delta e_{b_\lambda}} = \frac{I_{\lambda_e}(\lambda, \theta, \phi)\cos\theta\,\Delta\Omega}{I_{b_\lambda}(\lambda, T)\cos\theta\,\Delta\Omega} \tag{2-1}$$

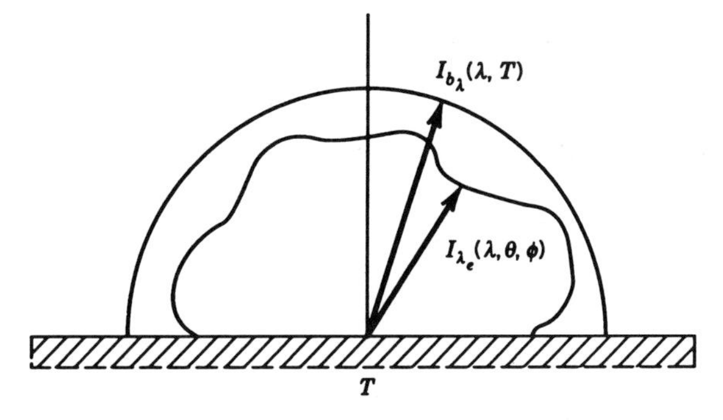

Figure 2-3 Emitted and blackbody intensity distributions.

The *spectral hemispherical emissivity* ε_λ is defined as the ratio of spectral hemispherical emitted flux to spectral hemispherical blackbody flux.

$$\varepsilon_\lambda = \frac{q_{\lambda_e}}{e_{b_\lambda}} = \frac{\int_{2\pi} I_{\lambda_e}(\lambda, \theta, \phi) \cos\theta \, d\Omega}{\pi I_{b_\lambda}(\lambda, T)} \tag{2-2}$$

Similarly, the *total directional emissivity* ε' is defined as the ratio of total directional emitted flux to total directional blackbody flux,

$$\varepsilon' = \frac{\Delta q_e}{\Delta e_b} = \frac{\int_0^\infty I_{\lambda_e} \cos\theta \, \Delta\Omega \, d\lambda}{I_b \cos\theta \, \Delta\Omega} \tag{2-3}$$

and the *total hemispherical emissivity* ε is defined as the ratio of total hemispherical emitted flux to total hemispherical blackbody flux.

$$\varepsilon = \frac{q_e}{e_b} = \frac{\int_0^\infty \int_{2\pi} I_{\lambda_e} \cos\theta \, d\Omega \, d\lambda}{\pi I_b} \tag{2-4}$$

Solving Eq. (2-1) for I_{λ_e} and substituting into Eq. (2-2) yields Eq. (2-5a).

$$\varepsilon_\lambda = \frac{1}{\pi} \int_{2\pi} \varepsilon'_\lambda \cos\theta \, d\Omega \tag{2-5a}$$

Similarly Eqs. (2-3) and (2-4) yield Eq. (2-5b).

$$\varepsilon = \frac{1}{\pi} \int_{2\pi} \varepsilon' \cos\theta \, d\Omega \tag{2-5b}$$

Equations (2-5a, b) demonstrate that integration of directional emissivity over the hemisphere above a surface yields the hemispherical emissivity. A similar relationship between spectral and total emissivities can be obtained as follows. Solving Eq. (2-1) for I_{λ_e} and substituting into (2-3) gives

$$\varepsilon'(T) = \frac{\int_0^\infty \varepsilon'_\lambda \pi I_{b_\lambda}(T)\, d\lambda}{\pi I_b(T)} = \frac{1}{\sigma T^4} \int_0^\infty \varepsilon'_\lambda e_{b_\lambda}(T)\, d\lambda \tag{2-6}$$

The integral in this expression can be further simplified by using the definition of the fractional function Eq. (1-44) to perform a variable transformation, replacing λ with $f(\lambda T)$.

$$\varepsilon'(T) = \int_0^1 \varepsilon'_\lambda \, df(\lambda T) \tag{2-7}$$

The integration in Eq. (2-7) can be readily performed if the variation of ε'_λ with wavelength is known. A reasonable estimate of the value of the integral can be obtained by approximating ε'_λ as a stair-step function with constant values over small, finite wavelength intervals and integrating numerically using a table of the fractional function. For more precise results, polynomial curve fits of $f(\lambda T)$ are available (see Appendix A), and the numerical integration can be performed by a digital computer.

Multiplying both sides of Eq. (2-7) by $(\cos\theta)/\pi$, integrating over 2π steradians, and using the relation between hemispherical and directional emissivity [Eq. (2-5a)] gives the following expression for the total hemispherical emissivity in terms of the spectral directional emissivity integrated over all directions and wavelengths.

$$\varepsilon(T) = \frac{1}{\pi} \int_0^1 \int_{2\pi} \varepsilon'_\lambda \cos\theta \, d\Omega \, df(\lambda T) \tag{2-8}$$

The most basic of the emissivities is the spectral directional emissivity ε'_λ. If ε'_λ is known for all wavelengths and all directions, any of the other emissivities can be obtained by integration. Similarly if the spectral hemispherical emissivity ε_λ is known for all wavelengths or the total directional emissivity ε' is known for all directions, the total hemispherical emissivity ε can be calculated by integrating over all wavelengths or all directions, respectively.

Example 2-1

Determine the total hemispherical emissivity of tungsten at 3000 and 3500 K using the approximate representation of the spectral hemispherical emissivity shown in Fig. 2-4. How does total emissivity vary with temperature?

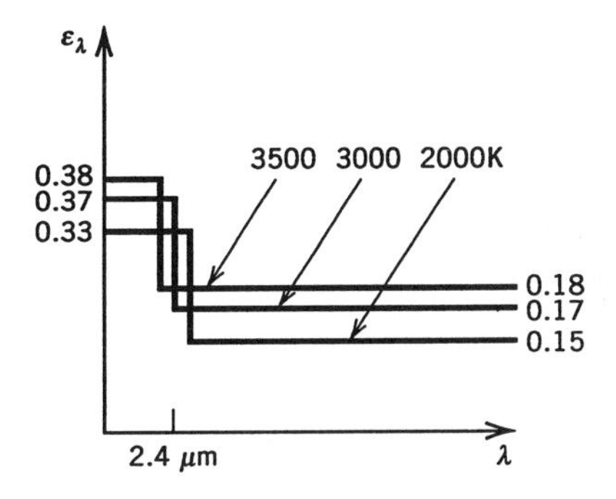

Figure 2-4 Spectral hemispherical emissivity of tungsten (approximate).

Combining Eqs. (2-5a) and (2-8) gives the following expression for total hemispherical emissivity as a function of spectral hemispherical emissivity.

$$\varepsilon(T) = \int_0^1 \varepsilon_\lambda(T)\, df(\lambda T)$$

According to Fig. 2-4, ε_λ is constant over wavelength except for a step change in value. The step change occurs at a cutoff wavelength of $\lambda_c = 2.4\ \mu\text{m}$, which corresponds to

$$\lambda_c T = (2.4\ \mu\text{m})(3000\ \text{K}) = 7200\ \mu\text{m K}$$

From Appendix A the fractional function is

$$f(7200\ \mu\text{m K}) = 0.82$$

Thus the total emissivity at 3000 K is

$$\varepsilon(3000\ \text{K}) = (0.37)(0.82) + (0.17)(1 - 0.82) = 0.33$$

At 3500 K the results are $\lambda_c T = 8400\ \mu\text{m K}$, $f(8400) = 0.87$ and

$$\varepsilon(3500\ \text{K}) = (0.38)(0.87) + (0.18)(1 - 0.87) = 0.35$$

Thus the total emissivity increases with temperature. This behavior is typical of metals. There are two reasons for this effect; one is the increase in the intrinsic spectral emissivity of the material with temperature and the other is the shift of the Planck function to shorter wavelengths as temperature increases. As the Planck function shifts to shorter wavelengths, the larger value of spectral emissivity is weighted more heavily.

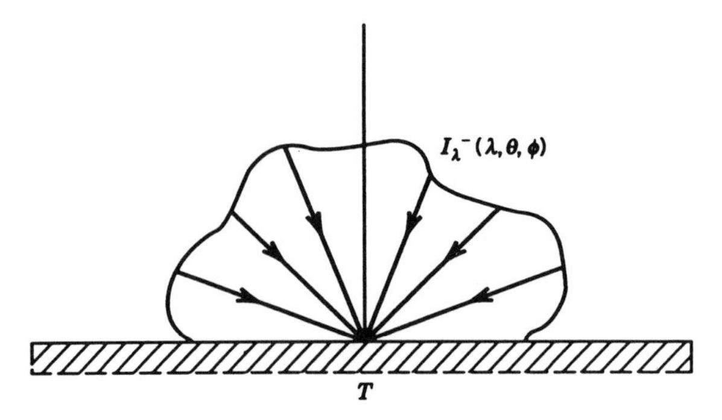

Figure 2-5 Incident intensity distribution.

ABSORPTIVITY

Consider a surface or the boundary of a participating medium characterized by temperature T with intensity distribution incident on the surface as depicted in Fig. 2-5. In general the incident intensity I_λ^- will vary with wavelength and direction.

The *spectral directional absorptivity* α'_λ is defined as the ratio of spectral absorbed flux at a given wavelength (λ) to the spectral directional (or partial) incident flux.

$$\alpha'_\lambda = \frac{\delta q_{\lambda_a}}{\Delta q_\lambda^-} = \frac{\delta q_{\lambda_a}}{I_\lambda^-(\lambda,\theta,\phi)\cos\theta\,\Delta\Omega} \tag{2-9}$$

The *spectral hemispherical absorptivity* α_λ is defined as the ratio of spectral absorbed flux to spectral hemispherical incident flux. It can also be thought of as a directional average of the spectral directional absorptivity, with the directional weighting function being the incident intensity.

$$\alpha_\lambda = \frac{q_{\lambda_a}}{q_\lambda^-} = \frac{\int_{2\pi} \delta q_{\lambda_a}}{\int_{2\pi} I_\lambda^-(\lambda,\theta,\phi)\cos\theta\,d\Omega} = \frac{\int_{2\pi} \alpha'_\lambda I_\lambda^- \cos\theta\,d\Omega}{\int_{2\pi} I_\lambda^- \cos\theta\,d\Omega} \tag{2-10}$$

The *total directional absorptivity* α' is defined as the ratio of total absorbed flux to total directional incident flux. It can also be thought of as a spectral average of the spectral directional absorptivity, with the weighting function again being the spectral incident intensity.

$$\alpha' = \frac{\delta q_a}{\Delta q^-} = \frac{\int_0^\infty \delta q_{\lambda_a}\,d\lambda}{\int_0^\infty I_\lambda^- \cos\theta\,\Delta\Omega\,d\lambda} = \frac{\int_0^\infty \alpha'_\lambda I_\lambda^-\,d\lambda}{\int_0^\infty I_\lambda^-\,d\lambda} \tag{2-11}$$

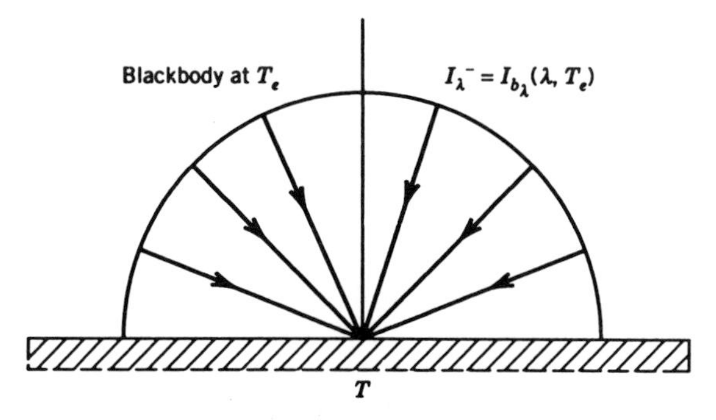

Figure 2-6 Incident intensity distribution for blackbody surroundings.

And the *total hemispherical absorptivity* α is defined as the ratio of total absorbed flux to total hemispherical incident flux with the additional interpretation of being a spectral and directional average of the spectral directional absorptivity.

$$\alpha = \frac{q_a}{q^-} = \frac{\int_0^\infty \int_{2\pi} \delta q_{\lambda_a} \, d\lambda}{\int_0^\infty \int_{2\pi} I_\lambda^- \cos\theta \, d\Omega \, d\lambda} = \frac{\int_0^\infty \int_{2\pi} \alpha'_\lambda I_\lambda^- \cos\theta \, d\Omega \, d\lambda}{\int_0^\infty \int_{2\pi} I_\lambda^- \cos\theta \, d\Omega \, d\lambda} \tag{2-12}$$

It should be noted that there is no directional information associated with the absorbed flux (i.e., partial directional vs. hemispherical) since all information regarding directionality is lost once the energy is absorbed and converted into thermal internal energy.

As was the case for emissivity, the most fundamental of the absorptivities is the spectral directional absorptivity. If α'_λ and the incident intensity are known for all directions and all wavelengths, any of the other absorptivities can be determined by integration. This process is illustrated by a special case that applies in many practical situations, namely, that of surroundings that are equivalent to a blackbody at temperature T_e.

Figure 2-6 illustrates a surface at temperature T subject to blackbody irradiation at temperature T_e. In this case the incident intensity is independent of direction, and the spectral distribution is given by the Planck function.

$$I_\lambda^- = I_{b_\lambda}(\lambda, T_e) \quad \text{(blackbody surroundings at } T_e\text{)} \tag{2-13}$$

The total hemispherical absorptivity as a function of spectral directional absorptivity is obtained by substituting (2-13) into (2-12) to give (2-14).

$$\alpha(T,T_e) = \frac{1}{\pi}\int_0^1 \int_{2\pi} \alpha'_\lambda(T)\cos\theta\, d\Omega\, df(\lambda T_e) \quad \text{(blackbody surroundings at } T_e) \tag{2-14}$$

Equation (2-14) shows that the total hemispherical absorptivity is a function of two temperatures (the surface and surroundings) whereas the total hemispherical emissivity is a function of only the surface temperature [see Eq. (2-8)]. The reason the environment temperature appears in absorptivity and not in emissivity is because the absorptivity depends on the incident radiation from the surroundings while the emissivity does not.

Kirchhoff's Law

In heat transfer analysis it is expedient, if justified, to assume that the emissivity and absorptivity of a surface or participating medium are equal. The equality between emissivity and absorptivity is known as Kirchhoff's law. The conditions when emissivity and absorptivity can be assumed equal vary depending on which spectral and directional combination is being considered.

For a system in thermodynamic equilibrium the emissivity and absorptivity are equal at the spectral directional level.

$$\varepsilon'_\lambda(\lambda,\theta,\phi,T) = \alpha'_\lambda(\lambda,\theta,\phi,T) \tag{2-15}$$

This relation applies to participating media in thermodynamic equilibrium, which includes opaque surfaces that are isothermal over the extent of surface layer which exchanges radiation with the surroundings. This condition is usually satisfied by highly opaque materials with high thermal conductivity such as metals. However some dielectric materials, particularly those with low thermal conductivity, such as ceramics, may not satisfy this condition. When the condition of thermodynamic equilibrium is satisfied, the equality in Eq. (2-15) holds for all wavelengths and all directions, but the value of emissivity (absorptivity) is not necessarily the same for all wavelengths and all directions. A surface whose emissivity varies with direction is referred to as a nondiffuse emitter-absorber. A surface whose emissivity varies with wavelength is referred to as a nongray emitter-absorber.

$$\varepsilon'_\lambda = \varepsilon'_\lambda(\theta,\phi) \quad \text{(nondiffuse emitter-absorber)} \tag{2-16}$$

$$\varepsilon'_\lambda = \varepsilon'_\lambda(\lambda) \quad \text{(nongray emitter-absorber)} \tag{2-17}$$

The spectral and directional characteristics of ε'_λ and α'_λ for a nondiffuse, nongray emitter-absorber are pictured in Fig. 2-7*a*.

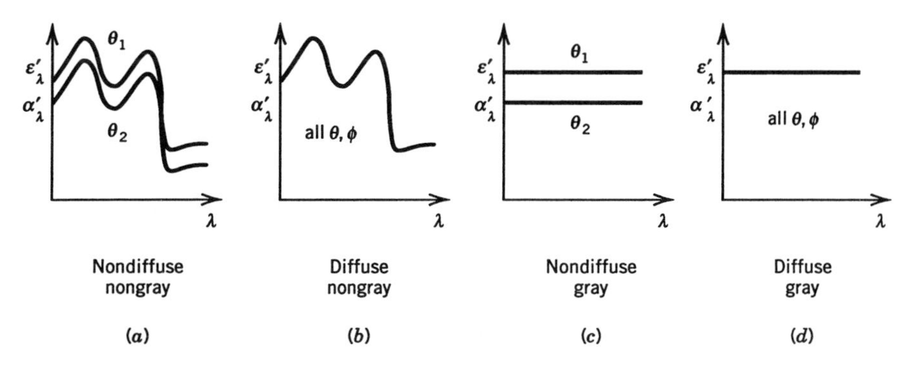

Figure 2-7 Spectral and directional variations of emissivity and absorptivity.

Kirchhoff's law can also be applied at other levels besides the spectral directional level if certain assumptions about the spectral and directional dependence of ε'_λ are satisfied. Consider a surface or medium whose spectral directional emissivity is independent of direction (Fig. 2-7*b*). Such a surface is known as a diffuse emitter-absorber.

$$\varepsilon'_\lambda \neq \varepsilon'_\lambda(\theta, \phi) \quad \text{(diffuse emitter-absorber)} \tag{2-18}$$

Substitution of Eq. (2-18) for ε'_λ into (2-5a), and the equivalent result for α'_λ into (2-10) gives the spectral hemispherical form of Kirchhoff's law, which is valid for a diffuse, nongray emitter-absorber.

$$\varepsilon_\lambda(\lambda, T) = \alpha_\lambda(\lambda, T) \quad \begin{array}{l}\text{(diffuse, nongray emitter-absorber} \\ \text{or diffuse irradiation)}\end{array} \tag{2-19}$$

Inspection of Eq. (2-10) reveals that Eq. (2-19) also results if the incident intensity is independent of direction (diffuse irradiation). The assumption of diffuse irradiation is one that is often justified in the calculation of radiative transfer between surfaces of an enclosure, and so it is quite common to invoke Eq. (2-19) even when the surfaces themselves are not perfectly diffuse emitter-absorbers.

Consider next a surface (or medium) whose spectral directional emissivity is independent of wavelength (Fig. 2-7*c*). Such a surface is known as a gray emitter-absorber.

$$\varepsilon'_\lambda \neq \varepsilon'_\lambda(\lambda) \quad \text{(gray emitter-absorber)} \tag{2-20}$$

Substitution of Eq. (2-20) for ε'_λ into (2-7), and the equivalent result for α'_λ

into (2-11) gives the total directional form of Kirchhoff's law, which is valid for a nondiffuse, gray emitter-absorber.

$$\varepsilon'(\theta, \phi, T) = \alpha'(\theta, \phi, T) \quad \text{(nondiffuse, gray emitter-absorber)} \quad (2\text{-}21)$$

Inspection of (2-11) shows that (2-21) also results if the incident intensity has a spectral distribution that is proportional to blackbody radiation at the same temperature as the surface (T). This case is so specialized, however, that it is of little practical interest. The main condition for invoking Eq. (2-21) is that of a gray emitting-absorbing surface.

Consider finally a surface (or medium) whose spectral directional emissivity is independent of both wavelength and direction (Fig. 2-7*d*). Such a surface is known as a diffuse, gray emitter-absorber.

$$\varepsilon'_\lambda \neq \varepsilon'_\lambda(\lambda, \theta, \phi) \quad \text{(diffuse, gray emitter-absorber)} \quad (2\text{-}22)$$

Substitution of Eq. (2-22) for ε'_λ into (2-8), and the equivalent result for α'_λ into (2-12) gives the total hemispherical form of Kirchhoff's law, which is valid for a diffuse, gray emitter-absorber.

$$\varepsilon(T) = \alpha(T) \quad \text{(diffuse, gray emitter-absorber or nondiffuse, gray emitter-absorber and diffuse irradiation)} \quad (2\text{-}23)$$

Inspection of Eq. (2-12) reveals that Eq. (2-23) also results if the surface is a gray emitter-absorber and the irradiation is diffuse. As noted previously, the assumption of diffuse irradiation is usually justified in the calculation of radiative heat transfer between surfaces of an enclosure, and so the validity of Eq. (2-23) is primarily determined by the validity of Eq. (2-20). For surfaces exhibiting nearly gray behavior, a total radiant interchange analysis can be performed on the enclosure using Eq. (2-23), and no spectral integration is required. For surfaces exhibiting significant nongray behavior, a spectral interchange analysis must be performed using Eq. (2-19), and the resulting spectral heat flux must then be integrated with respect to wavelength.

Example 2-2

Determine the total hemispherical absorptivity of tungsten at $T = 3000$ K subject to blackbody irradiation at $T_e = 1000$ K using the approximate representation of the spectral hemispherical emissivity shown in Fig. 2-4.

The total hemispherical absorptivity as a function of spectral hemispherical absorptivity can be written as

$$\alpha(T, T_e) = \int_0^1 \alpha_\lambda(T)\, df(\lambda T_e)$$

Since blackbody radiation is by definition diffuse, Kirchhoff's Law (2-19) holds, yielding

$$\alpha(T, T_e) = \int_0^1 \varepsilon_\lambda(T)\, df(\lambda T_e)$$

The step change in ε_λ occurs at $\lambda_c T_e$ = (2.4 μm)(1000 K) = 2400 μm K and the fractional function at this value is f(2400 μm K) = 0.14. Thus the value of the total absorptivity is

$$\alpha(3000\ K, 1000\ K) = (0.37)(0.14) + (0.17)(1 - 0.14) = 0.20$$

This value is less than the corresponding emissivity from Example 2-1 (0.33) due to the spectral distribution of incident radiation being more heavily weighted at wavelengths above 2.4 μm.

REFLECTIVITY

Consider again the surface or boundary of a participating medium characterized by temperature T with an incident intensity distribution as depicted in Fig. 2-5.

The *spectral directional reflectivity* ρ'_λ (also referred to as spectral *directional–hemispherical*) is defined as the ratio of spectral hemispherical reflected flux at a given wavelength (λ) to the spectral directional (or partial) incident flux.

$$\rho'_\lambda = \frac{\delta q_{\lambda_r}}{\Delta q_\lambda^-} = \frac{\int_{2\pi} I_{\lambda_r} \cos\theta_r\, d\Omega_r}{I_\lambda^-(\lambda, \theta, \phi) \cos\theta\, \Delta\Omega} \tag{2-24}$$

The *spectral hemispherical reflectivity* ρ_λ is defined as the ratio of spectral hemispherical reflected flux to spectral hemispherical incident flux. It can also be thought of as a directional average of the spectral directional reflectivity, with the directional weighting function being the incident intensity.

$$\rho_\lambda = \frac{q_{\lambda_r}}{q_\lambda^-} = \frac{\int_{2\pi} \delta q_{\lambda_r}}{\int_{2\pi} I_\lambda^-(\lambda, \theta, \phi) \cos\theta\, d\Omega} = \frac{\int_{2\pi} \rho'_\lambda I_\lambda^- \cos\theta\, d\Omega}{\int_{2\pi} I_\lambda^- \cos\theta\, d\Omega} \tag{2-25}$$

The *total directional reflectivity* ρ' (also referred to as total directional–hemispherical) is defined as the ratio of total hemispherical reflected flux to total directional incident flux (Fig. 2-8). It can also be thought of as a spectral average of the spectral directional reflectivity, with the weighting function again being the spectral incident intensity.

$$\rho' = \frac{\delta q_r}{\Delta q^-} = \frac{\int_0^\infty \delta q_{\lambda_r}\, d\lambda}{\int_0^\infty I_\lambda^- \cos\theta\, \Delta\Omega\, d\lambda} = \frac{\int_0^\infty \rho'_\lambda I_\lambda^-\, d\lambda}{\int_0^\infty I_\lambda^-\, d\lambda} \tag{2-26}$$

The *total hemispherical reflectivity* ρ is defined as the ratio of total hemispherical reflected flux to total hemispherical incident flux and has the additional interpretation of a spectral and directional average of the spectral directional reflectivity.

$$\rho = \frac{q_r}{q^-} = \frac{\int_0^\infty \int_{2\pi} \delta q_{\lambda_r}\, d\lambda}{\int_0^\infty \int_{2\pi} I_\lambda^- \cos\theta\, d\Omega\, d\lambda} = \frac{\int_0^\infty \int_{2\pi} \rho'_\lambda I_\lambda^- \cos\theta\, d\Omega\, d\lambda}{\int_0^\infty \int_{2\pi} I_\lambda^- \cos\theta\, d\Omega\, d\lambda} \tag{2-27}$$

It should be noted that there is no detailed information about the directional distribution of reflected energy in any of the preceding definitions. The notation δq_r refers to *hemispherical* reflected flux, not a directional (i.e., partial) flux. The differential part of the notation δ refers to the fact that the flux δq_r is taken to be that which is reflected (hemispherically) for only one direction of incidence. Integration of this quantity over solid angle $\int_{2\pi} \delta q_r$ refers to integration over incident angles, not reflected angles. Figures 2-8 and 2-9 illustrate the definitions of directional and hemispherical reflectivities, respectively.

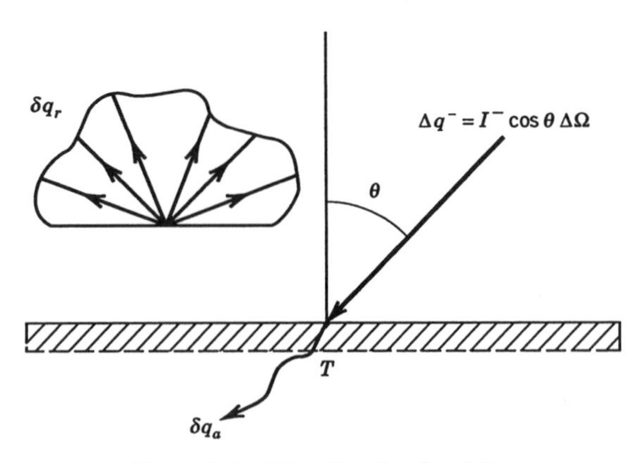

Figure 2-8 Directional reflectivity.

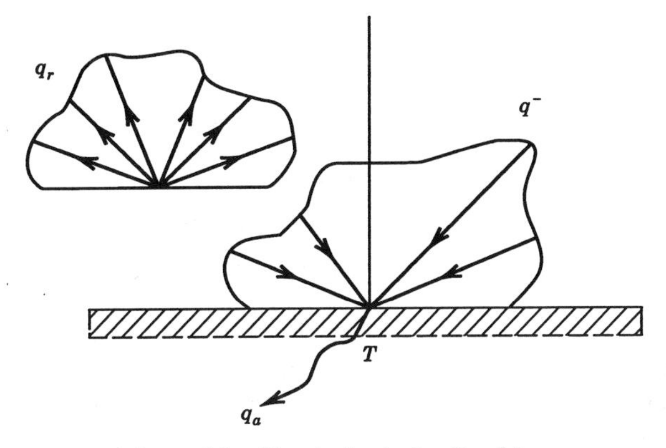

Figure 2-9 Hemispherical reflectivity.

As was the case for absorptivity, the reference flux for reflectivity is the incident flux, which, in general, is not diffuse and has a spectral distribution other than the Planck function. Therefore the total and hemispherical reflectivities (i.e., ρ_λ, ρ', and ρ) depend not only on the intrinsic material properties, as given by ρ'_λ, but also on the spectral and directional distribution of the incident intensity. For the special case of blackbody surroundings at temperature T_e (Fig. 2-6) the expressions for these integrated properties can be simplified through the use of the fractional function. The total hemispherical reflectivity can be written in terms of spectral directional reflectivity as shown in Eq. (2-28).

$$\rho(T, T_e) = \frac{1}{\pi}\int_0^1 \int_{2\pi} \rho'_\lambda(T) \cos\theta \, d\Omega \, df(\lambda T_e) \quad \text{(blackbody surroundings at } T_e\text{)} \tag{2-28}$$

Radiative Energy Balance at an Opaque Surface

The radiant flux incident upon an opaque surface (or an optically thick participating medium) must be either absorbed or reflected. This statement is true for both collimated and hemispherical incidence (see Figs. 2-8 and 2-9) and both spectrally and on a total basis.

$$\delta q_{\lambda_a} + \delta q_{\lambda_r} = \Delta q_\lambda^- \quad \text{(spectral directional)} \tag{2-29a}$$

$$q_{\lambda_a} + q_{\lambda_r} = q_\lambda^- \quad \text{(spectral hemispherical)} \tag{2-29b}$$

$$\delta q_a + \delta q_r = \Delta q^- \quad \text{(total directional)} \tag{2-29c}$$

$$q_a + q_r = q^- \quad \text{(total hemispherical)} \tag{2-29d}$$

Dividing Eqs. (2-29 a–d) by their right-hand sides gives the following set of equations.

$$\alpha'_\lambda + \rho'_\lambda = 1 \quad \text{(spectral directional)} \tag{2-30a}$$

$$\alpha_\lambda + \rho_\lambda = 1 \quad \text{(spectral hemispherical)} \tag{2-30b}$$

$$\alpha' + \rho' = 1 \quad \text{(total directional)} \tag{2-30c}$$

$$\alpha + \rho = 1 \quad \text{(total hemispherical)} \tag{2-30d}$$

These relations are expressions of a radiative energy balance at an opaque (optically thick) surface and are valid without restriction. By introducing the various forms of Kirchhoff's law, the following relations result, subject to the indicated restrictions.

$$\varepsilon'_\lambda + \rho'_\lambda = 1 \tag{2-31a}$$

$$\varepsilon_\lambda + \rho_\lambda = 1 \quad \text{(diffuse emitter-absorber or diffuse irradiation)} \tag{2-31b}$$

$$\varepsilon' + \rho' = 1 \quad \text{(gray emitter-absorber)} \tag{2-31c}$$

$$\varepsilon + \rho = 1 \quad \text{(diffuse, gray emitter-absorber or gray emitter-absorber and diffuse irradiation)} \tag{2-31d}$$

These relations, like the fundamental form of Kirchhoff's law, are subject to the condition that the material region exchanging radiation with the surroundings be in thermodynamic equilibrium. This condition is satisfied by metals at all wavelengths and by semiconductors in spectral regions where the extinction coefficient is large. The notable exception to this condition occurs in the case of dielectric materials, such as ceramics and glasses, which are usually characterized by low thermal conductivities and low extinction coefficients. The consequence of this combination of thermophysical properties is that the length scale for significant temperature change through the medium is comparable to the mean free photon path, and the isothermal requirement of Kirchhoff's law is not satisfied.

Bi-Directional Reflectivity

The preceding definitions of reflectivity have no provision for describing the directional distribution of the reflected energy, only the magnitude of the hemispherical reflected flux. To describe the directional distribution of reflected energy, a new reflectivity is defined that is a function of both the incident direction (θ, ϕ) and the reflected direction (θ_r, ϕ_r). The *spectral bi-directional reflectivity* $\rho''_\lambda(\lambda, \theta, \phi, \theta_r, \phi_r, T)$ is defined as the ratio of π times the reflected intensity over the incident partial flux.

$$\rho''_\lambda = \frac{\pi I_{\lambda_r}}{\Delta q_\lambda^-} = \frac{\pi I_{\lambda_r}(\theta_r, \phi_r, \lambda)}{I_\lambda^-(\theta, \phi, \lambda)\cos\theta\,\Delta\Omega} \tag{2-32}$$

A similar definition applies for the *total bi-directional reflectivity* $\rho''(\theta, \phi, \theta_r, \phi_r, T)$ using total instead of spectral intensities.

$$\rho'' = \frac{\pi I_r}{\Delta q^-} = \frac{\pi I_r(\theta_r, \phi_r)}{I^-(\theta, \phi)\cos\theta\,\Delta\Omega} \tag{2-33}$$

It is important to note that ρ''_λ and ρ'' are not ratios of the actual partial reflected flux to the incident partial flux. Such a definition may seem logically intuitive, given the similar form of all the preceding property definitions, but it would not be as convenient as the present definition. The present definition can still be interpreted, however, as the ratio of a hypothetical reflected flux to the incident partial flux. The hypothetical reflected flux referred to is the hemispherical flux that would be reflected if the reflected intensity field were diffuse with a magnitude of I_{λ_r} (i.e., πI_{λ_r}). The definition of ρ'' and its interpretation are depicted in Fig. 2-10.

Since ρ''_λ is not the ratio of the actual reflected flux to the incident flux, it is possible for its value to be greater than 1. This situation occurs in the case of smooth, mirror-like surfaces, which reflect specularly. The fact that the value of ρ''_λ is greater than 1 does not violate the balance of radiative energy at an interface. It must be remembered that the balance of radiative energy at an interface involves the directional properties (ρ'_λ and α'_λ) not ρ''_λ.

The directional reflectivity can be related to the bi-directional reflectivity by substituting Eq. (2-32) into (2-24).

$$\rho'_\lambda = \frac{\delta q_{\lambda_r}}{\Delta q_\lambda^-} = \frac{\int_{2\pi} I_{\lambda_r}\cos\theta_r\,d\Omega_r}{\Delta q_\lambda^-} = \frac{1}{\pi}\int_{2\pi} \rho''_\lambda \cos\theta_r\,d\Omega_r \tag{2-34a}$$

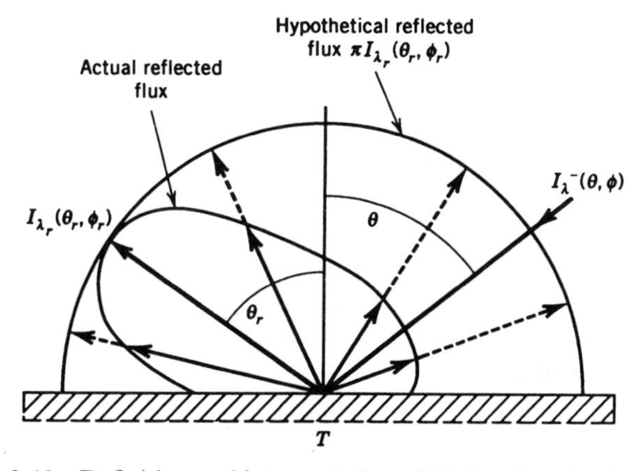

Figure 2-10 Definition and interpretation of bi-directional reflectivity.

$$\rho' = \frac{\delta q_r}{\Delta q^-} = \frac{\int_{2\pi} I_r \cos\theta_r \, d\Omega_r}{\Delta q^-} = \frac{1}{\pi}\int_{2\pi} \rho'' \cos\theta_r \, d\Omega_r \qquad \text{(2-34b)}$$

Equations (2-34a, b) show that the directional reflectivity is an integrated average of the bi-directional reflectivity over the reflectance angle for a given incidence angle. If the spectral bi-directional reflectivity is known for all wavelengths and all directions (both incident and reflected), any of the other reflectivities can be obtained by appropriate integration. Thus ρ''_λ is the most fundamental information about reflectivity that can be prescribed. As a practical matter, however, most tabulations of reflectivity are not given in terms of the bi-directional quantity since the amount of information involved is overwhelming. The detailed angular dependence of ρ'' is usually not necessary anyway.

Diffuse Reflection

There are two idealized cases of bi-directional reflectance that can be used to model the reflectance characteristics of real surfaces. One is that of a diffuse reflector and the other is a specular reflector. A *diffuse reflector* is defined as a surface or medium whose bi-directional reflectivity is independent of reflectance angle.

$$\rho''_\lambda \neq \rho''_\lambda(\theta_r, \phi_r) \quad \text{and} \quad \rho'' \neq \rho''(\theta_r, \phi_r)$$

$$\text{for a given } \theta, \phi \quad \text{(diffuse reflector)} \qquad \text{(2-35)}$$

In this case Eq. (2-34) indicates that the directional reflectivity and bi-directional reflectivity are numerically equivalent.

$$\rho'_\lambda = \rho''_\lambda \quad \text{and} \quad \rho' = \rho'' \quad \text{(diffuse reflector)} \qquad \text{(2-36)}$$

The reflected intensity distribution for a diffuse reflector is shown in Fig. 2-11. In the case of a diffuse reflector the actual reflected flux and the hypothetical reflected flux referred to earlier are identical as shown in Fig. 2-11.

Whether or not a surface is a diffuse reflector is an independent consideration from whether or not it is a diffuse emitter-absorber. Recall from Eq. (2-31a) that a definition of a diffuse emitter-absorber, which is equally valid with Eq. (2-18), is that of a surface whose directional reflectivity is independent of *incident* angle.

$$\rho'_\lambda \neq \rho'_\lambda(\theta, \phi) \quad \text{or} \quad \rho' \neq \rho'(\theta, \phi) \quad \text{(diffuse emitter-absorber)} \qquad \text{(2-37)}$$

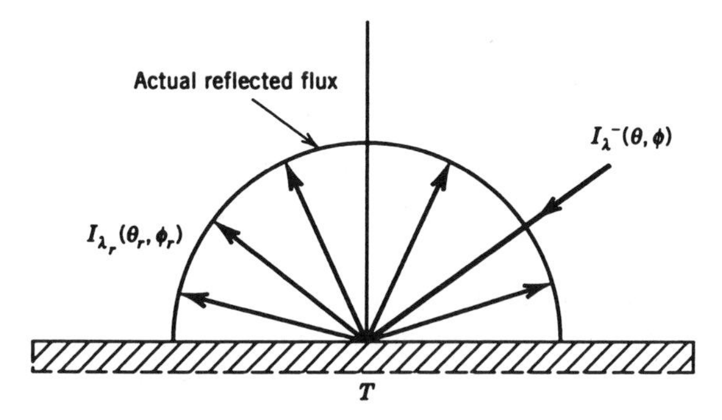

Figure 2-11 Diffuse reflector.

This case should not be confused with that of a diffuse reflector, whose bi-directional reflectivity is independent of *reflected* angle, for a given incident direction.

Specular Reflection

A *specular reflector* is a surface whose bi-directional reflectivity is zero for all angles of reflection except the specular angle ($\theta_r = \theta, \phi_r = \phi + \pi$), at which angle the bi-directional reflectivity goes to infinity. In other words, the fraction of energy incident from a given direction that is reflected is all reflected in a mirror-like fashion into a single direction (Fig. 2-12).

The bi-directional reflectivity for a specular reflector can be expressed through the use of the Dirac delta function.

$$\rho''_{\lambda} = \pi \rho'_{\lambda} \frac{\delta(\theta_r - \theta)\delta(\phi - \phi_r + \pi)}{\cos\theta \sin\theta} \tag{2-38}$$

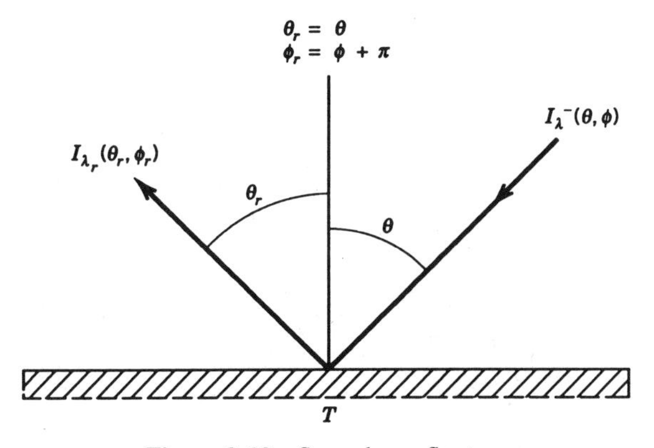

Figure 2-12 Specular reflector.

The Dirac delta function is defined as a function whose value is zero everywhere except where the argument goes to zero and there it goes to infinity.

$$\delta(x - x_0) = \begin{cases} \infty, & x = x_0 \\ 0, & x \neq x_0 \end{cases} \tag{2-39}$$

The Dirac function also has the property that when it is weighted against a continuous function $f(x)$ and integrated over the singular point x_0 the value of the integral takes on the value of the continuous function evaluated at x_0.

$$\int_0^\infty f(x)\,\delta(x - x_0)\,dx = f(x_0) \tag{2-40}$$

Equation (2-38) can be substituted into (2-34) to verify that it is the correct form of bi-directional reflectivity for a specular reflector. The term ρ'_λ appearing in (2-38) is simply the directional reflectivity.

$$\rho'_\lambda(\lambda, \theta, \phi, T) = \frac{\Delta q_{\lambda_r}}{\Delta q_\lambda^-} = \frac{I_{\lambda_r}(\theta_r, \phi_r, \lambda)\cos\theta_r\,\Delta\Omega_r}{I_\lambda^-(\theta, \phi, \lambda)\cos\theta\,\Delta\Omega} = \frac{I_{\lambda_r}(\theta_r, \phi_r, \lambda)}{I_\lambda^-(\theta, \phi, \lambda)} \tag{2-41}$$

Most real surfaces exhibit reflective behavior that is neither purely specular nor purely diffuse in character, but which is a combination of specular and diffuse behavior. One approach to modeling the reflectivity of real surfaces is to add a specular component and a diffuse component of reflectivity together (Fig. 2-13).

$$\rho'_\lambda = \rho'_{\lambda_s} + \rho'_{\lambda_d} \tag{2-42}$$

$$\rho''_\lambda = \pi\rho'_{\lambda_s}\frac{\delta(\theta_r - \theta)\,\delta(\phi - \phi_r + \pi)}{\cos\theta\sin\theta} + \rho'_{\lambda_d} \tag{2-43}$$

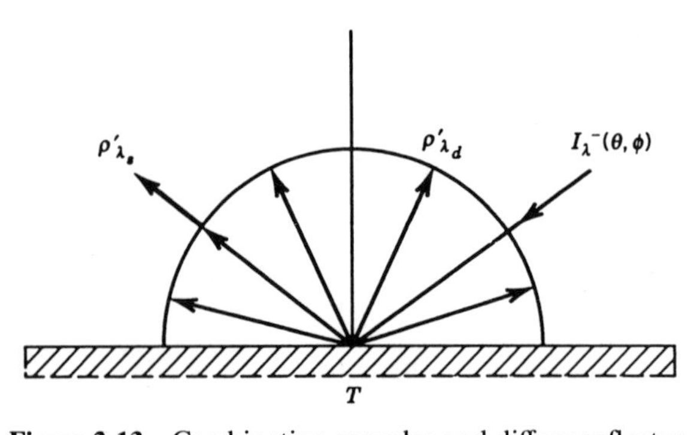

Figure 2-13 Combination specular and diffuse reflector.

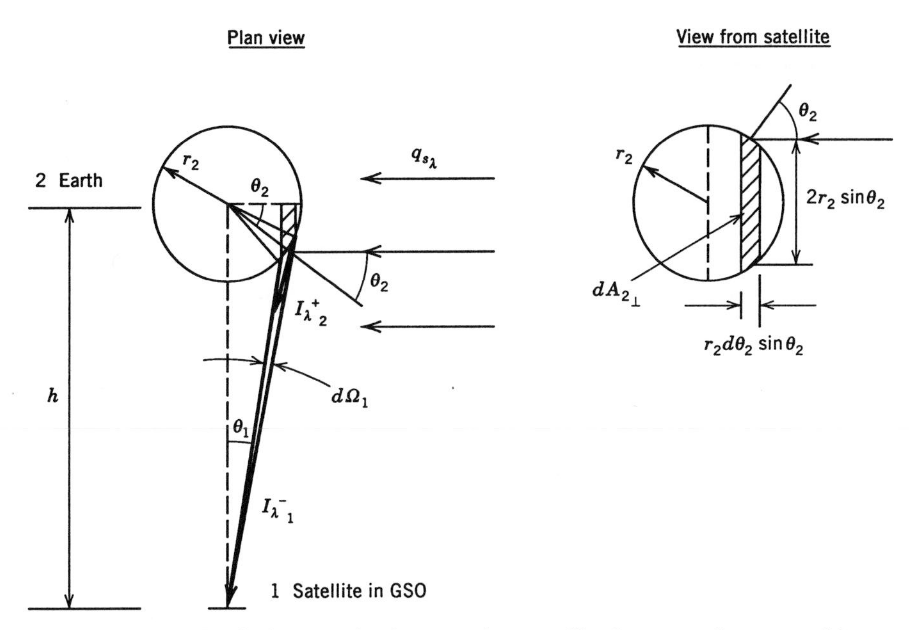

Figure 2-14 Radiative transfer from earth to satellite in geosynchronous orbit.

The primary reason for defining engineering radiative properties, such as emissivity, absorptivity and reflectivity, is so that radiative energy transfer between surfaces can be computed. Given the properties of a surface and the intensity field incident upon the surface, it is possible to calculate the radiative transfer to any other surface using just the property definitions introduced in this chapter and the definitions of intensity and flux introduced in Chapter 1. This procedure is illustrated in the following example.

Example 2-3

A satellite is in geosynchronous orbit (GSO) about the earth (Fig. 2-14). It may be assumed that the earth is a nongray, diffuse emitter and diffuse reflector with spectral hemispherical emissivity and reflectivity of ε_{λ_2} and ρ_{λ_2}. For the orientation shown (satellite 90° from direction of collimated solar incident flux, q_{s_λ}) obtain an expression for the spectral flux incident on the satellite from the earth $q_{\lambda_1}^-$ in terms of r_2, h, T_2, ε_{λ_2}, ρ_{λ_2}, and q_{s_λ}. Note that $h^2 \gg r_2^2$ for GSO.

The spectral incident flux is given by Eq. (1-13) as

$$q_{\lambda_1}^- = \int_{2\pi} I_{\lambda_1}^- \cos\theta_1 \, d\Omega_1 \tag{2-44}$$

Since $I_{\lambda_1}^-$ is nonzero only over the relatively small solid angle subtended by the earth and since $h^2 \gg r_2^2$ the $\cos\theta_1$ term can be set to 1. The intensity incident at 1 is equal to the intensity leaving 2, which is made up of the emitted and reflected contributions.

$$I_{\lambda_1}^- = I_{\lambda_2}^+ = I_{\lambda_{2r}} + I_{\lambda_{2e}} \tag{2-45}$$

For a diffuse reflector $\rho'_{\lambda_2} = \rho''_{\lambda_2}$ and for a diffuse emitter-absorber $\rho'_{\lambda_2} = \rho_{\lambda_2}$ and $\varepsilon'_{\lambda_2} = \varepsilon_{\lambda_2}$. Thus the definition of ε'_λ, Eq. (2-1), can be used to give the emitted intensity as

$$I_{\lambda_{2e}} = \varepsilon_{\lambda_2} I_{b_{\lambda 2}} \tag{2-46}$$

and the definition of ρ''_λ, Eq. (2-32), can be used to give the reflected intensity as

$$I_{\lambda_{2r}} = \frac{\rho_{\lambda_2}}{\pi}\cos\theta_2 I_{\lambda_2}^- \,\Delta\Omega = \frac{\rho_{\lambda_2}}{\pi}\cos\theta_2 q_{s_\lambda} \tag{2-47}$$

where θ_2 is the angle between the collimated solar flux and the normal to the earth's surface as shown in the figure.

The differential solid angle subtended by a uniformly illuminated strip on the earth's surface as viewed from the satellite is

$$d\Omega_1 = \frac{dA_{2\perp}}{h^2} = \frac{2r_2^2 \sin^2\theta_2 \, d\theta_2}{h^2} \tag{2-48}$$

The expression for the incident flux thus becomes

$$q_{\lambda_1}^- = \frac{2r_2^2}{h^2}\left(\varepsilon_{\lambda_2} I_{b_{\lambda 2}} \int_0^\pi \sin^2\theta_2 \, d\theta_2 + \frac{\rho_{\lambda 2}}{\pi} q_{s_\lambda} \int_0^{\pi/2} \sin^2\theta_2 \cos\theta_2 \, d\theta_2\right) \tag{2-49}$$

and since

$$\int_0^\pi \sin^2\theta_2 \, d\theta_2 = \frac{\pi}{2} \quad \text{and} \quad \int_0^{\pi/2} \sin^2\theta_2 \cos\theta_2 \, d\theta_2 = \frac{1}{3} \tag{2-50}$$

the result for the spectral incident flux at 1 is

$$q_{\lambda_1}^- = \frac{r_2^2}{h^2}\left(\varepsilon_{\lambda 2} e_{b_{\lambda 2}} + \frac{2\rho_{\lambda_2}}{3\pi} q_{s_\lambda}\right) \tag{2-51}$$

Thus there are two components of radiant flux emanating from the earth, terrestrial (low temperature) radiation and solar (high temperature) reflected radiation, which are incident on the satellite. It should be noted that there is usually a direct solar component, not considered in this example, which is also incident upon a satellite, which can impose a much larger heat load than that of the combined terrestrial and reflected solar fluxes considered in this example.

TRANSMISSIVITY

In all of the previous property definitions it was assumed that the emitting/absorbing/reflecting medium or surface was opaque and that there was no energy transmitted through a layer of some finite thickness. Now consider the case of a partially transmitting participating medium characterized by temperature T with finite thickness L, which is subject to irradiation.

The *spectral directional transmissivity* τ'_λ (also referred to as spectral *directional–hemispherical*) is defined as the ratio of spectral hemispherical transmitted flux at a given wavelength (λ) to the spectral directional (or partial) incident flux.

$$\tau'_\lambda = \frac{\delta q_{\lambda_t}}{\Delta q_\lambda^-} = \frac{\int_{2\pi} I_{\lambda_t} \cos\theta_t \, d\Omega_t}{I_\lambda^-(\lambda,\theta,\phi)\cos\theta\,\Delta\Omega} \tag{2-52}$$

The *spectral hemispherical transmissivity* τ_λ is defined as the ratio of spectral hemispherical transmitted flux to spectral hemispherical incident flux. It can also be thought of as a directional average of the spectral directional transmissivity, with the directional weighting function being the incident intensity.

$$\tau_\lambda = \frac{q_{\lambda_t}}{q_\lambda^-} = \frac{\int_{2\pi} \delta q_{\lambda_t}}{\int_{2\pi} I_\lambda^-(\lambda,\theta,\phi)\cos\theta\,d\Omega} = \frac{\int_{2\pi} \tau'_\lambda I_\lambda^- \cos\theta\,d\Omega}{\int_{2\pi} I_\lambda^- \cos\theta\,d\Omega} \tag{2-53}$$

The *total directional transmissivity* τ' (also referred to as total directional–hemispherical) is defined as the ratio of total hemispherical transmitted flux to total directional incident flux (Fig. 2-15). It can also be thought of as a spectral average of the spectral directional transmissivity, with the weighting function again being the spectral incident intensity.

$$\tau' = \frac{\delta q_t}{\Delta q^-} = \frac{\int_0^\infty \delta q_{\lambda_t}\,d\lambda}{\int_0^\infty I_\lambda^- \cos\theta\,\Delta\Omega\,d\lambda} = \frac{\int_0^\infty \tau'_\lambda I_\lambda^-\,d\lambda}{\int_0^\infty I_\lambda^-\,d\lambda} \tag{2-54}$$

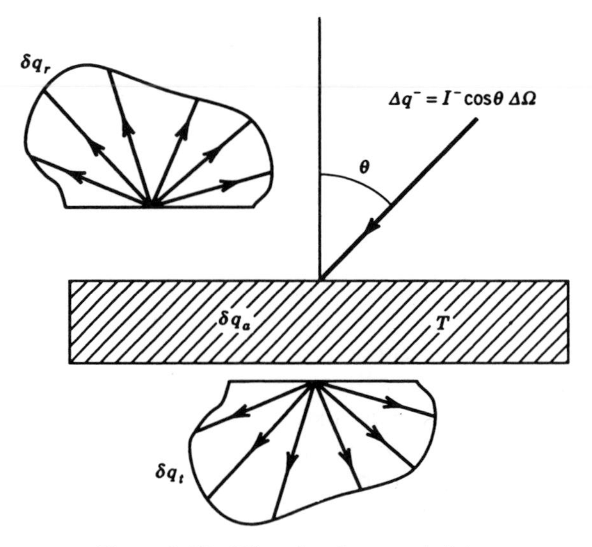

Figure 2-15 Directional transmissivity.

The *total hemispherical transmissivity* τ is defined as the ratio of total hemispherical transmitted flux to total hemispherical incident flux and has the additional interpretation of a spectral and directional average of the spectral directional transmissivity.

$$\tau = \frac{q_t}{q^-} = \frac{\int_0^\infty \int_{2\pi} \delta q_{\lambda_t} \, d\lambda}{\int_0^\infty \int_{2\pi} I_\lambda^- \cos\theta \, d\Omega \, d\lambda} = \frac{\int_0^\infty \int_{2\pi} \tau'_\lambda I_\lambda^- \cos\theta \, d\Omega \, d\lambda}{\int_0^\infty \int_{2\pi} I_\lambda^- \cos\theta \, d\Omega \, d\lambda} \tag{2-55}$$

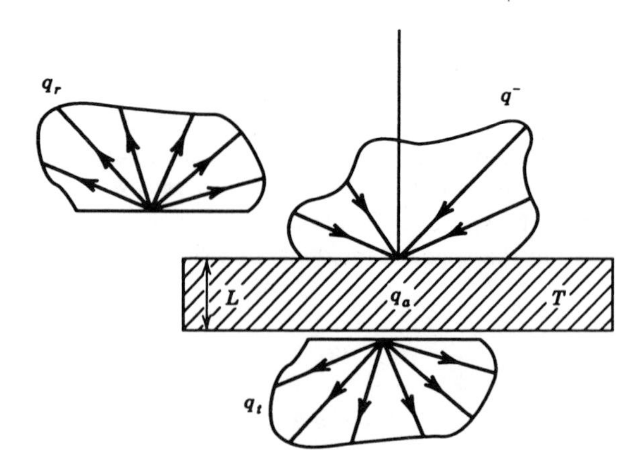

Figure 2-16 Hemispherical transmissivity.

Similar to the case of reflectivity, there is no detailed information about the directional distribution of transmitted energy in the preceding definitions of transmissivity. The notation δq_t refers to the *hemispherical* transmitted flux, not directional or partial flux. Figures 2-15 and 2-16 illustrate the definitions of directional and hemispherical transmissivities, respectively. It should also be noted that the transmissivity is a function of the thickness of the layer, L. Likewise, for a partially transmitting layer, all of the other properties (emissivity, absorptivity, and reflectivity) become functions of the layer thickness.

Radiative Energy Balance for a Transmitting Layer

The radiative energy balance for a transmitting layer is similar to that for an opaque surface except that the transmitted energy must be included.

$$\delta q_{\lambda_a} + \delta q_{\lambda_r} + \delta q_{\lambda_t} = \Delta q_\lambda^- \quad \text{(spectral directional)} \qquad \text{(2-56a)}$$

$$q_{\lambda_a} + q_{\lambda_r} + q_{\lambda_t} = q_\lambda^- \quad \text{(spectral hemispherical)} \qquad \text{(2-56b)}$$

$$\delta q_a + \delta q_r + \delta q_t = \Delta q^- \quad \text{(total directional)} \qquad \text{(2-56c)}$$

$$q_a + q_r + q_t = q^- \quad \text{(total hemispherical)} \qquad \text{(2-56d)}$$

Dividing Eqs. (2-56 a–d) by their right-hand sides gives the following set of equations.

$$\alpha'_\lambda + \rho'_\lambda + \tau'_\lambda = 1 \quad \text{(spectral directional)} \qquad \text{(2-57a)}$$

$$\alpha_\lambda + \rho_\lambda + \tau_\lambda = 1 \quad \text{(spectral hemispherical)} \qquad \text{(2-57b)}$$

$$\alpha' + \rho' + \tau' = 1 \quad \text{(total directional)} \qquad \text{(2-57c)}$$

$$\alpha + \rho + \tau = 1 \quad \text{(total hemispherical)} \qquad \text{(2-57d)}$$

These relations are an expression of a radiative energy balance on a partially transmitting layer and are valid without restriction. For a medium in thermodynamic equilibrium the various forms of Kirchhoff's law can be invoked to substitute emissivity for absorptivity in Eqs. (2-57 a–d), subject to the restrictions noted in Eqs. (2-31 a–d).

Bi-Directional Transmissivity

Since the preceding definitions of transmissivity have no means of describing the directional distribution of the transmitted energy, a new transmissivity is defined, analogous to the bi-directional reflectivity, which is a function of both the incident direction (θ, ϕ) and the transmitted direction (θ_t, ϕ_t). The *spectral bi-directional transmissivity* $\tau''_\lambda(\lambda, \theta, \phi, \theta_t, \phi_t, T, L)$ is defined as the ratio of π times the transmitted intensity over the incident partial flux.

$$\tau''_\lambda = \frac{\pi I_{\lambda_t}}{\Delta q_\lambda^-} = \frac{\pi I_{\lambda_t}(\theta_t, \phi_t, \lambda)}{I_\lambda^-(\theta, \phi, \lambda)\cos\theta\,\Delta\Omega} \tag{2-58}$$

A similar definition applies for the *total bi-directional transmissivity* $\tau''(\theta, \phi, \theta_t, \phi_t, T, L)$ with total instead of spectral intensities.

$$\tau'' = \frac{\pi I_t}{\Delta q^-} = \frac{\pi I_t(\theta_t, \phi_t)}{I^-(\theta, \phi)\cos\theta\,\Delta\Omega} \tag{2-59}$$

Like ρ''_λ and ρ'', τ''_λ and τ'' are not ratios of the actual transmitted flux to the incident flux, therefore it is possible for their values to be greater than 1. This situation occurs in the case of a clear window, which transmits specularly. The directional transmissivity can be related to the bi-directional transmissivity by substituting Eq. (2-58) into (2-52).

$$\tau'_\lambda = \frac{\delta q_{\lambda_t}}{\Delta q_\lambda^-} = \frac{\int_{2\pi} I_{\lambda_t}\cos\theta_t\,d\Omega_t}{\Delta q_\lambda^-} = \frac{1}{\pi}\int_{2\pi} \tau''_\lambda \cos\theta_t\,d\Omega_t \tag{2-60}$$

$$\tau' = \frac{\delta q_t}{\Delta q^-} = \frac{\int_{2\pi} I_t\cos\theta_t\,d\Omega_t}{\Delta q^-} = \frac{1}{\pi}\int_{2\pi} \tau'' \cos\theta_t\,d\Omega_t \tag{2-61}$$

Diffuse Transmission

A *diffuse transmitter* is a medium whose bi-directional transmissivity is independent of transmittance angle.

$$\tau''_\lambda \neq \tau''_\lambda(\theta_t, \phi_t) \quad \text{and} \quad \tau'' \neq \tau''(\theta_t, \phi_t) \quad \text{(diffuse transmitter)} \tag{2-62}$$

$$\tau'_\lambda = \tau''_\lambda \quad \text{and} \quad \tau' = \tau'' \quad \text{(diffuse transmitter)} \tag{2-63}$$

Examples of nearly diffuse transmitters are frosted glass and opal diffusers. The transmitted intensity distribution for a diffuse transmitter is shown in Fig. 2-17. Whether or not a medium is a diffuse transmitter is an independent consideration from whether or not it is a diffuse absorber-reflector. A diffuse absorber-reflector is a medium whose directional transmissivity is independent of *incident* angle.

$$\tau'_\lambda \neq \tau'_\lambda(\theta, \phi) \quad \text{or} \quad \tau' \neq \tau'(\theta, \phi) \quad \text{(diffuse absorber-reflector)} \tag{2-64}$$

This case is not to be confused with that of a diffuse transmitter, whose bi-directional transmissivity is independent of *transmitted* angle.

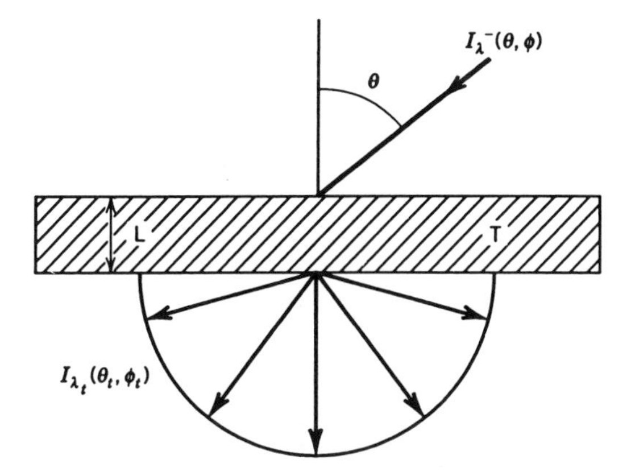

Figure 2-17 Diffuse transmitter.

Specular Transmission

A *specular transmitter* is a medium whose bi-directional transmissivity is zero for all angles of transmission except the specular angle, where the bi-directional transmissivity goes to infinity. In other words, the fraction of energy incident from a given direction that is transmitted is all transmitted as through a clear window into a single direction (Fig. 2-18). The bi-directional transmissivity for a specular transmitter can also be expressed through the use of the Dirac delta function

$$\tau_\lambda'' = \pi\tau_\lambda' \frac{\delta(\theta_t - \theta)\delta(\phi - \phi_t + \pi)}{\cos\theta \sin\theta} \tag{2-65}$$

where τ_λ' is simply the directional transmissivity.

$$\tau_\lambda'(\lambda,\theta,\phi,T,L) = \frac{\Delta q_{\lambda_t}}{\Delta q_\lambda^-} = \frac{I_{\lambda_t}(\theta_t,\phi_t,\lambda)\cos\theta_t\,\Delta\Omega_t}{I_\lambda^-(\theta,\phi,\lambda)\cos\theta\,\Delta\Omega} \tag{2-66}$$

Example 2-4

On an overcast day the earth's atmosphere can be modeled as a diffusely transmitting, plane-parallel layer with a collimated solar flux q_s incident on the top of the layer at an angle of θ_0 (Fig. 2-19). If the directional transmissivity of the atmosphere for the solar spectrum at this angle of incidence is τ', what is the solar intensity distribution incident at the surface of the earth? Neglect the fraction of q_s transmitted that undergoes multiple reflection between the earth and atmosphere.

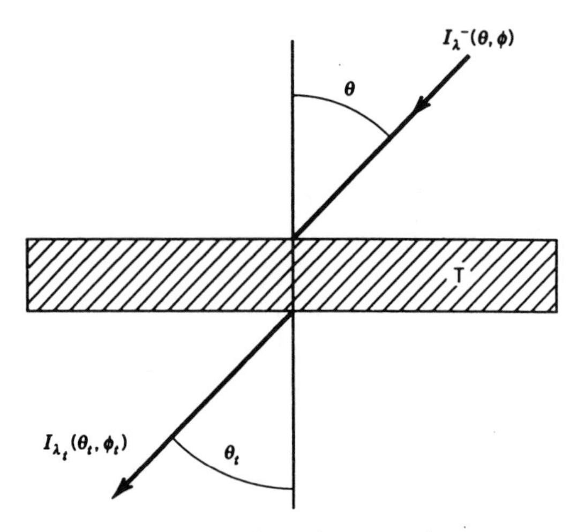

Figure 2-18 Specular transmitter.

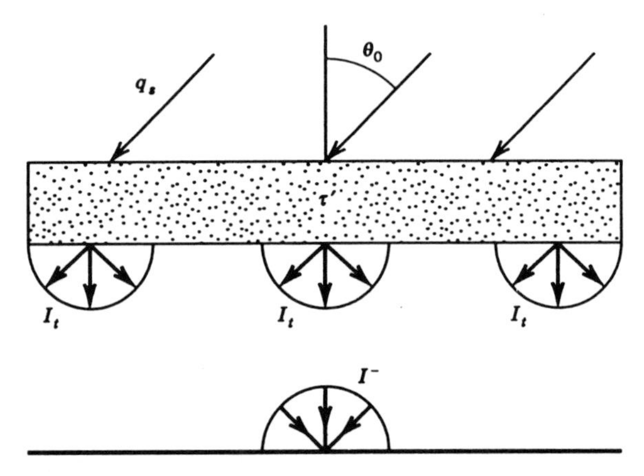

Figure 2-19 Diffusely transmitting layer of clouds.

The intensity incident at the earth's surface I^- is equal to the transmitted intensity leaving the cloud layer I_t. For a diffusely transmitting layer the transmitted intensity is independent of direction and thus $\tau' = \tau''$. Therefore the definition of τ'', Eq. (2-59), can be used to give the incident intensity as

$$I^- = I_t = \frac{\tau'}{\pi} \cos \theta_0 I^- \, \Delta\Omega = \frac{\tau'}{\pi} \cos \theta_0 q_s \qquad (2\text{-}67)$$

This example again illustrates how, with a knowledge of the incident intensity field and engineering radiative properties of a surface or medium, the radiative transfer to any other surface can be calculated using the property definitions and the definitions of intensity and flux.

RADIATIVE PROPERTIES OF OPAQUE SOLIDS

A large number of radiative property measurements have been made for various solid materials. These results have been reported in technical journals, company reports, and conference proceedings. Extensive compilations of some of these measurements can be found in Gubareff et al. [1960] and Touloukian [1970]. A limited tabulation of total emissivity is given in Table 2-1.

The values for radiative properties of solid surfaces depend on the thermodynamic state and the geometry of the surface. Temperature and composition (mole numbers) are the most important thermodynamic state variables influencing the radiative properties of a solid surface. Pressure is usually unimportant. Surface roughness has a strong influence on radiative properties. Thus it is important to try to ensure similarity of temperature, composition, and surface condition when adopting reported values of radiative properties for use in any heat transfer analysis.

For purposes of describing their radiative properties, solids are usually classified according to whether they are conductors (metals) or insulators (nonmetals). Metals usually have low total emissivities while nonmetals, if they are thick enough to be opaque, usually have high total emissivities, as can be seen from Table 2-1.

The key to understanding the total emissivity behavior of solids is the spectral emissivity, as shown in Fig. 2-20. Over the visible and infrared region, the spectral emissivity of metals usually decreases moderately with increasing wavelength while that for nonmetals usually increases, often rapidly in the near infrared region. As temperature increases, the spectral emissivity of metals usually increases slightly. (A similar generalization about the spectral emissivity of nonmetals is difficult to make.) The underlying physical reasons for these trends are discussed in Chapter 5.

The trends shown in Fig. 2-20 for spectral emissivity help explain the observed values of total emissivity. For metals the spectral emissivity is low over the entire visible and near infrared spectrum. Thus the total emissivity is also generally low. For nonmetals the spectral emissivity is usually high in the infrared region, where most of the spectral energy of the Planck function is located for temperatures between room temperature and several thousand degrees. Thus the total emissivity of nonmetals is usually high. As temperature increases the total emissivity of metals increases slightly because the energy distribution shifts to shorter wavelengths and because the intrinsic emissivity also increases. For nonmetals, as temperature increases the total emissivity usually decreases because of the shift in the Planck function toward shorter wavelengths.

TABLE 2-1 Total Emissivity Data

Material	300 K	500 K	800 K	1600 K
		Metals		
Aluminum				
Smooth, polished	0.04	0.05	0.08	0.19
Smooth, oxidized	0.11	0.12	0.18	
Rough, oxidized	0.2	0.3		
Anodized	0.9	0.7	0.6	0.3
Brass				
Highly polished		0.03		
Polished	0.1	0.1		
Oxidized	0.6			
Chromium				
Polished	0.08	0.17	0.26	0.40
Copper				
Polished	0.04	0.05	0.18	0.17
Oxidized	0.87	0.83	0.77	
Gold				
Highly polished	0.02		0.035	
Iron and steel				
Iron, polished	0.06	0.08	0.1	0.2
Iron, oxidized	0.6	0.7	0.8	
Stainless, polished	0.1	0.2		
Stainless, aged	0.5-0.8			
Mild steel, polished	0.1		0.3	
Mild steel, oxidized	0.8	0.8		
Lead				
Polished	0.05	0.08		
Oxidized	0.6	0.6		
Mercury				
Clean	0.1			
Magnesium				
Polished	0.07	0.13	0.18	0.24
Nichrome (wire)				
Clean	0.65		0.71	
Oxidized	0.95	0.98		
Nickel				
Polished	0.05	0.07		
Oxidized	0.4	0.5		
Platinum				
Polished	0.05		0.1	
Oxidized	0.07		0.1	
Silver				
Polished	0.01	0.02	0.03	
Oxidized	0.02		0.04	
Tin				
Polished	0.05			
Tungsten				
Filament	0.032	0.053	0.088	0.35 (3500 K)

TABLE 2-1 (*Continued*)

Material	300 K	500 K	800 K	1600 K
		Metals		
Zinc				
Polished	0.02	0.03		
Oxidized		0.1		
Galvanized	0.02-0.03			
		Nonmetals		
Aluminum oxide		0.7	0.6	0.4
Asbestos	0.95			
Asphalt	0.93			
Brick				
Alumina refractory			0.40	0.33
Fireclay	0.9		0.8	0.8
Kaolin insulating			0.70	0.53
Magnesite refractory	0.9			0.4
Red, rough	0.9			
Silica	0.9		0.8	0.8
Concrete (rough)	0.94			
Glass	0.95	0.9	0.7	
Graphite	0.7			0.8
Ice (273 K)				
Smooth	0.97			
Rough	0.99			
Limestone	0.9	0.8		
Marble (white)	0.95			
Mica	0.75			
Paints				
Aluminized	0.3-0.6			
Most others (incl. white)	0.9			
Paper	0.90-0.98			
Porcelain (glazed)	0.92			
Pyrex	0.82	0.80	0.72	0.6
Quartz	0.9		0.6	
Rubber				
Hard	0.95			
Soft, gray, rough	0.86			
Sand (silica)	0.9			
Silicon carbide		0.9		0.8
Skin	0.95			
Snow	0.8-0.9			
Soil	0.93-0.96			
Rocks	0.88-0.95			
Teflon	0.85	0.92		
Vegetation	0.92-0.96			
Water (> 0.1 mm thick)	0.96			
Wood	0.8-0.9			

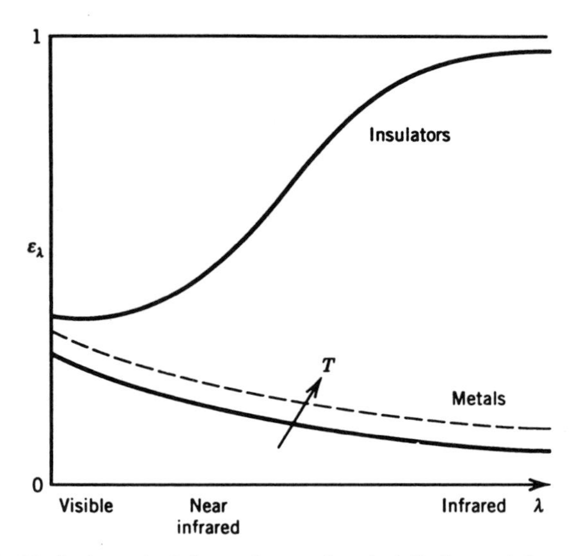

Figure 2-20 Typical spectral dependence of emissivity for metals and insulators.

The general trends for the directional distribution of emissivity for metals and nonmetals is shown in Fig. 2-21. Both metals and nonmetals exhibit diffuse emissive behavior in directions near the normal to the surface. As polar angle θ increases, metals and nonmetals exhibit opposite behavior; the emissivity of insulators tends to decrease uniformly, approaching zero at grazing angles ($\theta = \pi/2$) whereas the emissivity of metals increases, reaching a peak value before decreasing sharply to zero at $\theta = \pi/2$. These trends are corroborated by corresponding predictions of directional reflectivity from electromagnetic theory (see Chapter 4). Thus the hemispherical emissivity is usually slightly greater than the normal emissivity for metals while the opposite is true for nonmetals. As a practical matter, the difference between the normal and hemispherical emissivities is seldom enough to warrant a

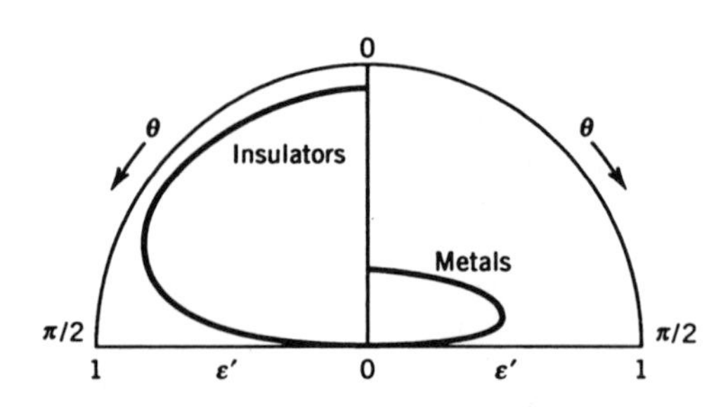

Figure 2-21 Typical distribution of directional emissivity for metals and insulators.

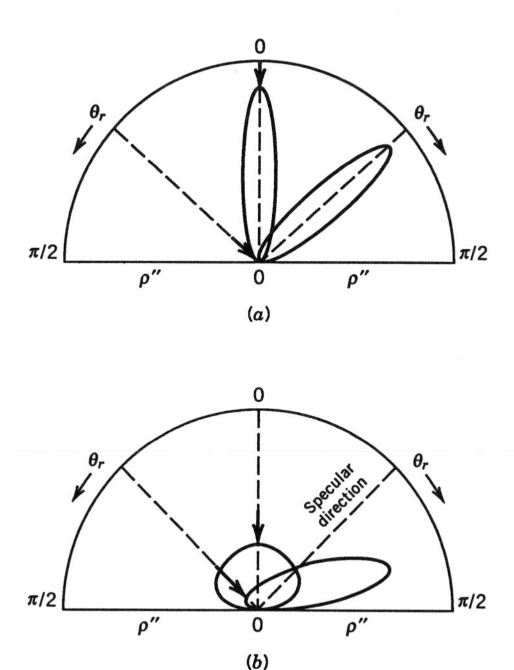

Figure 2-22 Typical variation of bi-directional reflectivity versus reflectance angle θ_r for normal ($\theta = 0$) and off-normal ($\theta \sim \pi/4$) incidence and for (*a*) relatively smooth surfaces $\sigma_{\text{rms}}/\lambda < 1$ and (*b*) relatively rough surfaces $\sigma_{\text{rms}}/\lambda > 1$.

distinction (the values in Table 2-1 may be taken to be either). The uncertainty associated with surface condition is usually larger than the difference due to directional integration. Occasionally azimuthal variations in emissivity are also important. For example, grooved surfaces exhibit azimuthally dependent properties [Hesketh et al., 1988].

One factor that strongly influences surface radiative properties is roughness. Surface roughness is usually characterized by the nondimensional parameter $\sigma_{\text{rms}}/\lambda$ where σ_{rms} is the root-mean-square surface roughness and λ is wavelength. For a surface with $\sigma_{\text{rms}}/\lambda \ll 1$ the surface is optically smooth and the results from electromagnetic theory for interaction of a plane wave with a plane interface hold (see Chapter 4). As σ_{rms} increases (with fixed λ), the hemispherical emissivity and absorptivity increase through the mechanism of multiple reflection between surface asperities (the blackbody cavity effect). This trend can be observed in the total emissivity values of Table 2-1. Multiple reflection associated with surface roughness also influences the bi-directional reflectivity as shown in Fig. 2-22. In Fig. 2-22, the typical variation of bi-directional reflectivity with reflected angle θ_r is shown for two angles of incidence, normal ($\theta = 0$) and off-normal ($\theta \sim \pi/4$). Figure

2-22*a* shows the trend for a relatively smooth surface ($\sigma_{\text{rms}}/\lambda < 1$) and Fig. 2-22*b* shows the trend for a relatively rough surface ($\sigma_{\text{rms}}/\lambda > 1$). The smooth surface (*a*) exhibits strong specular reflection for both angles of incidence as evidenced by strong peaks in the specular direction. In contrast the rough surface (*b*) exhibits relatively diffuse reflection, especially for normal incidence. Comparing the trends in (*a*) and (*b*) it can be seen that the effect of increasing roughness is to decrease ρ'' in the specular direction and increase ρ'' in most off-specular directions. It is also interesting that for rough surfaces with grazing incidence ($\pi/4 < \theta < \pi/2$) the peak value of ρ'' usually occurs at $\theta_r > \theta$. Several investigators have studied the effects of roughness on radiative properties [Beckman and Spizzichino, 1963; Porteus, 1963; Torrance and Sparrow; 1967]. Some theoretical relations have been proposed, but these effects are still difficult to predict and best results are obtained by recourse to measurements.

Oxidation is also an important factor that influences metal radiative properties. Oxidation affects radiative properties through two possible mechanisms. One mechanism is an increase in surface roughness. For an initially optically smooth surface, the buildup of an oxide layer can represent a considerable increase in surface roughness if the oxide layer is sufficiently thick. The other mechanism is a change in chemical composition. Both of these mechanisms generally result in an increase in total emissivity. This trend can be seen in some of the entries in Table 2-1.

Since the foregoing discussion has been restricted to opaque solid surfaces, it is well to consider what conditions are necessary to ensure opaqueness of a solid surface. Opaqueness is defined as negligible transmission through a given specimen and is thus strongly dependent on both the extinction coefficient of the material and the thickness of the specimen. Metals have very large extinction coefficients over the entire visible and infrared spectrum. As a result, most metal structures are opaque, except for very thin vapor deposited films (e.g., two-way mirrors). Nonmetallic solids generally have relatively small extinction coefficients, especially in the visible region and especially if the specimen is free of impurities and porosity. To be considered opaque some nonmetals (such as glass in the visible region) must acquire a thickness of several meters. For more information about the extinction coefficient and optical constants of metals and nonmetals see Chapters 4 and 5.

SIMPLE RADIATIVE TRANSFER

It has been previously stated in this chapter that the purpose of defining radiative surface properties is to be able to use these definitions and the definitions of intensity and flux to compute the net radiative transfer rate at a surface that is exchanging energy radiatively with other surfaces. This entails solving the so-called transfer problem. Solution of the transfer problem will

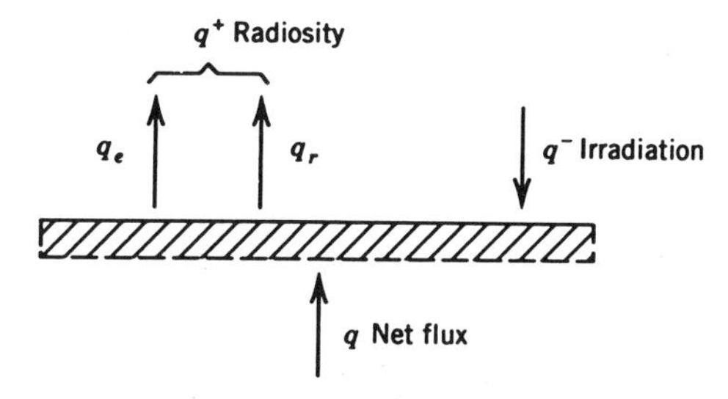

Figure 2-23 Net radiative flux at an opaque surface.

next be introduced formally by considering the net radiative flux at an opaque surface with unspecified incident radiation. Two simple cases will then be presented of incident radiation from a point source and from a large, isothermal surroundings.

Net Radiative Flux at an Opaque Surface

Consider an opaque surface that is exchanging energy radiatively with its surroundings. Energy is supplied by conduction to the bottom side of the interface (Fig. 2-23) and energy leaves the top side of the interface by radiation. Conduction or convection of energy away from the top side is temporarily assumed to be negligible.

The heat flux conducted to the interface is q. This flux must be equal to the net radiative flux away from the top of the interface, $q^+ - q^-$.

$$q = q^+ - q^- \tag{2-68}$$

By noting that the radiosity is the sum of the emitted flux and the reflected flux

$$q^+ = q_e + q_r \tag{2-69}$$

and using the result from the radiative energy balance at an opaque interface, Eq. (2-29d), the net flux can also be written as the difference between the emitted flux and the absorbed flux.

$$q = q_e - q_a \tag{2-70}$$

Using the definition of total hemispherical emissivity, Eq. (2-4), the emitted flux is given by

$$q_e = \varepsilon e_b(T_s) \tag{2-71}$$

where ε is given by Eq. (2-8) and T_s is the surface temperature. Using the definition of total hemispherical absorptivity, Eq. (2-12), the absorbed flux is given by

$$q_a = \alpha q^- \tag{2-72}$$

where q^- can be obtained from

$$q^- = \int_0^\infty \int_{2\pi} I_\lambda^- \cos\theta \, d\Omega \, d\lambda \tag{2-73}$$

and α can be obtained from (2-12). The net heat flux away from the surface is thus

$$q = \varepsilon \sigma T_s^4 - \alpha q^- \tag{2-74}$$

If heat convection from the surface is included in addition to radiation, the previous equations still hold. However, q given by (2-74) would represent only the net radiative heat flux q_{rad} and a convective contribution would have to be included to give the total net heat flux.

$$q = q_{\text{rad}} + q_{\text{conv}} \tag{2-75}$$

$$q_{\text{conv}} = h(T_s - T_\infty) \tag{2-76}$$

In order to determine the net radiative heat flux, it still remains to determine the incident hemispherical flux q^- and the absorptivity α, which requires a knowledge of the incident intensity distribution $I_\lambda^-(\theta, \phi)$. Determining $I_\lambda^-(\theta, \phi)$ in general requires a more detailed analysis of radiative transfer in the surroundings, which is the subject of subsequent chapters. However, there are two cases that commonly arise in which the incident flux in Eq. (2-73) can be easily determined without a detailed analysis of the surroundings. These two cases correspond to radiation incident on a flat or convex surface (1) from large, isothermal surroundings and (2) from a point source.

Incident Radiation from Large, Isothermal (Blackbody) Surroundings

Whenever a surface that is flat or convex exchanges radiation with surroundings that are much larger in area and reasonably isothermal, the surroundings will act like a blackbody and the radiation incident on the smaller surface will be blackbody radiation based on the temperature of the surroundings. This can be understood by recalling the concept of a hohlraum or isothermal enclosure from Chapter 1 and considering the small opening in the enclosure as if it were a surface at temperature T_s and the enclosure as a much larger surface at temperature T_e (Fig. 2-24).

Since a flat or convex surface cannot "see itself" and hence cannot exchange radiation with itself, the irradiation q^- comes entirely from the surroundings. The incident intensity I_λ^- is simply $I_{b_\lambda}(T_e)$. Thus Eq. (2-73) can

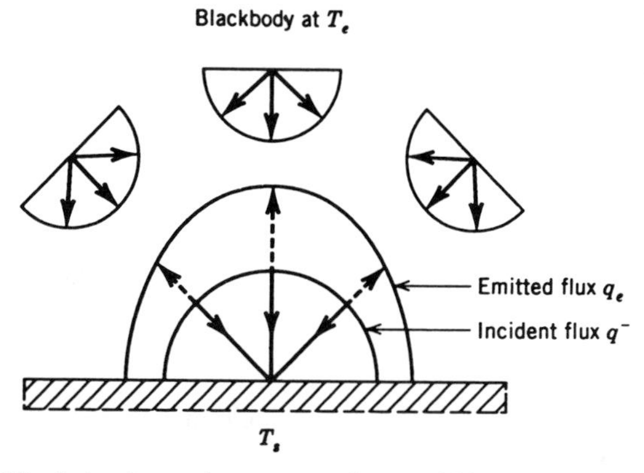

Figure 2-24 Interchange between surface and blackbody surroundings.

be evaluated to give the incident flux as the blackbody hemispherical flux based on the surroundings temperature T_e.

$$q^- = e_b(T_e) = \sigma T_e^4 \tag{2-77}$$

Substituting Eq. (2-77) into (2-74) gives the well-known expression for the net radiant heat flux from an opaque surface to blackbody surroundings.

$$q = \varepsilon\sigma T_s^4 - \alpha\sigma T_e^4 \quad \text{(nongray emitter-absorber)} \tag{2-78}$$

The absorptivity α can be determined using Eq. (2-14). The restrictions associated with Eq. (2-78) are:

1. The surroundings are much larger than the surface ($A_e \gg A_s$).
2. The surroundings are reasonably isothermal at T_e.
3. The surface A_s cannot see itself.

If, in addition to conditions 1 to 3 above, the surface is assumed to be a gray emitter-absorber, Eq. (2-23) applies, giving

$$q = \varepsilon\sigma\left(T_s^4 - T_e^4\right) \quad \text{(gray emitter-absorber)}. \tag{2-79}$$

Incident Radiation from a Point Source

Whenever a surface receives radiation from a point source* the incident flux will be collimated in a narrow solid angle $\Delta\Omega_0$ about a single direction θ_0 (see Fig. 2-25). If there is negligible radiation coming from the rest of the

*A point source is any surface whose projected area A is small relative to the square of the distance from the receiving surface s, i.e., $A \ll s^2$ (see Example 1-1).

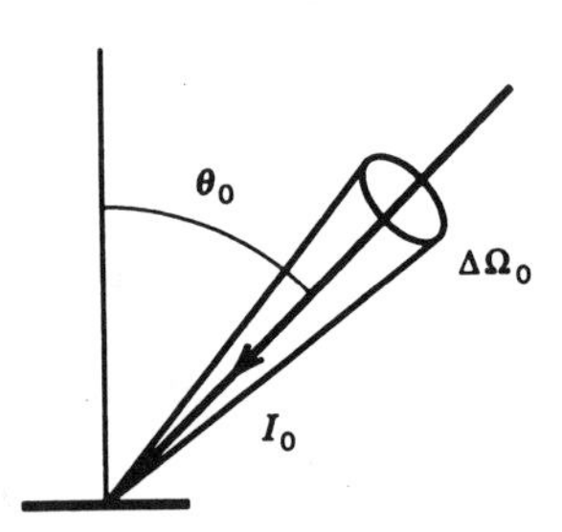

Figure 2-25 Collimated flux from a point source.

surroundings and the magnitude of the incident spectral intensity arriving from the point source is I_{λ_0}, the incident intensity distribution will be

$$I_\lambda^- = \begin{cases} I_{\lambda_0}, & \text{over } \Delta\Omega_0 \\ 0, & \text{all other directions} \end{cases} \tag{2-80}$$

Substituting (2-80) into (2-73) and (2-12) gives

$$q^- = \int_0^\infty \int_{\Delta\Omega_0} I_{\lambda_0} \cos\theta \, d\Omega \, d\lambda \tag{2-81}$$

and

$$\alpha = \frac{\displaystyle\int_0^\infty \int_{\Delta\Omega_0} \alpha'_\lambda I_{\lambda_0} \cos\theta \, d\Omega \, d\lambda}{\displaystyle\int_0^\infty \int_{\Delta\Omega_0} I_{\lambda_0} \cos\theta \, d\Omega \, d\lambda} \tag{2-82}$$

By definition, the solid angle subtended by a point source is very small, $\Delta\Omega_0 \ll 1$, and it is therefore usually reasonable to assume that θ is constant at θ_0 and α'_λ is constant at α'_{λ_0} over the interval $\Delta\Omega_0$ (Problem 2-12 is an exception to this rule). Furthermore, it is often reasonable to assume that the intensity from the source I_{λ_0} is also uniform over the interval $\Delta\Omega_0$ (Example 2-3 is an exception to this case). Thus, the incident flux becomes

$$q^- = I_0 \, \Delta\Omega_0 \cos\theta_0 \tag{2-83}$$

where

$$I_0 = \int_0^\infty I_{\lambda_0} \, d\lambda \tag{2-84}$$

and the absorptivity is

$$\alpha = \alpha'_0 = \frac{\displaystyle\int_0^\infty \alpha'_{\lambda_0} I_{\lambda_0} \, d\lambda}{I_0} \tag{2-85}$$

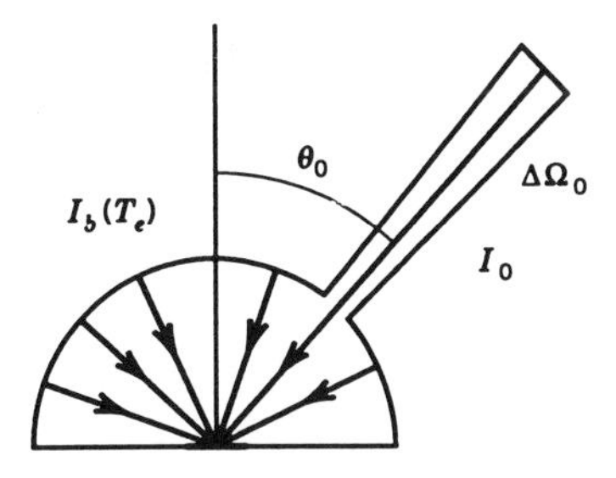

Figure 2-26 Incident radiation from point source and large isothermal surroundings.

The net radiative flux leaving the surface can then be obtained from Eq. (2-74) as

$$q = \varepsilon\sigma T_s^4 - \alpha'_0 I_0 \, \Delta\Omega_0 \cos\theta_0 \tag{2-86}$$

Example 2-5

A flat, opaque surface at T_s is exposed to a large isothermal environment at T_e as well as a point source at an angle of θ_0 relative to the normal to the surface (see Fig. 2-26). The solid angle subtended by the point source is $\Delta\Omega_0$ and the spectral intensity from the point source is I_{λ_0}. The spectral emissivity/absorptivity of the surface is $\varepsilon'_\lambda = \alpha'_\lambda$ for arbitrary direction and α'_{λ_0} for the direction of collimated incidence. Obtain an expression for the net radiant flux from the surface to the surroundings in terms of T_s, T_e, θ_0, $\Delta\Omega_0$, I_0, ε, α, and α'_0

Combining Eqs. (2-73) and (2-12), the absorbed flux can be written as

$$q_a = \alpha q^- = \int_0^\infty \int_{2\pi} \alpha'_\lambda I_\lambda^- \cos\theta \, d\Omega \, d\lambda \tag{2-87}$$

The incident intensity distribution is

$$I_\lambda^- = \begin{cases} I_{\lambda_0}, & \text{over } \Delta\Omega_0 \\ I_{b_\lambda}(T_e), & \text{all other directions} \end{cases} \tag{2-88}$$

Substituting the intensity distribution (2-88) into (2-87) gives

$$q_a = \int_0^\infty \int_{2\pi - \Delta\Omega_0} \alpha'_\lambda I_{b_\lambda}(T_e) \cos\theta \, d\Omega \, d\lambda + \int_0^\infty \int_{\Delta\Omega_0} \alpha'_\lambda I_{\lambda_0} \cos\theta \, d\Omega \, d\lambda \tag{2-89}$$

which can also be written as

$$q_a = \int_0^\infty \int_{2\pi} \alpha'_\lambda I_{b_\lambda}(T_e) \cos\theta \, d\Omega \, d\lambda + \int_0^\infty \int_{\Delta\Omega_0} \alpha'_\lambda \left[I_{\lambda_0} - I_{b_\lambda}(T_e) \right] \cos\theta \, d\Omega \, d\lambda \tag{2-90}$$

For a point source ($\Delta\Omega_0 \ll 1$) it is reasonable to assume that θ and α'_λ are constant over $\Delta\Omega_0$ at the values of θ_0 and α'_{λ_0}. Furthermore, if it is assumed that I_{λ_0} is constant over $\Delta\Omega_0$, the integration over $\Delta\Omega_0$ becomes trivial, giving

$$q_a = \int_0^\infty \int_{2\pi} \alpha'_\lambda I_{b_\lambda}(T_e)\cos\theta \, d\Omega \, d\lambda + \int_0^\infty \alpha'_{\lambda_0}\left[I_{\lambda_0} - I_{b_\lambda}(T_e)\right] d\lambda \, \Delta\Omega_0 \cos\theta_0 \tag{2-91}$$

By defining the total directional absorptivity as

$$\alpha'_0 = \frac{\int_0^\infty \alpha'_{\lambda_0}\left[I_{\lambda_0} - I_{b_\lambda}(T_e)\right] d\lambda}{I_0 - I_b(T_e)} \tag{2-92}$$

and the total hemispherical absorptivity as usual [i.e., Eq. (2-14)] the absorbed flux becomes

$$q_a = \alpha\sigma T_e^4 + \alpha'_0\left[I_0 - I_b(T_e)\right] \Delta\Omega_0 \cos\theta_0 \tag{2-93}$$

The net radiant flux from the surface to the surroundings is thus

$$q = \varepsilon\sigma T_s^4 - \left\{\alpha\sigma T_e^4 + \alpha'_0\left[I_0 - I_b(T_e)\right] \Delta\Omega_0 \cos\theta_0\right\} \tag{2-94}$$

These results can be further simplified in one of two ways depending on the relative magnitudes of I_0 and $I_b(T_e)$. If $I_0 \gg I_b(T_e)$, then $I_b(T_e)$ can be ignored relative to I_0 in Eqs. (2-92) through (2-94). If I_0 is the same order of magnitude as $I_b(T_e)$ or less, the collimated flux can be ignored completely since it was assumed that $\Delta\Omega_0 \ll 1$ and $\cos\theta_0$ is of order 1.

Example 2-6

During the lunar day, the temperature of the moon's surface can be reasonably estimated by assuming the lunar surface is adiabatic. Take the moon to be a diffuse, nongray absorber-emitter with a solar absorptivity of 0.90 and an emissivity of 0.97. Assume the sun is a blackbody emitter at 5785 K at a distance of 108 sun diameters (total incident flux, $q_s = 1360$ W/m^2).

(a) Determine the maximum temperature on the moon's surface.

(b) Determine the necessary angle between the sun's rays and the normal to the lunar surface (zenith angle θ; see Fig. 2-27) to experience a temperature of 60°F (289 K).

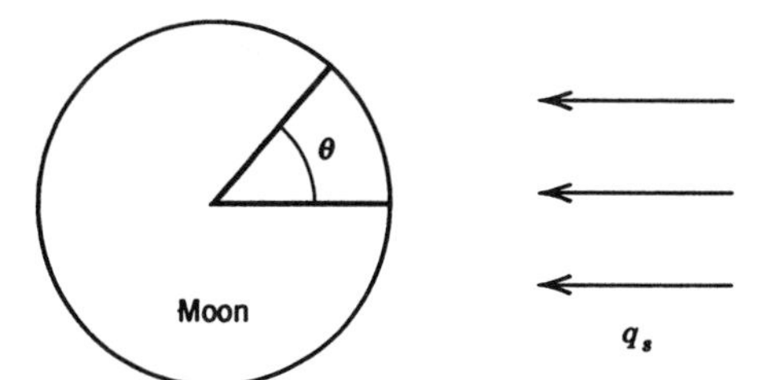

Figure 2-27 Angle between collimated solar flux and local normal to lunar surface.

Since the surface of the moon is locally adiabatic the emitted flux must equal the absorbed flux.

$$\varepsilon \sigma T^4 = \alpha q_s \cos\theta \qquad (2\text{-}95)$$

(a) The maximum temperature occurs at $\theta = 0$. Thus Eq. (2-95) can be solved for temperature to give

$$T_{\max} = \left[\frac{(0.9)(1360 \text{ W/m}^2)(1)}{(0.97)(5.67 \times 10^{-8} \text{ W/m}^2\text{K}^4)} \right]^{1/4} = 386 \text{ K} \qquad (2\text{-}96)$$

(b) For a temperature of 289 K, Eq. (2-95) can be solved for the zenith angle giving $\theta = 72°$.

This simple model gives a reasonable estimate of the lunar surface temperature during the lunar day. During the lunar night, energy stored in the lunar soil is conducted to the surface where it is re-emitted to deep space, thus preventing the surface temperature from falling to absolute zero.

SUMMARY

The purpose of defining surface radiative properties (emissivity, absorptivity, reflectivity, and transmissivity) is to be able to compute the net radiative heat flux at a surface that is exchanging energy radiatively with surrounding surfaces and participating media. In general these properties are functions of both wavelength and direction. Gray surfaces are those whose properties are independent of wavelength. Diffuse surfaces are those whose properties are independent of direction. Metals usually have low emissivities, low absorptivities, and high reflectivities. Nonmetals usually have high emissivities, high absorptivities, and low reflectivities. In general, determination of the net radiant flux at a surface requires an analysis of radiative transfer in the environment surrounding the surface. Two simple cases that do not require a complicated analysis of the surroundings are: (1) when the incident radiation is blackbody radiation from a large, isothermal surroundings and (2) when the incident radiation is collimated radiation from a point source.

REFERENCES

Beckmann P. and Spizzichino, A. (1963), *The Scattering of Electromagnetic Waves from Rough Surfaces*, Macmillan, New York.

Gubareff, G. G., Janssen, J. E., and Torborg, R. H. (1960), *Thermal Radiation Properties Survey*, Minneapolis, MN, Honeywell Research Center.

Hesketh, P. J., Gebhart, B., and Zemel, J. N. (1988), "Measurements of the Spectral and Directional Emission from Microgrooved Silicon Surfaces," *J. Heat Transfer*, Vol. 110, pp. 680–686.

Porteus, J. O. (1963), "Relation Between the Height Distribution of a Rough Surface and the Reflectance at Normal Incidence," *J. Opt. Soc. Am.*, Vol. 53, pp. 1394–1402.

Torrance, K. E., and Sparrow, E. M. (1967), "Theory for Off-Specular Reflectance from Roughened Surfaces," *J. Opt. Soc. Am.*, Vol. 57, pp. 1105–1114.

Touloukian, Y. S. (1970), *Thermal Radiative Properties*, Vols. 7–9, Plenum Publishing, New York.

PROBLEMS

1. Determine the total hemispherical emissivity of tungsten at 2000 K using the approximate representation of the spectral hemispherical emissivity shown in Fig. 2-4.

2. Consider a surface at temperature T subject to diffuse blackbody irradiation at temperature T_e. Write an integral expression for the following integrated absorptivities in terms of the given distributed ones.
   (a) α_λ in terms of α'_λ
   (b) α' in terms of α'_λ
   (c) α in terms of α'
   (d) α in terms of α_λ

3. A spectrally selective, nondiffuse, nongray emitting-absorbing surface at 500 K is subject to diffuse blackbody incident radiation at 1000 K. The spectral hemispherical emissivity of the surface at 500 K is 0.2 below 3 μm and 0.8 above 3 μm.
   (a) Calculate the total hemispherical emissivity. ($\varepsilon = 0.792$).
   (b) Calculate the total hemispherical absorptivity and comment on why $\alpha < \varepsilon$. ($\alpha = 0.636$)
   (c) Calculate the net radiant heat flux from the surface to the surroundings. ($q = -3.32\ \text{W/cm}^2$)

4. The spectral directional emissivity for a smooth dielectric material may be approximated as $\varepsilon'_\lambda(\theta) = \varepsilon'_{\lambda_n} \cos\theta$ where ε'_{λ_n} is the normal ($\theta = 0$) value of ε'_λ. Show that the spectral hemispherical emissivity is $\frac{2}{3}$ of the normal directional value.

5. A flat plate insulated on the bottom side is subject to a collimated solar flux of 1200 W/m^2 normal to the plate. The plate is a spectrally selective, diffuse emitter-absorber. Find the steady-state temperature of the plate, assuming the solar flux has the spectral distribution of a blackbody at 5785 K for the following two cases. Neglect convective heat transfer from the top of the plate and assume that the solar flux is the only contribution to the hemispherical incident flux.

(a) Spectral hemispherical emissivity equals 0.85 for wavelengths below 1.5 μm and 0.15 for wavelengths above. ($T_s = 573$ K)

(b) Spectral hemispherical emissivity equals 0.15 for wavelengths below 1.5 μm and 0.85 for wavelengths above. ($T_s = 276$ K)

6. On a clear day, a solar collector is irradiated by a collimated solar flux and a diffuse atmospheric flux. The magnitude of the collimated solar flux is $I_0\,\Delta\Omega_0 = 1000$ W/m^2 and the angle of incidence is $\theta_0 = 45°$. The effective blackbody sky temperature is 244 K and the effective blackbody solar temperature is 5785 K. It may be assumed that the collector behaves as an isothermal, nongray (spectrally selective), diffuse emitter-absorber at 320 K with a spectral, hemispherical emissivity as shown below. Calculate the net radiant heat flux absorbed by the collector. (568 W/m^2)

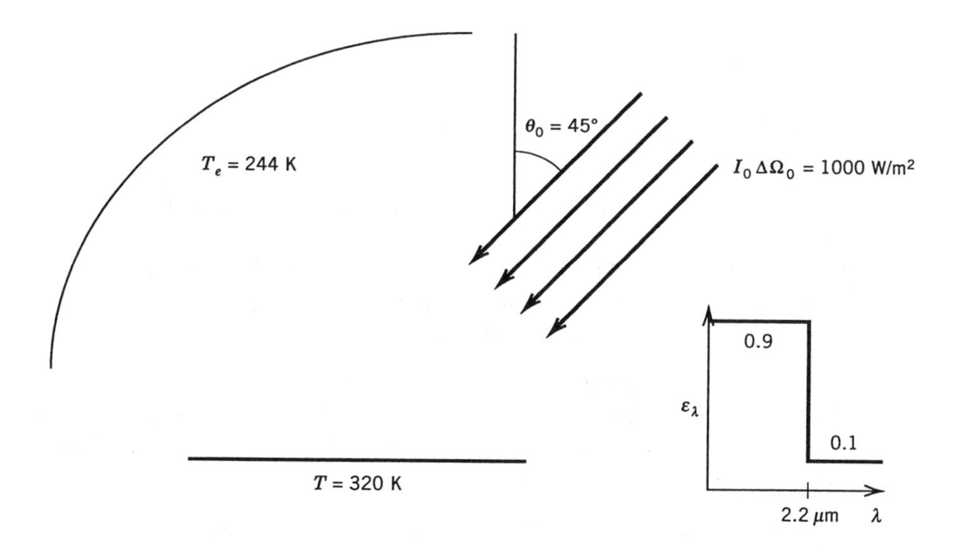

Figure P6

7. On a clear night with a full moon the total intensity reflected from the center of the moon's disc ($\theta = 0$) to the earth ($\theta_r = 0$) is 157 W/(m^2 sr). The collimated solar flux incident on the moon is 1390 W/m^2.

(a) Estimate the total bi-directional reflectivity of the moon's surface for normal incidence and normal reflectance. [$\rho''(0, 0) = 0.355$]

(b) If the observed brightness of the full moon from the earth is constant across the disc of the moon, what does that say about the variation of $\rho''(\theta = \theta_r, \phi = \phi_r)$ with θ?

(c) On a clear night with a quarter moon the total intensity reflected for normal incidence ($\theta = 0$) to the earth ($\theta_r = \pi/2$) is 31 W/(m^2 sr). Is the moon a diffuse reflector at this angle of incidence?

8. Consider radiative transfer from a small hole ΔA_1 in a furnace at $T_1 = 1000$ K to a small detector area ΔA_2 located at a distance s from the hole, such that $\Delta A_{1,2} \ll s^2$ (see Problem 1-5 and Example 1-1). Between the hole and detector is a spectrally selective filter ($\Delta A_f > \Delta A_{1,2}$), which is oriented normal to the line of sight between ΔA_1 and ΔA_2 with spectral directional transmissivity equal to 0.9 below 3 μm and zero above.

(a) Assuming the filter is a specular transmitter, show that the total power going from 1 to 2 is given by $C\tau'\sigma T_1^4$ where $C = \cos\theta_1 \cos\theta_2 \, \Delta A_1 \, A_2/(\pi s^2)$ and determine the value of τ'. ($\tau' = 0.246$)

(b) Assuming the filter is a diffuse transmitter, show that the total power going from 1 to 2 is $C_1\tau'\sigma T_1^4$ where

$$C_1 = \cos\theta_1 \cos\theta_2 \, \Delta A_1 \, \Delta A_2 \, \Delta A_f / \left(\pi^2 s_1^2 s_2^2\right)$$

and $s_{1,2}$ are the distances between the filter and $\Delta A_{1,2}$ ($\Delta A_f \ll s_{1,2}^2$).

9. Reconsider Problem 8 for the case where the filter is placed at either the detector or the hole (s_1 or $s_2 = 0$). For simplicity assume that $\Delta A_{f,1,2}$ are all parallel to each other, that ΔA_1 and ΔA_2 are normal to the line connecting them, and that the filter is a diffuse absorber-reflector (as well as a diffuse transmitter). Show that the power going from 1 to 2 is the same as it was in part (a) of Problem 2-8.

10. Show that if the filter of Problem 9 is placed halfway between the detector and the hole ($s_1 = s_2 = s/2$) and allowed to become infinite $\Delta A_f \to \infty$, the power going from 1 to 2 is $\frac{4}{3}$ of what it was in Problem 9. Also assume $\Delta A_{1,2} \ll (s/2)^2$.

11. Consider a small blackbody source at T_1, a detector, and a diffuse reflector as shown in the figure. Obtain an expression for the reflected

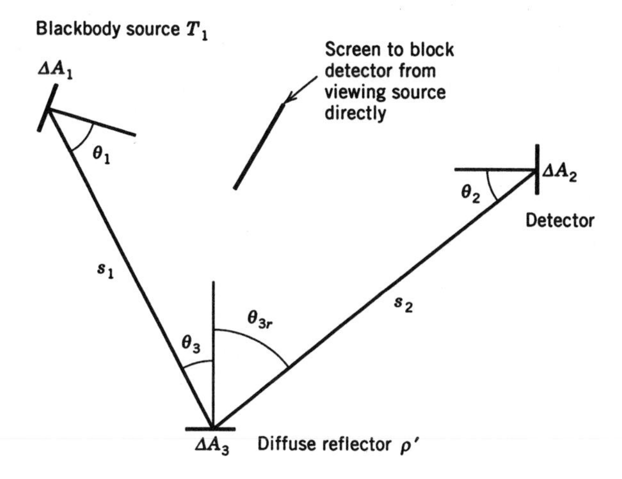

Figure P11

power from the source, which is incident on the detector (watts) Q_2^- in terms of the angles, distances, and properties shown in the figure .

12. For the satellite in geosynchronous orbit of Example 2-3, find the steady-state temperature of the satellite assuming no heat generation and for $r_2 = 3960$ miles, $h = 26{,}260$ miles, and $T_2 = 250$ K. The spectral hemispherical absorptivity of the earth $\alpha_{2\lambda}$ is 0.7 in the spectral region where the solar flux is concentrated ($\lambda < 5$ μm) and 1.0 in the spectral region where the earth's emission is located ($\lambda > 5$ μm). Treat the sun as a blackbody at 5785 K at a distance of 108 sun diameters from the satellite and earth. Treat the satellite as a two-sided, diffuse, gray flat plate. Hint: For the direct solar heat flux, $\cos\theta_1$ cannot be taken as constant over $\Delta\Omega_1$ for this orientation even though $\Delta\Omega_1 \ll 1$.

13. Take the sun to be a blackbody emitter and the earth to be a diffuse, gray reflector-absorber.

(a) Show that the temperature of the earth T_2 as a function of the temperature of the sun T_1, the radius of the sun r_1, and the distance between the sun and the earth r is $T_2 = T_1[r_1/(2r)]^{1/2}$.

(b) At a distance of 108 sun diameters away, what is the earth temperature for $T_1 = 5785$ K?

14. A bare thermocouple (A_1) is used to measure the temperature of a radiatively nonparticipating gas ($T_g = 2000$ K) flowing in a very long duct (A_3), with a wall temperature of $T_3 = 1000$ K. To reduce the radiation error in the measurement, a long tubular shield (A_2) is placed over the thermocouple in such a way that its presence causes minimal disturbance to the flow. The heat transfer coefficient to both the thermocouple and shield is 66 W/m^2 K. The thermocouple bead is a sphere 1 mm in diameter. Neglect conduction through the thermocouple lead wires. The shield is a cylinder 2 cm in diameter, 10 cm long (i.e., $A_3 \gg A_2 \gg A_1$). All surfaces are diffuse and gray with a thermocouple emissivity of $\varepsilon_1 = 0.1$ and a shield emissivity (both inside and outside) of $\varepsilon_{2o} = \varepsilon_{2i} = 0.1$.

(a) What is the temperature of the shield, T_2?

(b) What is the temperature of the thermocouple, T_1?

(c) Comment on the impact the shield emissivities ε_{2o} and ε_{2i} have on the accuracy of the temperature measurement.

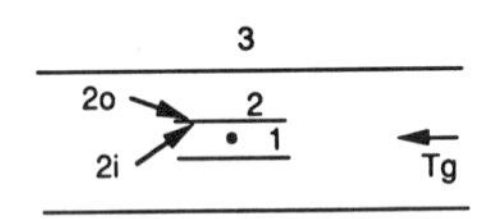

Figure P14

15. A thin, cylindrical aluminum wire (diameter = d) is to be ignited by irradiating it with a collimated flux $I_0 \Delta\Omega_0$ perpendicular to the axis of the wire. Assuming the wire is a gray, diffuse absorber-emitter ($\varepsilon = 0.05$) with constant properties and neglecting conduction heat losses from the ends of the wire, estimate the minimum flux $I_0 \Delta\Omega_0$ required to raise the aluminum wire to its ignition temperature of $T = 2320$ K. The surroundings is at $T_e = T_\infty = 300$ K with a convective heat transfer coefficient of $h = 10$ W/m^2 K.

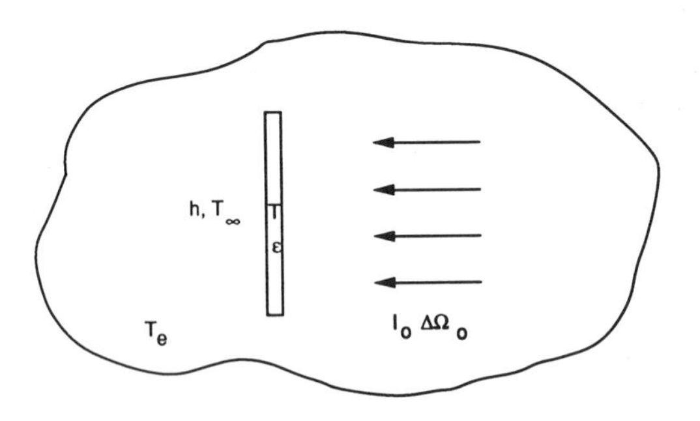

Figure P15

16. A surface at T_1 is exposed to large isothermal surroundings at T_2. The relative Planck curves for the surface (T_1) and surroundings (T_2) are shown in the figure along with the spectral hemispherical absorptivity of the surface. The notation $e_{b_{\lambda\max}}$ means the maximum value of e_{b_λ} at that temperature.

(a) Which is larger T_1 or T_2?

(b) Which is larger α_1 or ε_1?

(c) Is the net radiant heat flux at 1 positive or negative (positive q means net heat from surface to surroundings) or is there insufficient information given to determine this?

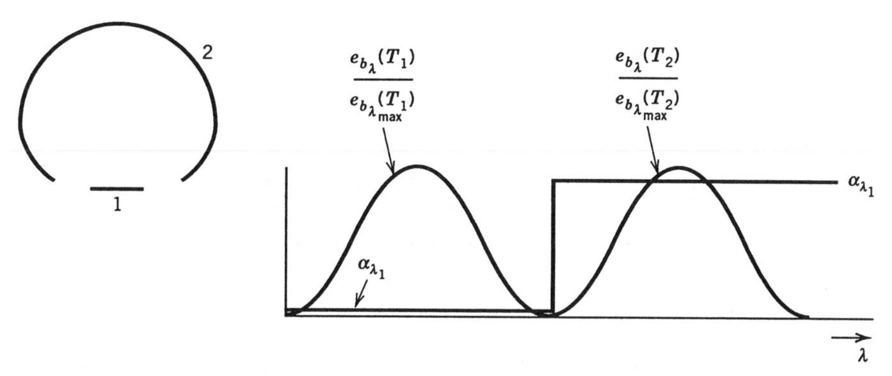

Figure P16

17. A translucent envelope often surrounds a point source of light to diffuse the light, i.e., reduce the intensity from the source. Examples of this include the frosted glass envelope that surrounds the tungsten filament of an ordinary light bulb and a translucent globe that covers a yard lamp light bulb. As a model of this situation, consider a spherical point source of radius r_1 surrounded concentrically by a spherical, diffusely absorbing, reflecting, and transmitting envelope with radius r_2 ($r_2 \gg r_1$). Because of the relative dimensions of the source and envelope, absorption and reflection by the source can be neglected. By using the definitions of bi-directional reflectivity and transmissivity and by considering the multiple reflections within the envelope:

(a) Obtain an expression for the intensity leaving the outside of the envelope I_2 in terms of the intensity emitted by the source I_1, the transmissivity and absorptivity of the envelope τ_2 and α_2, and the dimensions r_1 and r_2.

(b) Show that if the envelope absorptivity is zero, the illuminating flux from the envelope at a distance $L \gg r_2$ is the same as that which would emanate from the source without the envelope.

18. Consider a spherical surface (radius R) subject to collimated radiation with flux $I^- \Delta\Omega$. Show that the power absorbed by the surface is given by the flux times the projected area times the hemispherical absorptivity for diffuse incidence. Assume that α' is uniform over the spherical surface but the surface is not necessarily a diffuse absorber ($\alpha' \neq \alpha$).

CHAPTER 3

DIFFUSE SURFACE TRANSFER

This chapter deals with the topic of radiative transfer between surfaces that can be characterized as diffuse reflectors and diffuse emitter-absorbers. From Chapter 2 it may be recalled that a diffuse reflector is a surface whose bi-directional reflectivity is independent of reflectance angle, and a diffuse emitter-absorber is a surface whose directional reflectivity (absorptivity or emissivity) is independent of incidence angle. Although these assumptions may seem overly restrictive, the approach developed based on these assumptions is actually valid for predicting radiative heat transfer even between nondiffuse reflecting surfaces in many situations. In this chapter it is also assumed that the medium between surfaces is nonparticipating radiatively.

DIFFUSE, GRAY ENCLOSURE TRANSFER

In Chapter 2 the net radiative heat flux leaving a particular surface (denoted by subscript i) was shown to be the difference between the flux emitted by the surface and the flux absorbed by the surface.

$$q_i = \varepsilon_i \sigma T_i^4 - \alpha_i q_i^- \tag{3-1}$$

The most difficult part of determining the net radiative flux is determining the irradiation.

$$q_i^- = \int_{2\pi} I_i^- \cos\theta_i \, d\Omega_i \tag{3-2}$$

In Chapter 2 two special simple cases are considered in which the otherwise difficult integration of Eq. (3-2) is made trivial. The first special case is that of large, isothermal surroundings. For such a case the term I_i^- in Eq. (3-2) is constant and equal to blackbody intensity.

$$q_i^- = I_b \int_{2\pi} \cos\theta_i \, d\Omega_i \quad (I_i^- = I_b = \text{constant over } 2\pi) \tag{3-3}$$

Therefore the irradiation is, by definition, the blackbody hemispherical flux e_b. The second special case is that of collimated incident radiation, such as radiation from a point source. For this case the $\cos\theta_i$ term is usually constant over the small solid angle $\Delta\Omega_i$ for which I_i^- is nonzero, giving

$$q_i^- = I_i^- \cos\theta_i \int_{\Delta\Omega_i} d\Omega_i = I_i^- \cos\theta_i \, \Delta\Omega_i$$

$$(I_i^- \cos\theta_i \sim \text{constant over } \Delta\Omega_i \ll 2\pi) \tag{3-4}$$

While many practical radiative transfer problems fall into one of these two limiting cases, there are also many situations where the radiative transfer in the surroundings viewed by a surface of interest is more complicated, involving multiple sources of radiation at different temperatures. In this chapter the more general situation will be considered where the surroundings of a given surface cannot be described as either a blackbody (i.e., large and isothermal) or as a point source. To aid in this analysis it is useful to consider the concept of an enclosure.

An *enclosure* is defined as the set of all surfaces that either directly or indirectly contribute to the radiation incident on a given surface. According to this definition an enclosure is defined relative to a particular surface. As an example consider surface A_i in Fig. 3-1, which is exposed to radiation coming from all directions in the hemisphere above the surface. The enclo-

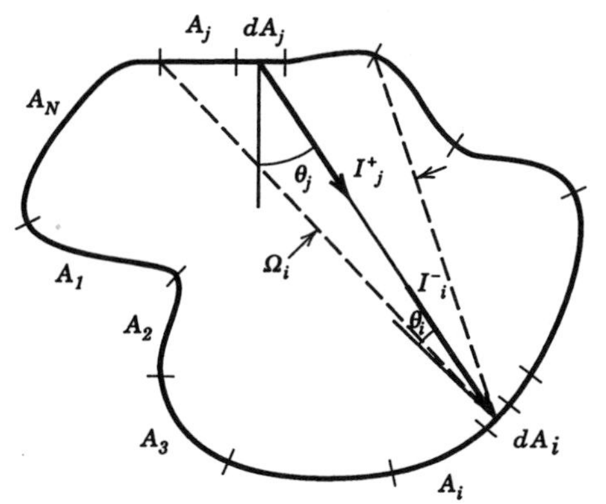

Figure 3-1 Enclosure concept.

sure is defined as the set of surfaces $1, 2, 3, \ldots, N$, which either directly or indirectly contribute to the irradiation on A_i. In particular, it should be noted that surface A_1, which cannot directly "see" surface A_i, is still part of the enclosure because it can emit or reflect radiation to other surfaces (such as A_j), which can reflect radiation to A_i. Thus A_1 indirectly contributes to the irradiation on A_i.

It is also important to realize that the surfaces of an enclosure do not have to be physical surfaces but can be imaginary surfaces that are assigned effective properties. Often a physical enclosure is only partially closed. In such cases it is expedient to complete the enclosure by defining an imaginary surface at the opening with an effective reflectivity of zero and an effective emissivity/absorptivity of one.

VIEW FACTOR

Definition

Consider the case where the intensity incident on a surface A_i at a particular location dA_i on the surface is constant over a finite solid angle Ω_i that is neither very small nor very large.

$$I_i^- = \text{constant over } \Omega_i, \qquad \Delta\Omega_i \ll \Omega_i < 2\pi \tag{3-5}$$

Equation (3-5) represents a condition that is intermediate between the two limiting cases already discussed [i.e., Eqs. (3-3) and (3-4)]. From Eq. (3-2) the flux incident at dA_i is

$$q_i^- = \sum_i \int_{\Omega_i} I_i^- \cos\theta_i \, d\Omega_i = \sum_i I_i^- \int_{\Omega_i} \cos\theta_i \, d\Omega_i \tag{3-6}$$

Since the medium inside the enclosure is assumed to be nonparticipating, the intensity leaving A_j equals the intensity incident on A_i, $I_i^- = I_j^+$. Furthermore, the integration with respect to solid angle can be replaced by one with respect to area in Eq. (3-6) using the small solid angle relation,

$$d\Omega_i = \frac{dA_j \cos\theta_j}{s^2} \tag{3-7}$$

giving,

$$q_i^- = \sum_j \pi I_j^+ \int_{A_j} \frac{\cos\theta_i \cos\theta_j}{\pi s^2} \, dA_j \tag{3-8}$$

The radiant power incident over the entire surface A_i is obtained by integrating Eq. (3-8) over the surface to give

$$Q_i^- = \int_{A_i} q_i^- \, dA_i = \int_{A_i} \sum_j \pi I_j^+ \int_{A_j} \frac{\cos\theta_i \cos\theta_j}{\pi s^2} \, dA_j \, dA_i \tag{3-9}$$

Let the assumption of Eq. (3-5) be extended to assume that $I_j^+ = I_i^-$ is diffuse and uniform

$$I_i^- = I_j^+ = \frac{q_j^+}{\pi} = \text{constant over both } A_i \text{ and } A_j \tag{3-10}$$

such that, upon interchanging the order of integration and summation, Eq. (3-9) becomes

$$Q_i^- = \int_{A_i} q_i^- \, dA_i = \sum_j q_j^+ \int_{A_i} \int_{A_j} \frac{\cos\theta_i \cos\theta_j}{\pi s^2} \, dA_j \, dA_i \tag{3-11}$$

The double integral in Eq. (3-11) is a purely geometric quantity that can be evaluated for any surface once the geometry of the enclosure is specified. The double integral is replaced by the following definition:

$$A_j F_{ji} = \int_{A_i} \int_{A_j} \frac{\cos\theta_i \cos\theta_j}{\pi s^2} \, dA_j \, dA_i \tag{3-12}$$

giving

$$Q_i^- = \int_{A_i} q_i^- \, dA_i = \sum_j q_j^+ A_j F_{ji} \tag{3-13}$$

The quantity F_{ji} is the radiation *view factor*. The view factor F_{ji} represents the fraction of *uniform*, *diffuse* radiation leaving surface A_j that is intercepted by surface A_i. The first subscript in the view factor corresponds to the surface from which radiant energy is leaving while the second subscript corresponds to the surface upon which the energy is incident. The assumptions inherent in the definition of F_{ji} are that the radiosity leaving surface A_j is uniform (independent of location on A_j) and diffuse (independent of direction).

Reciprocity

An important property of the view factor is the reciprocity relation. This relation follows from the definition of F_{ji} in Eq. (3-12). Since the right-hand side of (3-12) is symmetric in i and j (that is, i and j can be interchanged without changing the value of the double area integral), it follows that

$$A_i F_{ij} = A_j F_{ji} \tag{3-14}$$

This reciprocity relation is useful for determining unknown view factors from known ones. Substituting the reciprocity relation into Eq. (3-13) and dividing by A_i gives

$$q_i^- = \sum_{j=1}^{N} q_j^+ F_{ij} \tag{3-15}$$

where q_i^- now denotes the average flux over A_i. Equation (3-15) is sometimes a source of confusion to students because q_j^+ represents flux *leaving* surface j, yet F_{ij} refers to the fraction of diffuse, uniform energy going from surface i to surface j. The apparent conflict is resolved by remembering that the areas have been canceled out using reciprocity and that fluxes, not energy transfer rates, are involved in Eq. (3-15). If (3-15) is put back on a basis of power (watts) rather than flux (W/m^2), the correctness of the result is verified.

The general definition for the view factor Eq. (3-12) simplifies considerably when one (or both) of the areas in question is small relative to the distance between them. For example, when one area (A_i) is small ($A_i = \Delta A_i \ll s^2$), Eq. (3-12) simplifies to

$$A_j F_{ji} = \Delta A_i F_{ij} = \Delta A_i \int_{A_j} \frac{\cos\theta_i \cos\theta_j}{\pi s^2}\, dA_j \tag{3-16}$$

When both areas are small ($A_{i,j} = \Delta A_{i,j} \ll s^2$), Eq. (3-16) can be further simplified to

$$\Delta A_j F_{ji} = \Delta A_i F_{ij} = \frac{\cos\theta_i \cos\theta_j \Delta A_j\, \Delta A_i}{\pi s^2} \tag{3-17}$$

The product of area and view factor $\Delta A_i F_{ij}$ is the same geometric quantity encountered in Example 1-1 (then referred to as C) when the flux incident from a point source was introduced.

Summation Condition

In addition to the reciprocity relation, there is another useful relation for determining unknown view factors from known ones, called the summation condition. Consider diffuse, uniform radiative flux leaving area A_i and going to composite area A_{j+k}, which is composed of two areas A_j and A_k ($A_{j+k} = A_j + A_k$) as shown in Fig. 3-2. Conservation of energy dictates that

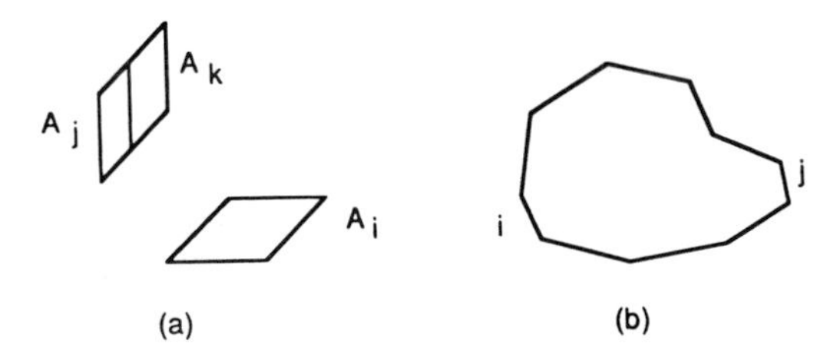

Figure 3-2 Summation relations for F_{ij}.

the fraction of diffuse, uniform radiation leaving A_i going to A_{j+k} is equal to the fraction going from A_i to A_j plus the fraction going from A_i to A_k.

$$F_{i-(j+k)} = F_{i-j} + F_{i-k} \quad \text{(partial summation condition)} \tag{3-18}$$

It is important to note that in general the reverse condition (a common mistake) is not true,

$$F_{(j+k)-i} \neq F_{j-i} + F_{k-i} \tag{3-19}$$

as can be readily verified by substituting the reciprocity relations into (3-18). More generally the partial summation condition can be written as

$$F_{i-\Sigma j} = \sum_j F_{i-j} \tag{3-20}$$

where the summation is made over the second subscript corresponding to the surface that is receiving, not sending, radiation.

If the summation in (3-20) is extended to include the entire 2π hemisphere above surface A_i, that is, every surface A_j which can be "seen" by A_i, then the view factor on the left-hand side of (3-20) is by definition unity.

$$1 = \sum_{\text{all } j} F_{ij} \quad \text{(total summation condition)} \tag{3-21}$$

Equation (3-21) indicates that the fraction of diffuse, uniform radiosity leaving surface i going to all possible surfaces j is one. In using either the partial or total summation condition, it is important not to forget the possibility of a surface exchanging radiation with itself, as represented by the term F_{ii}.

Determination Methods

Various methods for determining the geometrical view factor have been developed over the years. The advent of inexpensive, high speed computers

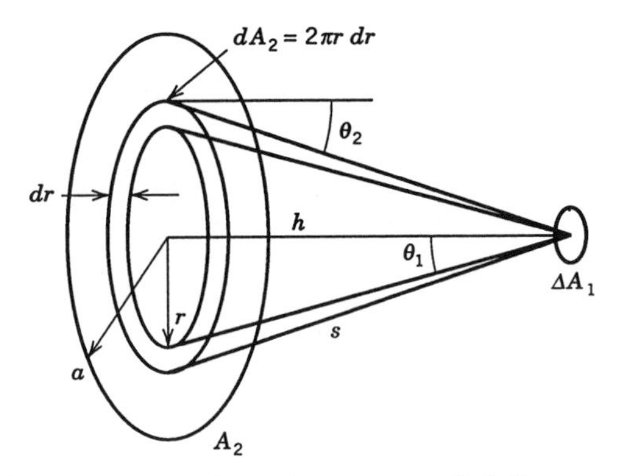

Figure 3-3 View factor between parallel discs.

has made some methods more advantageous than others and has made some methods virtually obsolete. Some of the methods still considered to be viable are direct integration, algebraic manipulation using summation and reciprocity, and numerical methods (e.g., Monte Carlo analysis).

Direct Integration

The most basic approach for determining the view factor of a particular geometric configuration is to evaluate the integral definition Eq. (3-12) directly. This has been done for many different configurations. A limited set of view factors is given in Appendix B. More extensive compilations can be found in Siegel and Howell [1988]. Before resorting to direct integration, it is advisable to consult a table of known view factors. The direct integration method is illustrated by the following examples.

Example 3-1

Consider transfer between a differential area ΔA_1 and a parallel circular disc $A_2 = \pi a^2$ whose center lies on the common normal line between ΔA_1 and A_2 as shown in Fig. 3-3.
(a) Find the view factor F_{12}.
(b) How does the flux incident on ΔA_1 vary with distance from A_2 in the limits $h \ll a$ and $h \gg a$?

(a) The appropriate equation for obtaining F_{ij} is Eq. (3-16) with $i = 1$ and $j = 2$. Setting the differential element $dA_2 = 2\pi r\, dr$ and noting that $\cos\theta_1 = \cos\theta_2 = h/(h^2 + r^2)^{1/2}$ and $s = (h^2 + r^2)^{1/2}$ Eq. (3-16) gives

$$F_{12} = 2h^2 \int_0^a \frac{r\, dr}{(h^2 + r^2)^2} = \frac{a^2}{h^2 + a^2} \tag{3-22}$$

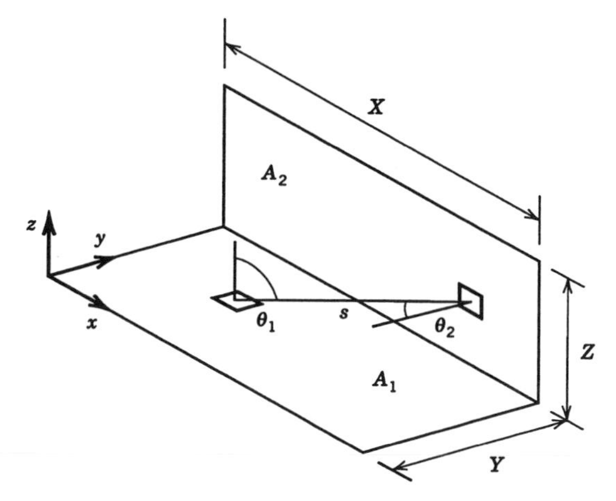

Figure 3-4 View factor between two adjacent walls.

(b) In the limit $h \ll a$, F_{12} goes to one as would be expected intuitively, and the flux incident on ΔA_1 varies as h^0 (a constant). In the limit $h \gg a$, $F_{12} \rightarrow a^2/h^2$, which is the same expression that would be obtained from Eq. (3-17) for two differential areas. In the latter case ($h \gg a$) the flux incident on ΔA_1 varies as h^{-2} and the disc looks like a point source to ΔA_1.

Example 3-2

Figure 3-4 shows a configuration consisting of two adjacent plane walls $A_1 = XY$ and $A_2 = XZ$ that are at right angles to each other. Write the integral expression for the view factor F_{12}.

The distance between a line connecting a differential element on A_1 to an element on A_2 is

$$s^2 = (x_2 - x_1)^2 + (Y - y_1)^2 + z_2^2 \tag{3-23}$$

and the cosines of the polar angles in terms of the Cartesian coordinates x, y, and z are

$$\cos\theta_1 = \frac{z_2}{s} \quad \text{and} \quad \cos\theta_2 = \frac{Y - y_1}{s} \tag{3-24}$$

Substituting (3-23) and (3-24) into (3-12) gives

$$F_{12} = \frac{1}{\pi XY}\int_0^Z \int_0^Y \int_0^X \int_0^X \frac{z_2(Y - y_1)\,dx_1\,dx_2\,dy_1\,dz_2}{\left[(x_2 - x_1)^2 + (Y - y_1)^2 + z_2^2\right]^2} \tag{3-25}$$

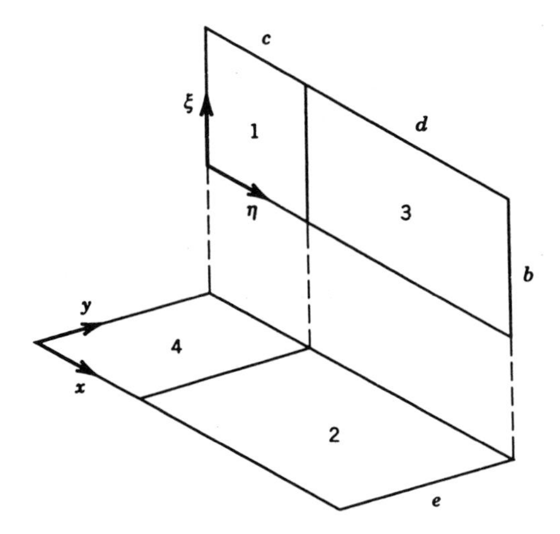

Figure 3-5 View factor between separated walls.

Example 3-3

Consider the configuration of Fig. 3-5. By writing the integral expressions for the view factors F_{12} and F_{34} show that the symmetry relation $A_1F_{12} = A_3F_{34}$ holds.

The view factor F_{12} in terms of the integral definition is

$$A_1F_{12} = \frac{1}{\pi}\int_{y=0}^{e}\int_{x=c}^{c+d}\int_{\eta=0}^{c}\int_{\xi=0}^{b}(\ )\,dy\,dx\,d\eta\,d\xi \tag{3-26}$$

where the argument ( ) is the same as in (3-25). By interchanging the order of integration and noting the equivalence of the x and η coordinates, this expression can be rewritten as

$$A_3F_{34} = \frac{1}{\pi}\int_{y=0}^{e}\int_{x=0}^{c}\int_{\eta=c}^{c+d}\int_{\xi=0}^{b}(\ )\,dy\,dx\,d\eta\,d\xi \tag{3-27}$$

thus establishing the symmetry relation

$$A_1F_{12} = A_3F_{34} \tag{3-28}$$

View Factor Algebra

The most expedient method for determining unknown view factors is to use the reciprocity, summation, and symmetry conditions to define unknown view factors in terms of known ones that can be obtained from tables. The basis

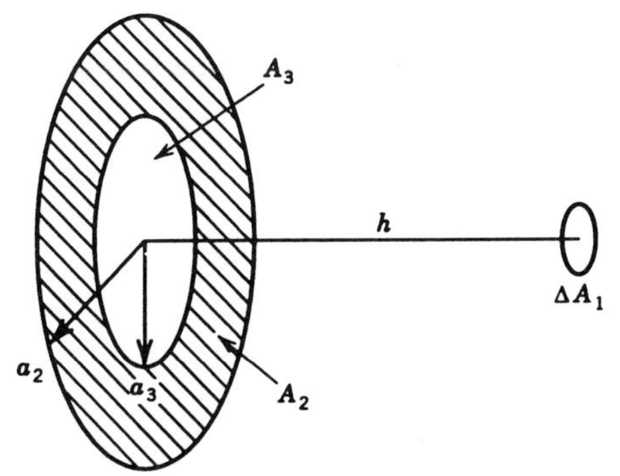

Figure 3-6 View factor between a differential area and a ring.

for applying this method is to recognize when either overlapping or adjacent areas define new view factors in terms of known ones. This method is called view factor algebra. The following examples illustrate this method.

Example 3-4

Figure 3-6 shows a ring shape composed of the space between a small circular disc $A_3 = \pi a_3^2$ and a larger concentric circular disc $A_2 = \pi(a_2^2 - a_3^2)$. Write the view factor F_{12} in terms of the result of Example 3-1.

Using the partial summation principle and the result of Example 3-1, the view factor from differential area ΔA_1 to the ring is

$$F_{1-2} = F_{1-(2+3)} - F_{1-3} = \frac{a_2^2}{h^2 + a_2^2} - \frac{a_3^2}{h^2 + a_3^2} \tag{3-29}$$

Example 3-5

Figure 3-7 shows two adjacent, perpendicular walls subdivided into many smaller sections. Show how to use flux algebra (reciprocity, summation, and symmetry) to find the unknown view factor F_{1-2}.

Since the view factor for two adjacent, perpendicular walls was discussed in Example 3-2, view factors of this type will be considered as known,

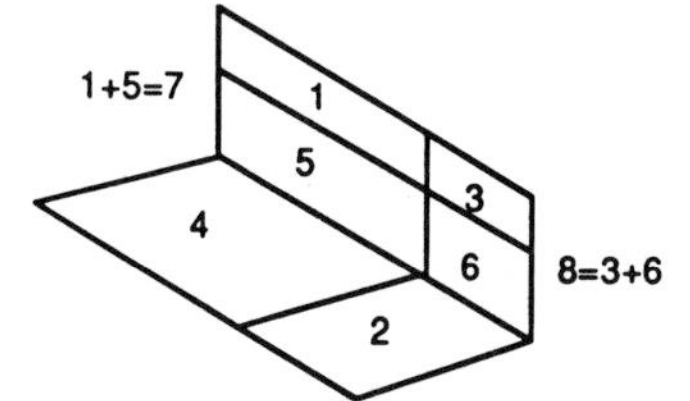

Figure 3-7 View factor between separated and diagonally opposed walls.

tabulated quantities and indicated by a star. The strategy will be indicated as follows, omitting the algebraic details. First reciprocity is used once to write

$$F_{1-2} = \frac{A_2}{A_1} F_{2-1} = \frac{A_2}{A_1} [F_{2-7} - F_{2-5}] \tag{3-30}$$

and again to give

$$F_{2-7} = \frac{A_7}{A_2} F_{7-2} \quad \text{and} \quad F_{2-5} = \frac{A_5}{A_2} F_{5-2} \tag{3-31}$$

Then, focusing on obtaining F_{7-2}, summation can be used to write

$$\begin{aligned} F^*_{(4+2)-(7+8)} &= F_{(4+2)-7} + F_{(4+2)-8} \\ &= \frac{A_7}{A_{4+2}} F_{7-(4+2)} + \frac{A_8}{A_{4+2}} F_{8-(4+2)} \\ &= \frac{A_7}{A_{4+2}} (F^*_{7-4} + F_{7-2}) + \frac{A_8}{A_{4+2}} (F_{8-4} + F^*_{8-2}) \end{aligned} \tag{3-32}$$

which can be solved for F_{7-2} in terms of known view factors F^* after recognizing that $F_{8-4} = A_7 F_{7-2}/A_8$ from the previous symmetry result obtained in Example 3-3. Then a similar strategy can be used to find F_{5-2} in terms of known view factors (i.e., $F^*_{(4+2)-(3+6)} = \ldots$) and the results substituted back into (3-30) and (3-31).

String Rule

In situations where surfaces interchanging energy radiatively are very long in one direction (i.e., two-dimensional), a simple formula called the string rule can be used to determine the view factors. Consider radiative transfer between two surfaces A_1 and A_2, which are very long in the direction perpendicular to the page as pictured in Fig. 3-8. An enclosure can be completed by drawing imaginary surfaces or strings from the endpoints on

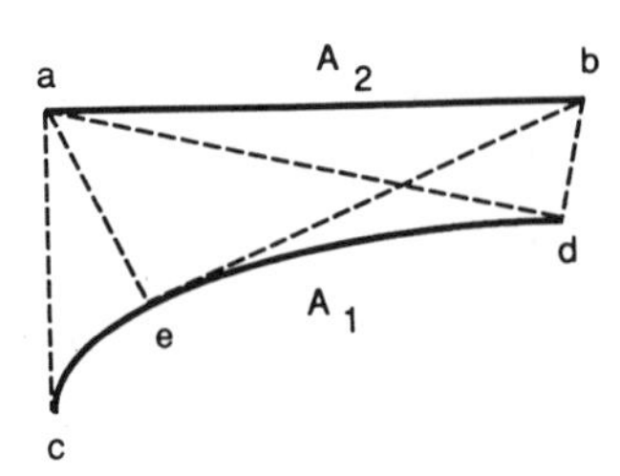

Figure 3-8 String rule geometry.

A_2 to the corresponding endpoints on A_1 (i.e., a to c and b to d). It is also imagined that crossed strings are stretched from the endpoints of A_2 to the opposite endpoints of A_1 (i.e., a to d and b to c). If necessary the crossed strings are imagined to be wrapped around the surface, such as at point e on A_1. The view factors between 1 and 2 are given in terms of the crossed and uncrossed string lengths by the following formula.

$$A_2F_{21} = A_1F_{12} = \frac{ad + bc - (bd + ac)}{2}$$

$$= \frac{\Sigma \text{ crossed strings} - \Sigma \text{ uncrossed strings}}{2} \tag{3-33}$$

This is known as *Hottel's string rule* [Hottel, 1954] and is valid for surfaces that are two-dimensional. It is valid even if parts of surface A_1 cannot be viewed from some points on A_2 (as e-c cannot be viewed from point b in Fig. 3-8). In the case of a partially obstructed view the strings are simply imagined to stretch around the obstruction. Even the uncrossed strings can be stretched around obstructions if necessary.

Numerical Methods

With the advent of high speed, low cost digital computers and computer graphics, numerical methods have increased in importance. Advances have been made in computer graphics that have almost direct application to radiative heat transfer and, in particular, view factor determination. The calculations necessary to simulate illumination and thus generate realistic computer images are essentially the same as those necessary to determine view factors. This similarity has been recognized by many researchers [Rushmeier, et al., 1990]. A detailed discussion of these methods is beyond the scope of this book. However, a limited discussion is given in Chapter 6 in relation to the Monte Carlo ray tracing method.

DIFFUSE ENCLOSURE TRANSFER EQUATIONS

The first step in formulating a diffuse surface transfer problem is to define the enclosure. This includes deciding how many different surfaces make up the enclosure. The number of nodes or surfaces that make up an enclosure is generally kept as small as possible, for computational simplicity. The minimum number necessary depends on how uniform the radiosity leaving the various surfaces is. Since the uniformity of radiosity leaving a surface is most strongly related to the uniformity of temperature, the number of nodes or surfaces necessary is generally determined by the uniformity of temperature. Surfaces that are close together (and therefore are subject to the same irradiation) and are also similar in temperature can be grouped together and considered as a single node or surface. If a surface has a significant temperature gradient across the face of it, then it must be divided into smaller surfaces in order to retain accuracy in the analysis. This requirement of uniform radiosity over any given node is directly related to the definition of view factor [recall Eq. (3-10)].

The other assumption made in the derivation of the view factor was that the radiosity leaving a surface was diffuse [Eq. (3-10)]. This condition can be met in two ways. One way to guarantee diffuse radiosity is for the surface to be a diffuse emitter and a diffuse reflector. The other way is for the surface to be a diffuse emitter and for the irradiation to be diffuse (in which case it doesn't matter even if the surface reflects specularly). Most surfaces are, to a good approximation, diffuse emitters. Furthermore, in an enclosure there are usually many multiple reflections such that the irradiation on any given surface is relatively diffuse. Therefore the analysis that follows generally yields accurate results for enclosures with specularly reflecting surfaces as well as diffusely reflecting surfaces.

Radiosity Formulation

Consider again the enclosure of Fig. 3-1 consisting of N gray emitting surfaces where the medium between the surfaces is assumed to be nonparticipating. Since it is assumed that either the surfaces are diffuse reflectors or the irradiation is diffuse, Kirchhoff's law at the hemispherical level can be applied [see Eq. (2-19)].

$$\varepsilon_i = \alpha_i = 1 - \rho_i \tag{3-34}$$

First the case will be considered where it is assumed that the temperatures (and thus the blackbody hemispherical emissive fluxes) of all the surfaces are known. The equation for the radiosity leaving surface A_i is

$$q_i^+ = \varepsilon_i e_{b_i} + \rho_i q_i^- \quad \text{(radiosity)} \tag{3-35}$$

The equation for the irradiation on surface A_i was obtained in Eq. (3-15) as

$$q_i^- = \sum_{j=1}^{N} q_j^+ F_{ij} \quad \text{(irradiation)} \tag{3-36}$$

Substituting (3-36) into (3-35) gives a system of N algebraic equations where $i = 1$ to N, in terms of the unknown radiosities.

$$q_i^+ = \varepsilon_i e_{b_i} + \rho_i \sum_{j=1}^{N} q_j^+ F_{ij} \quad \left(\text{prescribed } e_{b_i}(T_i) - \text{radiosity formulation}\right) \tag{3-37}$$

If the emissivities are independent of temperature, the system of equations is linear and can be solved for the radiosities by standard numerical techniques (Gauss-Seidel iteration, etc.) Once the radiosities have been obtained, the net radiative flux can be calculated from any one of the following three equivalent expressions:

$$q_i = q_i^+ - q_i^- \tag{3-38}$$

$$q_i = \varepsilon_i(e_{b_i} - q_i^-) \tag{3-39}$$

$$q_i = \frac{\varepsilon_i}{\rho_i}(e_{b_i} - q_i^+) \quad (\varepsilon_i \neq 1) \tag{3-40}$$

In some cases a prescribed net radiative heat flux is a more appropriate boundary condition than prescribed temperature. A specific example is a refractory surface where the net radiative heat flux can usually be taken as zero.

If the net radiative heat flux q_i is prescribed instead of temperature, the equation for radiosity for that surface (3-37) contains a new unknown, e_{b_i} and the system of equations cannot be solved as formulated. This problem can be overcome by using an equation that eliminates the unknown blackbody emissive flux e_{b_i} in the ith equation in favor of the known quantity q_i. Such an equation can be obtained by substituting the irradiation equation (3-36) into the net flux equation (3-38).

$$q_i = q_i^+ - \sum_j q_j^+ F_{ij} \quad \text{(prescribed } q_i \text{ -- radiosity formulation)} \tag{3-41}$$

Equation (3-41) or (3-37), depending on the prescribed boundary condition, is written for each surface of the enclosure. This set of equations constitutes a set of N linear algebraic equations for the N unknown radiosities. Once the radiosities are obtained, the unknown temperature or net heat flux can be obtained from (3-40). Generally either $T_i(e_{b_i})$ or q_i is prescribed for any given surface and the other is solved for. It should be noted that q_i cannot be arbitrarily prescribed for all surfaces because energy conservation must also be satisfied. In most cases, however, temperature would be prescribed for at least one of the surfaces.

NETWORK ANALOGY

Diffuse, Gray Enclosure Transfer

The network analogy from electrical circuit theory is a useful tool for solving the simultaneous algebraic equations of transfer in diffuse enclosures when only two or three surfaces are involved. It was first applied to this problem by

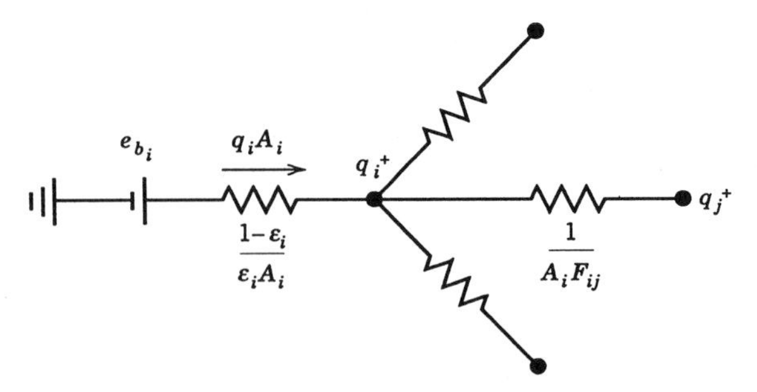

Figure 3-9 Network analogy of diffuse radiative transfer in enclosures.

Oppenheim [1956]. The analogy can be developed by starting with the radiosity equation for a prescribed temperature surface.

$$q_i^+A_i = \varepsilon_i e_{b_i} A_i + \rho_i A_i \sum_j q_j^+ F_{ij} \tag{3-42}$$

Subtracting $\varepsilon_i A_i q_i^+$ from both sides of (3-42) and multiplying the left-hand side by $\sum_j F_{ij}$ (which equals one) gives

$$\frac{e_{b_i} - q_i^+}{(1-\varepsilon_i)/\varepsilon_i A_i} = \sum_j \frac{q_i^+ - q_j^+}{1/A_iF_{ij}} = q_iA_i \tag{3-43}$$

Equation (3-43) can be interpreted in terms of the equivalent electrical circuit shown in Fig. 3-9 where e_{b_i} and q_i^+ are voltages, q_iA_i is current, and the terms in the denominator involving ε_i and F_{ij} are resistances. The effect of the emissivity resistance is to reduce the actual radiosity leaving the surface below the maximum possible blackbody flux. The smaller the emissivity the bigger the resistance and the smaller the radiosity. The effect of the view factor resistance is to reduce the net power transferred between two surfaces. The smaller the view factor the bigger the resistance and the less the heat transfer rate.

Based on Fig. 3-9 and Eq. (3-43) it is tempting to interpret the term

$$\frac{q_i^+ - q_j^+}{1/A_iF_{ij}}$$

as the net radiant heat transfer between surfaces i and j. However, as pointed out by Tao and Sparrow [1985], this interpretation is incorrect because it does not account for indirect transfer from i to j via other surfaces (through reflection as well as absorption and re-emission). It is

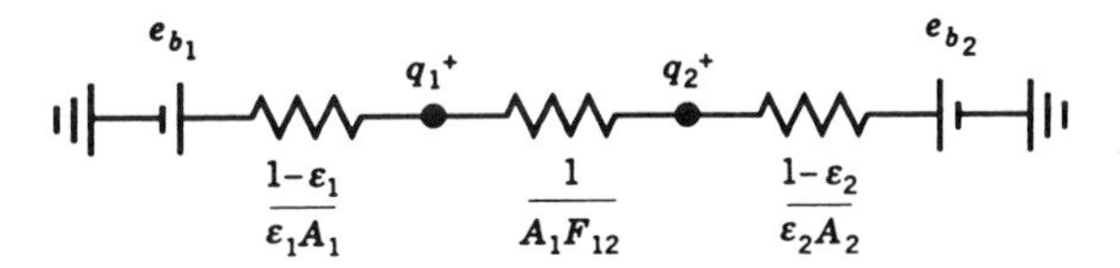

Figure 3-10 Two-node (source-sink) network.

possible to define a transfer factor that represents the net radiant heat transfer between two surfaces. Such a factor is referred to as a transfer factor [Edwards, 1981] or script-F factor [Hottel, 1954]. Gebhart [1961, 1971] also presented a method that is equivalent to the script-F approach. The radiative transfer formulation according to the script-F transfer factor is equivalent to the formulation presented here and is not discussed further.

Two Nodes: Source (1), Sink (2)

Consider transfer from a hot, isothermal surface (T_1) that is viewed by a single cold, isothermal surface (T_2). The network for a two-node source-sink system is shown in Fig. 3-10. Node 1 is the source and node 2 is the sink ($T_1 > T_2$). By adding the resistances together in series, the net radiative transfer from the source $q_1 A_1$ (which equals the net transfer to the sink, $-q_2 A_2$) is

$$A_1 q_1 = \frac{e_{b_1} - e_{b_2}}{\dfrac{1-\varepsilon_1}{\varepsilon_1 A_1} + \dfrac{1}{A_1 F_{12}} + \dfrac{1-\varepsilon_2}{\varepsilon_2 A_2}} \tag{3-44}$$

One Flat or Convex Node

If one of the nodes of a two-node system (e.g., node 1) is not concave such that $F_{11} = 0$ and $F_{12} = 1$, Eq. (3-44) can be further simplified to give

$$A_1 q_1 = \frac{e_{b_1} - e_{b_2}}{\dfrac{1}{\varepsilon_1 A_1} + \dfrac{1-\varepsilon_2}{\varepsilon_2 A_2}} \quad \text{or} \quad q_1 = \frac{\varepsilon_1 (e_{b_1} - e_{b_2})}{1 + \dfrac{\varepsilon_1 A_1}{\varepsilon_2 A_2}(1-\varepsilon_2)} \tag{3-45}$$

An example of a two-node system that fits this description is a pair of infinite, parallel flat plates. Setting $A_1 = A_2$ in Eq. (3-45) gives

$$q_1 = \frac{e_{b_1} - e_{b_2}}{\dfrac{1}{\varepsilon_1} + \dfrac{1}{\varepsilon_2} - 1} \quad \text{(parallel flat plates)} \tag{3-46}$$

One Convex, Small Node

A further specialization of the two-node, source-sink problem occurs if both $F_{11} = 0$ and $A_1 \ll A_2$ such that

$$\frac{\varepsilon_1 A_1}{\varepsilon_2 A_2}(1 - \varepsilon_2) \ll 1$$

In this case the net radiative flux from (3-45) reduces to

$$q_1 = \varepsilon_1(e_{b_1} - e_{b_2}) \tag{3-47}$$

The right-hand side of Eq. (3-47) can be interpreted as the emitted flux minus the absorbed flux. Since ε_1 is equivalent to α_1 for a gray surface subject to diffuse irradiation, e_{b_2} can be interpreted as the flux incident on A_1, which means that A_2 acts like a blackbody with respect to A_1. This is the same result as that obtained in Eq. (2-79) with the same restrictions and conditions.

Example 3-6

Consider a two-dimensional rectangular slot in an isothermal, diffuse, gray wall as an enclosure (Fig. 3-11). The slot length is L and the width is W. Treat the two side walls with length L and the bottom surface with width W as a single node (i.e., two-node network—source, sink). Obtain the effective emissivity-absorptivity of the slot as a function of the aspect ratio L/W and wall emissivity ε_2.

The effective emissivity of the slot is defined as the flux incident on surface 1 from the internal surfaces 2 divided by the hemispherical blackbody flux based on T_2 when the surroundings are at 0 K ($T_1 = 0$).

$$\varepsilon_{\text{eff}} = \left.\frac{q_1^-}{e_{b_2}}\right|_{e_{b_1}=0} \tag{3-48}$$

Since the surroundings are black at 0 K, no energy can be emitted from the surroundings into the slot and none of the energy escaping from the slot can be reflected back into the slot. Therefore $q_1^+ = 0$ and from Eq. (3-38) the

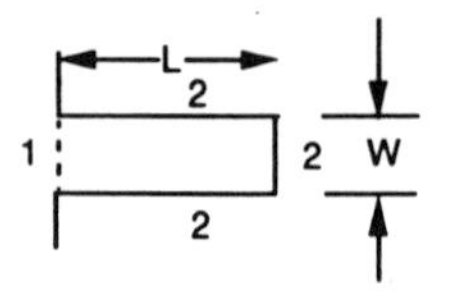

Figure 3-11 Isothermal slot.

irradiation at 1 is equal to minus the net flux at 1.

$$q_1^- = -q_1 \tag{3-49}$$

The net flux can be obtained for this two-node system from Eq. (3-45) by setting $\varepsilon_1 = 1$ and $e_{b_1} = 0$ giving

$$q_1 = \frac{-e_{b_2}}{1 + \dfrac{A_1}{A_2}\dfrac{(1-\varepsilon_2)}{\varepsilon_2}} \tag{3-50}$$

Thus the effective emissivity is

$$\varepsilon_{\text{eff}} = \frac{1}{1 + \dfrac{A_1}{A_2}\dfrac{(1-\varepsilon_2)}{\varepsilon_2}} \tag{3-51}$$

Setting $A_1 = W$ and $A_2 = W + 2L$ as the areas per unit depth gives

$$\varepsilon_{\text{eff}} = \frac{1}{1 + \dfrac{1}{1 + 2L/W}\dfrac{(1-\varepsilon_2)}{\varepsilon_2}} \tag{3-52}$$

The effective emissivity is plotted as a function of L/W for three values of emissivity 0.25, 0.50, and 0.75 in Fig. 3-12.

In the limit as $L/W \to 0$ the effective emissivity approaches the correct limit of ε_2. However, in the limit as $L/W \to \infty$ the effective emissivity approaches 1, independent of the value of ε_2, which is not correct. The value of ε_{eff} should approach a constant whose value is less than 1 and depends on the value of ε_2. The discrepancy in this two-node formulation is due to the assumption of uniform radiosity over the entire interior surface 2. For large

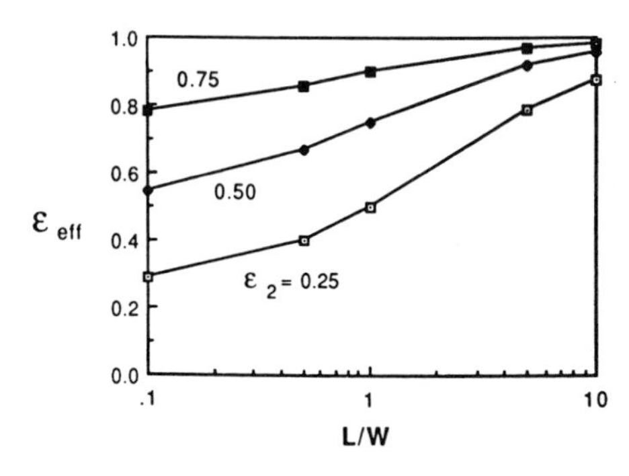

Figure 3-12 Effective emissivity of slot by two-node approximation.

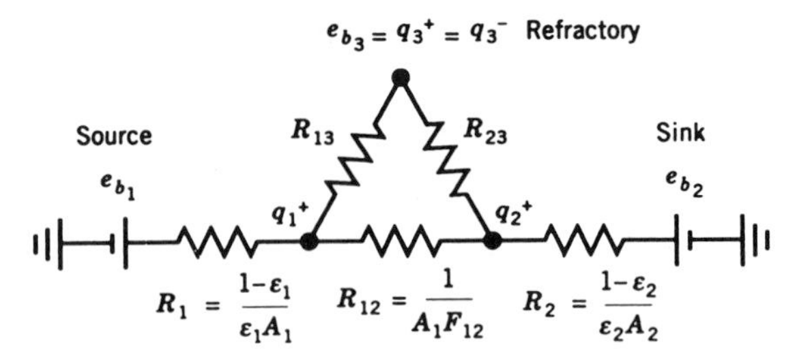

Figure 3-13 Three-node (source, sink, refractory) network.

aspect ratios it is perhaps intuitively obvious that the radiosity at surface 2 near the opening will be less than the radiosity deep in the slot. Deep in the slot the radiosity will approach the isothermal cavity value of e_{b_2}. An exact solution of this problem can be carried out by subdividing surface 2 into a large number of nodes approaching infinity (see Problem 3-3) or by using Monte Carlo techniques, as demonstrated in Chapter 6.

Three Nodes: Source (1), Sink (2), Refractory (3)

Many practical engineering systems can be modeled as three node source, sink, refractory systems. An example is a high temperature heat treating furnace. The network for a source, sink, refractory system is shown in Fig. 3-13. The refractory surface is represented by a floating potential since the net radiant heat transfer is zero. Since $q_3 = q_3^+ - q_3^- = \varepsilon_3(e_{b_3} - q_3^-) = 0$, the blackbody hemispherical flux, radiosity, and irradiation at an adiabatic, gray surface are all equal (assuming ε_3 is not zero). The emissivity of the refractory ε_3 doesn't enter the problem and can be taken as either 1 (perfect emitter) or 0 (perfect reflector) as convenient. (Note, however, that if the emissivity is taken as zero, $\varepsilon_3 = \alpha_3 = 0$, the blackbody flux e_{b_3} is indeterminate.)

Circuits of the type shown in Fig. 3-13 can be reduced using Y-Δ and Δ-Y transformations as shown in Fig. 3-14 [Eqs. (3-53) and (3-54)] and the rules for adding resistances in series and parallel.

$$R_1 = \frac{R_{12}R_{13}}{R_{12} + R_{23} + R_{13}} \tag{3-53}$$

$$R_{12}^{-1} = \frac{R_1^{-1}R_2^{-1}}{R_1^{-1} + R_2^{-1} + R_3^{-1}} \tag{3-54}$$

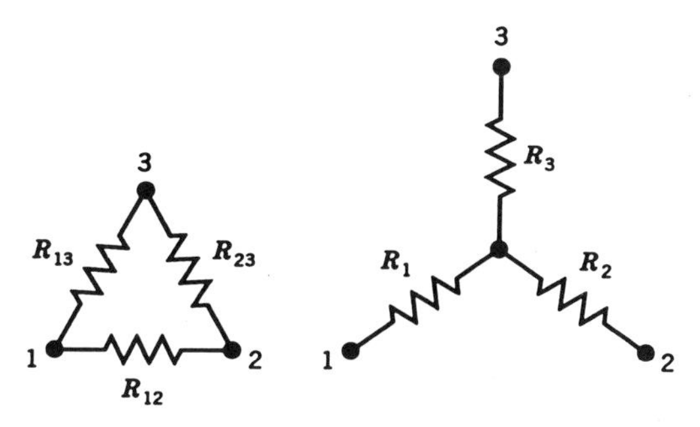

Figure 3-14 Delta-wye transformations.

Using the rules for adding resistances in series and parallel, the circuit in Fig. 3-13 can be reduced to a single equivalent resistance R_{12}''' between e_{b_1} and e_{b_2}.

$$A_1 q_1 = -A_2 q_2 = \frac{e_{b_1} - e_{b_2}}{R_{12}'''} \tag{3-55}$$

$$R_{12}''' = R_1 + R_2 + \frac{1}{\dfrac{1}{R_{12}} + \dfrac{1}{R_{13} + R_{23}}} \tag{3-56}$$

$$R_i = \frac{1 - \varepsilon_i}{\varepsilon_i A_i} \tag{3-57}$$

$$R_{ij} = \frac{1}{A_i F_{ij}} \tag{3-58}$$

and 1 = source, 2 = sink, and 3 = adiabatic node.

The same result can be derived without the aid of the network analogy by solving the radiosity and irradiation equations for a three-node system. The network is merely a convenient tool for solving low order systems of linear algebraic equations. For systems that are more complicated than the examples given here, the network may be useful for the physical insight it gives into the problem; but for solving the equations, it is recommended that a computer and equation solver package be used.

To complete this problem, the temperature of the refractory surface still needs to be determined. A strategy for finding $e_{b_3}(T_3)$ is to find q_1^+ and q_2^+ from Eq. (3-40), then find the branch "current" through R_{12} and subtract it from $q_1 A_1$ to give the branch "current" through R_{13}. Then the blackbody flux e_{b_3} can be determined from Ohm's law.

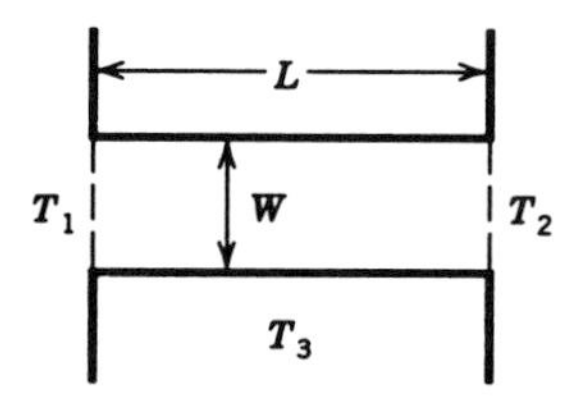

Figure 3-15 Furnace wall crack.

Example 3-7

Consider a two-dimensional rectangular crack in a furnace wall as an enclosure (Fig. 3-15). The wall thickness (crack length) is L and the crack width is W. The refractory wall material is diffuse and gray. The temperature inside and outside the furnace are T_1 and T_2, respectively.

(a) Assuming that the radiosity leaving the wall is diffuse and uniform (i.e., isothermal wall, three-node network), obtain the dimensionless heat transfer $q/(e_{b_1} - e_{b_2})$ as a function of the aspect ratio L/W.
(b) What is the dimensionless wall temperature, $(e_{b_3} - e_{b_2})/(e_{b_1} - e_{b_2})$?
(c) If $T_1 = 2000$ K and $T_2 = 300$ K, what is the wall temperature T_3?

(a) The crack can be modeled as an enclosure by treating the openings as black surfaces at the temperature of the surroundings. Using Eqs. (3-55) through (3-58) for a source-sink-refractory with $\varepsilon_1 = 1, \varepsilon_2 = 1$, $A_1 = A_2$, and $F_{23} = F_{13} = 1 - F_{12}$ gives the dimensionless heat flux as

$$\frac{q_1}{e_{b_1} - e_{b_2}} = \frac{F_{12} + 1}{2} \tag{3-59}$$

The string rule can be used to give the view factor F_{12} as

$$F_{12} = \sqrt{1 + (L/W)^2} - L/W \tag{3-60}$$

giving

$$\frac{q_1}{e_{b_1} - e_{b_2}} = \frac{1}{2}\left[\sqrt{1 + (L/W)^2} - L/W + 1\right] \tag{3-61}$$

In the limit as $L/W \to 0$, Eq. (3-61) predicts that $q_1/(e_{b_1} - e_{b_2}) \to 1$, which is the correct result, but in the limit as $L/W \to \infty$, Eq. (3-61) gives $q_1/(e_{b_1} - e_{b_2}) \to \frac{1}{2}$, which is not the correct result. The correct result is zero. The reason for the discrepancy is the assumption of uniform radiosity (i.e., uniform temperature) across the refractory surface, which is implicit in treating the refractory as a single node. To improve the accuracy of the result, the refractory surface should be divided into more nodes.

(b) The temperature of the refractory surface can be determined by referring to the circuit diagram of Fig. 3-13. This circuit represents the present problem if the gray surface resistances are set to zero, $R_1 = R_2 = 0$. The "current" (net heat transfer) from 1 to 3 is equal to the "current" (net heat transfer) from 3 to 2 since 3 is adiabatic.

$$\frac{e_{b_1} - e_{b_3}}{R_{13}} = \frac{e_{b_3} - e_{b_2}}{R_{23}} \tag{3-62}$$

By symmetry ($F_{23} = F_{13}$, $A_1 = A_2$) the view factor resistances are equal, $R_{23} = R_{13}$, and therefore the "voltage drop" from e_{b_1} to e_{b_2} is equally divided between the two resistances.

$$e_{b_3} = \frac{e_{b_1} + e_{b_2}}{2} \quad \text{or} \quad \frac{e_{b_3} - e_{b_2}}{e_{b_1} - e_{b_2}} = \frac{1}{2} \tag{3-63}$$

(c) For $T_1 = 2000$ K and $T_2 = 300$ K, the temperature of the refractory surface (treated as a single node) can be calculated from (3-63) as $T_3 = 1682$ K.

GENERALIZED DIFFUSE, GRAY ENCLOSURE TRANSFER FORMULATION

When more than three surfaces are involved, closed-form algebraic solutions (including the network method) are impractical, and it is advisable to seek a numerical solution to the system of equations. Such a numerical solution could be based on the radiosity formulation of the equations described earlier. However, that formulation involves the unnecessary intermediate step of calculating the radiosities. The step can be eliminated by algebraically manipulating the governing equations to remove radiosity (this is left as an exercise). The resulting equations for unknown heat flux q_i and blackbody flux e_{b_i} are given as follows. For convenience (only) suppose that the surfaces are numbered so that those with prescribed $T_i(e_{b_i})$ are designated as $1 \le i \le M$ while those with prescribed q_i are designated $(M+1) \le i \le N$.

For unknown q_i $(1 \le i \le M)$

$$q_i = \sum_{j=1}^{M} \Lambda_{ij} e_{b_j} - \frac{\varepsilon_i}{1 - \varepsilon_i} \sum_{j=M+1}^{N} \psi_{ij} q_j \quad (1 \le i \le M) \tag{3-64}$$

For unknown e_{b_i} $(M+1 \le i \le N)$

$$e_{b_i} = \sum_{j=1}^{M} \psi_{ij} e_{b_j} + \sum_{j=M+1}^{N} \Phi_{ij} q_j \quad (M+1 \le i \le N) \tag{3-65}$$

where

$$\Phi_{ij} = \frac{1-\varepsilon_i}{\varepsilon_i}\delta_{ij} + \psi_{ij} \tag{3-66}$$

$$\Lambda_{ij} = \frac{\varepsilon_i}{1-\varepsilon_i}(\delta_{ij} - \psi_{ij}) \tag{3-67}$$

$$\psi_{ij} = \chi_{ij}^{-1} \tag{3-68}$$

$$\chi_{ij} = \begin{cases} \dfrac{\delta_{ij} - (1-\varepsilon_i)F_{ij}}{\varepsilon_i}; & 1 \le i \le M \text{ prescribed } e_{b_i} \\ \delta_{ij} - F_{ij}; & M+1 \le i \le N \text{ prescribed } q_i \end{cases} \tag{3-69}$$

$$\delta_{ij} = \begin{cases} 1 & (i = j) \\ 0 & (i \ne j) \end{cases} \quad \text{Kronecker delta} \tag{3-70}$$

These equations can be readily programmed on a digital computer. A FORTRAN program that solves this set of equations is listed in Appendix D. If it is desired to change the value of a boundary condition, only the new value needs to be substituted into (3-64) and (3-65). The matrix inversion does not need to be repeated unless the view factors or emissivities are changed. Note that the order of the prescribed T_i and q_i surfaces can be mixed as long as the correct equation is used to define χ_{ij}.

The key to setting up the χ_{ij} matrix is the view factor matrix F_{ij}. Improper definition of the F_{ij} matrix often leads to singular χ_{ij} matrices that cannot be inverted. To ensure that χ_{ij} is properly defined the F_{ij} matrix should be set up to guarantee that the reciprocity and total summation conditions are satisfied. Use of the reciprocity and summation conditions reduces the number of independent view factors that need to be specified from N^2 to

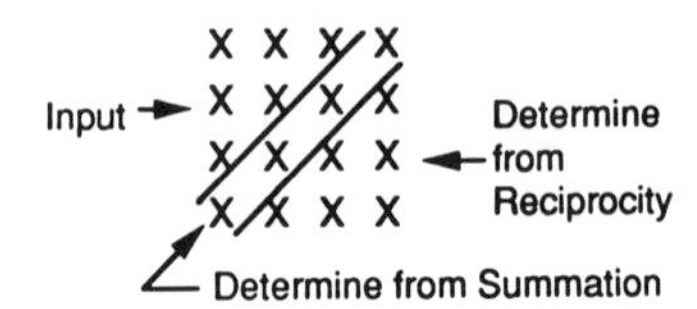

Figure 3-16 View factor matrix.

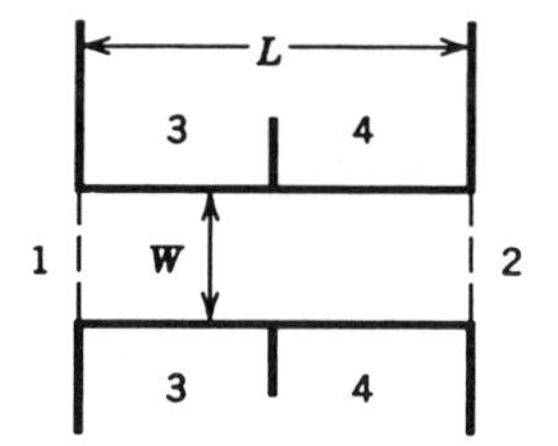

Figure 3-17 Four-node analysis of furnace wall crack.

$(N^2 - N)/2$. One way of visualizing this is to picture the F_{ij} matrix as an N by N array divided into an upper triangle of elements, a diagonal, and a lower triangle of elements (Fig. 3-16).

The diagonal consists of N elements and the triangular regions consist of $(N^2 - N)/2$ elements in each region. To set up the F_{ij} matrix it is only necessary to specify the elements in one of the triangular regions. The reciprocal elements in the other triangle can be calculated using reciprocity, and the diagonal elements can be calculated using the total summation condition. In this way the F_{ij} matrix will be defined consistent with the reciprocity and summation conditions, and numerical singularities in the inversion of χ_{ij} will be avoided.

Example 3-8

Repeat Example 3-7 and improve the accuracy by dividing the wall into two equal area, isothermal regions (3 and 4) as shown in Fig. 3-17.

(a) If $T_1 = 2000$ K and $T_2 = 300$ K, what are the wall temperatures T_3 and T_4 and what are the dimensionless wall temperatures,

$$\frac{e_{b_3} - e_{b_2}}{e_{b_1} - e_{b_2}} \quad \text{and} \quad \frac{e_{b_4} - e_{b_2}}{e_{b_1} - e_{b_2}} \quad \text{for } L/W = 20$$

(b) Calculate the dimensionless heat transfer $q/(e_{b_1} - e_{b_2})$ for $L/W = 20$.

The program difgray listed in Appendix D is used to solve this problem. The input file (difgray.inp) is listed in Table 3-1. On a per unit length basis perpendicular to the page, the areas of the enclosure are $A_1 = A_2 = W$ and $A_3 = A_4 = L$. Since the program works with areas, it is necessary to assume a length scale for this problem. A wall thickness of $L = 1$ m is assumed for this purpose. However, it should be noted that the results requested in (a) and (b) are functions of only the ratio L/W and not L and W individually. The emissivities of the refractory surfaces 3 and 4 are arbitrarily assigned to be one. The view factors can be calculated using the string rule. This is left as an exercise (see Problem 3-2).

The output file (difgray.out) is listed in Table 3-2. The unknown temperatures are $T_3 = 1805$ K and $T_4 = 1523$ K. The dimensionless temperatures are

$$\frac{e_{b_3} - e_{b_2}}{e_{b_1} - e_{b_2}} = 0.66 \quad \text{and} \quad \frac{e_{b_4} - e_{b_2}}{e_{b_1} - e_{b_2}} = 0.34$$

The dimensionless heat flux is

$$\frac{q}{e_{b_1} - e_{b_2}} = \frac{16370 \text{ W}}{(0.05 \text{ m}^2)(907200 - 459 \text{ W/m}^2)} = 0.361$$

These results are an improvement over those of Example 3-7, which predict a dimensionless heat flux of 0.51 for $L/W = 20$. Nevertheless, better accuracy could still be obtained through further subdivision of the adiabatic surfaces. In fact, it is possible to consider the limit of the interval of subdivision approaching zero. In that limit the number of subdivisions approaches infinity and the mathematical description of the problem becomes that of an integral equation. However, solution methods for such integral equations usually invoke approximations that are equivalent to reducing the problem

TABLE 3-1 Input File for Diffuse, Gray Enclosure Analysis (difgray.inp)

```
COMMENT TO BE WRITTEN ON HEADER OF OUTPUT FILE
Difgray-Example 3-8
NUMBER OF NODES
 4
        if Flag=1 ⇒ Prescribed Ti(K)
        if Flag=0 ⇒ Prescribed Qi(W)
   NODE #       AREA(m**2)       FLAG       Ti/Qi        Emissivity
     1,           0.05,           1,        2000.,           1.
     2,           0.05,           1,         300.,           1.
     3,           1.,             0,           0.,           1.
     4,           1.,             0,           0.,           1.
VIEW FACTORS, UPPER TRIANGLE
0.0250, 0.9501, 0.0249
        0.0249, 0.9501
                0.0463
```

TABLE 3-2 Output File for Diffuse, Gray Enclosure Analysis (difgray.out)

```
***Difgray - Example 3 - 8
View Factor Matrix:
NODE    1      2       3       4
  1   .0000  .0250   .9501   .0249
  2   .0250  .0000   .0249   .9501
  3   .0475  .0012   .9050   .0463
  4   .0012  .0475   .0463   .9050
           BB. Hem. Fluxes  Temperatures  Heat Transfer Rates
             (W / m**2)         (K)              (W)
           ---------------  ------------  -------------------
NODE 1      0.90720E + 06     2000.0         0.16370E + 05
NODE 2      0.45927E + 03      300.0        -0.16370E + 05
NODE 3      0.60219E + 06     1805.3         0.00000E + 00
NODE 4      0.30546E + 06     1523.5         0.00000E + 00
QSUM =      0.000000
```

back to a system of coupled algebraic equations. Thus the integral technique is not discussed here.

DIFFUSE, NONGRAY ENCLOSURE TRANSFER FORMULATION

The analysis for diffuse, gray enclosure transfer is also valid for nongray surfaces if the equations are written on a spectral basis. In terms of the radiosity formulation the equations are

$$q_{\lambda_i}^{+} = \varepsilon_{\lambda_i} e_{b\lambda_i} + \rho_{\lambda_i} \sum_j q_{\lambda_j}^{+} F_{ij} \quad \text{(radiosity)} \tag{3-71}$$

$$q_{\lambda_i}^{-} = \sum_j q_{\lambda_j}^{+} F_{ij} \quad \text{(irradiation)} \tag{3-72}$$

$$q_{\lambda_i} = q_{\lambda_i}^{+} - q_{\lambda_i}^{-} = \varepsilon_{\lambda_i}\left(e_{b\lambda_i} - q_{\lambda_i}^{-}\right) = \frac{\varepsilon_{\lambda_i}}{\rho_{\lambda_i}}\left(e_{b\lambda_i} - q_{\lambda_i}^{+}\right) \quad \text{(net flux)} \tag{3-73}$$

When the boundary conditions of the enclosure are all given in terms of prescribed temperature, the approach is very similar to that described in the gray case. The first step is to solve the above equations for the spectral heat fluxes q_{λ_i}. This must be done spectrally for each wavelength. Then the spectral heat fluxes can be integrated to give the total heat fluxes.

$$q_i = \int_0^\infty q_{\lambda_i}\, d\lambda \tag{3-74}$$

In the case where some of the surfaces are described by a prescribed net radiative heat flux instead of temperature, the analysis becomes more difficult. Now if the same approach that worked in the gray case of simply recasting the equations in terms of q_{λ_i} to solve for $e_{b\lambda_i}$ is tried, the resulting equation for spectral radiosity is

$$q_{\lambda_i}^+ = q_{\lambda_i} + \sum_j q_{\lambda_j}^+ F_{ij} \tag{3-75}$$

The problem is that the prescribed boundary condition is total, not spectral, net radiative flux (q_i not q_{λ_i}). For example, for a nongray refractory surface even though the total net radiative flux is zero ($q_i = 0$), at some wavelengths there would be a positive spectral net flux ($q_{\lambda_i} > 0$) and at other wavelengths there would be a negative spectral net flux ($q_{\lambda_i} < 0$) such that the average over all wavelengths was zero. One solution to this dilemma is to treat the problem as if the temperature was prescribed at all surfaces, guess the unknown T_i, solve for q_i, compare to the prescribed value, and iterate.

Example 3-9

A small tungsten filament wire (length = 2.1 cm; diameter = 0.1 cm) is radiatively dissipating 100 W at steady state to a large glass envelope at 500 K. Using the approximate properties for tungsten given in Fig. 2-4, estimate the temperature of the wire.

This problem fits the description of a two-node, source-sink problem with the source being the filament wire and the sink being the glass envelope. The prescribed boundary condition for the source (wire) is heat flux with a value of

$$q_1 = \frac{Q_1}{A_1} = \frac{100 \text{ W}}{\pi(0.1 \text{ cm})(2.1 \text{ cm})} = 152 \text{ W/cm}^2$$

and the prescribed boundary condition for the sink (envelope) is temperature $T_2 = 500$ K. Since the source has nongray properties (Fig. 2-4), a spectral analysis is needed. The appropriate expression for the net radiative flux is Eq. (3-47) written on a spectral basis.

$$q_{\lambda_1} = \varepsilon_{\lambda_1}(e_{b\lambda_1} - e_{b\lambda_2}) \tag{3-76}$$

Since $\varepsilon_{\lambda_1} = \alpha_{\lambda_1}$ (diffuse irradiation) Eq. (3-76) can also be written as

$$q_{\lambda_1} = \varepsilon_{\lambda_1} e_{b\lambda_1} - \alpha_{\lambda_1} e_{b\lambda_2} \tag{3-77}$$

Integrating over wavelength gives

$$q_1 = \varepsilon_1 \sigma T_1^4 - \alpha_1 \sigma T_2^4 \tag{3-78}$$

where

$$\varepsilon_1 = \frac{1}{\sigma T_1^4} \int_0^\infty \varepsilon_{\lambda_1} e_{b\lambda}(T_1)\, d\lambda = \int_0^1 \varepsilon_{\lambda_1}\, df(\lambda T_1) \tag{3-79}$$

$$\alpha_1 = \frac{1}{\sigma T_2^4} \int_0^\infty \alpha_{\lambda_1} e_{b\lambda}(T_2)\, d\lambda = \int_0^1 \alpha_{\lambda_1}\, df(\lambda T_2) \tag{3-80}$$

This is the same result as Eq. (2-78) for transfer between a small, nongray surface and blackbody surroundings. Solution of Eq. (3-78) can be accomplished by guessing T_1, calculating ε_1 (as demonstrated in Example 2-1), and iterating until $q_1 = 152\ \mathrm{W/cm^2}$. Convergence is hastened by assuming (and later confirming) that the absorbed flux is negligible, since $T_2^4 \ll T_1^4$. For a temperature of $T_1 = 3000$ K, Example 2-1 gives $\varepsilon_1 = 0.33$ and Eq. (3-78) gives

$$q_1 = (0.33)(5.67)(3)^4 = 152\ \mathrm{W/cm^2}$$

Thus $T_1 = 3000$ K is the steady-state temperature of the wire.

Example 3-10

Two nongray, parallel, infinite flat plates are exchanging energy by thermal radiation. Both plates have the same semigray spectral emissivity shown in Fig. 3-18, assumed to be independent of temperature. Write an expression for the total, net radiative flux between the two plates.

The equation for the net spectral flux from Eq. (3-46) is

$$q_{\lambda_1} = \frac{e_{b\lambda_1} - e_{b\lambda_2}}{\dfrac{1}{\varepsilon_{\lambda_1}} + \dfrac{1}{\varepsilon_{\lambda_2}} - 1} \tag{3-81}$$

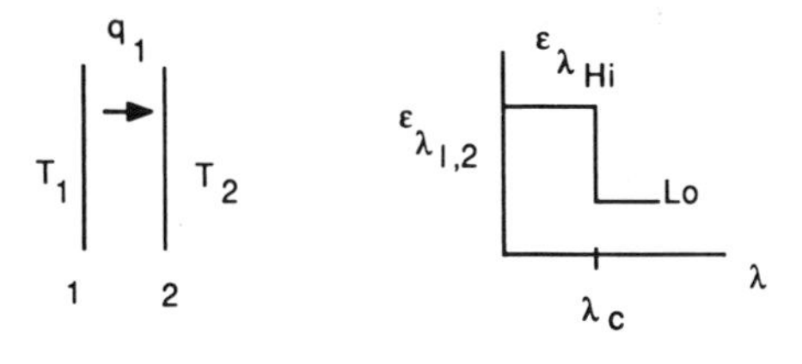

Figure 3-18 Nongray parallel plates.

Integrating (3-81) over all wavelengths gives

$$q_1 = \int_0^{\infty} \frac{e_{b\lambda}(T_1) - e_{b\lambda}(T_2)}{\dfrac{1}{\varepsilon_{\lambda_1}} + \dfrac{1}{\varepsilon_{\lambda_2}} - 1} d\lambda$$

$$= \int_0^{\infty} A_\lambda e_{b\lambda}(T_1)\, d\lambda - \int_0^{\infty} A_\lambda e_{b\lambda}(T_2)\, d\lambda \tag{3-82}$$

where for simplicity the spectral function A_λ has been defined as

$$A_\lambda \equiv \frac{1}{\dfrac{1}{\varepsilon_{\lambda_1}} + \dfrac{1}{\varepsilon_{\lambda_2}} - 1} \tag{3-83}$$

Using the definition of the fractional function in (3-82) gives

$$q_1 = e_{b_1} \int_0^1 A_\lambda\, df(\lambda T_1) - e_{b_2} \int_0^1 A_\lambda\, df(\lambda T_2) \tag{3-84}$$

$$= \sigma T_1^4 \{ A_{\lambda \mathrm{Hi}} f(\lambda_c T_1) + A_{\lambda \mathrm{Lo}} [1 - f(\lambda_c T_1)] \}$$
$$- \sigma T_2^4 \{ A_{\lambda \mathrm{Hi}} f(\lambda_c T_2) + A_{\lambda \mathrm{Lo}} [1 - f(\lambda_c T_2)] \} \tag{3-85}$$

If T_1 and T_2 are prescribed, then Eq. (3-85) can be solved directly for the unknown heat flux q_1. If q_1 is prescribed, then Eq. (3-85) can be solved implicitly for either T_1 or T_2.

Generalized Diffuse, Nongray Enclosure Transfer

The diffuse, nongray enclosure problem can also be generalized to eliminate the intermediate radiosity calculation as was done in the gray enclosure problem. The approach is to apply the generalized diffuse, gray formulation spectrally, on a band-by-band basis and treat each surface as if the temperature were prescribed. On a spectral basis the net radiative flux can be written from (3-64) as

$$q_{\lambda_i} = \sum_{j=1}^{N} \Lambda_{\lambda_{ij}} e_{b\lambda_j} \tag{3-86}$$

Integrating this expression over a constant property band $\Delta\lambda_k = \lambda_{k+1} - \lambda_k$ gives

$$q_{ik} = \sum_{j=1}^{N} \Lambda_{ijk} e_{b_j} [f(\lambda_{k+1} T_j) - f(\lambda_k T_j)] \tag{3-87}$$

where

$$\Lambda_{ijk} = \frac{\varepsilon_{ik}}{1 - \varepsilon_{ik}}(\delta_{ij} - \psi_{ijk}) \tag{3-88}$$

$$\psi_{ijk} = \chi_{ijk}^{-1} \tag{3-89}$$

$$\chi_{ijk} = \frac{\delta_{ij} - (1 - \varepsilon_{ik})F_{ij}}{\varepsilon_{ik}} \tag{3-90}$$

and the subscript i in ε_{ik} refers to the ith surface and k refers to the kth spectral band. Adding up the contribution from all M bands gives the total net radiative heat flux for the ith surface.

$$q_i = \sum_{k=1}^{M} q_{ik} \tag{3-91}$$

If the prescribed boundary condition for a surface is actually heat flux instead of temperature, the procedure would be to compare the calculated flux with the known flux, guess a new temperature, and iterate. A FORTRAN program for solving the nongray enclosure equations is listed in Appendix E. A sample problem is given in Example 3-11, which illustrates the format of the input and output files.

Example 3-11

Repeat Example 3-8 using a nongray analysis. Use five spectral bands delimited at 1, 2, 3, and 4 μm. Use constant (gray) emissivities and the temperatures obtained in Example 3-8 for the adiabatic surfaces to check consistency with the gray analysis. Comment on the spectral distribution of the heat flux at the various surfaces.

The program input file is listed in Table 3-3 and the output file is listed in Table 3-4. The results are consistent with those of the gray analysis in that the total heat transfer rate at surfaces 1 and 2 (16,370 W) is the same as that found in Example 3-8. Also surfaces 3 and 4 are adiabatic (within round-off error). The results for spectral heat flux are labeled PARTIAL Q. These results show that the heat flux at all wavelengths is positive (into the cavity) for surface 1 (source) and negative (out of the cavity) for surface 2 (sink). For surfaces 3 and 4 (refractories) the net heat flux is negative (into the surface) at short wavelengths (< 2 μm) and positive (out of the surface) at long wavelengths (> 2 μm). This trend indicates that the predominantly short wavelength energy that is radiated into the enclosure at surface 1 is converted into long wavelength energy as it is transferred through the enclosure to exit at surface 2. This conversion takes place via absorption and re-emission at the refractory surfaces.

Example 3-12

A satellite is in low earth orbit (LEO) as shown in Fig. 3-19. The altitude of a satellite in LEO is on the order of 100 to 200 miles (150 to 300 km). At this altitude the earth (which has a radius of 6378 km) still appears more or less as an infinite flat plane when viewed from the satellite. The earth can be treated as a nongray, diffuse emitter-absorber and diffuse reflector at $T_2 =$ 250 K with a spectral, hemispherical absorptivity of 0.7 below 5 μm and 1.0 above 5 μm. The satellite surfaces can be treated as adiabatic, diffuse, gray emitter-absorbers. The incident solar flux has a magnitude 1360 W/m^2 and

TABLE 3-3 Input File for Nongray Enclosure Analysis (nongray.inp)

```
COMMENT TO BE WRITTEN ON HEADER OF OUTPUT FILE
nongray - Example 3 - 11
NUMBER OF NODES
  4
NODE #               NODE AREA          TEMPERATURE(K)
  1,                   0.05,                2000.
  2,                   0.05,                 300.
  3,                   1.,                  1805.3
  4,                   1.,                  1523.5
NUMBER OF SPECTRAL BANDS
  5
SPECTRAL BAND LIMITS,μm
  1.0,2.0,3.0,4.0
    NOTE: Emissivity 1 corresponds to spectral
        band from 0μm to 1st band limit, etc.
NODE #,  EMISSIVITY 1,  EMISSIVITY 2,...
  1,       1.0,     1.0,      1.0, 1.0, 1.0
  2,       1.0,     1.0,      1.0, 1.0, 1.0
  3,       1.0,     1.0,      1.0, 1.0, 1.0
  4,       1.0,     1.0,      1.0, 1.0, 1.0
VIEW FACTORS, UPPER TRIANGLE
  0.0250,  0.9501,  0.0249
           0.0249,  0.9501
                    0.0463
```

TABLE 3-4 Output File for Nongray Enclosure Analysis (nongray.out)

```
***nongray - Example 3 - 11
-----------------------------------
View Factor Matrix:
NODE    1       2       3       4
 1    .0000 .0250   .9501   .0249
 2    .0250 .0000   .0249   .9501
 3    .0475 .0012   .9050   .0463
 4    .0012 .0475   .0463   .9050
-----------------------------------
      SPECTRAL RESULTS
      --------
LAMBDA(min)     LAMBDA(max)        NODE        PARTIAL Q
 (micron)        (micron)         NUMBER          (W)
 --------        --------         -----       ----------
 .0000            1.0000            1         0.18776E+04
                                    2        -0.31301E+03
                                    3        -0.78971E+03
                                    4        -0.77484E+03
1.0000            2.0000            1         0.82164E+04
                                    2        -0.46547E+04
                                    3        -0.71981E+03
                                    4        -0.28419E+04
2.0000            3.0000            1         0.36388E+04
                                    2        -0.47215E+04
                                    3         0.62679E+03
                                    4         0.45600E+03
3.0000            4.0000            1         0.13524E+04
                                    2        -0.27149E+04
                                    3         0.39588E+03
                                    4         0.96666E+03
4.0000           infinity           1         0.12822E+04
                                    2        -0.39658E+04
                                    3         0.49287E+03
                                    4         0.21907E+04
--------------------------
      TOTAL RESULTS
      -------
                 BB. Hem.                             Heat
                 Fluxes          Temperatures    Transfer Rates
                 (W / m**2)          (K)              (W)
               ---------------       ----             ----
NODE 1          0.90720E+06        2000.0          0.16367E+05
NODE 2          0.45927E+03         300.0         -0.16370E+05
NODE 3          0.60226E+06        1805.3          0.60165E+01
NODE 4          0.30546E+06        1523.5         -0.33381E+01
Qsum =           .000002
```

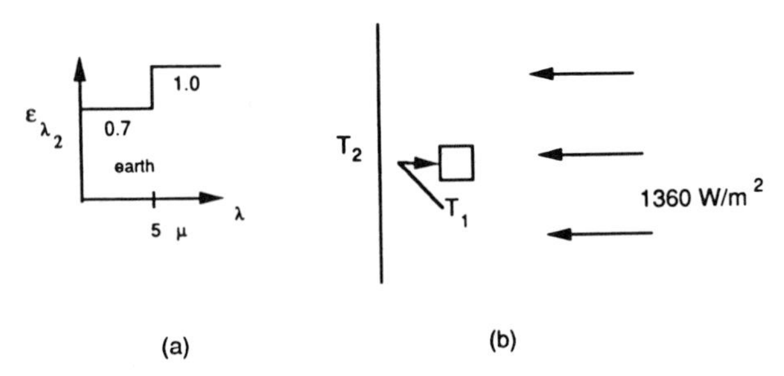

Figure 3-19 (*a*) Earth spectral hemispherical emissivity and (*b*) satellite in LEO.

the spectral distribution of a blackbody at 5785 K. Determine the temperature of a surface parallel to and facing the earth, T_1.

This problem is an example of a situation where the generalized enclosure formulation is not applicable (due to the collimated solar radiation). However, the view factor concept is still useful. The temperature of surface 1 can be determined from an energy balance on the surface. Since surface 1 is adiabatic ($q_1 = q_{r_1} + q_{c_1} = 0$) and there can be no conductive flux in outer space ($q_{c_1} = 0$), the net radiative flux is also zero ($q_{r_1} = 0$). Thus the emitted flux and absorbed flux are equal.

$$\varepsilon_1 \sigma T_1^4 = \alpha_1 q_1^- \tag{3-92}$$

Since surface 1 is assumed to be a diffuse, gray absorber-emitter, the total hemispherical absorptivity and emissivity are equal, $\varepsilon_1 = \alpha_1$. Furthermore, since the earth appears as an infinite flat plate and is assumed to be a diffuse reflector, the flux leaving the earth toward the satellite is uniform and diffuse. Thus the view factor can be used to express the incident flux at surface 1 in terms of the radiosity leaving surface 2 as

$$q_1^- = q_2^+ F_{12} \tag{3-93}$$

where

$$q_2^+ = \varepsilon_2 \sigma T_2^4 + \rho_2 q_2^- \tag{3-94}$$

The view factor from the satellite to the earth can be approximated as one, $F_{12} = 1$. Since most of the emitted energy from the earth is above 5 μm, the total emissivity of the earth is approximately one, $\varepsilon_2 = 1$. Similarly most of the solar energy is below 5 μm and the total reflectivity is $\rho_2 = 0.3$. Using $q_2^- = 1360$ W/m^2, Eq. (3-92) can be solved for the surface temperature of the satellite $T_1 = 325$ K. The temperature of the sides parallel to the solar rays can be determined using the same analysis but setting $F_{12} = 0.5$. The resulting temperature, 273 K, is the lowest achieved by any surface. The hottest temperature would be on the surface facing the sun, which would reach 393 K.

SUMMARY

Many engineering heat transfer problems can be solved by the diffuse surface interchange method. The problem of diffuse surface interchange reduces to a geometric problem of determining the view factors between the various surfaces. The view factor F_{ij} is the fraction of uniform, diffuse radiation leaving surface A_i that is intercepted by surface A_j. The mathematical solution of the transfer problem consists of solving a system of linear algebraic equations that represent the multiple reflection of radiation between the various surfaces.

The diffuse enclosure analysis is based on the assumption of *diffuse* and *uniform* radiosity leaving each surface. The uniform radiosity assumption can be satisfied to within the desired degree of accuracy simply by dividing large surfaces into many smaller surfaces. Subdivision has a greater effect when a large nonisothermal surface is divided into many smaller, more nearly isothermal surfaces; but there is also an improvement in accuracy realized upon subdividing an already isothermal surface into smaller surfaces.

The diffuse radiosity assumption is usually satisfied for enclosures with "regular" or "box-like" geometries even if some specularly reflecting surfaces are present because the effect of many multiple reflections within the enclosure is to make the irradiation nearly diffuse anyway. For "irregular" geometries (e.g., long, narrow cracks, etc.) with smooth surfaces it may be necessary to use a nondiffuse analysis, such as Monte Carlo or a mirror-image technique. But for the majority of practical problems the diffuse analysis will give satisfactory results.

REFERENCES

Edwards, D. K. (1981), *Radiation Heat Transfer Notes*, Hemisphere, New York.

Gebhart, B. (1961), "Surface Temperature Calculations in Radiant Surroundings of Arbitrary Complexity for Gray, Diffuse Surface Radiation," *Int. J. Heat Mass Transfer*, Vol. 3, No. 4, pp. 341–346.

Gebhart, B. (1971), *Heat Transfer*, 2nd ed., McGraw-Hill, New York.

Hottel, H. C. (1954), "Radiant Heat Transmission," in *Heat Transmission*, 3rd ed., W. H. McAdams, Ed., McGraw-Hill, New York, Chap. 4.

Oppenheim, A. K. (1956), "Radiative Analysis by the Network Method," *Trans. ASME*, Vol. 78, pp. 725–735.

Rushmeier, H. E., Baum, D. R., and Hall, D. E. (1990), "Accelerating the Hemi-Cube Algorithm for Calculating Radiation Form Factors," in *Radiation Heat Transfer: Fundamentals and Applications*, ASME HTD Vol. 137, T. F. Smith, M. F. Modest, A. M. Smith, and S. T. Thynell, Eds., presented at AIAA/ASME Thermophysics Conference, June, pp. 45–52.

Siegel, R. and Howell, J. R. (1988), *Thermal Radiation Heat Transfer*, 2nd ed., Taylor, Francis, Hemisphere, New York.

Tao, W. Q. and Sparrow, E. M. (1985), "Ambiguities Related to the Calculation of Radiant Heat Exchange Between a Pair of Surfaces," *Int. J. Heat Mass Transfer*, Vol. 28, No. 9, pp. 1786–1787.

PROBLEMS

1. Derive the generalized form of the diffuse enclosure transfer equations (3-64 and 3-65) from Eqs. (3-35) through (3-40).

2. Show how to calculate the view factors for Example 3-8 and execute the program to obtain the solution.

3. Repeat Example 3-6 and improve the accuracy by dividing the slot into four regions as shown in the figure below. Plot the effective emissivity/absorptivity of the slot as a function of L/W for $L/W = 0.1$ to 10 and for wall emissivities of 0.25, 0.50, and 0.75.

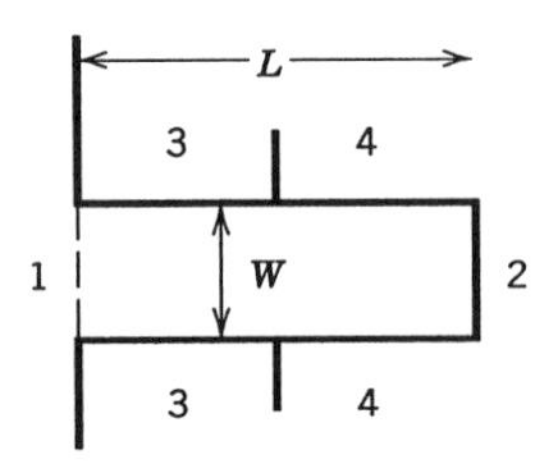

Figure P3

4. A long, radiant heating tube at $T_1 = 1000$ K with emissivity of 0.9 is at the center of a square, refractory-walled furnace (diffuse, gray, and adiabatic). One wall of the furnace is covered by a load consisting of a

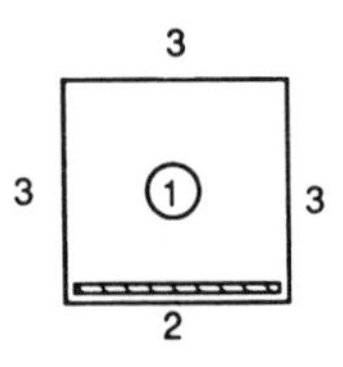

Figure P4

flat plate at $T_2 = 500$ K with emissivity 0.2. The furnace wall is one meter on a side and the tube diameter is $0.3/\pi$ meters. Calculate:

(a) the radiative heat transfer rate to the load (watts)

(b) the temperature of the refractory walls, T_3

5. Repeat Problem 3-4 for the case of a load consisting of a bank of closely spaced cylindrical tubes parallel to the radiant tube.

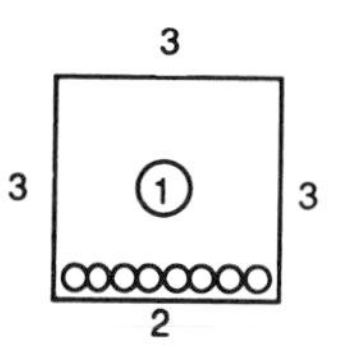

Figure P5

6. Consider the earth to be a diffuse, graybody at temperature T_1 with emissivity ε_1. A diffuse, gray satellite at T_2 is in low earth orbit on the dark side of the earth.

(a) Neglecting any heat generation on board the satellite, and assuming steady state, show that the temperature $T_2 = T_1(\varepsilon_1 F_{21})^{1/4}$, where due to scale considerations the view factor is approximately one half.

(b) Calculate the satellite temperature for $\varepsilon_1 = 1.0$ and $T_1 = 250$ K.

7. Two pipes carrying steam and cold water run parallel to each other through a large room. The hot pipe is at 120°C, the cold pipe is at 5°C, and the room is at 25°C. Both pipes are diffuse and gray with emissivities of 0.5. Compute the net radiant heat transfer from the hot pipe and the net radiant heat transfer to the cold pipe if both pipes have an outer diameter of 10 cm and the clearance between them is 2 cm.

8. A thin sheet of glass is used on one side of a long (2-D) greenhouse shaped like an equilateral triangle. (See next page.) The irradiation on the glass is comprised of the solar flux $q_s = 1100\ \mathrm{W/m^2}$, the atmospheric flux $q_{\mathrm{atm}} = 250\ \mathrm{W/m^2}$, and the irradiation from the inside surfaces q_i^-. The glass may be assumed to be totally transparent (i.e., $\tau_\lambda = 1$) for $\lambda <$ 1 μm and opaque and nonreflective (i.e., $\tau_\lambda = 0$; $\alpha_\lambda = 1$) for $\lambda < 1\ \mu$m. The solar flux is parallel to the back wall and intersects the floor and the glass at an angle of 30° to the normal. The back wall and floor are both black and adiabatic. The back wall may be assumed to be at the same temperature as the air inside the greenhouse T_i. The convective heat transfer coefficients at the floor, inside of the glass, and outside of the glass, respectively, are $h_f = 2\ \mathrm{W/m^2\ K}$, $h_i = 10\ \mathrm{W/m^2\ K}$, and $h_0 = 55$ $\mathrm{W/m^2\ K}$. The outside air is at $T_0 = 297$ K. Assuming steady-state conditions, write energy balance equations for the glass, inside air, and floor and solve for the glass, inside air, and floor temperatures $T_{g,i,f}$.

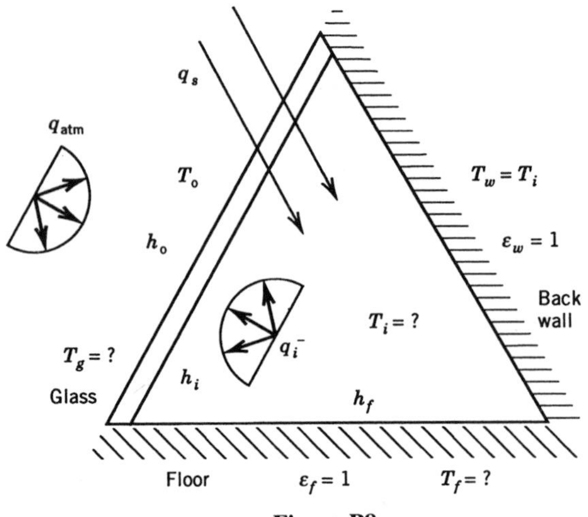

Figure P8

9. A flat plate with surface area on one side of A_1 ($\ll r_2^2$) is in orbit about the earth at an altitude of z. Its orientation is parallel to the surface of the earth. Show how the view factor F_{12} from the plate to the earth as a function of $\beta = z/r_2$ can be obtained from the result of Example 3-1 [Eq. (3-22)].

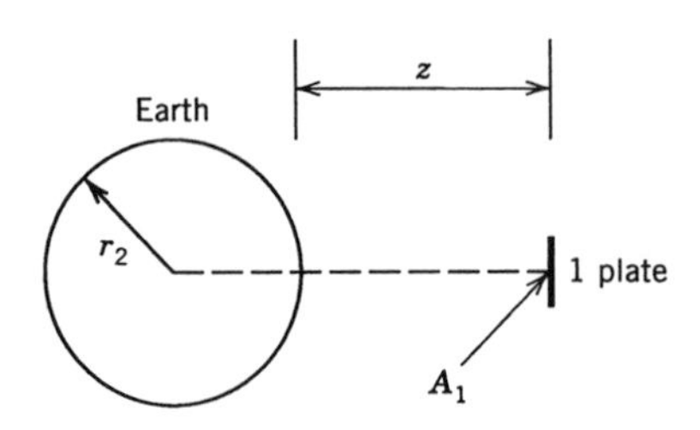

Figure P9

10. A flat plate with surface area A_1 (one side) is in orbit about the earth at an altitude of one earth radius ($z = r_2$). Its position is directly between the earth and the sun and its orientation is parallel to the solar flux. The view factor from one side of the plate to the earth is

$$F_{12} = \frac{1}{\pi}\left[\tan^{-1}\left(\frac{1}{\sqrt{H^2 - 1}}\right) - \frac{\sqrt{H^2 - 1}}{H^2}\right] \quad \text{where } H = 1 + \frac{z}{r_2}$$

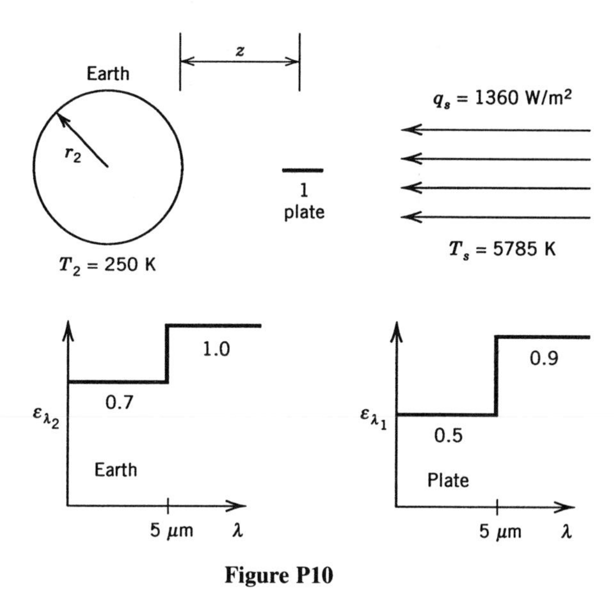

Figure P10

Make the following assumptions:

The sun is a blackbody emitter at a temperature of 5785 K and a distance of 216 sun radii ($216r_s$) from the earth.

The earth may be treated as a nongray, diffuse absorber-emitter with spectral, hemispherical emissivity ε_{λ_2} as given above.

The spectral, bi-directional reflectivity of the earth for normal incidence and normal reflectance equals the spectral, hemispherical reflectivity ($\rho''_{\lambda_2}(0,0) = \rho_{\lambda_2}$).

The plate may also be treated as a nongray, diffuse absorber-emitter with spectral, hemispherical emissivity ε_{λ_1} as given above.

The reflected intensity of solar radiation from the earth as viewed from the plate is uniform over the earth's surface.

What is the steady-state temperature of the plate T_1?

11. Reconsider Problem 4 with nongray instead of gray properties. Assume that the load is a diffuse, nongray emitter-absorber with a spectral hemispherical emissivity of 0.2 below 2 μm and 0.1 above 2 μm. Assume that the refractory wall is also a diffuse, nongray emitter-absorber with a spectral hemispherical emissivity of 0.9 below 3 μm and 0.7 above 3 μm. The tube is a diffuse, gray emitter-absorber with emissivity of 0.9, the same as in Problem 4. The temperatures of the tube and load are also the same as those in Problem 4. Determine the temperature of the refractory surface and the heat transfer rate to the load per unit depth.

12. A satellite is in low earth orbit (LEO) as shown in Fig. 3-19. The earth can be treated as a nongray, diffuse emitter-absorber and diffuse reflector at $T_2 = 250$ K with a spectral, hemispherical absorptivity of 0.7 below 5 μm and 1.0 above 5 μm as in Example 3-12. The satellite surfaces can be treated as diffuse, gray emitter-absorbers. The incident solar flux has a magnitude of 1360 W/m^2 and the spectral distribution of a blackbody at 5785 K. Internal heat transfer within the satellite maintains all surfaces at the same temperature T_1. Thus, as a whole, the satellite is adiabatic but individual surfaces may not be adiabatic. Treating the satellite as a cube with one face perpendicular to the solar flux, determine the temperature of the satellite, T_1.

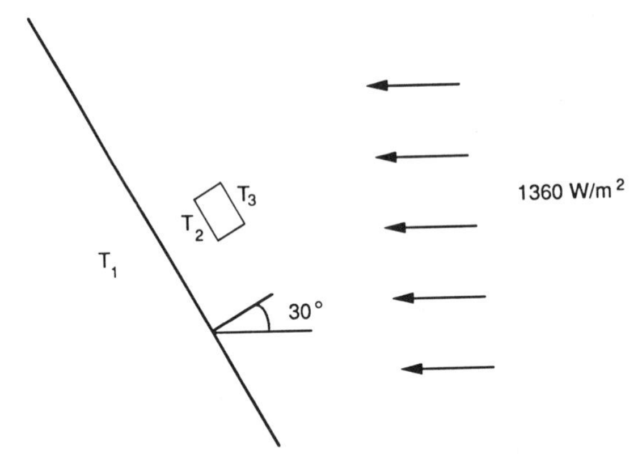

Figure P12

13. A satellite is in low earth orbit (LEO) as shown in the figure. The angle between the local normal to the earth's surface and the sun's rays is 30°. The earth can be treated as a nongray, diffuse emitter-absorber and diffuse reflector at 250 K with a spectral, hemispherical absorptivity of 0.7 below 5 μm and 1.0 above 5 μm. The satellite surfaces can be treated as adiabatic, diffuse, gray emitter-absorbers. The incident solar flux has a magnitude of 1360 W/m^2 and the spectral distribution of a blackbody at 5785 K.

(a) Determine the steady-state temperature of a surface of the satellite parallel to and facing away from the earth (T_3).

(b) Determine the steady-state temperature of a surface of the satellite parallel to and facing toward the earth (T_2).

14. Consider a row of tubes ($\varepsilon_2 = 0.9$) at $T_2 = 500$ K bounded on one side by a refractory wall ($Q_3 = 0$) and on the other side by a large furnace at

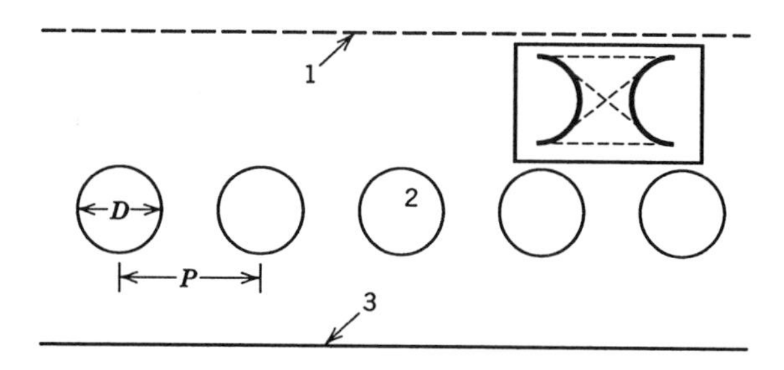

Figure P14

$T_1 = 1000$ K as shown in the figure. The pitch-to-diameter ratio of the tubes is $r = P/D = 2$.

(a) What is the heat flux to the tubes and

(b) the temperature of the refractory wall?

(Hint: Assuming there are enough tubes in the row that end effects are negligible, the view factor F_{22} can be obtained by considering only the transfer between two neighboring tubes and using the string rule as shown in the inset.)

CHAPTER 4

ELECTROMAGNETIC THEORY RESULTS

In 1864 James Clerk Maxwell published the now famous electromagnetic theory that provided a link between electrical, magnetic, and optical properties of matter. Many of the results from electromagnetic theory can be usefully applied to engineering radiative transfer. For example, the solution of Maxwell's equations for the interaction of a plane wave with a plane boundary separating two optically dissimilar media can be used to predict the reflectivity, absorptivity, and transmissivity of optically smooth surfaces and films. Also, the interaction of a plane wave with a spherical boundary between optically dissimilar media can be used to predict the scattering and absorption of radiation by particles.

In this chapter the solution of Maxwell's equations that describe the propagation of a linearly polarized, monochromatic plane wave will be presented. The fundamental optical properties of a material that govern the propagation of electromagnetic waves will be introduced and the interaction of a plane wave with a plane boundary will be discussed. The interaction of a plane wave with a sphere is discussed in Chapter 9.

MAXWELL'S EQUATIONS

The basis for electromagnetic theory is Maxwell's equations. In general vector notation these equations are written as follows [Bohren and Huffman,

1983; Born and Wolf, 1970]:

$$\nabla \times \mathbf{E} = -\frac{\partial \mathbf{B}}{\partial t} \tag{4-1}$$

$$\nabla \times \mathbf{H} = \frac{\partial \mathbf{D}}{\partial t} + \mathbf{j} \tag{4-2}$$

$$\nabla \cdot \mathbf{D} = \rho_e \tag{4-3}$$

$$\nabla \cdot \mathbf{B} = 0 \tag{4-4}$$

where

$$\mathbf{D} = \varepsilon_0 \mathbf{E} + \mathbf{P} \tag{4-5}$$

and

$$\mathbf{B} = \mu_0(\mathbf{H} + \mathbf{M}) \tag{4-6}$$

The vector (boldface) and scalar field variables appearing in Eqs. (4-1) through (4-6) are

$\mathbf{E}$ = electric field vector
$\mathbf{H}$ = magnetic field vector
$\mathbf{D}$ = electric displacement
$\mathbf{B}$ = magnetic induction
$\mathbf{j}$ = free charge current density
$\mathbf{P}$ = electric polarization (electric dipole moment per unit volume)
$\mathbf{M}$ = magnetization (magnetic dipole moment per unit volume)
ρ_e = electric charge density

In (4-5) and (4-6) ε_0 and μ_0 are the permittivity and permeability, respectively, of free space.

Maxwell's equations are augmented by the following set of constitutive relations for isotropic media:

$$\mathbf{j} = \sigma_e \mathbf{E} \tag{4-7}$$

$$\mathbf{P} = \varepsilon_0 \chi \mathbf{E} \tag{4-8}$$

$$\mathbf{B} = \mu \mathbf{H} \tag{4-9}$$

where the material properties are

σ_e = electrical conductivity ($= 1/r_e$)
r_e = electrical resistivity
ε = permittivity
μ = magnetic permeability
χ = electric susceptibility

These properties are all scalars according to the isotropic material assumption. Combining Eqs. (4-5) and (4-8) gives

$$\mathbf{D} = \varepsilon_0(1 + \chi)\mathbf{E} \tag{4-10}$$

which can also be written as

$$\mathbf{D} = \varepsilon\mathbf{E} \tag{4-11}$$

when the permittivity is defined as

$$\varepsilon = \varepsilon_0(1 + \chi) \tag{4-12}$$

The permittivity is a constant that is related to the force acting between two point charges. Coulomb's law states that when two charges e_1 and e_2 are separated by a distance r in a nonconducting, dielectric medium, the force between them is proportional to the magnitude of each charge and inversely proportional to the square of the distance between them.

$$F \sim \frac{e_1 e_2}{r^2} \tag{4-13}$$

The proportionality constant can be quantified by considering the case where the intervening medium is a vacuum. In that case, the force is defined by the equality

$$F_0 = \frac{1}{4\pi\varepsilon_0}\frac{e_1 e_2}{r^2} \tag{4-14}$$

where the vacuum permittivity is

$$\varepsilon_0 = 8.85 \times 10^{-12}\frac{\mathrm{C}^2}{\mathrm{N\,m}^2}$$

If the intervening medium is not a vacuum (but still a dielectric) the force is less than F_0 as given by

$$F = \frac{1}{4\pi\varepsilon}\frac{e_1 e_2}{r^2} < F_0, \qquad \varepsilon > \varepsilon_0 \tag{4-15}$$

where ε is the permittivity of the nonvacuum medium. Another commonly used property is the *relative permittivity* or *dielectric constant* defined as the ratio of the permittivity of the medium to that of a vacuum.

$$\varepsilon_r = \frac{\varepsilon}{\varepsilon_0} \tag{4-16}$$

The magnetic permeability is a constant related to the magnetic flux generated by a current flowing through a wire. Ampere's law states that the magnetic flux generated by a current (i) flowing through a wire is proportional to the magnitude of the current and inversely proportional to the distance from the wire.

$$B \sim \frac{i}{r} \tag{4-17}$$

The proportionality constant can again be defined by considering the case of a vacuum. In that case the magnitude of the flux is given by the equality

$$B = \frac{\mu_0}{2\pi}\frac{i}{r} \tag{4-18}$$

where the vacuum magnetic permeability is

$$\mu_0 = 4\pi \times 10^{-7}\,\frac{\text{Weber}}{\text{amp m}} \tag{4-19}$$

For most materials the magnetic permeability is equal to that of a vacuum $\mu = \mu_0$, and from Eqs. (4-6) and (4-9) the magnetization is zero, $\mathbf{M} = 0$.

Linearly Polarized Plane Wave

The full set of Maxwell's equations is quite complicated and the solutions are also quite complex. However, the essential results, which are pertinent to engineering radiative transfer, can be obtained by considering the simplified case of a linearly polarized plane wave.

For a linearly polarized plane wave propagating in the x direction, the electric vector has only one nonzero component and that component (E_y) is only a function of one spatial coordinate and time.

$$\mathbf{E} = \mathbf{E}(x, t) \qquad \text{(plane wave)} \tag{4-20}$$

$$E_y = E_y(x, t), \quad E_z = 0 \qquad \text{(linearly polarized)} \tag{4-21}$$

By substituting Eqs. (4-5) through (4-9) and (4-20) and (4-21) into (4-1) through (4-4) Maxwell's equations can now be simplified considerably. The resulting equation for the electric vector for zero net charge density $\rho_e = 0$ is

$$\frac{\partial^2 E_y}{\partial x^2} = \mu\varepsilon \frac{\partial^2 E_y}{\partial t^2} + \mu\sigma_e \frac{\partial E_y}{\partial t} \tag{4-22}$$

which is the equation for a damped wave.

Nonabsorbing Medium

For a medium that is characterized by a zero electric conductivity (or infinite resistivity), the last term on the right-hand side of Eq. (4-22) vanishes and the resulting governing equation is

$$\frac{\partial^2 E_y}{\partial x^2} = \mu\varepsilon \frac{\partial^2 E_y}{\partial t^2} \tag{4-23}$$

If the permittivity is a pure real number, Eq. (4-23) is the undamped wave equation. The classical solution to this equation is

$$E_y(x, t) = f\left(-x + \frac{t}{\sqrt{\mu\varepsilon}}\right) + g\left(x + \frac{t}{\sqrt{\mu\varepsilon}}\right) \tag{4-24}$$

where f and g are arbitrary functions determined by boundary conditions. The second function g can be discarded because it is physically unrealistic (the wave would be propagating forward in space in the negative time domain). The form of the physically interesting solution f is found by assuming that E_y is either periodic in x at time $t = 0$ or periodic in time at $x = 0$. The former approach is illustrated in Fig. 4-1, which is representative of the instantaneous spatial distribution of the electric vector for polychromatic radiation.

Once it is assumed that E_y is periodic (but not necessarily harmonic), Fourier analysis can be employed to resolve the complex periodic function f into its harmonic spectral components. Each of the spectral components represents monochromatic radiation at a particular wavelength or frequency.

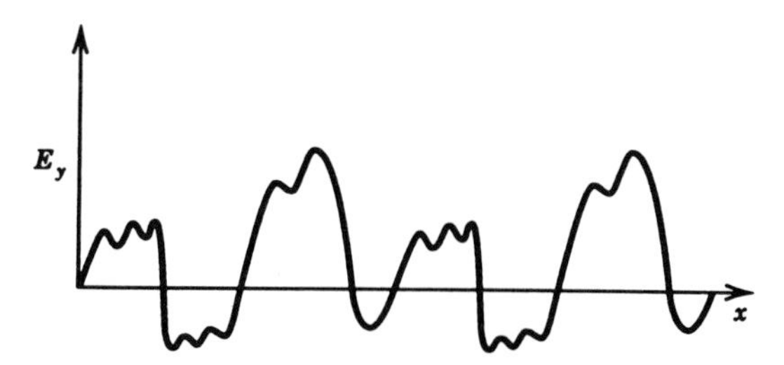

Figure 4-1 Instantaneous spatial distribution of polychromatic wave.

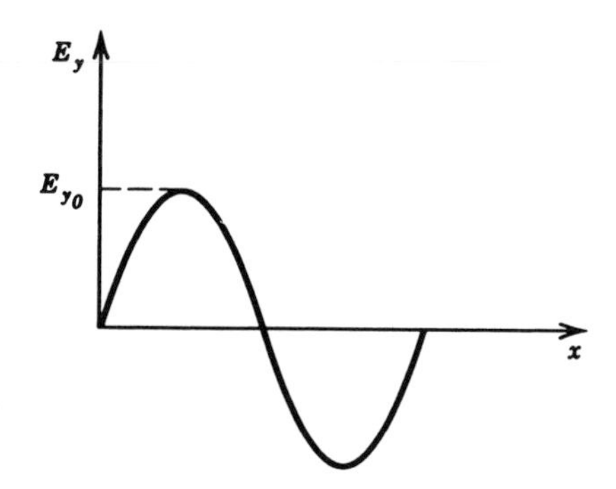

Figure 4-2 Instantaneous spatial distribution of monochromatic wave.

The contribution at one such wavelength can be written using the exponential function with a pure complex argument as in Eq. (4-25).

$$E_y = E_{y_0} \text{Re}\left\{ \exp\left[i\frac{2\pi}{\lambda}\left(-x + \frac{t}{\sqrt{\mu\varepsilon}} \right) \right] \right\} \tag{4-25}$$

Equation (4-25) is the mathematical representation of a sinusoidal traveling wave with an amplitude of E_{y_0} and a wave speed* of

$$c = \frac{1}{\sqrt{\mu\varepsilon}} \quad \text{(wave speed)} \tag{4-26}$$

The instantaneous spatial distribution of the electric vector for monochromatic radiation is shown in Fig. 4-2. By defining the refractive index of a

*In vacuum, the wave speed is $c_0 = 1/\sqrt{\mu_0\varepsilon_0} = 3.0 \times 10^8$ m/s. The fact that Maxwell was able to predict the measured speed of light from electromagnetic constants was remarkable evidence of the validity of the theory.

medium as the ratio of the speed of light in a vacuum to that in the medium

$$n = \frac{c_0}{c} = \frac{\lambda_0}{\lambda} \tag{4-27}$$

and the circular frequency as

$$\omega = \frac{2\pi c_0}{\lambda_0} = \frac{2\pi c}{\lambda} \tag{4-28}$$

Eq. (4-25) can be rewritten as

$$E_y = E_{y_0} \mathrm{Re}\left\{ \exp\left[i\omega\left(-\frac{nx}{c_0} + t \right) \right] \right\} \tag{4-29}$$

This result for the electric vector for a linearly polarized plane wave demonstrates that electromagnetic energy propagation in a medium with zero conductivity ($\sigma_e = 0$) and a pure real permittivity (or susceptibility) can be described as undamped, traveling, transverse, vector waves. There is also a magnetic vector solution that goes along with the electric vector. The nonzero component of the magnetic vector in this example occurs in the z direction and is given by the following relation, which can be obtained from Maxwell's equations.

$$H_z = \frac{n}{\mu c_0} E_y \tag{4-30}$$

Thus the electric vector and the magnetic vector oscillate in phase and at right angles to each other, with an amplitude ratio as given by (4-30). Figure 4-3 shows the relationship of the oscillating electric and magnetic vectors of a linearly polarized plane electromagnetic wave.

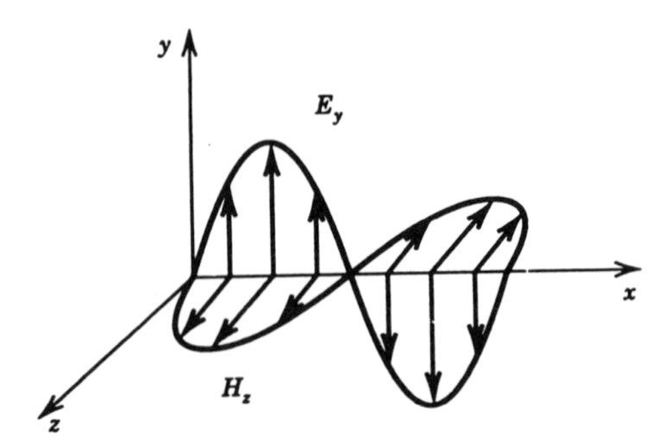

Figure 4-3 Linearly polarized plane wave.

Absorbing Medium

For a medium with a finite electrical conductivity the last term on the right-hand side of Eq. (4-22) is not zero. Having established already by assumption that the time dependence of E_y is $\exp(i\omega t)$, Eq. (4-22) can be written for finite σ_e as

$$\frac{\partial^2 E_y}{\partial x^2} = \mu\tilde{\varepsilon}\frac{\partial^2 E_y}{\partial t^2} \tag{4-31}$$

where the *complex permittivity* is defined as

$$\tilde{\varepsilon} = \varepsilon_0(1 + \tilde{\chi}) - i\frac{\sigma_e}{\omega} \tag{4-32}$$

Alternatively, the *complex refractive index*, defined analogous to Eq. (4-27) as

$$\tilde{n} = n - ik = \sqrt{\frac{\mu\tilde{\varepsilon}}{\mu_0\varepsilon_0}} \tag{4-33}$$

can be used as a phenomenological coefficient instead of $\tilde{\varepsilon}$.* The solution to Eq. (4-31) can then be obtained from the comparable solution for the nonabsorbing case, Eq. (4-29), by simply substituting the complex refractive index for n, giving

$$E_y = E_{y_0}\mathrm{Re}\left\{\exp\left[i\omega\left(-\frac{nx}{c_0} + t\right)\right]\exp\left[-\frac{\omega kx}{c_0}\right]\right\} \tag{4-34}$$

Since the argument of the second exponential term in Eq. (4-34) is a real number, the amplitude of the wave (which is initially E_{y_0} at $x = 0$) decays exponentially due to the influence of k. Thus the imaginary part of the refractive index (also referred to as the *absorption index*) accounts for damping or absorption of the wave as it travels through the medium. Together, n and k are referred to as *optical constants*.

Another form of phenomenological coefficient commonly used is the *complex dielectric constant*.

$$\tilde{\varepsilon}_r = \tilde{n}^2 = \frac{\mu\tilde{\varepsilon}}{\mu_0\varepsilon_0} = \varepsilon_r' - i\varepsilon_r'' = (1 + \tilde{\chi}) - i\frac{\sigma_e}{\omega\varepsilon_0} \tag{4-35}$$

As stated earlier, for most materials (metals and dielectrics) the magnetic permeability is equal to that of a vacuum $\mu = \mu_0$, thus $\tilde{\varepsilon}_r$ can also be thought

*If the argument of Eq. (4-24) had been chosen as $x - t/\sqrt{\mu\varepsilon}$ then $\tilde{\varepsilon} = \varepsilon + i(\sigma_e/\omega)$ and $\tilde{n} = n + ik$ would be used.

of as a complex relative permittivity, after the fashion of Eq. (4-16). The relationships between the real and imaginary parts of $\tilde{\varepsilon}_r$ and $\tilde{n}^2$ can be obtained from Eq. (4-35) as follows:

$$n^2 = \frac{\varepsilon_r'}{2}\left[1 + \sqrt{1 + \left(\frac{\varepsilon_r''}{\varepsilon_r'}\right)^2}\,\right] \tag{4-36}$$

$$k^2 = \frac{\varepsilon_r'}{2}\left[-1 + \sqrt{1 + \left(\frac{\varepsilon_r''}{\varepsilon_r'}\right)^2}\,\right] \tag{4-37}$$

$$\varepsilon_r' = 1 + \chi' = n^2 - k^2 \tag{4-38}$$

$$\varepsilon_r'' = \chi'' - \frac{\sigma_e}{\omega\varepsilon_0} = 2nk \tag{4-39}$$

These relations are important because they establish a connection between the optical properties and the electrical properties of a material. Nonabsorbing materials are characterized by zero electrical conductivity, $\sigma_e(= 1/r_e) = 0$ and a pure real susceptibility ($\chi'' = 0$), thus giving $n = c_0/c$ and $k = 0$. Absorbing materials have a finite, nonzero electrical conductivity and/or an imaginary component of susceptibility and therefore a finite, nonzero absorption index. The imaginary component of the susceptibility is associated with absorption by bound charged particles (such as bound electrons or ions) and the conductivity is associated with absorption by free charge carriers (such as conduction electrons). Depending on the nature of the material, either bound or free charges (or both) can contribute to the absorption of radiative energy at a particular wavelength. In metals, free electrons are generally responsible for absorption of radiation from zero frequency (dc) up through the visible region (10^{14} Hz), while bound electrons are responsible for absorption in the visible and ultraviolet regions in both metals and dielectrics. Dielectrics can also absorb infrared radiation through the oscillating dipole associated with lattice vibration. Thus, a dielectric that is nonabsorbing at frequencies near dc does not remain so over the entire frequency spectrum. As a result, the values of n and k (the so-called optical constants) are not at all constant and are in general quite dependent on frequency or wavelength, $\tilde{n} = \tilde{n}(\lambda)$. Appendix C contains a table of n and k as a function of wavelength for several common materials including metals, semiconductors, and dielectrics. The subject of the variation of the values of n and k with frequency is called *dispersion theory* and is discussed in Chapter 5.

Energy Propagation

The energy flux carried by a propagating electromagnetic wave is related to the Poynting vector. The Poynting vector **S** is the cross product of the electric

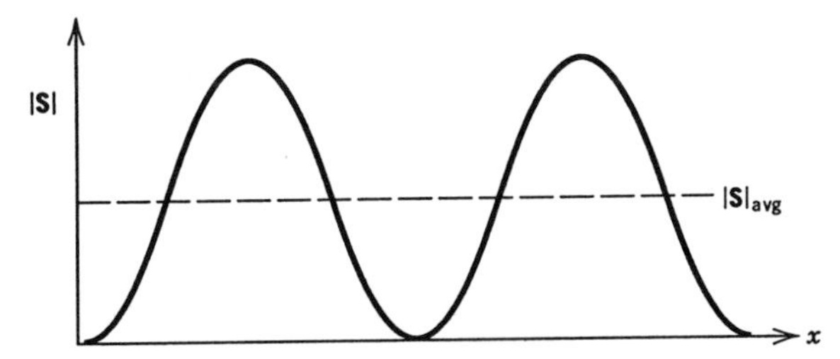

Figure 4-4 Instantaneous spatial distribution of magnitude of Poynting vector for nonabsorbing medium.

and magnetic vectors.

$$\mathbf{S} = \mathbf{E} \times \mathbf{H} \tag{4-40}$$

Since the electric vector and magnetic vector are at right angles to each other, the magnitude of the Poynting vector is simply the product of the magnitudes of these two vectors.

$$|\mathbf{S}| = |\mathbf{E}||\mathbf{H}| \tag{4-41}$$

Recalling Eqs. (4-30) and (4-34), and assuming $k \ll n$, the magnitude of $\mathbf{S}$ can be written as

$$|\mathbf{S}| = \frac{n}{\mu c_0} E_y^2 = \frac{n}{\mu c_0} E_{y_0}^2 \left[\mathrm{Re}\left\{ \exp\left[i\omega\left(-\frac{nx}{c_0} + t \right) \right] \right\} \right]^2 \exp\left[-\frac{2\omega kx}{c_0} \right] \tag{4-42}$$

A schematic representation of the instantaneous magnitude $|\mathbf{S}|$ for a nonabsorbing medium is shown in Fig. 4-4. Since it is the cross product of two perpendicular transverse vector waves (**E** and **H**), the Poynting vector is a longitudinal vector wave with its only nonzero component in the direction of propagation.

The average magnitude $|\mathbf{S}|_{\text{avg}}$ can be obtained from the instantaneous magnitude $|\mathbf{S}|$ by using the integral

$$\frac{1}{t_0} \int_0^{t_0} \sin^2(\omega t + \delta)\, dt = \frac{1}{2} \tag{4-43}$$

to average $|\mathbf{S}|$ over the period t_0 and obtain

$$|\mathbf{S}|_{\text{avg}} = \frac{n}{2c_0\mu} E_{y_0}^2 \exp\left[-\frac{2\omega kx}{c_0} \right] \tag{4-44}$$

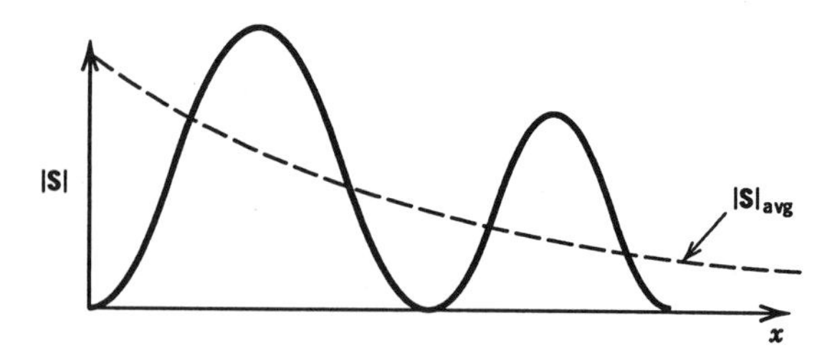

Figure 4-5 Instantaneous spatial distribution of magnitude of Poynting vector for absorbing medium.

The instantaneous and average magnitudes of the Poynting vector for an absorbing medium are shown in Fig. 4-5.

As indicated by Eq. (4-44), in an absorbing medium the average energy flux of a plane wave is attenuated exponentially with distance by absorption. The flux is constant if the medium is nonabsorbing. The attenuation factor can also be written in terms of wavelength instead of frequency.

$$\frac{|\mathbf{S}|_{\text{avg}}}{|\mathbf{S}|_{\text{avg}_0}} = \exp\left[-\frac{4\pi kx}{\lambda_0}\right] \tag{4-45}$$

The coefficient of x in Eq. (4-45) is defined as the *absorption coefficient* for a homogeneous medium.

$$K_{a\lambda} = \frac{4\pi k}{\lambda_0} \tag{4-46}$$

Example 4-1

Room temperature aluminum at 1.06 μm (Nd-YAG laser wavelength) has optical constants of $n = 1.2$ and $k = 10$ (see Appendix C). What is the characteristic length scale for attenuation of laser energy at this wavelength in aluminum?

The characteristic length scale for attenuation of radiation in a homogeneous medium is given by Eqs. (4-45) and (4-46) as the inverse of the absorption coefficient.

$$\frac{1}{K_{a\lambda}} = \frac{\lambda_0}{4\pi k} = \frac{1.06\ \mu\text{m}}{4\pi(10)} = 8.4\ \text{nm}$$

Thus the length scale for attenuation of radiation in aluminum, as in metals generally, is less than a hundredth of a wavelength. In general this length scale is referred to as the *mean free photon pathlength*. In metals it is also referred to as the *skin depth*.

Relation between Poynting Vector and Intensity

The average magnitude of the Poynting vector for a plane wave, as derived in the previous section, is the monochromatic radiant power per unit area perpendicular to the direction of propagation. In the nomenclature of Chapter 1 this quantity is the product of intensity and solid angle.

$$|\mathbf{S}|_{\text{avg}} = \frac{\text{power}}{\Delta A_{\perp}\ \Delta\lambda} = I_{\lambda}\,\Delta\Omega \tag{4-47}$$

In a nonattenuating medium the intensity is constant in the direction of propagation, whereas in an absorbing medium the intensity decreases in the direction of propagation according to the relation

$$I_{\lambda} = I_{\lambda_0}\exp[-K_{a\lambda}x] \tag{4-48}$$

where I_{λ_0} is the initial intensity at $x = 0$ and I_{λ} is the intensity after propagating a distance x. If a plane wave is incident on a plane surface at an angle θ (Fig. 4-6), the contribution to the flux in the direction normal to the surface (i.e., partial flux) can be written, based on Eqs. (4-47) and (1-15) as

$$\Delta q_{\lambda} = I_{\lambda}\cos\theta\,\Delta\Omega = |\mathbf{S}|_{\text{avg}}\cos\theta \tag{4-49}$$

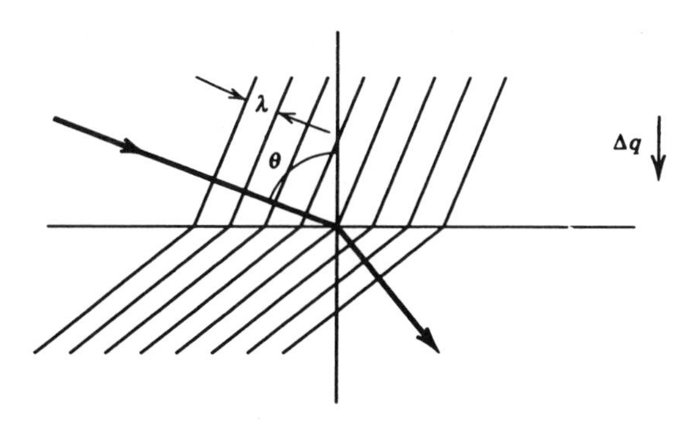

Figure 4-6 Flux at an interface from a plane wave.

Interaction of a Plane Wave with a Plane Boundary

Most of the important engineering radiative transfer results that can be obtained from electromagnetic theory are derived from the interaction of a plane wave with a plane boundary separating two semi-infinite media with different optical properties. In particular the reflectivity and transmissivity of an optically smooth interface can be obtained by applying the appropriate continuity and conservation conditions for the vector waves at a plane boundary.

Consider two adjacent media separated by a plane boundary as shown in Fig. 4-7. Medium 1 is nonabsorbing with real refractive index n_1 and medium 2 is absorbing with complex refractive index $n_2 - ik_2$.

A plane wave with electric vector $\mathbf{E}_i$ is obliquely incident on medium 2 from medium 1 at an angle of θ_1. The interaction of the incident wave with medium 2 produces a specularly reflected wave $\mathbf{E}_r$ that propagates back through medium 1 and a specularly transmitted wave $\mathbf{E}_t$ that propagates into medium 2. The *plane of incidence* is defined as the plane containing the normal to the boundary **N** and the direction of propagation of the incident wave. By definition of specular reflection and transmission, the plane of incidence also contains the directions of propagation of the reflected and transmitted waves.

To apply the interface continuity conditions it is necessary to resolve the various vector waves (**E**, **H**, **D**, **B**, and **j**) into components. Since plane waves are being considered, there are only two components of each of the vectors. These two components are referred to as a perpendicular polarization (or

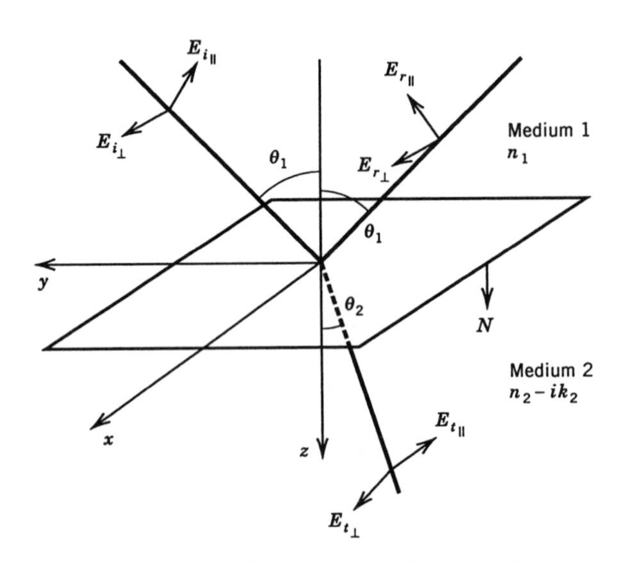

Figure 4-7 Plane wave at a plane interface.

component) and a parallel polarization (or component). The definitions of the parallel and perpendicular components of polarization are made with respect to the electric vectors $\mathbf{E}_{i,r,t}$. This procedure is illustrated in Fig. 4-7.

The *parallel* component of the electric vector $E_{\parallel}$ is the component that oscillates parallel to the plane of incidence. Other names for this component are the TM (transverse magnetic) mode and p polarization (from the German parallel). The *perpendicular* component of the electric vector $E_{\perp}$ is the component that oscillates perpendicular to the plane of incidence. Other names for this component are the TE (transverse electric) mode and s polarization (from the German word meaning perpendicular, *senkrecht*).

$$\parallel = p = \text{TM} = \text{parallel to plane of incidence}$$
$$\perp = s = \text{TE} = \text{perpendicular to plane of incidence}$$

According to these definitions of polarization, $E_{\perp}$ is always parallel to the plane interface, because it is perpendicular to the plane of incidence, which is perpendicular to the interface.

The arrows labeled $E_{\parallel}$ and $E_{\perp}$ in Fig. 4-7 represent the positive directions for the components of the electric vector. These directions do not necessarily coincide with the directions of the electric vectors themselves. By convention one always looks against the direction of propagation of the wave [Bennett and Bennett, 1978]. Thus the positive direction of $E_{\parallel}$ appears to undergo a phase change of 180° upon reflection while the positive direction of $E_{\perp}$ remains unchanged. The actual directions of the electric vector components may be opposite those of the positive coordinates, depending on the magnitudes of n_2 and n_1. For example, if $n_2 > n_1$ both components $E_{\parallel}$ and $E_{\perp}$ undergo a phase change of 180° upon reflection. Thus, if the incident components were in the positive direction (coincident with the arrows shown in Fig. 4-7), the reflected parallel component would be in the positive direction, coincident with the arrow labeled $E_{r\parallel}$, while the reflected perpendicular component would be in the negative direction, opposite the arrow labeled $E_{r\perp}$.

The amplitude of the reflected and transmitted electric vectors is obtained by applying the following interface continuity conditions.

$$\mathbf{N} \times (\mathbf{E}_t - \mathbf{E}_r - \mathbf{E}_i) = 0 \quad \text{(continuity of tangential component of } \mathbf{E}) \tag{4-50}$$

$$\mathbf{N} \times (\mathbf{H}_t - \mathbf{H}_r - \mathbf{H}_i) = 0 \quad \text{(continuity of tangential component of } \mathbf{H}) \tag{4-51}$$

$$\mathbf{N} \cdot (\mathbf{B}_t - \mathbf{B}_r - \mathbf{B}_i) = 0 \quad \text{(continuity of magnetic flux)} \tag{4-52}$$

$$\mathbf{N} \cdot (\mathbf{j}_t - \mathbf{j}_r - \mathbf{j}_i) = -\frac{\partial}{\partial t}[\mathbf{N} \cdot (\mathbf{D}_t - \mathbf{D}_r - \mathbf{D}_i)] \quad \text{(conservation of electric charge)} \tag{4-53}$$

Equations (4-50) through (4-53) give a total of six scalar relations, two each from the vector equations (4-50) and (4-51) and one each from the scalar equations (4-52) and (4-53). (The vector equations only give two relations instead of three each because the coordinate system has been defined with the x axis perpendicular to the plane of incidence.)

Of the six relations that result from the continuity equations, four are called the Fresnel relations and one is referred to as Snell's law. The remaining relation simply states that the angles of incidence and reflectance are equal, as was already assumed.

FRESNEL RELATIONS

The Fresnel coefficients $\tilde{r}$ and $\tilde{t}$ are the ratios of the complex amplitudes of the reflected and transmitted electric vectors to the complex amplitude of the incident electric vector. Since these coefficients are complex there is both magnitude and phase change information contained in them.

$$\tilde{t}_{12_\parallel} = \frac{E_{t_{\parallel 0}}}{E_{i_{\parallel 0}}} = \frac{2n_1 \cos\theta_1}{\tilde{n}_2 \cos\theta_1 + n_1 \cos\tilde{\theta}_2} \tag{4-54}$$

$$\tilde{r}_{12_\parallel} = \frac{E_{r_{\parallel 0}}}{E_{i_{\parallel 0}}} = \frac{\tilde{n}_2 \cos\theta_1 - n_1 \cos\tilde{\theta}_2}{\tilde{n}_2 \cos\theta_1 + n_1 \cos\tilde{\theta}_2} \tag{4-55}$$

$$\tilde{t}_{12_\perp} = \frac{E_{t_{\perp 0}}}{E_{i_{\perp 0}}} = \frac{2n_1 \cos\theta_1}{n_1 \cos\theta_1 + \tilde{n}_2 \cos\tilde{\theta}_2} \tag{4-56}$$

$$\tilde{r}_{12_\perp} = \frac{E_{r_{\perp 0}}}{E_{i_{\perp 0}}} = \frac{n_1 \cos\theta_1 - \tilde{n}_2 \cos\tilde{\theta}_2}{n_1 \cos\theta_1 + \tilde{n}_2 \cos\tilde{\theta}_2} \tag{4-57}$$

Snell's law gives the angle of transmission in terms of the optical constants. The interpretation of Snell's law can be easily understood when medium 2 is nonabsorbing. In that case n_2 and θ_2 are both pure real numbers and θ_2 is the angle of transmission. Snell's law says that radiation always bends toward the normal to the interface when passing from a lower index medium to a higher index medium (Fig. 4-8).

$$n_1 \sin\theta_1 = \tilde{n}_2 \sin\tilde{\theta}_2 \qquad \text{(Snell's law)} \tag{4-58}$$

When $k_2 > 0$, then θ_2 is complex and does not have the simple interpretation of the angle of transmission. The true angle of transmission can be determined from complex algebraic manipulation, but that discussion is beyond the scope of the present treatment (see [Bohren and Huffman, 1983]).

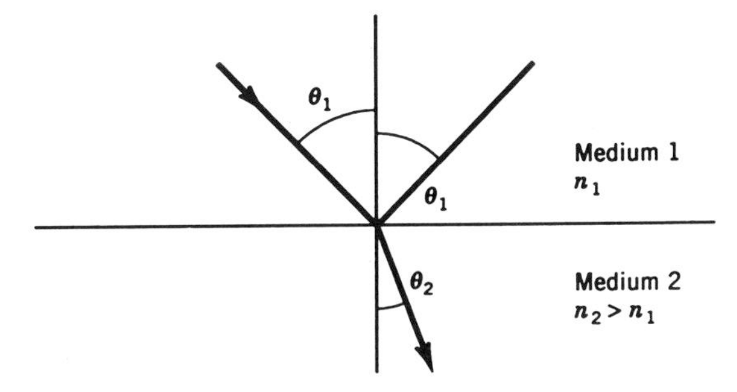

Figure 4-8 Illustration of Snell's law.

Evaluation of Interface Reflectivity and Transmissivity

From the results obtained in Eq. (4-49) the spectral flux approaching or leaving the interface can be written as

$$\Delta q = I \cos\theta \, \Delta\Omega = |\mathbf{S}|_{\text{avg}} \cos\theta \tag{4-59}$$

where the λ subscript has been omitted for simplicity. Substituting the result from (4-44) into (4-59), the incident, reflected, and transmitted fluxes can be written as

$$\Delta q^- = \frac{n_1}{2c_0\mu_1} E_{i_0}^2 \cos\theta_1 \quad \text{(incident flux)} \tag{4-60}$$

$$\Delta q_r = \frac{n_1}{2c_0\mu_1} E_{r_0}^2 \cos\theta_1 \quad \text{(reflected flux)} \tag{4-61}$$

$$\Delta q_t = \frac{\tilde{n}_2}{2c_0\mu_2} E_{t_0}^2 \cos\tilde{\theta}_2 \quad \text{(transmitted flux)} \tag{4-62}$$

The interface reflectivity is the ratio of the reflected flux to the incident flux

$$\mathcal{R} = \frac{\Delta q_r}{\Delta q^-} = \frac{E_{r_0}^2}{E_{i_0}^2} = |\tilde{r}_{12}|^2 \quad (\text{interface reflectivity } \parallel \text{ or } \perp) \tag{4-63}$$

and similarly for the interface transmissivity

$$\mathcal{T} = \frac{\Delta q_t}{\Delta q^-} = \frac{E_{t_0}^2 \cos\tilde{\theta}_2}{E_{i_0}^2 \cos\theta_1} \frac{\tilde{n}_2}{n_1} \frac{\mu_1}{\mu_2} = |\tilde{t}_{12}|^2 \frac{\cos\tilde{\theta}_2}{\cos\theta_1} \frac{\tilde{n}_2}{n_1}$$

$$(\text{interface transmissivity } \parallel \text{ or } \perp) \tag{4-64}$$

From conservation of electromagnetic energy at the interface the sum of the interface reflectivity and transmissivity must equal one.

$$\mathcal{R} + \mathcal{T} = 1 \tag{4-65}$$

To obtain a useful result, the Fresnel relations must be substituted into Eqs. (4-63) and (4-64) and the complex algebra carried out. This can be facilitated by defining a new complex variable $u - iv$,

$$u - iv = \tilde{n}_2 \cos \tilde{\theta}_2 = \tilde{n}_2 \sqrt{1 - \sin^2 \tilde{\theta}_2} = \sqrt{\tilde{n}_2^2 - n_1^2 \sin^2 \theta_1} \tag{4-66}$$

where Snell's law has been used in (4-66). Equating real and imaginary parts to solve for u and v gives

$$2u^2 = \left(n_2^2 - k_2^2 - n_1^2 \sin^2 \theta_1\right) + \sqrt{\left(n_2^2 - k_2^2 - n_1^2 \sin^2 \theta_1\right)^2 + 4n_2^2 k_2^2} \tag{4-67}$$

$$2v^2 = -\left(n_2^2 - k_2^2 - n_1^2 \sin^2 \theta_1\right) + \sqrt{\left(n_2^2 - k_2^2 - n_1^2 \sin^2 \theta_1\right)^2 + 4n_2^2 k_2^2} \tag{4-68}$$

The perpendicular and parallel interface reflectivities can then be expressed as

$$\mathscr{R}_\perp = \frac{\left(n_1 \cos \theta_1 - u\right)^2 + v^2}{\left(n_1 \cos \theta_1 + u\right)^2 + v^2} \tag{4-69}$$

$$\mathscr{R}_\parallel = \frac{\left[\left(n_2^2 - k_2^2\right)\cos \theta_1 - n_1 u\right]^2 + \left[2n_2 k_2 \cos \theta_1 - n_1 v\right]^2}{\left[\left(n_2^2 - k_2^2\right)\cos \theta_1 + n_1 u\right]^2 + \left[2n_2 k_2 \cos \theta_1 + n_1 v\right]^2} \tag{4-70}$$

The corresponding interface transmissivities can be calculated from Eq. (4-65).

For normal incidence ($\theta_1 = 0$), Eqs. (4-67) and (4-68) simplify to give $u = n_2$ and $v = k_2$. This result, when substituted into (4-69) and (4-70), gives the interface reflectance for normal incidence.

$$\mathscr{R}_n = \frac{\left(n_2 - n_1\right)^2 + k_2^2}{\left(n_2 + n_1\right)^2 + k_2^2} \tag{4-71}$$

For normal incidence the normal to the boundary and the direction of incidence are colinear and the plane of incidence is not defined. Therefore there is no distinction between the components of polarization and both (4-69) and (4-70) reduce to the same expression (4-71) as $\theta_1 \to 0$.*

*It is interesting to note, however, that at $\theta = 0$, the Fresnel reflection coefficients from Eqs. (4-55) and (4-57) differ by a sign. This is due to the fact that, according to sign convention, the positive direction for the parallel component changes direction upon reflection while the perpendicular component does not (as shown in Fig. 4-7). This, together with the fact that there is an actual phase change of either 180° ($n_2 > n_1$) or 0° ($n_2 < n_1$), which is the same for both components, results in an apparent phase change of 180° in the perpendicular component and no apparent phase change for the parallel component. Thus the sign difference.

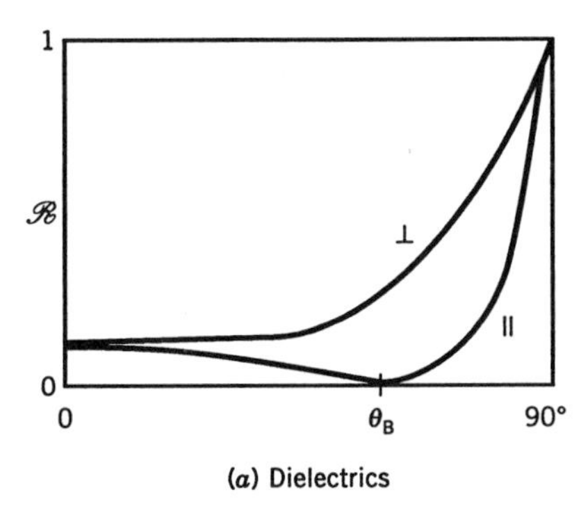

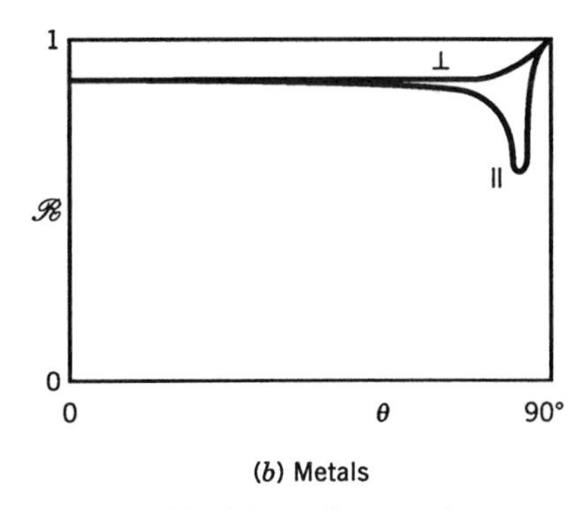

Figure 4-9 Fresnel interface reflectivity vs. angle of incidence ($n_2 > n_1$).

Another simplification of the general expressions occurs when medium 2 is weakly absorbing or nonabsorbing, $k_2 \ll n_2$. In this case the expressions for u and v reduce to $u^2 = n_2^2 - n_1^2\sin^2\theta_1$ and $v = 0$ and the angle of propagation into medium 2 is given by the pure real angle θ_2 as calculated from Snell's law (4-58).

Having derived a new type of reflectivity (namely the interface reflectivity of an optically smooth surface $\mathscr{R}$), the question arises as to the relationship between $\mathscr{R}$ and the various reflectivities defined in Chapter 2. Is $\mathscr{R}$ the bi-directional, directional, or hemispherical reflectivity? The answer is that $\mathscr{R}$ corresponds to ρ'_λ, the directional spectral reflectivity. This can be confirmed by comparing the definition of $\mathscr{R}$ from Eq. (4-63) with the definition of ρ'_λ for a specular reflecting surface in Eq. (2-41). The present case of an optically smooth surface fits the definition of a specular reflector as described in Chapter 2. That is, for collimated incidence, all of the reflected energy is collimated and reflected into the specular angle ($\theta_r = \theta$). The bi-directional reflectivity is given by the Dirac delta function as discussed in Chapter 2. Even though the subscript λ has not been attached to $\mathscr{R}$, it should be emphasized that $\mathscr{R}$ is a spectral quantity through the spectral dependence of the optical constants n and k.

The Fresnel interface reflectivities as given by Eqs. (4-69) and (4-70) are plotted in Fig. 4-9 for the usual case of $n_2 > n_1$.

Comparing the reflectivity of dielectrics (e.g., glass, water, etc.) in nonabsorbing spectral regions with metals, the behavior is seen to be quite different. For normal incidence, dielectrics are characterized by low reflectivities (4–6%) while polished metals have very high reflectivities (90–99%). As the incidence angle approaches 90° both components of polarization for both types of materials approach a reflectivity of 1. Thus, a still pond of water is a rather inefficient mirror when looking straight on at your reflection, but it makes an excellent mirror of a distant scene reflected at near grazing angles. The same can be said for a clear piece of glass. A metal-coated piece of glass (or mirror), on the other hand, is an excellent reflector at nearly all angles.

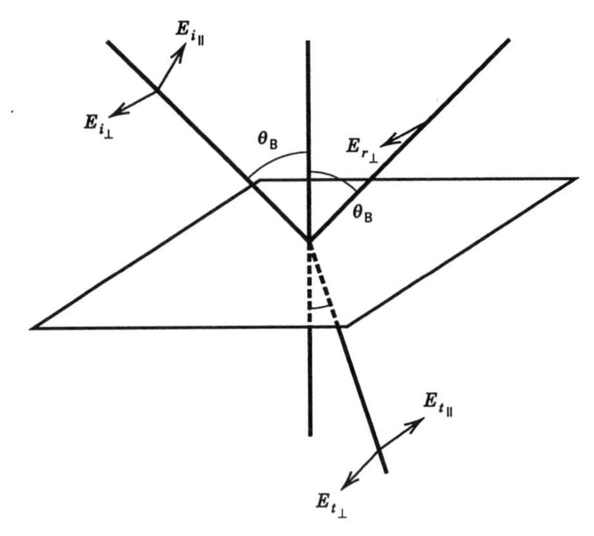

Figure 4-10 Polarization of light by reflection at Brewster angle.

The two components of polarization also behave differently as a function of incident angle. It can be seen from Fig. (4-9) that perpendicular reflectivity increases monotonically from normal to grazing incidence, while parallel reflectivity goes through a minimum. For nonabsorbing dielectrics, the value of $\mathscr{R}_\parallel$ reaches a minimum value of zero at an angle called the Brewster* angle θ_B.

$$\theta_B = \tan^{-1}\left(\frac{n_2}{n_1}\right) \quad (k_2 \to 0) \tag{4-72}$$

For this angle of incidence, any parallel polarized radiation incident on a dielectric will be completely transmitted through the interface. The reflected wave will have only a perpendicular component and that only if the incident wave has a perpendicular component (see Fig. 4-10). This does not imply that the perpendicular component is totally reflected at the Brewster angle, but only that the parallel component is not reflected at all.

The principle of polarization of light upon reflection near the Brewster angle is used to advantage in the design of Polaroid sunglasses. Most of the reflected light, which appears as unwanted glare, is perpendicularly polarized due to reflection from smooth horizontal surfaces. Polaroid sunglass lenses are made of long polymer molecules preferentially oriented to absorb perpendicular polarized light but transmit parallel polarized light. Thus the

*The Brewster angle is named for Sir David Brewster (1781–1868), a Scottish physicist whose pioneering work in crystallography led to the founding of the science of optical mineralogy.

polarized light is selectively absorbed by the glasses, removing the annoying glare.

The Brewster angle for typical nonabsorbing dielectrics in the visible region ($n_2 = 1.3$ to 1.8) in air ($n_1 = 1$) is between 50° and 60°. For metals, $\mathscr{R}_{\parallel}$ also goes through a minimum, but the value does not go to zero and the angle where the minimum occurs is closer to 90° (typically 80° to 87°).

For heat transfer calculations, the distinction between components of polarization is usually not important. In the first place, heat transfer surfaces usually are not optically smooth, and therefore the reflectivity is not determined from the Fresnel relations. Additionally, if multiple internal reflections are present from many different directions, as in an enclosure, the effect of polarization is averaged out even for a surface that is optically smooth. In this case the spectral reflectivity can be calculated assuming that the incident radiation is randomly polarized by taking the average of the two components of polarization.

$$\rho'_{\lambda} = \frac{\rho'_{\lambda_{\parallel}} + \rho'_{\lambda_{\perp}}}{2} \tag{4-73}$$

Example 4-2

What are the normal, spectral reflectivity and absorptivity of a polished, optically smooth aluminum surface at room temperature in air at 1.06 μm (Nd-YAG laser wavelength)?

From Appendix C, $n = 1.2$ and $k = 10$. Using Eq. (4-71) the reflectivity is given as

$$\rho'_{\lambda_n} = \frac{(1.2 - 1)^2 + 10^2}{(1.2 + 1)^2 + 10^2} = 0.954$$

The characteristic length for absorption in the aluminum was calculated in Example 4-1 as 84 angstroms. Thus for any thickness of aluminum greater than a few hundred angstroms, the transmissivity will be zero and the absorptivity can be calculated as

$$\alpha'_{\lambda_n} = 1 - \rho'_{\lambda_n} = 0.046$$

These high reflectivity and low absorptivity values are typical of clean, polished metals at all wavelengths throughout the visible, infrared, and longer wavelength regions.

Total Internal Reflection and the Critical Angle

The previous discussion of interface reflectivity has focused on the usual case of $n_2 > n_1$, where radiation is incident from a lower index nonabsorbing

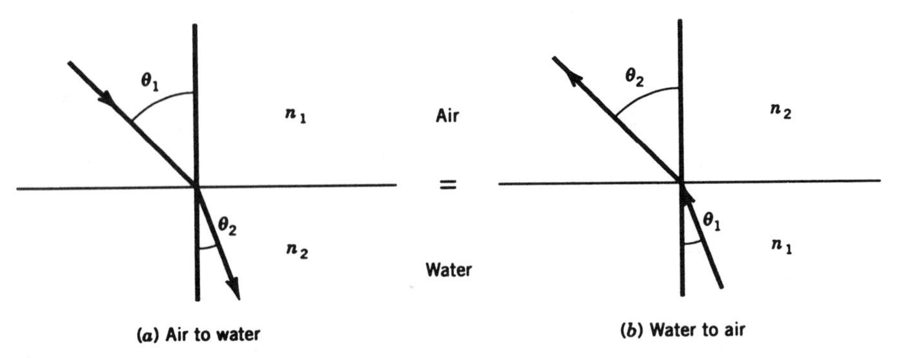

Figure 4-11 Reciprocity of Fresnel reflectivity.

medium on a higher index medium. Now the opposite case will be considered of radiation incident from a higher index medium on a lower index medium ($n_2 < n_1$). Both media will be assumed to be nonabsorbing.

For the sake of discussion it will be assumed that the diagram of $\mathscr{R}$ for a nonabsorbing dielectric medium in Fig. 4-9 ($n_2 > n_1$) corresponds to the reflectivity of light incident from air ($n_1 = 1$) on water ($n_2 = 1.33$) in the visible region.

Now consider the case of light incident from water ($n_1 = 1.33$) on air ($n_2 = 1$). Snell's law says that if n_1 and n_2 are interchanged the angle of incidence and transmittance (θ_1 and θ_2) also interchange. Furthermore, the Fresnel relations [Eqs. (4-55) and (4-57)] indicate that if n_1 and n_2 are interchanged and θ_1 and θ_2 are interchanged, the magnitude of the reflectivity will remain the same.

$$|\tilde{r}_{12}|^2 = |\tilde{r}_{21}|^2 \rightarrow \mathscr{R}_{12} = \mathscr{R}_{21} \tag{4-74}$$

This reciprocity condition is illustrated in Fig. 4-11, which shows that the reflectivity of light incident on water from air equals the reflectivity of light incident on air from water if the angles of incidence and transmittance are interchanged.

Now consider the situation as the angle of incidence from water to air (θ_1 in Fig. 4-11*b*) is increased. Eventually a critical angle is reached θ_c where the transmitted direction is parallel to the interface ($\theta_2 = 90°$) (Fig. 4-12). From Snell's law the critical angle is

$$\theta_c = \sin^{-1}\left(\frac{n_2}{n_1}\right) \tag{4-75}$$

According to the reciprocity relation Eq. (4-74), the reflectivity at this condition is equal to the reflectivity when light is incident at 90° from air on

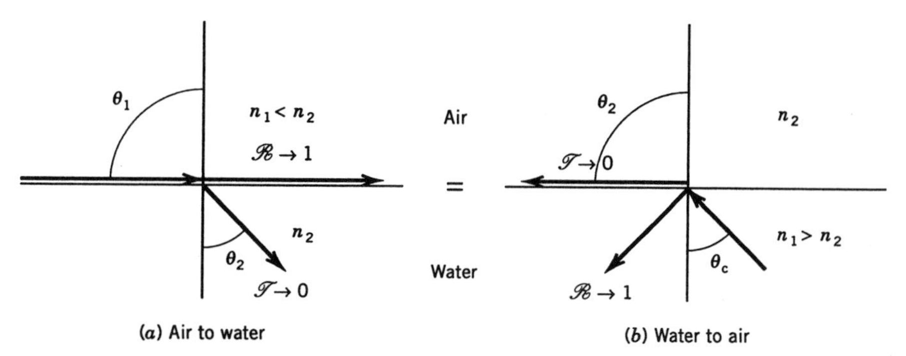

Figure 4-12 Illustration of critical angle and total internal reflection.

water, namely the reflectivity is 1 (see Fig. 4-9*a*). Moreover, for all angles of incidence greater than the critical angle ($\theta > \theta_c$), the reflectivity is 1. This condition is referred to as total internal reflection and can be readily observed in a swimming pool. This is also the principle of operation of fiber optics. The reflectivity for this situation as a function of incident angle is shown in Fig. 4-13.

A phenomenon closely related to total internal reflection is the mirage. When the layer of air near the ground is heated by the sun such that a sufficient temperature (and hence density and refractive index) gradient exists near the ground, light rays incident on the ground at angles near 90° can be refracted so that instead of reaching the ground they bend upward and propagate away from the ground. To an observer this often appears as a dark patch (often mistaken for water) against a lighter background. The dark patch is actually the distorted image of the dark blue sky (or some other dark object) in the distance above the horizon.

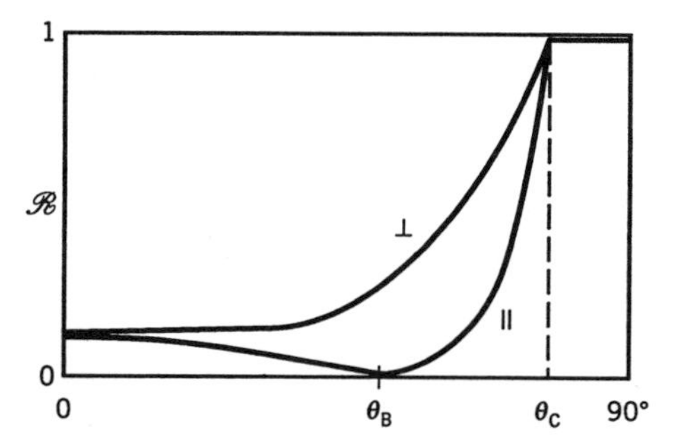

Figure 4-13 Fresnel interface reflectivity for dielectrics with total internal reflection ($n_2 < n_1$).

Intensity Change through an Interface

When a beam of collimated radiation passes through an interface, there is a change in intensity that arises from two effects. One effect is the transmissivity loss, which is simply the fact that some of the energy is converted into reflected energy. The other effect that changes the intensity is the change in solid angle $\Delta\Omega$ due to refraction.

With reference to the air-water system of Fig. 4-11*a*, the energy that is incident from the entire hemispherical region above the water is compressed into the solid angle defined by the critical angle as the radiation passes into the water. Ignoring reflection losses, this refraction has the effect of compressing the same energy into a smaller solid angle, which increases the intensity upon passing from a lower index medium into a higher index medium (Fig. 4-14). The overall change in intensity upon passing through an interface is determined by the relative contributions from these two effects and can be determined as follows.

From Eq. (4-64) the interface transmissivity can be written in terms of the incident and transmitted intensities as

$$\mathscr{T} = \frac{\Delta q_t}{\Delta q^-} = \frac{I_2\, \Delta\Omega_2 \cos\theta_2}{I_1\, \Delta\Omega_1 \cos\theta_1} \tag{4-76}$$

By substituting Snell's law and the differentiated form of Snell's law,

$$n_1 \cos\theta_1\, d\theta_1 = n_2 \cos\theta_2\, d\theta_2 \tag{4-77}$$

into (4-76) the change in intensity can be expressed as

$$\frac{I_2}{I_1} = \mathscr{T}\left(\frac{n_2}{n_1}\right)^2 \tag{4-78}$$

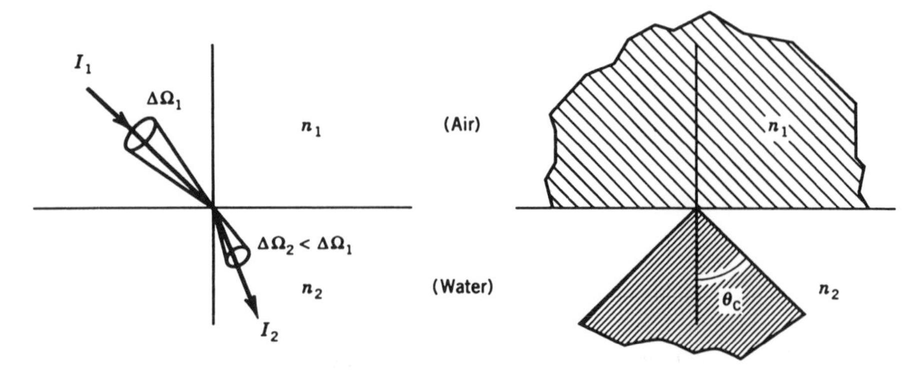

Figure 4-14 Intensity change due to refraction.

which clearly shows how the two effects, reflection and refraction, change the intensity upon passing through an interface. The transmissivity is always less than or equal to one and thereby reduces the intensity, whereas the refraction term (n_2/n_1) can be less than or greater than one, depending on which direction the rays are passing. Thus the intensity can actually increase upon passing through an interface from a lower index medium into a higher index medium, in spite of the reflectance loss.

THIN FILM OPTICS (INTERFERENCE EFFECTS)

Many engineering surfaces are covered with a film or coating that modifies the reflectivity of the substrate. Examples are oxide layers, ceramic coatings, and paint films. Assuming the layer can be modeled as homogeneous and planar, its reflectivity can be predicted by combining the results for two interfaces as pictured in Fig. 4-15.

Consider a plane wave propagating through a nonabsorbing medium (medium 1) that is incident on an absorbing film (medium 2). At the first interface part of the energy will be converted into reflected energy and part will be transmitted through the interface. The transmitted energy will be exponentially absorbed as the wave propagates through the film. Of the energy that reaches the second interface part will be converted into reflected energy and the rest will be transmitted into the absorbing substrate (medium 3). Of the energy reflected from the second interface part will reach the top interface where part will be reflected and the rest transmitted. The part that is transmitted back into medium 1 will interfere, either constructively or destructively, with the wave reflected at the first interface. The energy that is

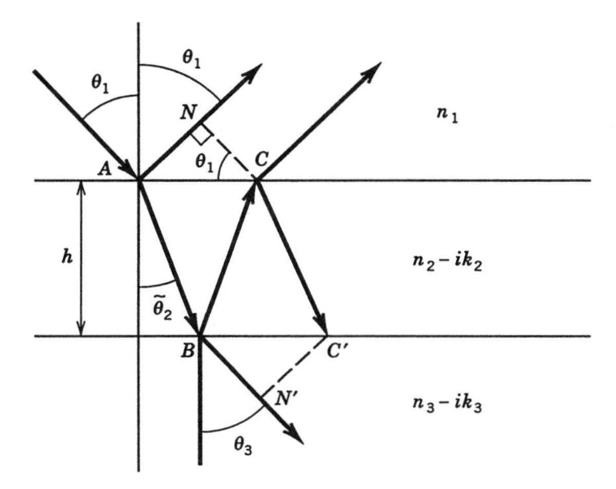

Figure 4-15 Multiple reflection in a thin film with interference.

trapped inside the film will be reflected forth and back until it is either absorbed in the film or transmitted out. By combining all the contributions, the overall reflectivity for the film-substrate can be determined.

From the preceding discussion it is evident that there are two issues involved that must be addressed: multiple reflection and interference (or phase change) effects. Interference effects will be considered first. It can be shown that the phase difference between the wave arriving at point N in Fig. 4-15 after being reflected at point A and the wave arriving at point C after being reflected at point B is the frequency ω times the time difference between the arrival of the two waves.

$$\omega(\text{time difference}) = 2\pi\nu 2\tilde{n}_2 h \cos\tilde{\theta}_2/c_0 = 4\pi\tilde{n}_2 h \cos\tilde{\theta}_2/\lambda_0 \quad (4\text{-}79)$$

Dividing by two gives the phase difference per transit $\tilde{\beta}$ as

$$\tilde{\beta} = \frac{2\pi\tilde{n}_2 h \cos\tilde{\theta}_2}{\lambda_0} = \frac{2\pi h}{\lambda_0}(u - iv) \quad (4\text{-}80)$$

The same result is obtained if the analysis is based on the waves arriving at points N' and C'.

The real part of $\tilde{\beta}$ contains information about the change in phase angle that occurs as the wave passes once through the film. The imaginary part of $\tilde{\beta}$ gives the attenuation of the wave amplitude for one pass through the film.

Overall Film Reflectivity

The amplitude of the wave reflected from the film-substrate system can be determined by adding the contributions from all the multiple reflections emerging from the film. In the following notation the tilde is omitted, but it should be remembered that the amplitude ratios (e.g., r_{12}) are complex numbers as given by the Fresnel relations.

$$r_{\text{film}} = r_{12} + t_{12}r_{23}t_{21}e^{-i2\beta} + t_{12}r_{23}r_{21}r_{23}t_{21}e^{-i4\beta} + \cdots \quad (4\text{-}81)$$

This series (except for the first term) is a geometric progression with a ratio of $r_{23}r_{21}e^{-i2\beta}$. Using the formula for the sum of a geometric progression

$$\text{Sum} = \frac{1}{1 - \text{ratio}} \quad (4\text{-}82)$$

the film amplitude ratio can be written as

$$r_{\text{film}} = r_{12} + \frac{t_{12}t_{21}r_{23}e^{-i2\beta}}{1 - r_{23}r_{21}e^{-i2\beta}} \quad (4\text{-}83)$$

Using the following results, which can be derived from the Fresnel relations,

$$r_{21} = -r_{12} \tag{4-84}$$

$$r_{12}^2 + t_{12}t_{21} = 1 \tag{4-85}$$

$$t_{21_\perp} = 2 - t_{12_\perp} \qquad \frac{n_1}{\tilde{n}_2} t_{21_\parallel} = 2 - \frac{\tilde{n}_2}{n_1} t_{12_\parallel} \tag{4-86}$$

the film amplitude ratio can also be written as

$$r_{\text{film}} = \frac{r_{12} + r_{23}e^{-i2\beta}}{1 + r_{12}r_{23}e^{-i2\beta}} \tag{4-87}$$

The overall film reflectance would then be given by the magnitude of the amplitude ratio squared.

$$\mathscr{R}_{\text{film}} = |r_{\text{film}}|^2 \tag{4-88}$$

Film Transmissivity

The transmissivity of the film can be determined by adding up all the contributions transmitted through the film as follows.

$$t_{\text{film}} = t_{12}t_{23}e^{-i\beta} + t_{12}r_{23}r_{21}t_{23}e^{-i3\beta} + \cdots \tag{4-89}$$

$$t_{\text{film}} = \frac{t_{12}t_{23}e^{-i\beta}}{1 + r_{12}r_{23}e^{-i2\beta}} \tag{4-90}$$

The film transmissivity is then given as

$$\mathscr{T}_{\text{film}} = |t_{\text{film}}|^2 \frac{n_3 \cos\theta_3}{n_1 \cos\theta_1} \tag{4-91}$$

It should be noted here that for an absorbing film, $\mathscr{T}_{\text{film}}$ cannot simply be calculated as $1 - \mathscr{R}_{\text{film}}$ as was the case for a single interface due to the unknown fraction of energy absorbed by the film.

If the film and substrate are both nonabsorbing, then the film reflectivity and transmissivity are given by the following relatively "simple" expressions.

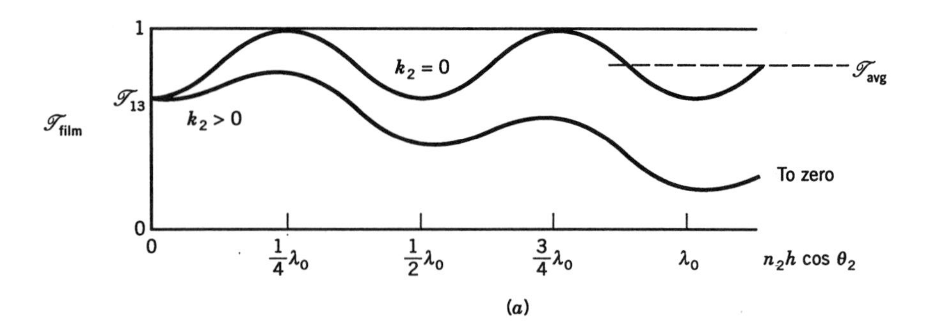

(a)

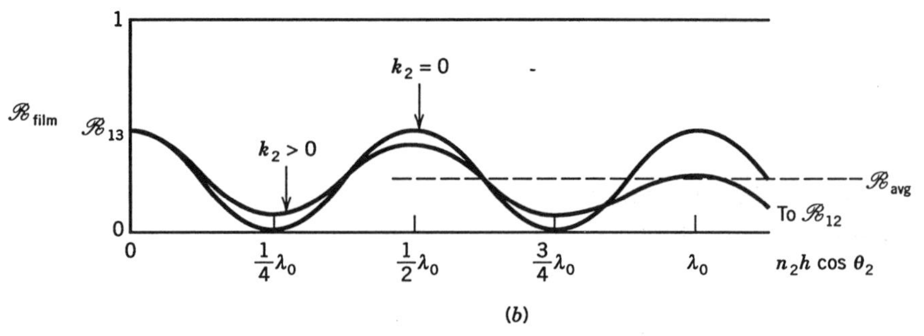

(b)

Figure 4-16 Thin film (*a*) transmissivity and (*b*) reflectivity with interference ($k_3 = 0$, $n_2 = \sqrt{n_3 n_1}$, $n_3 > n_2$).

(The corresponding expressions for an absorbing film on a nonabsorbing substrate can be found in Born and Wolf [1970].)

$$\mathcal{R}_{\text{film}} = \frac{r_{12}^2 + r_{23}^2 + 2r_{12}r_{23}\cos(2\beta)}{1 + r_{12}^2 r_{23}^2 + 2r_{12}r_{23}\cos(2\beta)} \quad (k_2 = k_3 = 0) \tag{4-92}$$

$$\mathcal{T}_{\text{film}} = \frac{n_3 \cos\theta_3}{n_1 \cos\theta_1} \frac{t_{12}^2 t_{23}^2}{1 + r_{12}^2 r_{23}^2 + 2r_{12}r_{23}\cos(2\beta)} \quad (k_2 = k_3 = 0) \tag{4-93}$$

A plot of the film reflectivity and transmissivity is given in Fig. 4-16 as a function of $n_2 h \cos\theta_2$ for the case $k_3 = 0$, $n_2 = \sqrt{n_3 n_1}$ and $n_3 > n_2$.

At zero film thickness, the film transmissivity and reflectivity approach the single interface values, $\mathcal{T}_{13}$ and $\mathcal{R}_{13}$, respectively, as indicated in Fig. 4-16. For a nonabsorbing film ($k_2 = 0$) both the reflectivity and transmissivity oscillate harmonically and undamped as $n_2 h \cos\theta$ increases. If there is

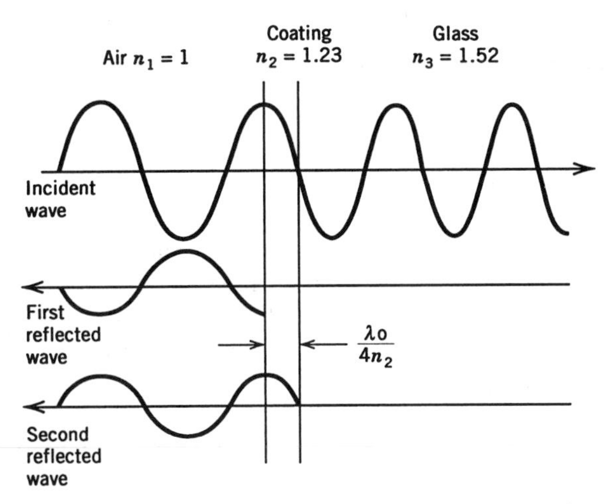

Figure 4-17 Destructive interference in antireflection coatings.

absorption in the film, the film transmissivity approaches zero, and the film reflectivity approaches the value of reflectivity for a single interface $\mathscr{R}_{12}$ as the value of $n_2 h \cos \theta_2$ increases.

The first minimum in $\mathscr{R}_{\text{film}}$ occurs when the value of $n_2 h \cos \theta_2$ is $\lambda_0/4$. For normal incidence this minimum occurs when the film thickness is $h = \lambda_0/(4n_2)$ or one quarter of a wave thick. The value of $\mathscr{R}_{\text{film}}$ at the minimum is zero for the particular refractive index selected, $n_2 = \sqrt{n_3 n_1}$. This fact is the basis for the design of antireflection coatings for lenses and other optics.

As pictured in Fig. 4-17, the wave reflected at the first interface and the wave reflected at the second interface will be exactly 180° out of phase if the film is one quarter of a wave thick (as measured in the film). Furthermore if n_2 is selected to be $\sqrt{n_3 n_1}$, all the reflected waves (including multiple internal reflections) will have amplitudes such that they all destructively interfere giving zero net reflectivity.

For air ($n_1 = 1$) and glass ($n_3 = 1.52$) the optimum coating index is $n_2 = 1.23$. However, durable coating materials with an index this low have not been found. A material whose index is reasonably close to 1.23 is magnesium fluoride with $n = 1.38$. Most antireflection coatings are made of MgF_2. The film reflectivity for MgF_2 on glass at 0.55 μm is about 1.5%. By increasing the number of layers the overall reflectivity can be reduced to less than 0.05% if necessary. This analysis, however, has been applied at only one wavelength, 0.55 μm (green). For other wavelengths away from the design wavelength the reflectivity will not be as low. For example, the wavelengths at the ends of the visible spectrum (blue at 0.4 μm and red at 0.7 μm) will have a slightly higher reflectivity, resulting in a slightly purplish appearance, which characterizes MgF_2 coated optics.

Example 4-3

Most metals in the far infrared, microwave, and radio wave regions can be treated as a perfect reflector ($n_3 = k_3 \rightarrow \infty$). Such a surface is to be made nonreflective by coating it with an absorptive material with optical constants $1.7 - ik_2$. Investigate the optimum coating thickness for various values of k_2 (10^{-3}, 10^{-2}, 10^{-1}, 1, and 10) by calculating the layer reflectivity as a function of $\pi h/\lambda_0$.*

The equations needed to calculate the layer reflectivity are (4-88), (4-87), (4-80), (4-58), (4-57), and (4-55). These equations can be combined using complex algebra to obtain a rather complicated expression for the layer reflectivity (see [Born and Wolf, 1970; Potter, 1985]). As an alternative a FORTRAN compiler with complex operations can also be used. A listing of such a code is given in Table 4-1. To evaluate $\cos \tilde{\theta}_i (i = 2 \text{ or } 3)$, Snell's law [Eq. (4-58)] is used along with the identity $\cos^2\tilde{\theta}_i + \sin^2\tilde{\theta}_i = 1$.

The results for normal incidence are plotted in Fig. 4-18. For all values of k_2, at $\pi h/\lambda_0 \ll 0.1$, the reflectivity approaches the value of the bare substrate, 1. For most values of $k_2 (< 10)$ the reflectivity undergoes a first minimum near $\pi h/\lambda_0 = 0.5$, which corresponds to $h/n_2\lambda_0 = 0.25$ or approximately one-quarter wave in the coating. This minimum is due to destructive interference as discussed in connection with Fig. 4-17. As h/λ_0 increases, the reflectivity undergoes alternating maxima and minima as the wave reflected from the coating experiences alternating cycles of constructive and destructive interference. Another effect that can be seen in Fig. 4-18 is that as h/λ_0 increases the mean value of reflectivity decreases, due to increasing absorption of energy in the coating. The larger the value of k_2 the faster the mean value falls due to stronger absorption in the coating. The reflectivity approaches that of a semi-infinite coating at a value of $\pi h/\lambda_0$ of approximately $1/k_2$. This follows from Eqs. (4-80) and (4-90), which indicate that absorption of energy in the coating goes like $\exp(-2\pi hk_2/\lambda_0)$. For $k_2 \ll n_2 = 1.7$ this limiting value of reflectivity is $(1.7 - 1)^2/(1.7 + 1)^2 = 0.067$ from Eq. (4-71). As can be seen from Fig. 4-18 there are a number of ways to minimize the reflectivity. For example, with $k_2 = 0.1$, the layer thickness could be selected to correspond to the fourth minimum at $\pi h/\lambda_0 = 3.3$ to give a reflectivity of 0.001. Alternatively, a minimum reflectivity could also be achieved with a thinner coating by designing at the first

*It should be apparent from Eq. (4-80) that the layer properties are a function of the ratio h/λ_0 only and not the individual values h and λ_0. This does not mean that the layer properties are independent of wavelength, however, as the optical constants are still functions of wavelength.

TABLE 4-1 FORTRAN Listing for Layer Reflectivity

```
c    program for calculating fresnel reflectance
c    for an absorbing layer (n2,k2)
c    on a substrate (n3,k3), including interference
     real n1,n2,k2,n3,k3,lambda0
     complex refr1,refr2,refr3,ct1,ct2,ct3,r12perp,r12para,r23perp
    1,r23para,rfperp,rfpara,beta,i,rperp,rpara
     i = cmplx(0.,1.)
     n1 = 1.
     n2 = 1.7
     k2 = 0.001
     n3 = 1.e3
     k3 = 1.e3
     refr1 = cmplx(n1,0)
     refr2 = cmplx(n2,- k2)
     refr3 = cmplx(n3,- k3)
 5   write (9,10)
10   format( / 1x,'input theta1(deg),h / lam0')
     read (9,*) theta1,hlam
     if (theta1.lt.0.) stop
     theta1 = theta1*3.1415926 / 180.
     ct1 = cmplx(cos(theta1),0.)
     ct2 = csqrt(1.- (n1 / refr2)**2*sin(theta1)**2)
     ct3 = csqrt(1.- (refr2 / refr3*cs1qrt(1.- ct2**2))**2)
     beta = 2.*3.1415926*hlam*refr2*ct2
     r12perp = rperp(ct1,refr1,ct2,refr2)
     r12para = rpara(ct1,refr1,ct2,refr2)
     r23perp = rperp(ct2,refr2,ct3,refr3)
     r23para = rpara(ct2,refr2,ct3,refr3)
     rfperp = (r12perp + r23perp*cexp( - i*2.*beta)) /
    1(1.+ r12perp*r23perp*cexp( - i*2.*beta))
     rfpara = (r12para + r23para*cexp( - i*2.*beta)) /
    1(1.+ r12para*r23para*cexp( - i*2.*beta))
     rfpe = cabs(rfperp)**2
     rfpa = cabs(rfpara)**2
     write (9,30) rfpe,rfpa
30   format (1x,'Rperp =',f8.4,3x,'Rpara =',f8.4)
     pause
     goto 5
     end
     function rpara(cti,rni,ctj,rnj)
     complex cti,rni,ctj,rnj
     rpara = (rnj*cti - rni*ctj) / (rnj*cti + rni*ctj)
     return
     end
     function rperp(cti,rni,ctj,rnj)
     complex cti,rni,ctj,rnj
     rperp = (rni*cti - rnj*ctj) / (rni*cti + rnj*ctj)
     return
     end
```

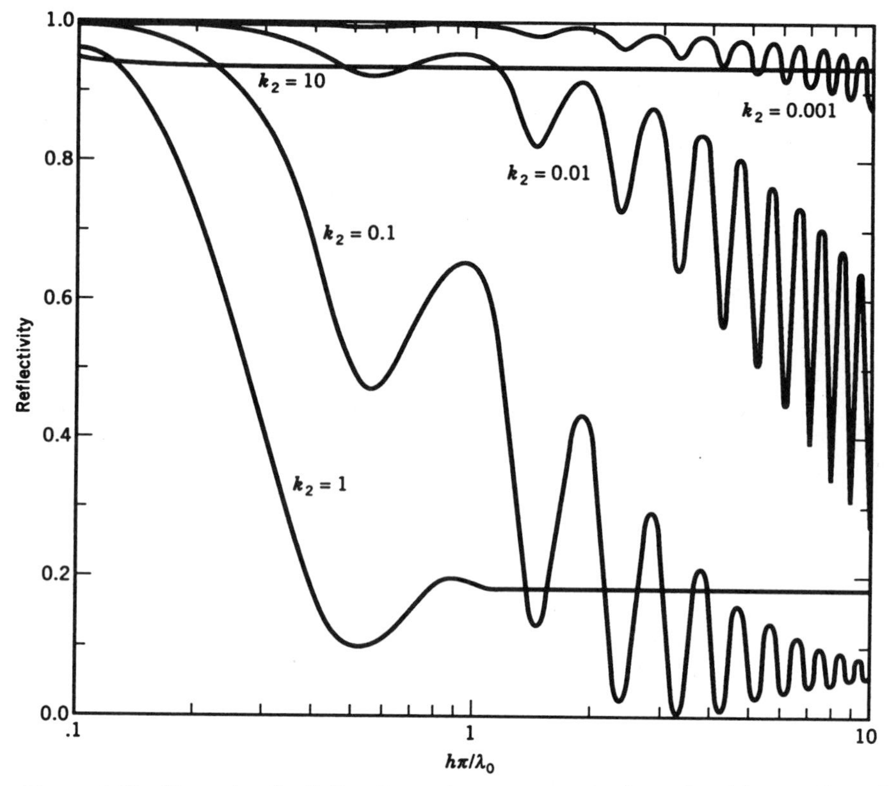

Figure 4-18 Normal reflectivity of a system composed of an absorbing coating of thickness h on a perfectly reflective substrate in air ($n_1 = 1$, $k_1 = 0$, $n_2 = 1.7$, $n_3 = k_3 \to \infty$).

minimum near $\pi h/\lambda_0 = 0.5$ and using a value of k_2 between 0.1 and 1 (see Problem 4-10).

Average Film Reflectivity and Transmissivity

The preceding discussion has focused on the effects of interference. These effects are only noticeable when the film thickness is on the order of the wavelength. When $h \gg \lambda_0$ interference effects usually average out due to naturally occurring variations in n_2, h, λ_0, and θ_2. As h becomes very large relative to λ_0, the phase angle parameter β [Eq. (4-80)] becomes very large and for very large β, a small change in n_2, h, λ_0, or θ_2 is sufficient to produce a difference of π in β which results in a cycle of interference. For example, it can be shown from the definition of β, Eq. (4-80), that the change in wavelength necessary to produce an increment of π in β is

$$\Delta\lambda_0 = \frac{\lambda_0^2}{2n_2 h \cos\theta_2 + \lambda_0} \tag{4-94}$$

For $\lambda_0 = 0.5\ \mu\text{m}$, $n_2 = 1.5$, $h = 1$ mm, and $\theta_2 = 0$ the increment in wavelength $\Delta\lambda_0$ is 0.083 nm, which is unresolvable by most commercial spectrometers. Therefore, when $h \gg \lambda_0$, or broadband or diffuse irradiation is involved, it is appropriate to use average values.

By rewriting the film transmissivity from Eq. (4-93) as

$$\mathscr{T}_{\text{film}} = \frac{n_3 \cos\theta_3}{n_1 \cos\theta_1} \frac{t_{12}^2 t_{23}^2}{\left(1 + r_{12}^2 r_{23}^2\right)\left(1 + \dfrac{2r_{12}r_{23}\cos(2\beta)}{1 + r_{12}^2 r_{23}^2}\right)} \tag{4-95}$$

the integral

$$\frac{1}{2\pi}\int_0^{2\pi} \frac{dx}{1 + A\cos x} = \frac{1}{\left(1 - A^2\right)^{1/2}} \tag{4-96}$$

can be used to obtain the average film transmissivity $\mathscr{T}_{\text{avg}}$ as

$$\mathscr{T}_{\text{avg}} = \frac{1}{2\pi}\int_0^{2\pi} \mathscr{T}_{\text{film}}\, d(2\beta) = \frac{n_3\cos\theta_3}{n_1\cos\theta_1} \frac{t_{12}^2 t_{23}^2}{1 - r_{12}^2 r_{23}^2} \tag{4-97}$$

When media 1 and 3 have the same index $(n_1 = n_3)$

$$\frac{n_3\cos\theta_3}{n_1\cos\theta_1} = 1$$

and $\mathscr{T}_{\text{avg}}$ reduces to

$$\mathscr{T}_{\text{avg}} = \frac{\left(1 - r_{12}^2\right)^2}{1 - r_{12}^4} = \frac{\left(1 - \mathscr{R}_{12}\right)^2}{1 - \mathscr{R}_{12}^2} = \frac{1 - \mathscr{R}_{12}}{1 + \mathscr{R}_{12}} \quad (n_1 = n_3) \tag{4-98}$$

ENERGY METHOD

When interference effects can be neglected, the average film reflectivity and transmissivity can be obtained in a more simple manner than that just illustrated. By the energy method, a unit of incident flux is tracked through the multiple reflections inside the film without keeping track of phase.

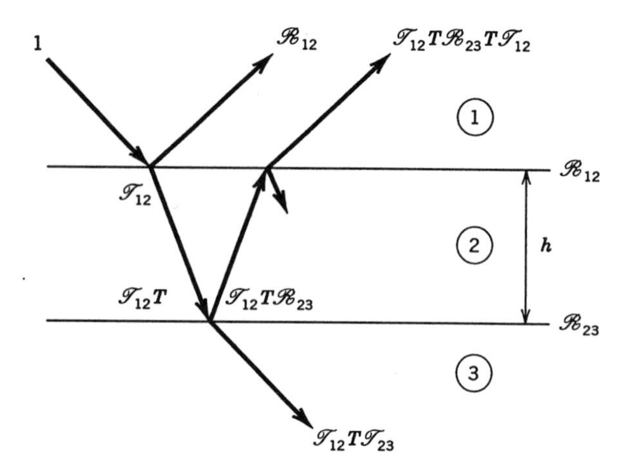

Figure 4-19 Multiple reflection in a thick layer without interference.

Consider again a plane wave propagating through a nonabsorbing medium (1) which is incident on an absorbing film (2) (see Fig. 4-19). A fraction of the energy $\mathcal{R}_{12}$ will be reflected at the interface and the rest, $\mathcal{T}_{12} = 1 - \mathcal{R}_{12}$, will be transmitted through the interface. The transmitted energy will be exponentially absorbed as the wave propagates through the film. The internal transmittance T can be calculated from either

$$T = \exp\left(-\frac{4\pi k_2 h}{\lambda_0 \cos\theta_2}\right) \quad \text{for} \quad (k_2 \ll n_2, \text{any } \theta_2) \qquad (4\text{-}99)$$

where θ_2 is calculated from Snell's law, or

$$T_n = \exp\left(-\frac{4\pi k_2 h}{\lambda_0}\right) \quad \text{for} \quad (\theta_2 = 0, \text{any } k_2) \qquad (4\text{-}100)$$

The energy that reaches the second interface is $T(1 - \mathcal{R}_{12})$. Of that fraction, $T(1 - \mathcal{R}_{12})(1 - \mathcal{R}_{23})$ will be transmitted into the substrate (3) and the rest, $T(1 - \mathcal{R}_{12})\mathcal{R}_{23}$, will be reflected. This process continues with the energy at the interface being divided into reflected and transmitted portions and energy traversing the film being exponentially absorbed. By combining all the contributions that leave the top interface and all the contributions that leave the bottom interface, the overall reflectivity and transmissivity for the film-substrate can be determined.

$$\mathscr{R}_{avg} = \mathscr{R}_{12} + (1 - \mathscr{R}_{12})T\mathscr{R}_{23}T(1 - \mathscr{R}_{12}) + (1 - \mathscr{R}_{12})T\mathscr{R}_{23}T\mathscr{R}_{12}T\mathscr{R}_{23}T(1 - \mathscr{R}_{12}) + \cdots \tag{4-101}$$

$$\mathscr{T}_{avg} = (1 - \mathscr{R}_{12})T(1 - \mathscr{R}_{12}) + (1 - \mathscr{R}_{12})T\mathscr{R}_{23}T\mathscr{R}_{12}T(1 - \mathscr{R}_{23}) + \cdots \tag{4-102}$$

Using the formula for the sum of a geometric progression, Eq. (4-82), these series can be written as

$$\mathscr{R}_{avg} = \mathscr{R}_{12} + \frac{(1 - \mathscr{R}_{12})^2\mathscr{R}_{23}T^2}{1 - \mathscr{R}_{12}\mathscr{R}_{23}T^2} \tag{4-103}$$

$$\mathscr{T}_{avg} = \frac{(1 - \mathscr{R}_{12})(1 - \mathscr{R}_{23})T}{1 - T^2\mathscr{R}_{12}\mathscr{R}_{23}} \tag{4-104}$$

The fraction of incident flux absorbed by the film would be $1 - \mathscr{R}_{avg} - \mathscr{T}_{avg}$. If media 1 and 3 are the same, then $\mathscr{R}_{12} = \mathscr{R}_{23}$. If in addition the film is nonabsorbing, then $T = 1$ and $\mathscr{T}_{avg}$ reduces to

$$\mathscr{T}_{avg} = \frac{1 - \mathscr{R}_{12}}{1 + \mathscr{R}_{12}} \tag{4-105}$$

which is the same result obtained in Eq. (4-98).

The energy method can be easily used to extend the results for a single layer to multiple layers. The strategy is to let the reflectivity and transmissivity of a single layer become the new effective properties of an interface. By combining two such interfaces the problem reduces to the one just solved. For example, consider two layers that are separated by a nonabsorbing medium (Fig. 4-20). The transmissivity and reflectivity of the two layers are $\mathscr{T}_{1,2}$ and $\mathscr{R}_{1,2}$, respectively. From the previous derivation the transmissivity and reflectivity of the system of two layers can be written as follows:

$$\mathscr{T}_{syst} = \frac{\mathscr{T}_1\mathscr{T}_2}{1 - \mathscr{R}_1\mathscr{R}_2} \tag{4-106}$$

$$\mathscr{R}_{syst} = \mathscr{R}_1 + \frac{\mathscr{R}_2\mathscr{T}_1^2}{1 - \mathscr{R}_1\mathscr{R}_2} \tag{4-107}$$

This process can be generalized by considering a complex system of N layers with transmissivity $\mathscr{T}_N$ and reflectivity $\mathscr{R}_N$. The transmissivity and reflectivity

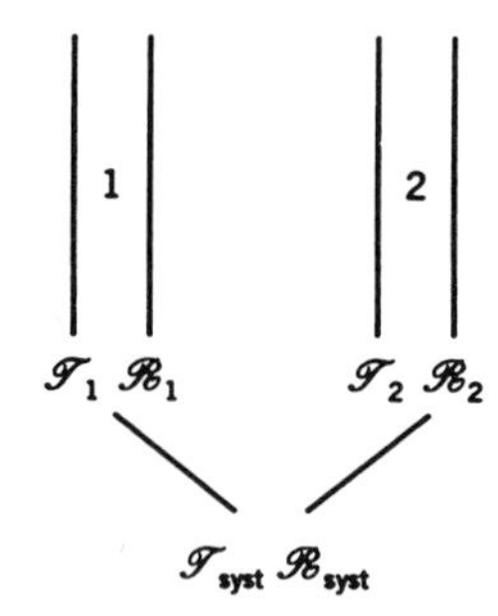

Figure 4-20 System of two layers.

of the new system obtained by adding one more layer (with properties $\mathscr{T}_1$ and $\mathscr{R}_1$) are

$$\mathscr{T}_{N+1} = \frac{\mathscr{T}_1 \mathscr{T}_N}{1 - \mathscr{R}_N \mathscr{R}_1} \tag{4-108}$$

$$\mathscr{R}_{N+1} = \mathscr{R}_1 + \frac{\mathscr{R}_N \mathscr{T}_1^{\,2}}{1 - \mathscr{R}_1 \mathscr{R}_N} \tag{4-109}$$

As the number of layers N becomes very large, even for a system composed of nonabsorbing layers ($\mathscr{T}_i + \mathscr{R}_i = 1$), the system transmissivity will approach zero and the system reflectivity will approach a constant. That is, the addition of one more layer will not change the reflectivity much. For a system composed of a large number of identical layers (with properties $\mathscr{T}_1$ and $\mathscr{R}_1$), the asymptotic value of $\mathscr{R}_N$, as N becomes very large, is

$$\mathscr{R}_N = \frac{\left(1 + \mathscr{R}_1^2 - \mathscr{T}_1^{\,2}\right) - \left[\left(1 + \mathscr{R}_1^2 - \mathscr{T}_1^{\,2}\right)^2 - 4\mathscr{R}_1^2\right]^{1/2}}{2\mathscr{R}_1} \tag{4-110}$$

Perfect Emitter-Absorber (Blackbody) Behavior

In Chapter 1, blackbody behavior was considered from the standpoint of an isothermal cavity or hohlraum. It was pointed out that any opaque material manufactured into the geometry of a cavity with a small opening and maintained at constant temperature behaved like a blackbody. This approach uses geometrical effects to create a perfect emitter-absorber from any opaque material, whether it be intrinsically a good or poor absorber. The mechanism of multiple internal reflection is responsible for the blackbody behavior. In

Chapter 2, the concept was introduced of a material that would be intrinsically a perfect emitter and absorber without utilizing the cavity effect. Such a surface would have an emissivity/absorptivity of one and reflectivity of zero without taking advantage of multiple reflections. It is now possible to consider what combination of optical constants is necessary to produce a material that would be a perfect absorber and emitter.

Consider a nonscattering, absorbing, emitting, isothermal layer of material with optical constants $n - ik$, in air, as pictured in Fig. 4-19. Equations (4-103) and (4-104) give

$$\mathscr{T}_{\text{avg}} = \frac{(1 - \mathscr{R}_{12})^2 T}{1 - T^2 \mathscr{R}_{12}^2} \tag{4-111}$$

and

$$\mathscr{R}_{\text{avg}} = \mathscr{R}_{12} + \frac{(1 - \mathscr{R}_{12})^2 \mathscr{R}_{12} T^2}{1 - \mathscr{R}_{12}^2 T^2} \tag{4-112}$$

where $\mathscr{R}_{12}$, the interface reflectivity, is the same at both surfaces since the same medium (air) is on both sides of the layer. Conservation of energy gives the directional spectral absorptivity of the layer for an arbitrary direction of incidence as

$$\alpha'_\lambda = 1 - \mathscr{R}_{\text{avg}} - \mathscr{T}_{\text{avg}} = \frac{(1 - \mathscr{R}_{12})(1 - T)}{1 - \mathscr{R}_{12} T} \tag{4-113}$$

For normal incidence, the interface reflectivity, from Eq. (4-71), is

$$\mathscr{R}_{12_n} = \frac{(n-1)^2 + k^2}{(n+1)^2 + k^2} \tag{4-114}$$

and the internal transmittance, from Eq. (4-100), is

$$T_n = \exp\left[-\frac{4\pi k h}{\lambda_0} \right] \tag{4-115}$$

which gives the normal layer absorptivity/emissivity as

$$\alpha'_{\lambda_n} = \varepsilon'_{\lambda_n} = \frac{\left[\dfrac{4n}{(n+1)^2 + k^2}\right]\left[1 - \exp\left(-\dfrac{4\pi kh}{\lambda_0}\right)\right]}{1 - \left[\dfrac{(n-1)^2 + k^2}{(n+1)^2 + k^2}\right]\exp\left(-\dfrac{4\pi kh}{\lambda_0}\right)} \tag{4-116}$$

For a perfect, nonabsorbing dielectric the internal transmittance is one ($T_n = 1$) and the emissivity is zero ($\varepsilon'_{\lambda_n} = 0$). For an opaque layer ($k \neq 0$, $h \gg \lambda_0$) the internal transmittance is zero ($T_n = 0$) and the emissivity/absorptivity reduces to one minus the single interface reflectivity, $\varepsilon'_{\lambda_n} = 1 - \mathscr{R}_{12_n}$.

For a material to be a good emitter-absorber (without changing its geometry to increase multiple reflections), two requirements must be satisfied; the material must be both opaque (that is, nontransparent) and nonreflective. In order to be opaque the internal transmittance of the layer must approach zero ($T \to 0$), which requires that k must be an order of magnitude or more larger than $\lambda_0/(4\pi h)$. However the value of k cannot be indefinitely large, because in order to be nonreflective the interface reflectivity $\mathscr{R}_{12}$ must also approach zero ($\mathscr{R}_{12} \to 0$), which requires that k must be an order of magnitude or more smaller than n, which must be approximately one. Combined, these conditions give the following restrictions for a material to be a good emitter-absorber.

$$\frac{\lambda_0}{4\pi h} \ll k \ll n \sim 1 : (\varepsilon'_\lambda \to 1) \tag{4-117}$$

Metals are generally poor emitter-absorbers because k and n are both large (see Appendix C), giving $\mathscr{R}_{12} \to 1$. At the other extreme, nonabsorbing materials, such as pure, stoichiometric oxides (e.g., Al_2O_3, MgO, SiO_2, etc.) are poor emitter-absorbers in the visible and near infrared region because k is too small, giving $T \to 1$. The best emitting and absorbing materials are "weakly" absorbing (lossy) dielectrics or poor conductors with k on the order of 10^{-3} to 10^{-1} and n close to one, such as carbon black. Substoichiometric oxides (e.g., Al_2O_{3-x}, MgO_{1-x}) are also good emitter-absorbers, sometimes exhibiting a dark, black appearance when viewed in natural light at low temperature. Nonabsorbing dielectric materials can be impregnated with small concentrations of very small metal particles as impurities to make the material a good absorber (e.g., platinum black and gold black). Application of

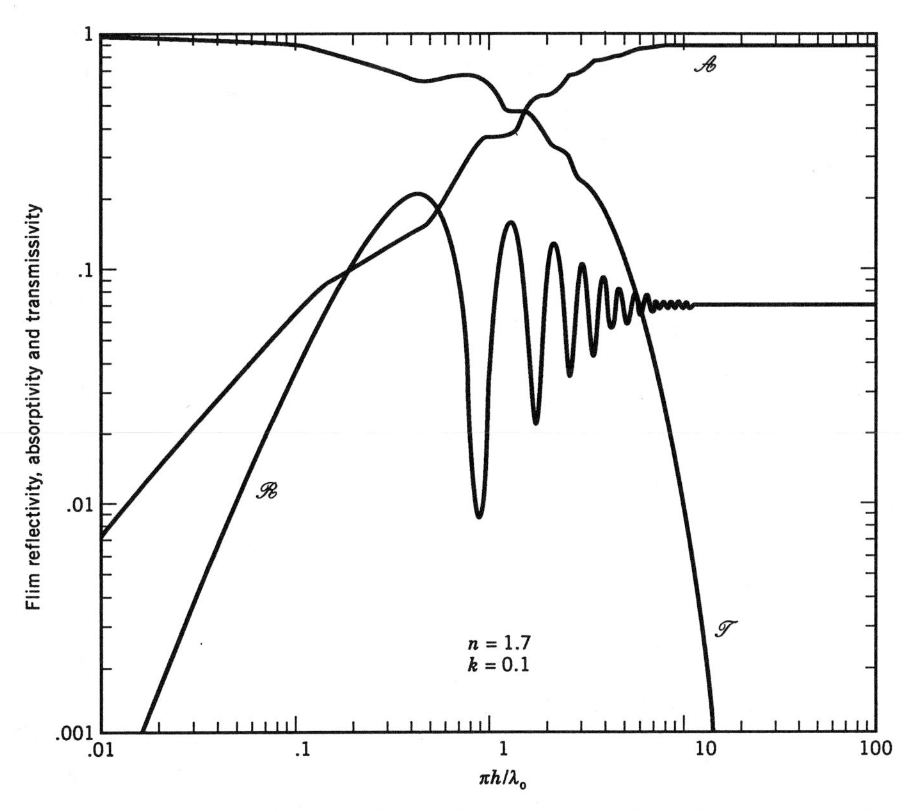

Figure 4-21 Normal reflectivity, absorptivity and transmissivity of a plane layer with $n = 1.7$ and $k = 0.1$ (including interference).

this principle is made in the design of stealth aircraft, which are built mostly from composites made of carbon fiber-reinforced resin, a radar absorbing material.

As an example of a highly absorbing material, the normal absorptivity, reflectivity, and transmissivity of a plane layer in air with $n = 1.7$ and $k = 0.1$ are shown in Fig. 4-21, with interference effects included. In accordance with Eq. (4-117), the layer is opaque and highly absorbing (i.e., black) when the thickness parameter $x = \pi h/\lambda_0$ is large, $x > 1/k = 10$. At $x < 1/(100k) = 0.1$ the layer becomes transmitting.

For thin layers with n on the order of 1 and $k \ll n$ an interesting feature occurs, which can be observed in Fig. 4-21. For $x \ll 1$ it can be shown (see Problem 4-13) that the film reflectivity is proportional to x^2 while the film absorptivity is proportional to x. Thus as x decreases the layer reflectivity decreases faster than the absorptivity and absorption, if present at all (i.e.,

$k \neq 0$), eventually dominates over reflection. The critical value of x where the crossover occurs can be shown to be given by $x_c = k/(n-1)^2$ for $k \ll (n-1) \ll 1$ (in Fig. 4-21, $x_c \approx 0.1/0.7^2 = 0.2$). Similar behavior is also exhibited in the scattering and absorption behavior of small particles (see Rayleigh scattering in Chapter 9).

SUMMARY

The principal results obtained in this chapter are the Fresnel reflectivity relations. The assumptions involved in their derivation are (1) a plane incident wave and (2) plane interface.

With regard to the plane wave assumption (1) it was shown in Chapter 1 (Example 1-1) that spherical waves from a point source behave locally like plane waves. Even in situations of thermal radiative transfer between two surfaces of finite area, a differential area on one surface acts like a point source to a differential area on the other surface. Thus, in most heat transfer calculations (except when the spacing between surfaces becomes of the order of a wavelength) the plane wave approximation is a valid one.

The plane interface assumption (2) is valid for optically smooth surfaces. More precisely, this means that the root-mean-square roughness of a surface (σ_{rms}) must be much less than the incident wavelength for the surface to be considered a plane interface $\sigma_{\text{rms}} \ll \lambda$. Depending on the value of σ_{rms} this condition may place a restriction on the wavelengths for which the Fresnel relations can be applied. But other than this restriction there is no other inherent assumption that would invalidate the electromagnetic theory results even in the ultraviolet region.

As a practical matter, however, it should be noted that most heat transfer surfaces can be treated as optically smooth only in the far infrared region and beyond. Optically smooth surfaces in the visible region can sometimes be achieved with painstaking graded sandpaper treatment and diamond paste polishing. And to achieve an optically smooth surface in the ultraviolet region requires ultra-high vacuum deposition. For an unpolished surface the actual reflectivity in the visible and near infrared region will be less than that for a smooth surface due to the influence of multiple reflections between surface asperities (i.e., blackbody cavity effect) [Beckmann and Spizzichino, 1963; Bass and Fuks, 1979; Majumdar and Tien, 1990].

REFERENCES

Bass, F. G. and Fuks, I. M. (1979), *Wave Scattering from Statistically Rough Surfaces*, Pergamon Press, Oxford.

Beckmann P. and Spizzichino, A. (1963), *The Scattering of Electromagnetic Waves from Rough Surfaces*, Macmillan, New York.

Bennett, J. M. and Bennett, H. E. (1978), "Polarization," in *Handbook of Optics*, W. G. Driscoll, Ed., McGraw-Hill, New York, Chap. 10.

Bohren, C. F. and Huffman, D. R. (1983), *Absorption and Scattering of Light by Small Particles*, John Wiley & Sons, New York.

Born, M. and Wolf, E. (1970), *Principles of Optics*, Pergamon Press, Oxford.

Majumdar, A. and Tien, C. L. (1990), "Reflection of Radiation by Rough Fractal Surfaces," in *Radiation Heat Transfer: Fundamentals and Applications*, ASME HTD Vol. 137, T. F. Smith, M. F. Modest, A. M. Smith, and S. T. Thynell, Eds., presented at AIAA/ASME Thermophysics Conference, June.

Potter, R. F. (1985), "Basic Parameters for Measuring Optical Properties," in *Handbook of Optical Constants of Solids*, E. D. Palik, Ed., Academic Press, New York.

PROBLEMS

1. Aluminum oxide has the following optical constants at $\lambda = 1.06\ \mu$m:

	n	k
Solid, void-free (nonscattering, 300 K)	1.75	6×10^{-8}
Liquid (2320 K)	1.62	1×10^{-3}

(a) Plot the interface reflectivity (both polarization components) versus incident angle from 0° to 90° for an optically smooth sample of solid, void-free, aluminum oxide.

(b) What would be the spectral directional emissivity at 0°, 60°, and 85° for an opaque sample of the solid material? (Is the approximation of Problem 2-4 a good one?)

(c) How thick would a sample of each of these materials (both solid and liquid) need to be for the sample to be opaque at 1.06 μm? (Take 0.1% normal transmission as the criterion for opaqueness.)

2. The optical constants for aluminum are given at two common industrial laser wavelengths:

	λ (μm)	n	k
Nd:YAG	1.06	1.2	10
CO_2	10.6	27	93

(a) Calculate the normal spectral absorptivity of a polished aluminum sample for each laser. All other things being equal (e.g., incident power, focused spot size, etc.), which laser would be more suitable for processing aluminum?

(b) Plot spectral directional reflectivity (both polarization components) versus incident angle from 0° to 90° for 1.06 μm.

(c) What would be the angle of incidence for maximum absorptivity (at 1.06 μm) of collimated, unpolarized incident radiation and what would be the corresponding absorptivity?

(d) What would be the angle of incidence for maximum absorbed flux (at 1.06 μm) of collimated, unpolarized incident radiation and what would be the corresponding absorptivity?

3. A metal is coated with a lossy dielectric layer. The optical constants of the metal and layer are, respectively, $1.2 - i10$ and $1.75 - i10^{-3}$ at 1.06 μm. Make a plot of the spectral directional reflectivity (both polarization components) versus incident angle (0° to 90°) at 1.06 μm for a 1, 10, and 100 μm thick layer. Neglect interference effects.

4. Consider emission at 1.06 μm by a metal that forms an absorbing oxide. To simplify the analysis, assume the oxide and metal are homogeneous layers with a distinct interface between them. Assume the optical constants of the metal are $0.40 - i7.0$ and for the oxide, $1.75 - i0.1$. Assuming the oxide layer and metal are isothermal, make a plot of the normal spectral directional emissivity as a function of oxide layer thickness. Let the layer thickness range from being thin enough that the metal dominates to being thick enough that the oxide becomes effectively semi-infinite. Neglect interference effects.

5. Reconsider the situation of Problem 4 including interference effects. Compare these results with the results of Problem 4. Why do the two methods give diverging results as the layer thickness goes to zero? (Hint: show that as $h/\lambda_0 \to 0$ and $T \to 1$, $\mathscr{R}_{avg}$ does not in general approach $\mathscr{R}_{13}$.) Comment on the reason for this difference between $\mathscr{R}_{avg}$ and $\mathscr{R}_{film}$.

6. The optical constants of three common materials at room temperature in the visible region ($\lambda = 0.5$ μm) are:

	n	k
Glass	1.5	1×10^{-7}
Silicon	4	0.05
Aluminum	1	7

Neglecting interference, what is the normal spectral absorptivity of a 1 mm thick, optically smooth sample of each of these materials in air?

7. The optical constants of copper at room temperature in the visible region are:

λ (μm)	n	k
0.40	1.18	2.21
0.48	1.15	2.50
0.56	0.826	2.60
0.65	0.214	3.67

Calculate the normal spectral reflectivity of an optically smooth, opaque sample of copper at these wavelengths in the visible region and discuss the implication of these results with regard to the color of copper.

8. The optical constants of gold at room temperature in the visible region are:

λ (μm)	n	k
0.40	1.66	1.96
0.48	1.24	1.80
0.56	0.306	2.88
0.65	0.166	3.15

Calculate the normal spectral reflectivity of an optically smooth, opaque sample of gold at these wavelengths in the visible region and discuss the implication of these results with regard to the color of gold. Does this calculation indicate anything about the spectral sensitivity of the retina?

9. Silicon, a material that is opaque at visible wavelengths, becomes transparent for wavelengths above 1 μm. For a 1 mm thick, optically smooth sample of silicon, what is the transmissivity, reflectivity, and absorptivity at $\lambda = 1.06$ μm where $n = 3.5$ and $k = 6 \times 10^{-5}$?

10. Design a millimeter-wave (94 GHz) radar absorbing coating for a titanium airframe by specifying (a) the coating thickness and (b) the desired effective coating absorption index k_2 subject to the following parameters. The coating matrix is epoxy resin with a dielectric constant of 3 ($n_2 = 1.7$). The titanium d.c. electrical resistivity is 5.5×10^{-6} Ω-cm, corresponding to optical constants $n_3 = k_3 = 1300$ (see Hagen-Rubens equation in Chapter 5). Design for minimum reflectivity with minimum coating thickness (i.e., design at the first minimum in reflectivity).

11. In Fig. 4-21, at a value of $\pi h/\lambda_0$ corresponding to one-quarter wavelength in the layer [$h = \lambda_0/(4n)$] the layer reflectivity goes through an extremum that is a maximum instead of a minimum (as was the case in Fig. 4-16*b*). Explain the difference. (Hint: What is the significance of the sign change that occurs in the Fresnel coefficients when n_1 and n_2 are interchanged?) What would Fig. 4-17 look like for $n_3 < n_2$?

12. Carry out a simplified derivation of Eqs. (4-79) and (4-80) by making the (overly restrictive) assumption that $k_2 \ll n_2$ such that the speed of wave propagation in medium 2 is $c_0/\tilde{n}_2$ and the angle of propagation is $\tilde{\theta}_2$.

13. Derive simple analytic expressions for the normal reflectivity ($\mathscr{R}_f$), transmissivity ($\mathscr{T}_f$), and absorptivity ($\mathscr{A}_f$) of a thin, absorbing layer in air when $x = \pi h/\lambda_0 \ll 1$ and $k \ll (n - 1) \ll 1$. Obtain an expression for the critical value of x, above which $\mathscr{R}_f > \mathscr{A}_f$ and below which $\mathscr{R}_f < \mathscr{A}_f$. Check the results by showing that $\mathscr{A}_f$ is proportional to k (as $k \to 0$, $\mathscr{A}_f \to 0$) and that as $n \to 1$, $\mathscr{R}_f \to 0$, and $\mathscr{T}_f = 1 - \mathscr{A}_f$.

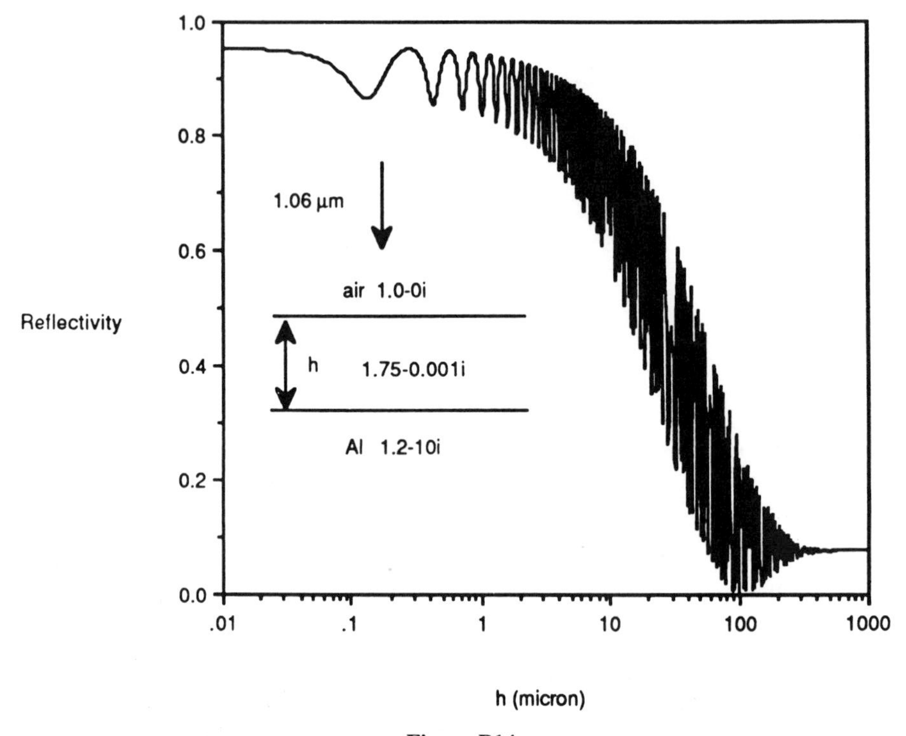

Figure P14

14. An aluminum surface is to be made nonreflective to 1.06 μm laser radiation by coating it with an absorptive material with optical constants $1.75 - i0.001$. Investigate the optimum coating thickness (acceptably low reflectivity with minimum thickness) by calculating the layer reflectivity as a function of coating thickness. Ignore interference effects and compare the results with the accompanying plot which includes interference. What is the magnitude of the error incurred as thickness goes to zero by neglecting interference?

15. Consider the problem of determining n and k from measured values of normal reflectivity, R_n, for an opaque, optically smooth surface (in air) by investigating the allowable values of n and k for measured values of $R_n = 0.01$, 0.1 and 0.5. Is this an effective method for determining k when $k < 0.1$? Comment on the uniqueness of the value of n for a given k. What value of n gives the maximum allowable value of k for a given R_n?

CHAPTER 5

CLASSICAL DISPERSION THEORY

Chapter 4 demonstrates how the engineering radiative properties for optically smooth surfaces can be predicted from electromagnetic theory if the optical constants n and k are known. These "constants" n and k are fundamental properties of a given material and are usually strongly dependent on wavelength. A material (e.g., glass) may be nonabsorbing at certain wavelengths (visible) and absorbing at other wavelengths (infrared or ultraviolet). The study of how to predict the values of n and k from basic electromagnetic properties as a function of wavelength is the subject of dispersion theory.

Dispersion theory can be applied to both gases and condensed matter. However, gases are usually not characterized in terms of the spectral absorption index k but rather in terms of band parameters that have been smoothed over a narrow spectral region (the radiative properties of gases is treated in detail in Chapter 8). The present discussion is primarily concerned with the application of dispersion theory to solids and liquids.

The study of dispersion theory is strongly related to the fields of quantum theory and solid-state physics. These modern physical theories, however, are overly complicated for engineering purposes. The object of this chapter is to develop a general understanding of why various materials interact with radiation of different wavelengths the way they do. To meet this objective it is sufficient to avoid the implications of quantum physics and consider dispersion theory in terms of the classical model of a damped oscillator, which was developed in the early days of electron theory (circa 1880) by Drude, Lorentz, and Zener [see Slater and Frank, 1947; Bohren and Huffman, 1983].

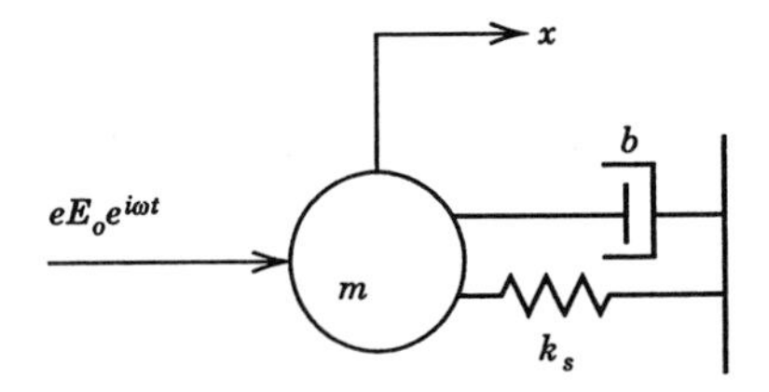

Figure 5-1 Damped oscillator model.

DAMPED OSCILLATOR (LORENTZ) MODEL

Most of the qualitative features of the spectral dependence of n and k can be described by the relatively simplistic mechanical model of a damped oscillator. Although it is not a completely accurate model of the microscopic structure of matter, this model gives results that are surprisingly in good agreement with the more rigorous predictions of quantum theory. The key assumptions of the damped oscillator model (Fig. 5-1) are as follows.

Assumptions

1. The material is assumed to be composed of charged particles (electrons or ions) held in equilibrium positions by Hooke's law forces.
2. A damping force acts on the particles that is proportional to the velocity of the particle (linear damping).
3. These forces are isotropic.

Under the action of an incident electromagnetic field a charged particle will be displaced from its equilibrium position ($x = 0$) by the force of the electric field acting on the charged particle. The magnitude of this driving force is the product of the charge on the particle (e) and the magnitude of the electric vector (E_x).

$$\text{Driving force} = eE_x$$

A spring force, which is representative of interatomic forces, acts to restore the particle to its equilibrium position. This force is proportional to the displacement and acts in a direction opposite to the displacement.

$$\text{Spring force} = -k_s x$$

A dissipative damping force acts to resist the motion of the particle. This force models the irreversible degradation of (ordered) electromagnetic energy to (disordered) internal energy (i.e., absorption).

$$\text{Damping force} = -b\dot{x}$$

The force balance on the particle is obtained by summing the forces and setting the result equal to the particle acceleration

$$m\ddot{x} = -b\dot{x} - k_s x + eE_x \tag{5-1}$$

where m is the mass of the particle. Upon rearranging, the force balance becomes

$$m\ddot{x} + b\dot{x} + k_s x = eE_x \tag{5-2}$$

The oscillating electric vector given in complex variable notation is

$$E_x = E_0 e^{i\omega t} \tag{5-3}$$

Letting $b = \gamma m$ and $k_s = \omega_0^2 m$ the equation of motion* for the particle becomes

$$\ddot{x} + \gamma\dot{x} + \omega_0^2 x = \frac{eE_0}{m} e^{i\omega t} \tag{5-4}$$

where γ is referred to as the damping factor or *relaxation frequency* and ω_0 is the natural or *resonant frequency* of the oscillator.

A particular (or steady-state) solution can be obtained by assuming that the displacement has the form

$$x(t) = Ae^{i\omega t} \tag{5-5}$$

Substituting this assumed form of the solution into the differential equation gives the following relation:

$$-A\omega^2 + i\gamma\omega A + A\omega_0^2 = \frac{eE_0}{m} \tag{5-6}$$

which can be solved for the amplitude A as

$$A = \frac{eE_0}{m} \cdot \frac{1}{\omega_0^2 - \omega^2 + i\gamma\omega} \tag{5-7}$$

Thus the solution for the motion of the charged particle is

$$x(t) = \frac{eE_0}{m} \frac{e^{i\omega t}}{\omega_0^2 - \omega^2 + i\gamma\omega} \tag{5-8}$$

*This formulation ignores the influence of the induced electric field, which is created by the polarization of the charged particles in the material. The induced field effect is included at the end of the chapter with only a minor change in the equation of motion.

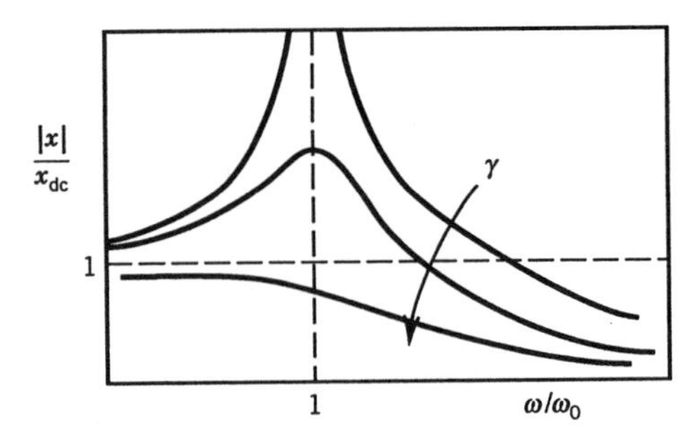

Figure 5-2 Amplitude of damped oscillator.

This solution indicates that the charged particle oscillates with the same frequency as the external field ω, and that the amplitude of the oscillation depends on the value of ω relative to ω_0. A plot of the amplitude of the displacement as a function of frequency is shown in Fig. 5-2.

The phase shift between the input (driving force eE_x) and output (x) is shown in Fig. 5-3.

This solution is a classical result from vibration theory for a mass, spring, dashpot system subjected to forced oscillations. The maximum amplitude response occurs when the driving frequency is close to the natural frequency of the oscillator ($\omega = \omega_0$). The maximum energy dissipation or absorption also occurs near $\omega = \omega_0$. Little energy absorption occurs for driving frequencies that are far from the resonant frequencies of the material. This interpretation leads to the notion of absorption bands occurring at certain frequencies near resonant frequencies of a material.

To relate this model to the optical constants n and k, the polarization **P** is introduced. It may be recalled from Chapter 4 that the polarization is a measure of the dipole moment per unit volume. For the one-dimensional plane wave under consideration the x component of the polarization is

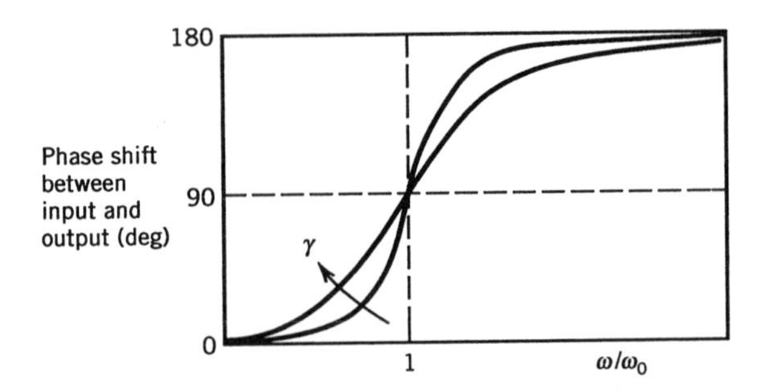

Figure 5-3 Phase shift of damped oscillator.

given by

$$P_x = Nex \quad \text{where} \quad N = \frac{\text{number of oscillators}}{\text{volume}} \tag{5-9}$$

The polarization is also related to the optical properties through Eqs. (4-8), (4-38), and (4-39) as

$$\mathbf{P} = \tilde{\chi}\varepsilon_0\mathbf{E} = (\tilde{\varepsilon}_r - 1)\varepsilon_0\mathbf{E} \tag{5-10}$$

where

$$\tilde{\varepsilon}_r \equiv \tilde{\varepsilon}/\varepsilon_0 = \varepsilon_r' - i\varepsilon_r'' = \tilde{n}^2 = (n - ik)^2 \tag{5-11}$$

is the complex dielectric constant, which is equal to the square of the complex refractive index. It should be noted that the term from Eq. (4-39) involving conductivity σ_e has been dropped to be consistent with the convention that σ_e represents the free charge contribution while the imaginary susceptibility χ'' represents the bound charge contribution. Combining Eqs. (5-8) through (5-11) gives

$$P_x = (\tilde{\varepsilon}_r - 1)\varepsilon_0 E_x = Nex = \frac{Ne^2 E_x}{m(\omega_0^2 - \omega^2 + i\gamma\omega)} \tag{5-12}$$

which can be solved for the square of the complex refractive index

$$\varepsilon_r' - i\varepsilon_r'' = (n - ik)^2 = 1 + \frac{\omega_p^2}{\omega_0^2 - \omega^2 + i\gamma\omega} \tag{5-13}$$

where the *plasma frequency* ω_p is defined as

$$\omega_p = \sqrt{\frac{Ne^2}{m\varepsilon_0}} \tag{5-14}$$

To account for multiple oscillators with various resonant frequencies, Eq. (5-13) can be written as a sum, assuming that the effect on $\tilde{\varepsilon}_r$ is additive.

$$\varepsilon_r' - i\varepsilon_r'' = (n - ik)^2 = 1 + \sum_j \frac{\omega_{p_j}^2}{\omega_{0_j}^2 - \omega^2 + i\gamma_j\omega} \tag{5-15}$$

Separating (5-15) into real and imaginary parts gives

$$\varepsilon_r' = n^2 - k^2 = 1 + \sum_j \frac{\omega_{p_j}^2\left(\omega_{0_j}^2 - \omega^2\right)}{\left(\omega_{0_j}^2 - \omega^2\right)^2 + \gamma_j^2\omega^2} \tag{5-16a}$$

$$\varepsilon_r'' = 2nk = \sum_j \frac{\omega_{p_j}^2\gamma_j\omega}{\left(\omega_{0_j}^2 - \omega^2\right)^2 + \gamma_j^2\omega^2} \tag{5-16b}$$

The parameters in these equations are the plasma frequencies ω_{p_j}, the oscillator frequencies ω_{0_j}, and the relaxation frequencies γ_j. If these parameters are known, the real and imaginary components of the complex dielectric constant can be calculated from Eqs. (5-16). The refractive and absorption indices can then be calculated from the dielectric constants using the following equations:

$$n = \left[\frac{\varepsilon_r' + \sqrt{\varepsilon_r'^2 + \varepsilon_r''^2}}{2}\right]^{1/2} \tag{5-17}$$

$$k = \left[\frac{-\varepsilon_r' + \sqrt{\varepsilon_r'^2 + \varepsilon_r''^2}}{2}\right]^{1/2} \tag{5-18}$$

The general trends for n and k given by Eqs. (5-16) through (5-18) are plotted in Fig. 5-4 and the corresponding normal reflectivity is shown in Fig. 5-5.

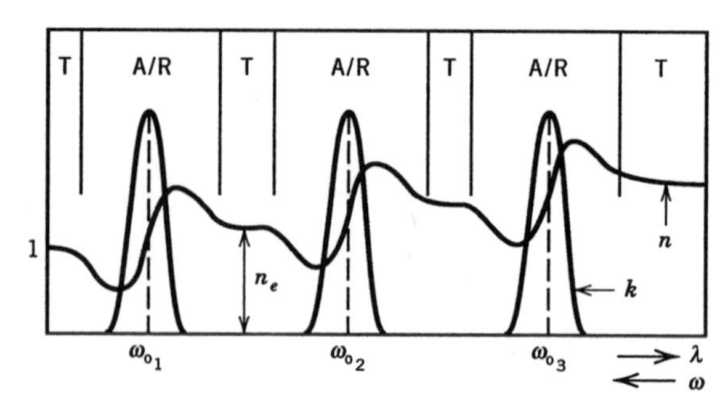

Figure 5-4 Optical constants of nonmetallic solids and liquids.

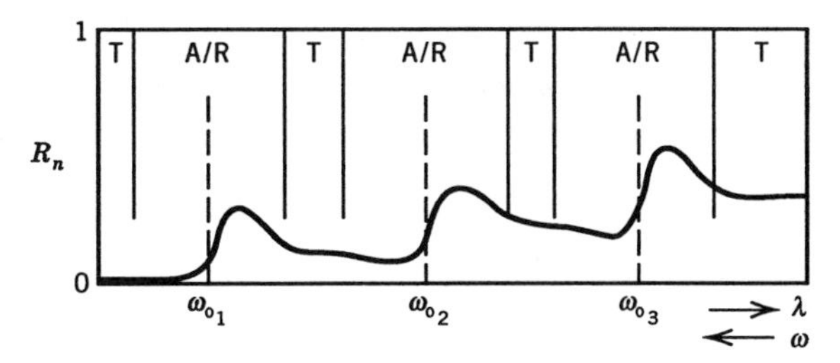

Figure 5-5 Normal reflectivity of nonmetallic solids and liquids.

General Behavior of *n* and *k* for Lorentz Oscillators

Inspection of Eqs. (5-16) shows that the imaginary optical constants (ε_r'' and k) are zero when the frequency is far from a resonant frequency and nonzero only when the frequency is in the neighborhood of a resonant frequency. This behavior is also evident in Fig. 5-4. Frequencies in the vicinity of the resonant frequencies where the value of k is nonzero define spectral regions where the interaction of the material with electromagnetic radiation is dominated by absorptive/reflective (A/R) behavior. These absorption/reflection regions are separated from each other by transparent (T) regions where the absorption index goes to zero. Figure 5-4 shows the optical constants for a material consisting of three independent oscillators separated from each other by transparent spectral regions. The width of the absorption band of an independent oscillator at the half peak point is equal to the relaxation frequency γ_j. It is not uncommon in some materials for two oscillator frequencies to be sufficiently close to each other (and/or the bandwidths sufficiently wide) that the bands are not independent and no transparent region separates the region of influence of one oscillator from the other.

Upon passing through an absorption region, the value of n initially decreases, then increases, and then decreases again approaching a new (nearly constant) value. The final value of n on the low frequency side of the oscillator is always higher than the value on the high frequency side. Upon passing through many oscillators the value of n gradually increases due to the incremental increase from each oscillator whereas the value of k always goes through a local maximum near the oscillator frequency and then approaches zero between oscillators.

Optical Constants of Nonmetals

At very high frequencies ($\omega > \omega_{0_1}$ in Fig. 5-4) all nonmetallic solids (and most metals for that matter) have a pure real refractive index with a magnitude of one. In terms of the mechanical oscillator model the interpretation of this behavior is that there are no charged particles (either bound or unbound) that can respond to the high frequency oscillations of x-ray radiation. With reference to Fig. 5-2, the driving frequency is much larger than the nearest resonant frequency, and the amplitude of the displacement of the charged particles is essentially zero.

As the frequency decreases toward ω_{0_1} in Fig. 5-4, most materials will begin to exhibit interband (bound electron) energy state transitions that will result in absorption of radiation. Because the mass of electrons is so small, the natural frequency of oscillation of bound electrons is relatively large $[\omega_0 = (k_s/m)^{1/2}]$ and generally falls in the ultraviolet or visible range (the color of objects in the visible region is usually due to bound electronic energy transitions).

At still lower frequencies (ω_{0_2} and ω_{0_3} in Fig. 5-4) most nonmetals exhibit lattice vibrations with associated oscillating dipoles, which gives rise to absorption bands in the infrared region. These absorption bands are characterized by much lower resonant frequencies than those for bound electrons due to the fact that the mass of atoms is so much larger than that of electrons.

For a limited region of interest, a simplified form of the dispersion relations can be obtained, assuming the oscillators are far enough apart in frequency to act independently. For example, in the neighborhood of the frequency ω_{0_2}, ($\omega \approx \omega_{0_2}$) the term for the ω_{0_1} contribution will be constant with frequency (since $\omega_{0_1} \gg \omega$) and can be added to the leading 1 in Eq. (5-15) or (5-16a) and called n_e^2. The contribution from lower frequency oscillators, such as the term for ω_{0_3} would be negligible since $\omega_{0_3} \ll \omega_{0_2} \approx \omega$ and the summation sign can be dropped from Eqs. (5-16).

$$\varepsilon_r' = n^2 - k^2 = n_e^2 + \frac{\omega_{p2}^2\left(\omega_{0_2}^2 - \omega^2\right)}{\left(\omega_{0_2}^2 - \omega^2\right)^2 + \gamma_2^2\omega^2} \qquad \left(\omega_{0_3} \ll \omega \ll \omega_{0_1}\right) \quad \text{(5-19a)}$$

$$\varepsilon_r'' = 2nk = \frac{\omega_{p_2}^2\gamma_2\omega}{\left(\omega_{0_2}^2 - \omega^2\right)^2 + \gamma_2^2\omega^2} \qquad \left(\omega_{0_3} \ll \omega \ll \omega_{0_1}\right) \quad \text{(5-19b)}$$

Optical Behavior of Glass

Figure 5-4 is a qualitative representation of the optical constants of any insulator, including glass. The first oscillator represents bound electron absorption in the ultraviolet region. Between the first and second oscillators is a region of transmission. For glass this region extends from approximately 0.4 to 2.5 μm. In this transparent region the value of n for glass is nearly constant at a value of 1.5. (As indicated by the previous discussion, this value is determined by the influence of bound electrons, which have "natural frequencies" much higher than the driving frequency but which still exert an influence on the optical constants at lower frequencies.) In this transparent

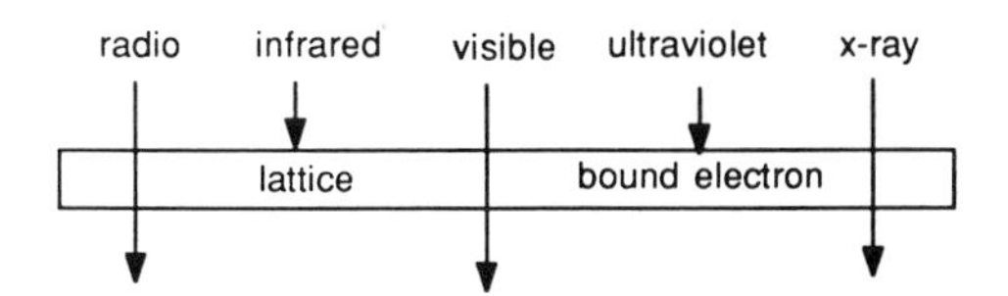

Figure 5-6 Interaction of glass with the electromagnetic spectrum.

region (as in any transparent region), the refractive index actually decreases slightly as wavelength increases. For example, the refractive index of fused silica glass for blue light (0.4 μm) is 1.47 and for red light (0.7 μm) it is 1.45. This type of variation of n with λ ($dn/d\lambda < 0$) is called *normal dispersion* and is responsible for the operation of a prism as well as chromatic aberration in lenses. Near the oscillator frequencies where absorption is occurring the opposite trend is observed and n increases with λ. This variation ($dn/d\lambda > 0$) is called *anomalous dispersion*.

The interaction of glass with radiation of various wavelengths is summarized in Fig. 5-6. Radio waves, visible light, and x-rays are all transmitted by glass while infrared radiation is absorbed by lattice vibrations, and ultraviolet light is absorbed by interband electronic transitions (bound electron oscillations). Other nonmetals exhibit behavior similar to that shown for glass in Fig. 5-6. Most nonmetals have a transparent window between the electronic absorption edge (also called the *fundamental band gap*) and the lattice vibration bands. For dielectrics the transparent window usually extends from the near infrared region down through the visible region. For semiconductor materials the lower wavelength absorption edge is usually shifted to higher wavelengths. For example silicon is opaque in the visible region and becomes transparent at about 1.5 μm [Edwards, 1985].

Influence of Oscillator Parameters

The three oscillator parameters are oscillator frequency ω_0, plasma frequency ω_p, and damping or relaxation frequency γ. The oscillator frequency is the location parameter. It determines the spectral location of the band. Its value depends on the resonant frequency of the material. Increasing or decreasing the oscillator frequency shifts the band structure to higher or lower frequencies.

The plasma frequency ω_p is the strength parameter. It determines whether the oscillator is relatively weak or strong. Figures 5-7 and 5-8 show the influence of the plasma frequency on n and k. As can be seen in Fig. 5-7, the magnitude of ω_p is an indicator of the area under the curve of k. As ω_p increases the value of k increases at all frequencies and the area under the curve increases. Figure 5-8 shows the influence of ω_p on n. When k is very

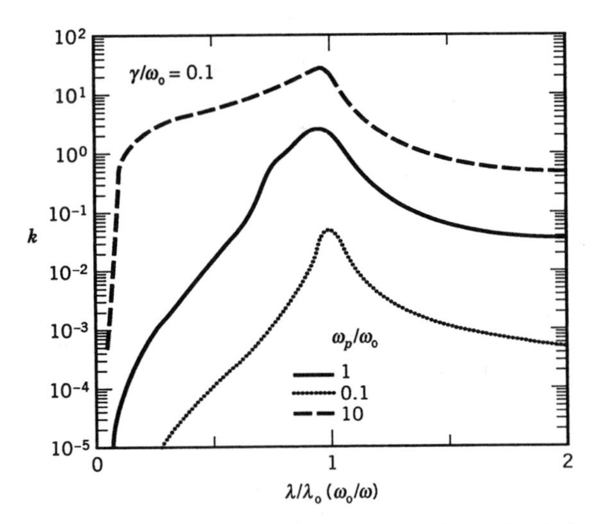

Figure 5-7 Effect of plasma frequency on absorption index.

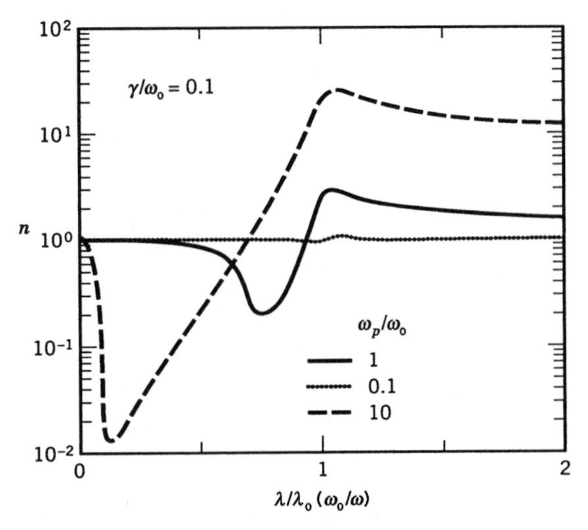

Figure 5-8 Effect of plasma frequency on refractive index.

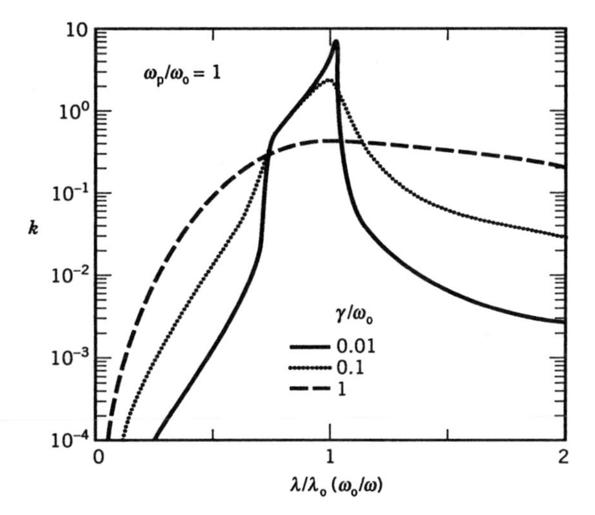

Figure 5-9 Effect of damping frequency on absorption index.

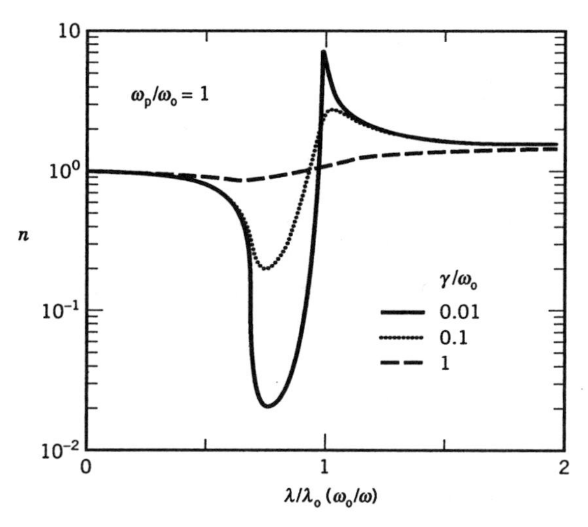

Figure 5-10 Effect of damping frequency on refractive index.

small (corresponding to $\omega_p/\omega_0 = 0.1$), it can be seen that n is relatively constant at a value of one. This case describes the condition of gases.

The damping frequency γ is the parameter that describes the width of the absorption band for a given strength. Figure 5-9 shows the influence of γ on k. As γ increases the width of the band increases with a relatively constant area under the curve. Figure 5-10 shows the corresponding influence of γ on n.

Example 5-1

The optical constants of glass (amorphous SiO_2) at room temperature ($T = 298$ K) in the region $0.4 < \lambda < 10$ μm can be modeled using the following dispersion parameters:

$$\eta_0 = 1063 \text{ cm}^{-1} \ (\lambda_0 = 9.4 \ \mu\text{m}) \qquad \eta_p = 900 \text{ cm}^{-1}$$

$$\gamma = 30 \text{ cm}^{-1} \qquad n_e = 1.46$$

Using these parameters calculate the following:

(a) The optical constants n and k (compare these with the experimental values given in Appendix C).
(b) The spectral normal absorptivity for a 5-mm-thick layer in the region $0.4 < \lambda < 10$ μm.
(c) The total normal absorptivity for a 5-mm-thick piece of glass exposed to solar radiation ($T_e = 5785$ K).

A FORTRAN computer code (Table 5-1) is used to determine the optical constants from the dispersion relations, Eqs. (5-17) through (5-19), at 0.1-μm intervals between 0.4 and 20 μm. The normal spectral absorptivity of the layer is calculated from the optical constants using Eq. (4-116), neglecting interference effects. The total absorptivity is obtained by integrating the spectral absorptivity using the fractional function.

$$\alpha_n'(T, T_e) = \int_0^1 \alpha_{\lambda n}'(T) \, df(\lambda T_e)$$

The results for the optical constants predicted by the single oscillator model using the given parameters are shown in Fig. 5-11 along with the experimental values from Appendix C. The primary oscillator at 9.4 μm is due to the vibration of Si — O — Si bridging oxygen. Several minor oscillators and another strong oscillator near 20 μm are not shown in the data of Fig. 5-11. From Fig. 5-11 it can be seen that the single oscillator model overpredicts the value of k in the low absorption wings away from the center

TABLE 5-1 FORTRAN Listing for Dispersion Relations

```
c  Program for calculating total normal absorptivity of a layer
c  of thickness h with optical properties described by a two
c  oscillator model using Eq. (4-116), ignoring interference.
c  Note: oscillator frequencies are effective values including
c  induced field effect.
c
   implicit real*8 (a-h,o-z)
   dimension alpha(196),wave(196),rn(196),rk(196)
   rlf=0.4              !first (lower) wavelength limit, μm
   rll=20.              !last (upper) wavelength limit, μm
   step=.1              !wavelength increment in μm
   l1=(rll-rlf)/step    !number of discrete wavelengths
   data pi/3.141592654d0/
   bolt=5.67d-12        !Boltzmann constant, W/cm**2 K**4
   c2=14388.0d0         !second radiation constant, μmK
   c0=2.99d14           !light speed, μm/s
c
c  input parameters
c
   h=5000.              !layer thickness, μm
   Te=5785.             !characteristic temperature of radiation, K
   w01=0.               !effective natural frequency of oscillator
c                        one, 1/s
   wp1=0.               !plasma frequency of oscillator one, 1/s
   gamma1=0.            !damping frequency of oscillator one, 1/s
   rne=1.46             !high frequency refractive index
   w02=2.*pi*c0*1.d-4*1063.    !natural frequency of oscillator
c                               two, 1/s
   wp2=2.*pi*c0*1.d-4*900.     !plasma frequency of oscillator
c                               two, 1/s
   gamma2=2.*pi*c0*1.d-4*30.   !damping frequency of oscillator
c                               two, 1/s
c
   rla=rlf              !wavelength, μm
c
c  loop on wavelength
c
   do 10 i=1,l1
   wave(i)=rla
   w=2.*pi*c0/rla       !circular frequency, 1/s
c
c  set up dispersion equations
c
   erp=rne*rne
  1+wp1*wp1*(w01*w01-w*w)/((w01*w01-w*w)**2+gamma1*gamma1*w*w)
  2+wp2*wp2*(w02*w02-w*w)/((w02*w02-w*w)**2+gamma2*gamma2*w*w)
    erpp=gamma1*w*wp1*wp1/((w01*w01-w*w)**2+gamma1*gamma1*w*w)
  1+gamma2*w*wp2*wp2/((w02*w02-w*w)**2+gamma2*gamma2*w*w)
```

TABLE 5-1 (*Continued*)

```
c
c  solve for optical constants, n and k
c
      rn(i) = dsqrt((erp + dsqrt(erp*erp + erpp*erpp))*.5)
      rk(i) = erpp*.5 / rn(i)
c
c  solve for interface transmissivity, tau1, internal
c  transmissivity, tint,
c  and layer spectral absorptivity, alpha(i)
c
      tau1 = 4.*rn(i) / ((rn(i) + 1.)**2 + rk(i)**2)
      tint = dexp( - 4.*pi*rk(i)*h / rla)
      alpha(i) = tau1*(1.- tint) / (1.- ((1.- tau1)*tint))
10 rla = rla + step
c
c  print spectral results
c
      open (unit = 56,file = 'alftot.dat',status = 'new')
      write (56,20)
20 format(1x, 'wavelength(micron)',10x, 'n',15x, 'k',15x, 'alflambda'/ )
   do 30 i = 1,l1
30 write (56,40) wave(i),char(9),rn(i),char(9),rk(i),char(9),
c  alpha(i)
40 format (1x,4(e15.4,a1))
c
c  integrate spectral absorptivity over wavelength to give total
c  absorptivity using fractional function. do in three parts:
c  below rlf,
c  between rlf and rll, and above rll. assume spectral absorptivity
c  is constant
c  at alpha(1) below rlf and constant at alpha(l1) above rll.
c
      v = c2 / ((rlf - (step / 2.0d0))*Te)
      alftot = alpha(1)*fract(v)
      v = c2 / ((rll + (step / 2.0d0))*Te)
      alftot = alftot + alpha(l1)*(1.0d0 - fract(v))
      rla = rlf
      do 50 i = 1,l1
   rless = rla - (step / 2.0d0)
   rmore = rless + step
   vless = c2 / (rless*Te)
   vmore = c2 / (rmore*Te)
   alftot = alftot + alpha(i)*(fract(vmore) - fract(vless))
50 rla = rla + step
   write(56,60) Te,alftot
60 format( / 1x, 'alphatot( ',f5.0,') = ',f5.3)
   end
c
```

TABLE 5-1 (*Continued*)

```
c  fractional function subroutine
c
      function fract(v)
      implicit real*8 (a-h,o-z)
      data pi/3.141592654d0/
      if(v.ge.2.0d0)then
      fract=0.
      do 5 m=1,100
      f1=(dexp(-m*v)/m**4)*(((m*v+3.0d0)*m*v+6.0d0)*m*v+
     1 6.0d0)
      if (dabs(f1).le.1.d-7)goto10
      fract=fract+f1
 5    continue
10    fract=fract*(15.d0/pi**4)
      else
      fract=1.0d0-(15.d0/pi**4)*v**3*(1.d0/3.d0-v/8.d0+v**2/
     1 6.d1-v**4/5040.d0+v**6/272160.d0-v**8/13305600.d0)
      endif
      return
      end
```

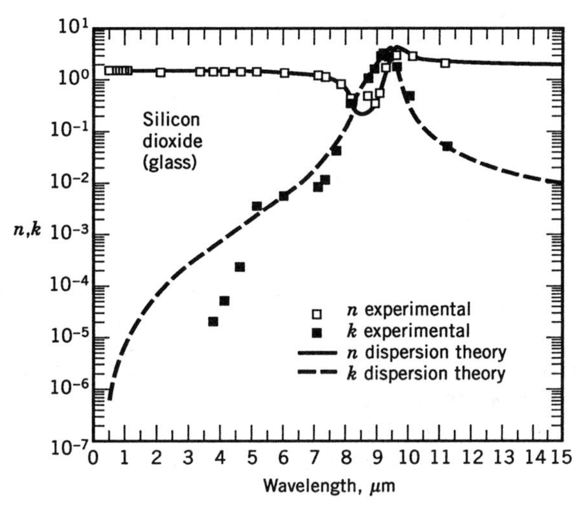

Figure 5-11 Optical constants of SiO_2 (glass) at room temperature (experimental values from Appendix C).

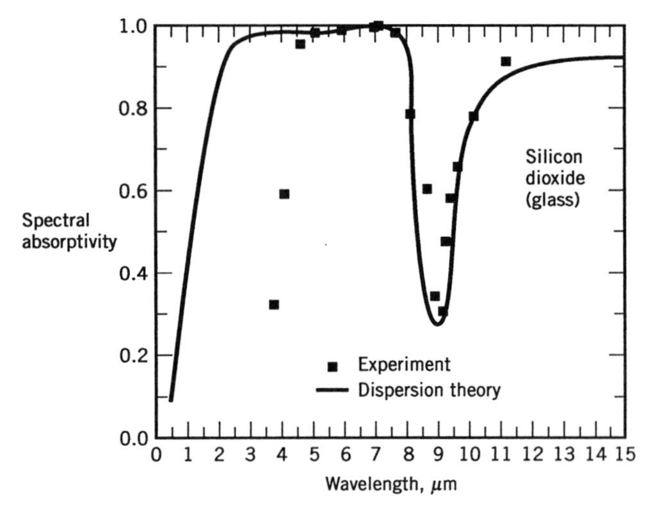

Figure 5-12 Spectral normal absorptivity of 5-mm-thick SiO_2 (glass) at room temperature.

frequency. This is a characteristic shortcoming of the classical (linear) oscillator theory. This shortcoming can be remedied by including temperature- and frequency-dependent functions for the damping term γ corresponding to anharmonic (nonlinear) material behavior.

The spectral absorptivities corresponding to both the theoretical and measured optical constants are shown in Fig. 5-12. It can be seen that at wavelengths below 4 μm the oscillator model greatly overpredicts the measured values. This is directly due to the overprediction of the k values in this region. It is important to note that the value of k in regions where k is small (such as $\lambda < 4$ μm) is extremely sensitive to impurities. Although the experimental values of n and k for SiO_2 presented in Appendix C represent, as nearly as can be determined, pure SiO_2, water and OH impurities as well as defect absorption can greatly influence the value of k in these regions. Commercial glass contains many impurities. Thus, commercial soda lime glass displays a cutoff near 2.5 μm instead of at 4 μm as indicated in Fig. 5-12. The total absorptivity calculated from the oscillator model is $\alpha_n' = 0.32$ while the value corresponding to the experimental values is 0.02. The experimental value is much lower than the prediction due to the difference in the short wavelength cutoff and the fact that most of the incident radiation is distributed at wavelengths below the cutoff.

Optical Constants of Metals

Metals and plasmas are characterized by the presence of many free charge carriers (electrons or ions). Free charge carriers, such as conduction electrons in metals, are not bound by restoring forces. The motion of conduction electrons under the influence of an oscillating electromagnetic field can be modeled by letting the spring constant k_s go to zero, which implies that the natural frequency goes to zero and the corresponding wavelength goes to infinity.

$$\omega_0 \to 0, \lambda_0 \to \infty \quad \text{(metals; free electrons—no restoring force)}$$

If the oscillator frequency is set to zero, Eq. (5-8) for the particle displacement becomes

$$x(t) = \frac{eE_0}{m} \frac{e^{i\omega t}}{-\omega^2 + i\gamma\omega} \tag{5-20}$$

and Eqs. (5-16) for the optical constants become

$$\varepsilon_r' = n^2 - k^2 = 1 - \sum_j \frac{\omega_{p_j}^2}{\omega^2 + \gamma_j^2} \tag{5-21}$$

$$\varepsilon_r'' = 2nk = \sum_j \frac{\omega_{p_j}^2 \gamma_j}{\omega^3 + \gamma_j^2 \omega} \tag{5-22}$$

Equations (5-21) and (5-22) demonstrate that the optical constants for metals exhibit continuous (not banded) spectra. Because free electrons are not bound by any restoring force, they are free to oscillate at any frequency, limited only by their own inertia at very high frequencies. The practical result is that metals are opaque to radiation at any frequency except very high energy ultraviolet and x-rays.

It is also possible to add the effect of bound electrons in metals to Eqs. (5-21) and (5-22). This is necessary if it is desired to accurately model the optical properties of metals at higher frequencies (near infrared through ultraviolet). For modeling the optical constants in just the infrared region it is usually sufficient to consider only the influence of conduction electrons, as discussed in the next section.

DRUDE THEORY

The Drude model assumes the existence of only one type of conducting or free electron with a relaxation frequency of γ. For a single unbound oscillator Eqs. (5-21) and (5-22) become

$$\varepsilon_r' = n^2 - k^2 = 1 - \frac{\omega_p^2}{\omega^2 + \gamma^2} \tag{5-23}$$

$$\varepsilon_r'' = 2nk = \frac{\gamma\omega_p^2}{\omega^3 + \gamma^2\omega} \tag{5-24}$$

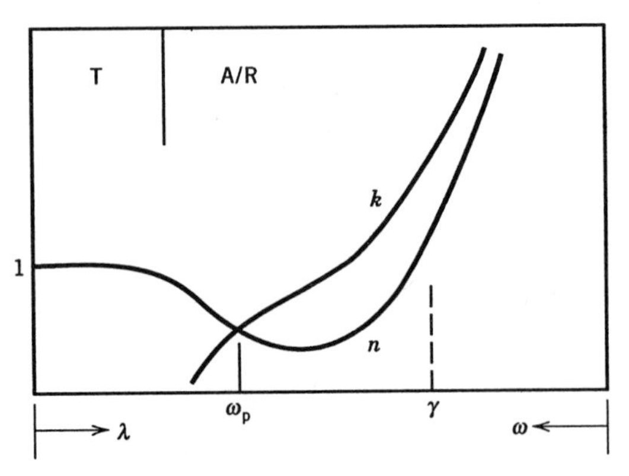

Figure 5-13 Optical constants for metallic solids and liquids (single Drude oscillator).

where the plasma frequency, defined in Eq. (5-14), is a measure of the number density of free electrons present. For most metals the plasma frequency is much greater than the relaxation frequency.

$$\omega_p \gg \gamma \quad \text{(metals)} \tag{5-25}$$

A typical plot of n and k for a single Drude oscillator is shown in Fig. 5-13 and the corresponding normal reflectivity in Fig. 5-14.

At high frequencies the value of n is close to one and k is essentially zero. With increasing wavelength, the value of n goes through a minimum and then increases monotonically. The value of k increases monotonically with

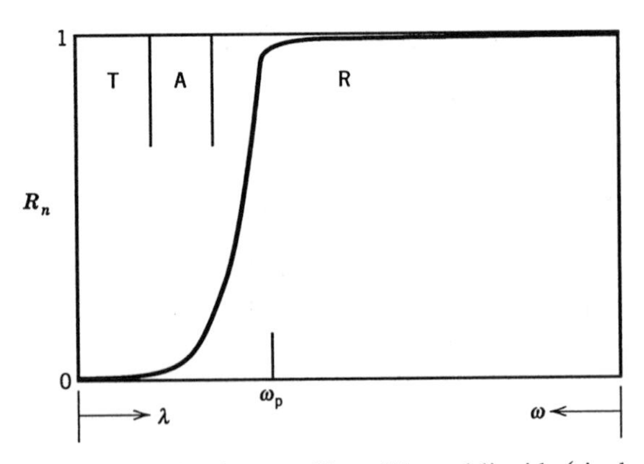

Figure 5-14 Normal reflectivity for metallic solids and liquids (single Drude oscillator).

wavelength. At frequencies above the plasma frequency ($\omega > \omega_p$) the value of k is less than that of n. At the plasma frequency ($\omega = \omega_p$) n and k are equal. At frequencies below the plasma frequency ($\omega < \omega_p$) the value of k is greater than n. The general behavior of n, k, and R_n in the various spectral regions, near, above and below the plasma frequency, can be summarized as follows:

$$\omega = \omega_p: n = k \ll 1 \quad \text{and} \quad R_n \to 1 \tag{5-26}$$

$$\omega \gg \omega_p: n > k \ (n \to 1, k \to 0) \quad \text{and} \quad R_n \to 0 \tag{5-27}$$

$$\omega \ll \omega_p: k > n \gg 1 \quad \text{and} \quad R_n \to 1 \tag{5-28}$$

These trends can be readily shown by considering the limiting behavior of Eqs. (5-23) and (5-24) and using the relation $\omega_p \gg \gamma$, which is true for most metals. This derivation is left as an exercise.

This analysis shows that metals undergo a rapid transformation in optical behavior at the plasma frequency. Below the plasma frequency the metal is a good reflector. Just above the plasma frequency the metal rapidly changes from a good reflector to a transparent material. For most metals the plasma frequency is in the far ultraviolet region.

Self-Extinction of a Laser-Induced Plasma

One interesting practical implication of this phenomenon of rapid transition from being a transmitter to a reflector is the self-extinction of a laser-induced plasma formed next to a target. Unlike the free electrons in a metal, the number of free charge carriers in a plasma is not constant but increases with the temperature of the plasma. Therefore the plasma frequency of a plasma is a variable that depends on the temperature (or degree of ionization) of the plasma. If the power density of a laser focused on a target is large enough, the plasma formed can become reflective and prevent the laser energy from being absorbed by either the target or the plasma, thus extinguishing the plasma. The condition for the plasma to remain nonreflective is for the plasma frequency ω_p to remain below the laser frequency ω.

$$\omega > \omega_p = \sqrt{\frac{Ne^2}{m\varepsilon_0}} \tag{5-29}$$

This condition places an upper limit on the number density of free electrons.

$$N < \frac{\omega^2 m\varepsilon_0}{e^2} \tag{5-30}$$

If this condition is not satisfied and $\omega < \omega_p$, then $R_n \to 1$ and the plasma reflects the laser energy and stops generation of the plasma.

Equation (5-29) also has an important application in the transmission and reflection of radio waves at the ionosphere. The number density of free electrons in the ionosphere is approximately 10^6 cm^{-3}. This value corresponds to a plasma frequency of 10 MHz and a plasma wavelength of 30 m. Thus AM radio waves (1 MHz) are reflected by the ionosphere while FM radio waves (100 MHz) are transmitted. As a result AM radio signals can be received on the surface of the earth at greater distances from the transmitter than FM signals for equal signal strengths; on the other hand FM and higher frequencies must be used for communications between the ground and satellites.

Hagen-Rubens Equation

The description of metal optical properties given by the Drude theory is good qualitatively over the entire frequency spectrum. Near the visible region the influence of bound electrons must be included with the free electrons in the dispersion relations to retain quantitative accuracy. In the long wavelength region (radio, microwave, and infrared) the influence of bound electrons is negligible and the Drude theory is also quantitatively accurate. This means that at low frequencies the optical constants of metals can be predicted from dc electrical properties, as will be discussed in the following section.

In the limit of zero frequency an oscillating electric field E_x becomes a constant dc field $E_{x_{\mathrm{dc}}}$. For a constant external field, electrons will acquire a constant velocity according to the solution of Eq. (5-4) with $\omega_0 = 0$ and the acceleration term set to zero,

$$\dot{x}_{\mathrm{dc}} = \frac{eE_{x_{\mathrm{dc}}}}{m\gamma} \tag{5-31}$$

so that if N is number density of conducting electrons, the current is given by

$$j = Ne\dot{x} = \frac{Ne^2 E_{x_{\mathrm{dc}}}}{m\gamma} = \sigma_{e_{\mathrm{dc}}} E_{x_{\mathrm{dc}}} \tag{5-32}$$

which defines the dc conductivity as

$$\sigma_{e_{\mathrm{dc}}} = \frac{Ne^2}{m\gamma} = \frac{\omega_p^2 \varepsilon_0}{\gamma} \tag{5-33}$$

For an oscillating field it can be shown from Eqs. (5-8) and (5-32) that the complex ac conductivity is

$$\sigma_e = \frac{Ne^2}{m(\gamma + i\omega)} \tag{5-34}$$

However, for long wavelengths or low frequencies ($\omega \ll \gamma$) the conductivity is essentially constant at the dc value. Therefore the Drude dispersion relations for metals can be rewritten in terms of the dc conductivity as follows:

$$\varepsilon_r' = n^2 - k^2 = 1 - \frac{\sigma_{e_{dc}}\gamma}{\varepsilon_0} \cdot \frac{1}{\omega^2 + \gamma^2} \tag{5-35}$$

$$\varepsilon_r'' = 2nk = \frac{\sigma_{e_{dc}}\gamma}{\varepsilon_0} \frac{\gamma}{\omega^3 + \gamma^2\omega} \tag{5-36}$$

These relations (5-35) and (5-36) can be used to model the optical constants of metals in the infrared and longer wavelength regions with good accuracy. This requires the value of one parameter γ. The value of γ can be obtained from a reflectivity measurement at one wavelength.

A single relation that is even simpler than (5-35) and (5-36) (but also less accurate) can be obtained by eliminating the relaxation frequency parameter γ as follows. According to the order of magnitude argument $\omega \ll \gamma$, Eq. (5-36) can be simplified to give

$$2nk \approx \frac{\sigma_{e_{dc}}}{\varepsilon_0\omega} \tag{5-37}$$

In the limit of low frequency ($\omega \to 0$), $2nk$ becomes very large. However $n^2 - k^2$ does not become correspondingly large so the value of n approaches the value of k (to within an order of magnitude). Using the approximation that n and k are equal, Eq. (5-37) gives the following relation for n and k:

$$n \approx k \approx \sqrt{\frac{\sigma_{e_{dc}}}{2\omega\varepsilon_0}} = \sqrt{\frac{\lambda_0}{4\pi c_0\varepsilon_0 r_{e_{dc}}}} = \sqrt{\frac{0.003\lambda_0}{r_{e_{dc}}}} \tag{5-38}$$

where λ_0 is the vacuum wavelength in microns and $r_{e_{dc}}$ is the dc resistivity in Ω-cm. This relation is the Hagen-Rubens equation and is useful for predicting the infrared optical constants of metals to within an order of magnitude using the dc resistivity. This relation also gives an indication of the temperature dependence of the optical constants for metals in the infrared. For most metals $r_{e_{dc}}$ increases with temperature and the values of n and k therefore decrease as temperature increases. (With this much information and the results of Chapter 4 you should now be able to explain the trends shown for metals in Fig. 2-20.)

Example 5-2

Calculate the optical constants of aluminum at 298 K using the Drude model with the following parameters:

$$\omega_p = 1.97 \times 10^{16}\ \mathrm{s}^{-1} \quad \gamma = 1.0 \times 10^{14}\ \mathrm{s}^{-1}$$

Compare the predicted optical constants and normal reflectivity over the region $0.4 < \lambda < 20\ \mu$m with the values given for aluminum in Appendix C.

The program listed in Table 5-1 can be used by setting the oscillator frequency to zero. The resulting optical constants are shown in Fig. 5-15 and the reflectivities are given in Fig. 5-16. The optical constants and reflectivity values are well matched by the Drude model in the region $\lambda > 2\ \mu$m. However, there is an interband (bound electron) transition at 0.8 μm that causes the predicted values to deviate somewhat from the measured values for $\lambda < 2\ \mu$m. As a result of this interband transition, there is a local minimum in the reflectivity of aluminum near 0.8 μm. While the relative change in the reflectivity in this region is small, the change in absorptivity is very large, due to the inherently small values of absorptivity. The reflectivity changes from 0.99 at 10 μm to 0.87 at 0.8 μm, which is a decrease of 12%. However, the absorptivity changes from 0.01 at 10 μm to 0.13 at 0.8 μm, which represents an increase by a factor of 13. Thus, in laser processing of aluminum, Nd-YAG lasers (1.06 μm) are often preferred over more powerful CO_2 lasers (10.6 μm) due to the higher intrinsic absorptivity of aluminum at

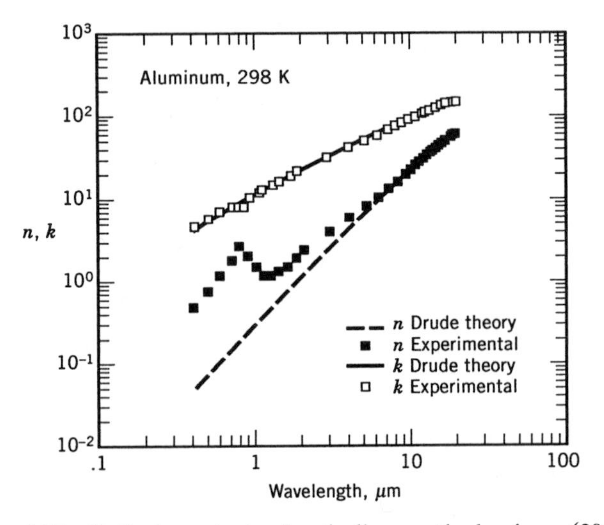

Figure 5-15 Optical constants of optically smooth aluminum (298 K).

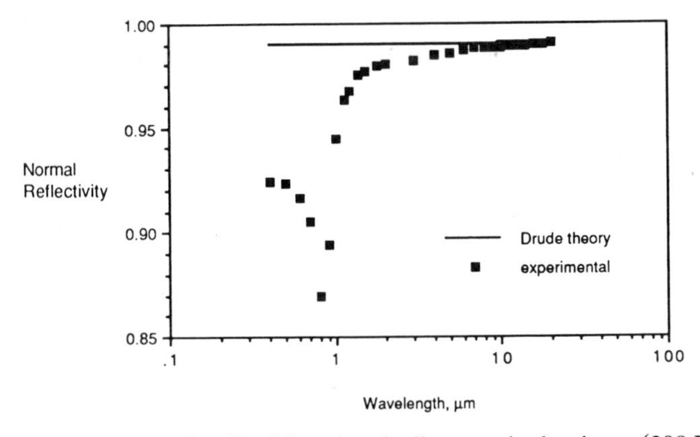

Figure 5-16 Normal reflectivity of optically smooth aluminum (298 K).

1.06 μm (see Problem 4-2). It should be noted that the parameters used in this example were selected specifically to match the data in the infrared region. If the ultraviolet region was of more interest, a different set of parameters could be used that would match better at short wavelengths but not as well at long wavelengths (see Problem 5-12).

Summary of Dispersion in Solids

There are two basic mechanisms of absorption in solids: electronic absorption and lattice vibration. Both of these mechanisms can be modeled by the classical damped oscillator. Electronic absorption can be modeled using both bound and unbound electrons (to model both interband and intraband transitions). And lattice vibration can be modeled using bound ions.

Figure 5-17 summarizes the various mechanisms of absorption that influence the optical properties of metals, semiconductors, and dielectrics over the various regions of the EM spectrum.

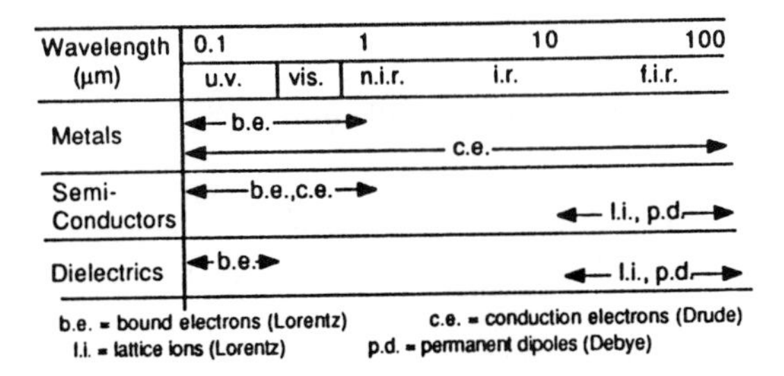

Figure 5-17 Summary of absorption mechanisms in solids and liquids.

In metals, intraband transitions (free conduction electrons) dominate absorption over the ultraviolet (u.v.), visible (vis.) and infrared (i.r.) regions. Interband transitions (bound electrons) also make a minor contribution in the u.v., vis., and n.i.r. (near infrared) regions.

In dielectrics, u.v. absorption is dominated by bound electrons while i.r. absorption is dominated by lattice vibration. Most pure dielectrics are opaque in the u.v. and have a transparent window somewhere between the u.v. and i.r. region.

Finally, semiconductors are a hybrid case between metals and dielectrics. In semiconductors there are no free conduction electrons so i.r. absorption is dominated by lattice vibration. However, semiconductors are characterized by electrons that can absorb and emit photons by changing energy states between the valence band and conduction band (bound-free transition). Since the conduction band side of the transition is unbounded, electrons undergoing this type of transition can emit or absorb photons of any frequency as long as the energy ($h\nu$) is big enough to overcome the band gap. Thus the opaque electronic absorption edge is usually extended to longer wavelengths for semiconductors compared with dielectrics.

Dispersion in Gases

As was mentioned at the beginning of this chapter, dispersion theory can also be used to model the behavior of gases. The dispersion in gases is also described by Eq. (5-15).

$$(n - ik)^2 = 1 + \sum_j \frac{N_j e^2}{m_j \varepsilon_0} \frac{1}{\omega_{0_j}^2 - \omega^2 + i\gamma_j \omega} \tag{5-15}$$

For gases, the number density of oscillating charged particles, which create an oscillating dipole moment N_j, is much smaller than that for condensed matter. As a result, the second term on the right-hand side of (5-15) is much smaller than one, and Eq. (5-15) can be written to an accurate approximation as

$$n - ik = 1 + \frac{1}{2} \sum_j \frac{N_j e^2}{m_j \varepsilon_0} \frac{1}{\omega_{0_j}^2 - \omega^2 + i\gamma_j \omega} \tag{5-39}$$

which can be separated into real and imaginary parts.

$$n = 1 + \frac{1}{2} \sum_j \frac{N_j e^2}{m_j \varepsilon_0} \frac{\omega_{0_j}^2 - \omega^2}{\left(\omega_{0_j}^2 - \omega^2\right)^2 + \gamma_j^2 \omega^2} \tag{5-40}$$

$$k = \frac{1}{2} \sum_j \frac{N_j e^2}{m_j \varepsilon_0} \frac{\gamma_j \omega}{\left(\omega_{0_j}^2 - \omega^2\right)^2 + \gamma_j^2 \omega^2} \tag{5-41}$$

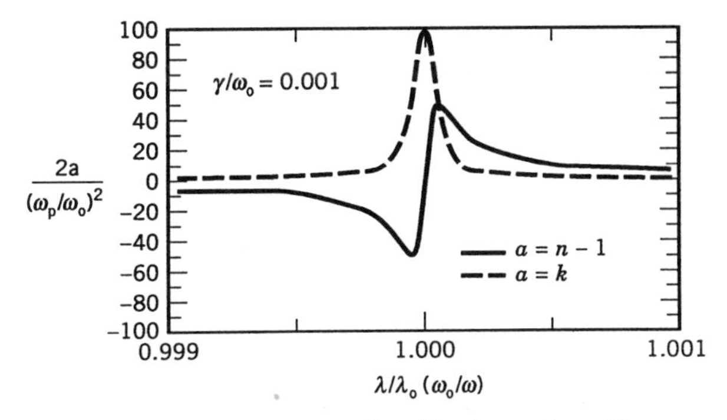

Figure 5-18 Optical constants for a gas line (single weak oscillator, $\omega_p/\omega_0 \ll 1$).

These relations are shown plotted in Fig. 5-18 for a single oscillator. The behavior with frequency is similar to the general behavior depicted in Fig. 5-4. As frequency decreases n goes through an anomalous dispersion in the neighborhood of a resonant frequency. However, because the magnitude of k is so small, the value of n for gases remains nearly constant at a value of one.

The spectral absorption coefficient can be obtained from the relation

$$K_{a_\lambda} = \frac{4\pi k}{\lambda} = \frac{2\omega k}{c_0} \tag{5-42}$$

where λ is understood to be wavelength in vacuum but the 0 subscript has been omitted to avoid confusion with the resonant frequency notation. (This point is academic anyway since the refractive index is essentially one for all gases.) Substituting (5-41) into (5-42) gives the spectral absorption coefficient in the neighborhood of a single independent oscillator.

$$K_{a_\lambda} = \frac{Ne^2}{m\varepsilon_0 c_0} \frac{\gamma\omega^2}{\left(\omega_0^2 - \omega^2\right)^2 + \gamma^2\omega^2} \tag{5-43}$$

Since the relaxation frequency γ for gases is relatively low, the only contribution from the right-hand side of (5-43) will occur when the frequency is very close to the resonant frequency and the approximation $\omega_0 + \omega \approx 2\omega$ can be used in (5-43) to give

$$K_{a_\lambda} = \frac{Ne^2}{m\varepsilon_0 c_0} \frac{\gamma}{4(\omega_0 - \omega)^2 + \gamma^2} \tag{5-44}$$

This relation is called a Lorentz profile, and it can also be derived from quantum theory for a polar gas molecule undergoing vibrational-rotational energy transitions. This relation is considered again in Chapter 8 where a more detailed discussion of gas radiation properties is given.

Induced Field Effect

When a dielectric material is subjected to a dc electrical field, the dipoles in the material become oriented as illustrated in Fig. 5-19. The orientation of the dipoles in the material sets up an induced field that is opposite the external field in polarity. In Fig. 5-19, E_0 is the external field strength, E_{ind} is the strength of the field that is produced by the orientation of the dipoles in the material, and E is the resulting net electric field produced in the material.

$$E = E_0 - E_{\text{ind}} \tag{5-45}$$

It can be shown [Slater and Frank, 1947] that the magnitude of the induced field is

$$E_{\text{ind}} = \frac{P_x}{3\varepsilon_0} \tag{5-46}$$

Thus the magnitude of the field inside the material is somewhat less than the external field due to the influence of the induced field. This same effect also occurs when an oscillating electric field is imposed on a dielectric material. The reduction in the internal field can be accounted for by using the reduced field strength E in place of the external field strength that was used in the development of the equations at the beginning of the chapter. If $E_0 - P_x/3\varepsilon_0$ is used in place of E_0, Eq. (5-4) gives

$$\frac{(n - ik)^2 - 1}{(n - ik)^2 + 2} = \frac{1}{3} \sum_j \frac{\omega_{p_j}^2}{\omega_{0_j}^2 - \omega^2 + i\gamma_j \omega} \tag{5-47}$$

or, using arguments outlined by Slater and Frank, [1947] it can be shown that

$$\varepsilon_r' - i\varepsilon_r'' = (n - ik)^2 = 1 + \sum_j \frac{\omega_{p_j}^2}{\Omega_{0_j}^2 - \omega^2 + i\gamma_j \omega} \tag{5-48}$$

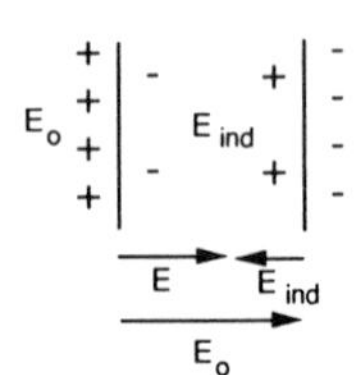

Figure 5-19 Effect of induced field.

where

$$\Omega_{0_j}^2 = \omega_{0_j}^2 - \frac{\omega_{p_j}^2}{3} \tag{5-49}$$

Equation (5-48) is identical to the original dispersion relation, Eq. (5-15), except that the resonant frequency has been shifted as indicated by Eq. (5-49). Since the resonant frequency is usually determined experimentally from spectroscopic measurements rather than from fundamental physical constants, this modification is of no real significance to the previous discussion. However, there is one important result that comes out of this modified theory.

By evaluating Eq. (5-47) at frequencies far from resonant frequencies, that is, in nonabsorbing or weakly absorbing regions ($k \ll n$), and recognizing that the plasma frequency squared is proportional to the oscillator number density N_j, which is proportional to the density of the material ($N \propto \rho$), the following relation is obtained:

$$\frac{n^2 - 1}{n^2 + 2}\frac{1}{\rho} = \text{constant} \quad \text{(liquids and solids, } n > 1\text{)} \tag{5-50}$$

This expression is known as the *Lorenz-Lorentz* equation and can be used to evaluate the variation of n with density for liquids and solids. Since the variation of density with temperature is usually well known for most liquids and solids, Eq. (5-50) can be used to predict the variation of n with temperature for nonabsorbing or weakly absorbing liquids and solids, as verified for Al_2O_3 solid propellant combustion particles by Parry and Brewster [1990]. It should also be noted that Eq. (5-50) or (5-47) can be summed over constituents to model the optical properties of mixtures in terms of the optical properties of the constituents, as illustrated by Goodwin and Mitchner for coal slags [1989].

For gases a simpler expression can be obtained directly from Eq. (5-40) (since the correction for the induced field is not important in gases) or by evaluating (5-50) for $n \approx 1$.

$$\frac{n - 1}{\rho} = \text{constant} \quad \text{(gases, } n \approx 1\text{)} \tag{5-51}$$

This result is known as the *Gladstone-Dale* equation and gives an excellent prediction of the variation of n with temperature for gases and vapors.

Example 5-3

Estimate the refractive index of ice at 1 μm from the value for water. Use the properties for liquid water given in Appendix C.

Evaluating Eq. (5-50) at the solid and liquid states gives

$$\frac{n_s^2 - 1}{n_s^2 + 2}\frac{1}{\rho_s} = \frac{n_l^2 - 1}{n_l^2 + 2}\frac{1}{\rho_l}$$

which can be solved to give

$$n_s = \sqrt{\frac{2C + 1}{1 - C}} \qquad C = \frac{\rho_s}{\rho_l}\frac{n_l^2 - 1}{n_l^2 + 2}$$

Substituting the values $\rho_s = 0.9168$, $\rho_l = 1.000$, and $n_l = 1.326$ gives $n_s = 1.296$, which agrees to within 0.6% of the value given for ice in Appendix C of 1.302. Thus, the refractive index of water decreases upon freezing because the density decreases (unlike most substances whose density increases upon freezing).

DEBYE RELAXATION

Many condensed materials, particularly liquids (e.g., water) consist of molecules that have a *permanent* dipole that can be oriented by an oscillating electromagnetic field to produce a net dipole moment. This orientation of the molecules, which takes place against the tendency toward random orientation due to thermal collisions with other molecules, gives rise to a mechanism for absorption of photons called Debye relaxation. The frequency of radiation that can be so absorbed is on the order of the thermal relaxation frequency of the molecules, which is typically in the microwave region. However, the bandwidth associated with thermal relaxation is relatively wide, such that the influence of Debye absorption often extends down to the infrared region. For this reason it is briefly mentioned here.

The dielectric constant for Debye relaxation can be derived by considering the return to equilibrium of a polarized region of matter under the influence of thermal collisions with surrounding particles. The results are

$$\varepsilon_r' = n^2 - k^2 = n_v^2 + \frac{n_D^2 - n_v^2}{1 + \omega^2\tau^2} \tag{5-52}$$

$$\varepsilon_r'' = 2nk = \frac{\omega\tau\left(n_D^2 - n_v^2\right)}{1 + \omega^2\tau^2} \tag{5-53}$$

where n_D is the refractive index on the low frequency side of the Debye region and n_v is the residual refractive index on the high frequency (lattice vibrational) side of the Debye region. The relaxation time τ is a parameter

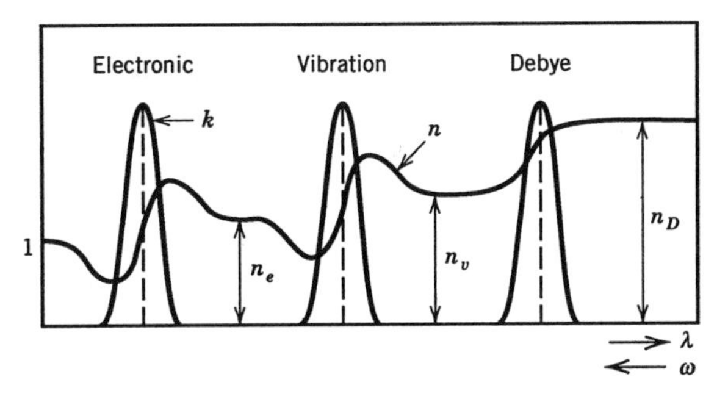

Figure 5-20 Optical constants of nonconductors including Debye relaxation.

that can be estimated from that of a sphere of diameter d in a fluid of viscosity μ as

$$\tau = \frac{\pi \mu d^3}{2k_B T} \tag{5-54}$$

where k_B is Boltzmann's constant.

The spectral behavior of n and k predicted by Eqs. (5-52) and (5-53) is fundamentally different than the results of the Lorentz model pictured in Fig. 5-4. The difference arises from the fact that in the Lorentz model a hypothetical restoring force tries to return the particle to its equilibrium position, which results in overshoot in the particle motion, whereas in the Debye model the restoring force is generated by random thermal collisions, which do not produce overshoot in the particle motion. As a result, n exhibits "underdamped" oscillatory behavior for the Lorentz model as seen in Fig. 5-4. However, the corresponding plot for the Debye model in Fig. 5-20 shows a monotonic, "overdamped" transition from n_v to n_D.

Optical Constants of Water

Water is one of the most pervasive and optically important substances on earth. It is also of pedagogical interest in that it exhibits Debye relaxation absorption in addition to Lorentzian oscillator absorption. The optical constants of both water and ice are listed in Appendix C, as taken from Irvine and Pollack [1968]. These properties are plotted in Figs. 5-21 and 5-22.

Figure 5-21 shows the absorption index for room temperature water and ice at 270 K. In the ultraviolet region, electronic transitions are responsible for absorption of light. In the visible region, a relatively nonabsorbing window appears. Between 1 and 10 μm intramolecular vibrations are primarily responsible for absorption. Various bands can be seen at 1.5, 2.0, 3.0, 4.7, and 6.0 μm, which are very similar between water and ice. These bands can

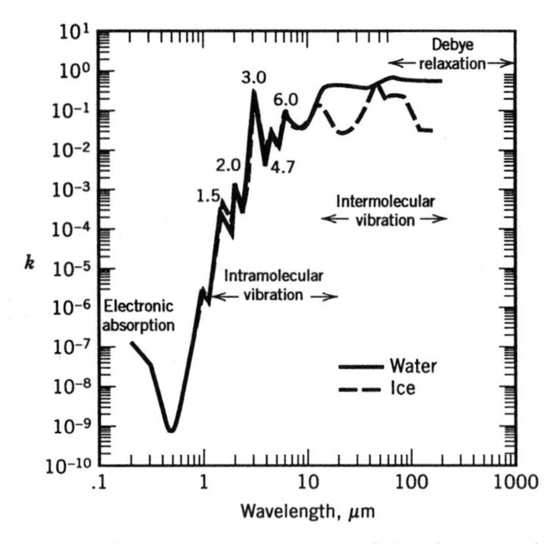

Figure 5-21 Absorption index of water (298 K) and ice (270 K).

be identified with the vibration modes of the free water vapor molecule (see Chapter 8 for a further discussion of these modes). The band wavelengths for the condensed states are slightly shifted from those listed in Chapter 8 for the free molecule due to the weak perturbing influence that intermolecular forces have on the intramolecular vibrations in the condensed states. Between 10 and 100 μm the absorption index of water and ice differ significantly. This difference is due to the difference in intermolecular forces between the solid and liquid state. In the liquid state intermolecular translational motion (62 μm) and intermolecular rotational motion (17 μm) are responsible for weak absorption bands, which appear on top of the short wavelength tail of the Debye absorption band. In the solid state the Debye absorption band does not appear due to the inability of the molecules to rotate freely. Thus the absorption index for ice is significantly lower than that for water over most of the region from 10 to 100 μm. As a result, the intermolecular rotation and translation bands in ice are more pronounced. These bands appear at 12 and 45 μm, respectively, in ice. The shift of the intermolecular band locations to lower wavelengths (higher frequencies) is due to the increased strength in the intermolecular bonds associated with freezing.

Figure 5-22 shows the corresponding refractive index for both water and ice. In the nonabsorbing visible region the refractive index of ice is a few percent below the value for water due to the density difference between ice and water, as was discussed in Example 5-3. Throughout the far infrared ($\lambda > 100$ μm), microwave, and radio wave regions, the refractive index of water remains significantly larger in magnitude than that for ice.

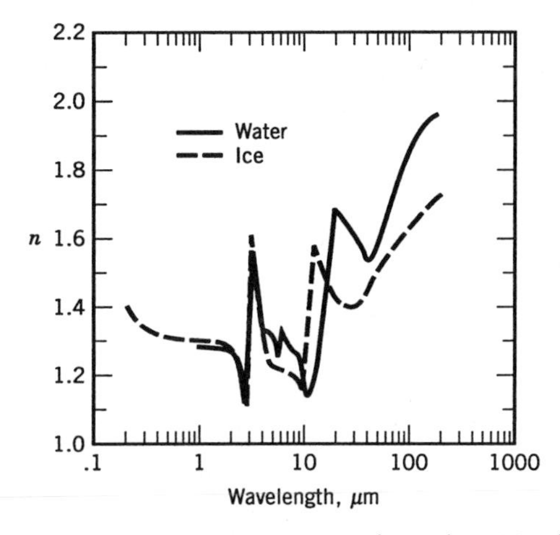

Figure 5-22 Refractive index of water (298 K) and ice (270 K).

The difference in the radio wave optical properties of ice and water has important implications in terms of weather forecasting and radar detection of storms. Water droplets possess a large radar back-scattering cross section due to a relatively large value of n, whereas ice particles have a much smaller back-scattering cross section. As a result rainstorms are very easy to detect with radar while ice and hail storms are more difficult. Another important application of long wavelength optical properties of water is microwave cooking. Most cooking and heating of food in microwave ovens is a result of Debye absorption by liquid water.

Determination of *n* and *k* from Normal Reflectivity Measurements

The oscillator model offers a useful method for describing the optical constants of most materials given that an appropriate set of oscillator parameters can be determined. The question is how to determine a set of oscillator parameters. A relatively easy method is to measure some quantity that can be predicted from n and k using electromagnetic theory results over a range of frequencies. Normal reflectance from a smooth surface is a quantity often selected.

$$\mathfrak{R} = \frac{(n-1)^2 + k^2}{(n+1)^2 + k^2} \tag{5-55}$$

Then, the spectral variation of the measured values of $\mathfrak{R}$ (such as that depicted in Fig. 5-5 or 5-14) can be curve fit using a least squares criterion. The fit is obtained by minimizing the error between the measured and theoretical values of reflectance, $\Sigma_i[\Re_{i,\mathrm{exp}} - \Re_{i,\mathrm{theo}}]^2$. The minimization may be accomplished using a computer program based on the method of steepest descent [Verleur, 1968]. It is usually possible to obtain a fit with a reasonable degree of accuracy if a sufficient number of oscillators are used [Spitzer and Kleinman, 1961]. Another method for determining optical constants from normal reflectivity measurements is Kramers-Kronig (or KK) analysis.

KRAMERS-KRONIG RELATIONS

From the preceding discussion it should be apparent that the optical constants n and k are not entirely independent of each other. At any particular spectral location where k increases due to an absorption mechanism, a corresponding "wiggle" (a region of anomalous dispersion) is induced in the spectrum of n. The residual value of n that emerges on the low frequency side of the absorption band is higher than residual value that n had going into the high frequency side of the anomalous dispersion. The net change in the residual value of n is related to the strength of the absorption band (i.e., the magnitude and width of the band). The stronger the band, the larger the residual change in n. The various absorption mechanisms that occur at different frequencies throughout the electromagnetic spectrum all contribute to the overall change that takes place in the value of n across the full spectrum. Across the full spectrum the value of n goes from 1 at very high frequencies to a relatively large value at very low frequencies (dc). For insulators this maximum value is finite and corresponds to the square root of the dc dielectric constant. For conductors this maximum value is unbounded, as can be seen from the Hagen-Rubens relation. For example, the dielectric constant of water owes its large value ($\varepsilon_r' = 78, n = 8.8$) at $\lambda \to \infty$ to the many absorption mechanisms that exist across the spectrum, including electronic absorption, intra- and intermolecular vibration, and Debye relaxation.

The relationship that exists between n and k is expressed mathematically in terms of equations known as the Kramers-Kronig relations [Kronig, 1926, Kramers, 1927, and Cardona, 1969]. These relations are a result of complex variable theory. The only physical assumption is that the response of the material to the incident radiation is linear and causal (i.e., no radiation is reflected until radiation is incident on the material). The Kramers-Kronig relations can be expressed in many different forms, depending on the choice of variables. In terms of the optical constants n and k the Kramers-Kronig relations are

$$n(\omega) = 1 + \frac{2}{\pi} P \int_0^\infty \frac{\omega' k(\omega')}{\omega'^2 - \omega^2} \, d\omega' \tag{5-56}$$

$$k(\omega) = \frac{2\omega}{\pi} P \int_0^\infty \frac{n(\omega')}{\omega'^2 - \omega^2} \, d\omega' \tag{5-57}$$

where P signifies the Cauchy principal value of the integral. These relations indicate that the value of n (or k) at any particular frequency is determined by the values of k (or n) at all frequencies, but most strongly by the values of k (or n) in the neighborhood of the frequency of interest. The latter observation follows from the fact that the major contribution to the integrals comes from spectral region where the frequency ω' is in the neighborhood of ω. If k is known over all frequencies, then n is also, and vice versa. These relations are useful for checking the validity of optical constant data that are reported for various materials. A valid set of data for n and k should satisfy the Kramers-Kronig relations.

The Kramers-Kronig relations may also be used as a method for determining optical constants from measurements of normal reflectivity. In order to use this method, it is useful to consider the Kramers-Kronig relations in a form that introduces the Fresnel interface reflectivity. If a complex, normal reflectivity is defined as discussed in Chapter 4 in connection with Eq. (4-63),

$$\Re(\omega) = |\tilde{r}_{12}|^2 = \left|\sqrt{\Re(\omega)}\, e^{i\phi(\omega)}\right|^2 \tag{5-58}$$

complex variable theory can be used to show that the optical constants are given by

$$n(\omega) = \frac{1 - \Re(\omega)}{1 + \Re(\omega) - 2\sqrt{\Re(\omega)}\cos\phi(\omega)} \tag{5-59}$$

$$k(\omega) = \frac{2\sqrt{\Re(\omega)}\sin\phi(\omega)}{1 + \Re(\omega) - 2\sqrt{\Re(\omega)}\cos\phi(\omega)} \tag{5-60}$$

where ϕ is the phase difference between the incident and reflected waves given by

$$\phi(\omega) = \frac{2\omega}{\pi} P \int_0^\infty \frac{\ln\sqrt{\Re(\omega')}}{\omega'^2 - \omega^2}\, d\omega' \tag{5-61}$$

This form of the Kramers-Kronig relations indicates that if the normal reflectivity can be measured over the entire frequency range, both n and k can be determined by integration.

Although Kramers-Kronig analysis is exact in theory, in practice there are problems. Foremost is the fact that measurements of normal reflectivity are required over all frequencies, which is not very feasible. Fortunately, it is often possible to extrapolate to zero and infinity. And since the contribution to the integral in Eq. (5-61) is small for ω' approaching zero and infinity when ω is intermediate, the error due to uncertainty in these regions is minimal anyway. A discussion of this and other difficulties associated with Kramers-Kronig analysis can be found in Schatz, et al. [1963], Bowlden and Wilmshurst

[1963], Gottlieb [1960], and Sanderson [1965]. Schemes that combine the oscillator fit method with KK analysis have also been tried with some success [Andermann et al., 1965].

SUMMARY

The results obtained in this chapter for the optical constants of solids, liquids, and gases are based on a very simplistic model. This model assumes that dielectrics are composed of charged particles (ions and electrons) that are held in equilibrium positions by linear restoring forces and subject to linear damping forces. Metals are modeled as being composed of conduction electrons that have no restoring forces, but are subject to damping forces. This model is surprisingly good in practice, considering what is now known from quantum theory about the real behavior of electrons and atoms.

According to quantum theory atoms, ions, electrons, and molecules have certain energy levels that they can occupy, and the absorption and emission of radiation of frequency ν is associated with the transition of these particles between energy states according to the relation

$$E_2 - E_1 = h\nu = \frac{h\omega}{2\pi} \tag{5-62}$$

By applying certain perturbation techniques to the quantum theory, however, it can be shown that a theory that is mathematically equivalent to the dispersion theory is obtained. For example, the frequency ω associated with a transition between energy states in the quantum theory is equivalent to the resonant frequency ω_0 of the classical oscillator model. Furthermore, the broadening of upper and lower energy states ($E_i = E_{i0} \pm \Delta E_i$) due to collisions between particles that arises in quantum theory is equivalent to the damping mechanism of the oscillator model. Thus the dispersion theory is of great practical significance in modeling optical properties of materials.

REFERENCES

Andermann, G., Caron, A., and Dows, D. A. (1965), *J. Opt. Soc. Am.*, Vol. 55, p. 1210.

Bowlden, H. J. and Wilmshurst, J. K. (1963), *J. Opt. Soc. Am.*, Vol. 53, p. 1073.

Cardona, M. (1969), in *Optical Properties of Solids*, S. Nudelman and S. S. Mitra, Eds., Plenum, New York, p. 137.

Edwards, D. F. (1985), "Silicon (Si)," in *Handbook of Optical Constants of Solids*, E. D. Palik, Ed., Academic Press, New York, pp. 547–569.

Goodwin, D. G. and Mitchner, M. (1989), "Infrared Optical Constants of Coal Slags: Dependence on Chemical Composition," *Journal of Thermophysics and Heat Transfer*, Vol. 3, No. 1, Jan., pp. 53–60.

Gottlieb, M. (1960), *J. Opt. Soc. Am.*, Vol. 50, p. 343.

Irvine, W. M. and Pollack, J. B. (1968), "Infrared Optical Properties of Water and Ice Spheres," *ICARUS*, Vol. 8, pp. 324–360.

Kronig, R. de L. (1926), *J. Opt. Soc. Am.*, Vol. 12, p. 547.

Parry, D. L. and Brewster, M. Q. (1991), "Optical Constants of Al_2O_3 Smoke in Propellant Flames," *J. Thermophysics and Heat Transfer*, Vol. 5, No. 2, pp. 142–149.

Sanderson, R. B. (1965), *J. Phys. Chem. Solids*, Vol. 26, p. 803.

Schatz, P. N., Maeda, S., and Kozima, K. (1963), *J. Chem. Phys.*, Vol. 38, p. 2658.

Slater, J. C. and Frank, N. H. (1947), *Electromagnetism*, Dover Publications Inc., New York.

Spitzer, W. G. and Kleinman, D. A. (1961), *Phys. Rev.*, Vol. 121, p. 1324.

Verleur, J. (1968), *J. Opt. Soc. Am.*, Vol. 58, p. 1356.

REFERENCES FOR FURTHER READING

Bohren, C. F. and Huffman, D. R., *Absorption and Scattering of Light by Small Particles*, John Wiley & Sons, New York, 1983.

Kramers, H. A., *Atti Congr. Intern. Fis. Como.*, Vol. 2, 1927, pp. 545–557 (reprinted in H. A. Kramers, "Collected Scientific Papers," North-Holland Publ., Amsterdam, 1956).

PROBLEMS

1. Using the assumption that the plasma frequency is much greater than the relaxation frequency show that the Drude dispersion relations (5-23) and (5-24) give the results shown in Eqs. (5-26) through (5-28) for metals.

2. Use the Hagen-Rubens relation to calculate the optical constants (n and k) and normal spectral directional absorptivity for optically smooth aluminum. Use the parameters given in Example 5-2. Compare your results with the measured values given in Appendix C and the values predicted by the Drude theory in Example 5-2. Comment on the accuracy of the two models in the various spectral regions, in terms of both n and k and absorptivity.

3. Upon devolatilization, coal experiences a significant increase in the number of free electrons, and the optical constants can be described by the Drude model. The optical constants of a particular type of coal char have been curve fit with the Drude model using the following optical

parameters:

$$N = 5.66 \times 10^{21}\ \text{cm}^{-3} \qquad \gamma = 2.31 \times 10^{15}\ \text{s}^{-1}$$

$$e = 1.6022 \times 10^{-19}\ \text{C} \qquad m = 9.1096 \times 10^{-31}\ \text{kg}$$

$$\varepsilon_0 = 8.85 \times 10^{-12}\ \text{C}^2/(\text{Nm}^2)$$

(a) Based on the given electron number density calculate the plasma frequency and compare it with the relaxation frequency. Based on this comparison comment on whether char is very "metallic" in nature.

(b) Calculate the normal total emissivity and absorptivity of an "optically smooth" opaque sample of char at 1000 K in a 2000-K blackbody environment.

(c) How would the actual emissivity/absorptivity values of a rough and porous (but opaque) char particle compare with the respective values from part (b)?

4. Assume that the optical properties of water in the region $2 < \lambda < 4\ \mu\text{m}$ can be approximated by an oscillator at 3 μm corresponding to O—H molecular stretching and one at $\lambda \ll 3\ \mu\text{m}$ corresponding to bound electronic transitions. By forcing the dispersion equations to match the measured values of n and k at a limited number of wavelengths determine the dispersion parameters and use these to obtain and plot n and k and normal spectral reflectivity at other wavelengths between 2 and 4 μm.

λ (μm)	2.0	2.4	3.0	3.2	4.0
n	1.304	1.276	1.351	1.509	1.349
k	.0011	.0008	0.259	0.094	0.0048

5. The refractive index of pure, solid aluminum oxide ($k_s \ll n_s$) at 2300 K is

λ (μm)	0.5	1	2	3	4	5	6
n_s (2300 K)	1.83	1.81	1.80	1.77	1.74	1.69	1.60

and the density is $\rho_s(2300\ \text{K}) = 3.73\ \text{g/cm}^3$. The density of liquid aluminum oxide ($T_{\text{mp}} = 2320$ K) is given by

$$\rho_l = 5.632 - 1.127 \times 10^{-3} T\ (\text{K})\ \text{in g/cm}^3$$

Estimate the refractive index of molten aluminum oxide n_l at 2320 K and 1 μm wavelength. (Note that $k_l \ll n_l$ even though $k_l \gg k_s$.)

6. The optical constants of liquid aluminum oxide at 3000 K produced by combustion of aluminum in solid propellants can be represented by two oscillators with the following effective parameters (which include induced field effects):

	Oscillator 1 (μm^{-1})	Oscillator 2 (cm^{-1})
η_0	7.17	171
η_p	9.45	940
γ	0.447	104

By approximating a particle of liquid Al_2O_3 with diameter d as an equivalent slab with thickness h, show the strong dependence of particle emissivity on size by calculating the total normal emissivity of a slab of molten Al_2O_3 at 3000 K for thicknesses of 1, 10, 100, and 1000 μm.

7. The infrared optical constants of silicon carbide (SiC) at room temperature can be represented by a single oscillator with the following parameters (which include induced field effects):

n_e	2.59
η_0	793 cm^{-1}
η_p	1440 cm^{-1}
γ	4.76 cm^{-1}

Calculate the total normal absorptivity of a layer of SiC at room temperature with blackbody incident radiation at 1500 K for two layer thicknesses, 1 and 5 mm.

8. The infrared optical constants of magnesium oxide (MgO) at room temperature can be represented by two oscillators with the following effective parameters (which include induced field effects):

	Oscillator 1 (cm^{-1})	Oscillator 2 (cm^{-1})
η_0	401	640
η_p	1030	135.8
γ	7.62	102.4

$$n_e = 1.73$$

Calculate the total normal absorptivity of a 1-mm-thick homogeneous layer of MgO at room temperature that is subject to blackbody irradiation at 1000 K.

9. Obtain limiting expressions for a single Lorentz oscillator that show that at frequencies well above the oscillator frequency ($\omega \gg \omega_0$) the refractive index approaches one and the absorption index varies as ω^{-3}

$$\omega \gg \omega_0: \; n \sim 1 - \frac{\omega_p^2}{2\omega^2} \qquad k \sim \frac{\gamma \omega_p^2}{2\omega^3}$$

and at frequencies well below the oscillator frequency ($\omega \ll \omega_0$) the refractive index approaches a constant greater than one and the absorption index varies as ω

$$\omega \ll \omega_0: \; n \sim \sqrt{1 + \frac{\omega_p^2}{\omega_0^2}} \qquad k \sim \frac{\gamma \omega_p^2 \omega}{2n\omega_0^4}$$

10. The optical constants of sapphire (crystalline Al_2O_3) at room temperature ($T = 298$ K) in the region $9 < \lambda < 20$ μm can be modeled using the following dispersion parameters.

$n_e = 1.77$

$\eta_{0_1} = 420\ \text{cm}^{-1}\ (\lambda_{0_1} = 24\ \mu\text{m}) \qquad \eta_{p_1} = 900\ \text{cm}^{-1} \qquad \gamma_1 = 6\ \text{cm}^{-1}$

$\eta_{0_2} = 575\ \text{cm}^{-1}\ (\lambda_{0_2} = 17\ \mu\text{m}) \qquad \eta_{p_2} = 870\ \text{cm}^{-1} \qquad \gamma_2 = 15\ \text{cm}^{-1}$

(a) Use these parameters to calculate the optical constants over $0.4 < \lambda < 20$ μm (compare with the experimental values given in Appendix C; use the average of the ordinary and extraordinary rays).

(b) Using both theoretical (dispersion) and experimental (Appendix C) optical constants, calculate the spectral normal absorptivity for a 5-mm-thick sapphire window in the region $0.4 < \lambda < 20$ μm, and

(c) the total normal absorptivity for a 5-mm-thick sapphire window exposed to blackbody radiation at $T_e = 1000$ K.

11. Suppose that the optical behavior of SiO_2 due to bound electron transitions is to be represented by a single equivalent Lorentz oscillator located somewhere in the ultraviolet region. In order for the refractive index in the visible region to take on the correct value, namely, $n_e = 1.46$, what should be the value of η_p/η_0 for the u.v. oscillator?

12. The Drude parameters used in Example 5-2 work well in modeling the optical properties of aluminum in the infrared region but not as well in the ultraviolet region. Find a set of parameters that is more appropriate (see Appendix C) for the region $0.04 < \lambda < 0.4$ μm. What is the plasma wavelength for aluminum at 298 K?

13. The microwave optical properties of room temperature water can be modeled using the Debye model and treating the water molecules as equivalent spheres of diameter 2.75 angstroms. Estimate the normal absorptivity, reflectivity, and transmissivity of a 5-mm-thick slab of water at various wavelengths between 0.01 and 100 cm. Assume the residual refractive index on the high frequency side of the Debye region is 2.30. The low frequency (dc) dielectric constant is 77.5. The viscosity of room temperature water is 0.01 poise (g/cm s) and Boltzmann's constant is 1.3806E-16 erg/K (erg = g cm^2/s^2). Neglect interference.

CHAPTER 6

MONTE CARLO SURFACE TRANSFER

Radiative transfer in enclosures with a nonparticipating medium can usually be modeled quite accurately using the diffuse transfer analysis of Chapter 3. If greater accuracy is needed for treating nonisothermal or otherwise nonuniform surfaces, the number of nodes can be increased. In the limit as the number of nodes becomes infinite and the area of each node approaches zero, the system of simultaneous algebraic equations yields integral equations (which are usually recast as simultaneous finite difference equations to be solved anyway). Thus the exact formulation of radiative transfer between surfaces takes the form of either integral equations or high order systems of algebraic equations. Although the solution of large systems of algebraic equations is routinely done on modern computers, the determination of all the geometric view factors that are needed as input parameters can be a prohibitive exercise in itself. Furthermore, under special circumstances where nondiffuse surface behavior or polarization effects are important, the methods of Chapter 3 are not applicable.

Another technique for solving radiative transfer in enclosures with a nonparticipating medium is the Monte Carlo method [Howell, 1966; Haji-Sheikh, 1988]. With the Monte Carlo method the treatment of nondiffuse surfaces or polarization [Edwards and Tobin, 1967] is conceptually no more difficult than that for diffuse surfaces and unpolarized radiation.

The idea of the Monte Carlo method is to use probability concepts to model stochastic physical events such as photon emission, reflection, and absorption. The paths of individual photon "bundles" are tracked beginning with emission at some location and ending with absorption at some other location. To determine which direction a photon bundle should go and whether it should be absorbed or reflected when it encounters a surface, random numbers are generated and compared with appropriate probability functions. There are two basic steps that constitute a Monte Carlo solution to a radiative transfer problem:

1. Assign discrete values of energy to photon bundles.
2. Write an algorithm to track the bundles and keep score of absorptions.

ASSIGNING DISCRETE VALUES OF ENERGY

The Monte Carlo model of radiative transport consists of discrete photon bundles that transport energy between various surfaces of an enclosure. To guarantee a proper balance of energy by radiative transport, it is important to establish a consistent and clear relationship between the power emitted and absorbed by the various surfaces of the enclosure and the energy transport associated with each conceptual photon bundle. The starting point for defining such a relationship is the equation for net radiative power from each surface into the enclosure.

$$q_i A_i = \varepsilon_i e_{b_i} A_i - \alpha_i q_i^- A_i \tag{6-1}$$

The power absorbed by the surface $\alpha_i q_i^- A_i$ is represented on a discrete basis as

$$\alpha_i q_i^- A_i = w S_i \tag{6-2}$$

where w is the energy per bundle and S_i is the number of bundles absorbed (per unit time) by surface A_i. Thus Eq. (6-1) becomes

$$q_i A_i = \varepsilon_i e_{b_i} A_i - w S_i \tag{6-3}$$

The energy per bundle w is defined as the power emitted from a given surface the A_i divided by the number of bundles emitted (per unit time) by the surface N_i as shown in Eq. (6-4).

$$w = \frac{\varepsilon_i e_{b_i} A_i}{N_i} = \frac{\varepsilon_1 e_{b_1} A_1}{N_1} = \text{constant} \tag{6-4}$$

According to this definition, the energy per bundle is maintained constant for all of the bundles regardless of which surface they are emitted from. It can also be seen from this definition that the number of bundles that should be emitted from any given surface varies as the emissive power of the surface. Usually at least one of the surfaces of any enclosure will have a prescribed temperature boundary condition, for example, T_1, and this surface can be used as a reference surface. The number of bundles to be emitted from the reference surface N_1 is arbitrary but should be large enough to achieve the desired statistical accuracy. The energy per bundle w is thus uniquely determined by Eq. (6-4) as are the number of bundles that should be emitted from all other prescribed temperature surfaces.

For surfaces with prescribed temperature, the solution scheme is relatively simple. The appropriate number of bundles (N_i) is emitted from each surface. Each bundle is tracked through the enclosure until it is absorbed at some surface. A count is kept of how many bundles are absorbed at each surface (S_i) and the unknown heat transfer rate is calculated from Eq. (6-3).

For surfaces with prescribed radiative heat flux the situation is somewhat more complicated than the previous case. For *gray* surfaces with prescribed radiative heat flux a solution scheme can be formulated based on Eq. (6-3). Upon substitution of Eq. (6-4), Eq. (6-3) can be solved for the number of bundles that should be emitted from A_i giving

$$N_i = S_i + \frac{q_i A_i N_1}{\varepsilon_1 e_{b_1} A_1} \tag{6-5}$$

Emission of photon bundles continues until Eq. (6-5) is satisfied for all prescribed heat flux surfaces. The temperature for these surfaces can then be calculated from Eq. (6-4). Unfortunately, the foregoing approach does not work for *nongray* surfaces with prescribed radiative heat flux even though Eq. (6-5) is still valid. The problem is that nongray surfaces require spectral analysis, which means that a wavelength must be assigned to each photon bundle. However the procedure of assigning a wavelength to the bundle [which is given by Eq. (6-13)] requires knowledge of the surface temperature, which is unknown. Therefore enclosures containing nongray surfaces with prescribed radiative heat flux prove to be especially difficult, as was the case in the analysis of Chapter 3. However, prescribed temperature surfaces, which are either gray or nongray, and prescribed heat flux surfaces, which are gray, can be readily treated by the method described above.

ALGORITHM TO TRACK BUNDLES

The algorithm to track bundles consists of the following steps:

1. Choose a location on the enclosure surface for emission of a photon bundle.
2. Choose the wavelength of the bundle if the enclosure properties are nongray.
3. Choose the direction of propagation of the bundle.
4. Determine the new location where the bundle intersects the enclosure.

5. Decide if the bundle is reflected or absorbed at the new location.
6. If the bundle is reflected go to step 3. If the bundle is absorbed keep score and go to step 1.
7. Repeat steps 1 to 6 for many bundles.

Steps 1, 2, 3 and 5 require more detailed discussion.

Choose a Location on the Surface for Emission of the Photon Bundle. The location on a particular surface of area A where a photon bundle should be emitted can be determined from probability laws that represent the emission process. The probability of emission from a differential area dA into a small solid angle $d\Omega = \sin\theta\, d\theta\, d\phi$ in a small wavelength interval $d\lambda$ is the ratio of the power emitted from dA into the $d\Omega$ and $d\lambda$ intervals divided by the total power emitted from A into all solid angles at all wavelengths.

$$\begin{array}{l}\text{Probability of emission}\\ \text{from } dA \text{ into } d\Omega\\ \text{and } d\lambda \text{ intervals}\end{array} = \frac{\text{power emitted from } dA \text{ into } d\Omega\, d\lambda \text{ intervals}}{\text{total power emitted from } A}$$

This probability is given by the product of a *probability distribution function* $P(\lambda, \theta, \phi, \mathbf{r})$ and the small intervals $d\theta\, d\phi\, d\lambda\, dA$ where $\mathbf{r}$ represents the position vector as shown in Fig. 6-1a.

$$P(\lambda, \theta, \phi, \mathbf{r})\, d\theta\, d\phi\, d\lambda\, d\mathbf{r} = \frac{\varepsilon'_\lambda(\lambda, \theta, \phi) I_{b_\lambda}(\lambda, T) \cos\theta \sin\theta\, d\theta\, d\phi\, d\lambda\, dA}{\int_A \int_0^\infty \int_0^{2\pi} \int_0^{\pi/2} \varepsilon'_\lambda I_{b_\lambda} \cos\theta \sin\theta\, d\theta\, d\phi\, d\lambda\, dA} \tag{6-6}$$

Assuming the temperature and spectral directional emissivity are uniform over A, the denominator can be integrated to give

$$P(\lambda, \theta, \phi, \mathbf{r})\, d\theta\, d\phi\, d\lambda\, d\mathbf{r} = \frac{\varepsilon'_\lambda(\lambda, \theta, \phi) I_{b_\lambda}(\lambda, T) \cos\theta \sin\theta\, d\theta\, d\phi\, d\lambda\, dA}{\varepsilon e_b(T) A} \tag{6-7}$$

The probability of emission from dA at any wavelength and in any direction can be obtained by integrating this expression over all directions and all wavelengths giving

$$P(\mathbf{r})\, d\mathbf{r} = \frac{dA}{A} \tag{6-8}$$

The probability of emission from a subset of A that is larger than dA but smaller than A can be obtained by integrating over the subset area ∂A.

$$R(\mathbf{r}) = \int_{\partial A} P(\mathbf{r})\, d\mathbf{r} = \frac{\partial A}{A} \tag{6-9}$$

Example 6-1

What is the probability of emission from a differential rectangular strip $dA = W\,dx$ and a finite rectangular strip $\partial A = Wx$ that is a subset of a rectangular surface $A = WL$ as pictured in Fig. 6-1b?

Using $dA = Wdx$, the evaluation of Eq. (6-8) for the rectangular geometry gives

$$P(x)\, dx = \frac{dx}{L} \tag{6-10}$$

and Eq. (6-9) gives

$$R(x) = \int_0^x P(x')\, dx' = \frac{x}{L} \tag{6-11}$$

When $x = 0$ the probability of emission from $\partial A = 0$ is zero and when $x = L$ the probability of emission from $\partial A = WL$ is one.

From the preceding example it can be seen that one way to determine the x location for emission from a rectangular surface would be to generate a uniformly distributed random number R_x between 0 and 1, set it equal to the

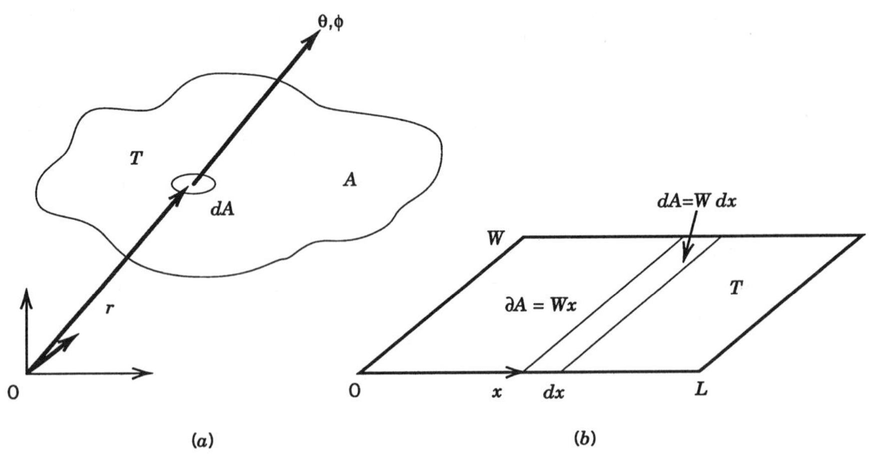

Figure 6-1 Probability of emission from (a) a small area dA and (b) a finite area ∂A.

cumulative probability function $R(x)$, and solve for x. For a large number of emission events a uniform distribution would be obtained. A simpler way to determine the x location for emission would be to uniformly distribute the emission locations over the area of the surface by subdividing the surface into many small equal-area elements and emitting the same number of bundles from each element.

Choose the Wavelength of the Bundle if the Enclosure Properties are Nongray. The probability laws discussed in the previous section can also be used to determine the wavelength that should be assigned to a photon bundle in the case of a spectral analysis involving nongray enclosure properties. The probability of emission from any location on a given isothermal surface in the interval $d\lambda$ and in any direction is the ratio of the spectral power emitted in all directions to the total power emitted in all directions.

$$\text{Probability of emission in } d\lambda = \frac{\text{power emitted in } d\lambda}{\text{total power emitted}}$$

This probability can be represented mathematically by integrating Eq. (6-6) over the entire area of the surface and over all directions giving

$$P(\lambda)\,d\lambda = d\lambda \int_A \int_0^{2\pi} \int_0^{\pi/2} P(\lambda, \theta, \phi, \mathbf{r})\,d\theta\,d\phi\,dA = \frac{\varepsilon_\lambda e_{b_\lambda}}{\varepsilon e_b}\,d\lambda \quad (6\text{-}12)$$

The probability of emission in the wavelength interval between 0 and λ (in any direction) is the integral of (6-12) between 0 and λ.

$$R(\lambda) = \int_0^\lambda P(\lambda')\,d\lambda' = \frac{1}{\varepsilon e_b}\int_0^\lambda \varepsilon_\lambda e_{b_\lambda}\,d\lambda' = \frac{1}{\varepsilon}\int_0^{f(\lambda T)} \varepsilon_\lambda\,df(\lambda' T) \quad (6\text{-}13)$$

$R(\lambda)$ is the cumulative probability function for emission in the wavelength band between 0 and λ. Its value ranges between 0 and 1. For a gray emitter (6-13) reduces to the fractional function.

$$R(\lambda) = f(\lambda T) \quad \text{(gray emitter)} \qquad (6\text{-}14)$$

The utility of using the cumulative probability function is that a pseudo-random number R_λ uniformly distributed between 0 and 1 can be easily generated by a computer and set equal to $R(\lambda)$ in Eq. (6-13). Then (6-13) can be solved for λ to determine the wavelength that should be assigned to the bundle. In this way the correct distribution of wavelengths will be assigned to the emitted bundles to correspond to the spectral distribution of the actual emitted energy. For example, if ε_λ is larger at certain wavelengths near $\lambda = \lambda_0$ (and/or $e_{b_\lambda}(T)$ peaks near λ_0), the choice of λ will be biased so that it is more likely to produce a value of λ near λ_0. One drawback of the present formulation [Eq. (6-14)] is that it requires implicit evaluation of the fractional function to obtain λT from a given random number. To overcome this problem Haji-Sheikh and Sparrow [1969] presented a polynomial curve fit for λT as a function of f that allows explicit evaluation of λT.

Choose the Direction of Propagation of the Bundle. To choose the direction of emission from a given differential area dA the probability of emission into a particular direction will be used. Two angles must be specified, polar angle θ and azimuthal angle ϕ. Consider first the probability of emission into a small polar solid angle in any azimuthal direction as pictured in Fig. 6-2

$$\text{Probability of emission into } d\theta = \frac{\text{power emitted into } d\theta}{\text{total power emitted}}$$

The probability density function $P(\theta)$ can be obtained by integrating (6-6) over all wavelengths, azimuthal angles, and area.

$$P(\theta)\,d\theta = d\theta \int_A \int_0^{2\pi} \int_0^{\lambda} P(\lambda, \theta, \phi, \mathbf{r})\,d\lambda\,d\phi\,dA = \frac{2\varepsilon' \cos\theta \sin\theta\,d\theta}{\varepsilon} \tag{6-15}$$

In performing the integration to obtain Eq. (6-15), it has been assumed that ε' is independent of ϕ. Otherwise the directional emissivity appearing in (6-15) represents an azimuthally averaged value of the directional emissivity.

The cumulative probability function is the integral of (6-15) with respect to θ between 0 and θ.

$$R(\theta) = \int_0^{\theta} P(\theta')\,d\theta' = \frac{1}{\varepsilon} \int_0^{\sin^2\theta} \varepsilon'(\theta')\,d(\sin^2\theta') \tag{6-16}$$

$R(\theta)$ represents the probability of emission at any wavelength in the direction interval between 0 and θ. Its value ranges between 0 and 1, depending on the value of θ. The strategy for determining the polar angle θ for emission is to generate a uniformly distributed random number R_θ between 0 and 1, set it

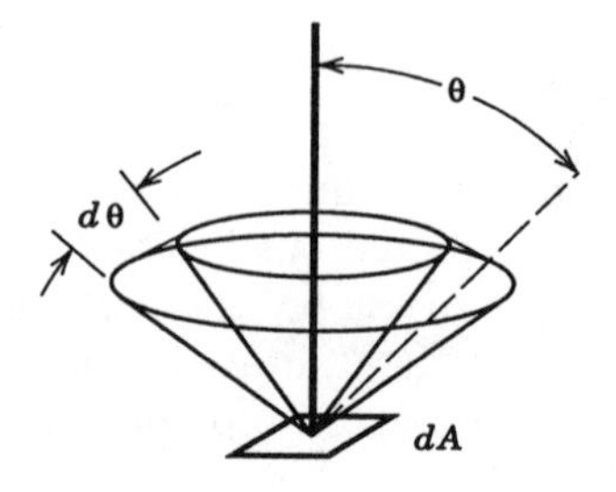

Figure 6-2 Probability of emission into small polar angle.

equal to $R(\theta)$ and solve for θ. For a diffuse emitter Eq. (6-16) reduces to

$$R(\theta) = \sin^2\theta \quad \text{(diffuse emitter)} \tag{6-17}$$

Similar arguments apply to the determination of the azimuthal angle ϕ. For most problems the directional emissivity will be independent of azimuthal angle and the cumulative probability for emission between 0 and ϕ is $\phi/2\pi$.

$$R(\phi) = \frac{\phi}{2\pi} \quad \text{(azimuthal symmetry)} \tag{6-18}$$

Equations (6-17) and (6-18) also apply for the determination of the reflectance angles for a surface that is a diffuse reflector.

Decide if Bundle Is Reflected or Absorbed. When a photon bundle intersects a surface of the enclosure, it is necessary to decide if the bundle is absorbed or reflected. To do this a random number is generated and compared with the directional absorptivity. The directional absorptivity is determined from the known emissivity for the particular direction of incidence.

$$\alpha'_{\lambda i}(\lambda, \theta, \phi) = \varepsilon'_{\lambda i}(\lambda, \theta, \phi) \tag{6-19}$$

This value is then compared with a uniformly distributed random number between zero and one (R) to determine if the bundle is absorbed or reflected.

$$\alpha'_{\lambda i} < R \quad \text{(reflected)}$$

$$\alpha'_{\lambda i} > R \quad \text{(absorbed—increment } S_i\text{)}$$

The Monte Carlo technique is best understood by considering specific examples. Thus the following illustrative example will be used to demonstrate the method. In Chapter 3 the heat transfer through a two-dimensional rectangular slot in a diffuse, gray refractory wall was solved using three- and four-node approximations (Example 3-7 and Problem 3-2). It was demonstrated that these low order approximations fail to predict that the heat flux goes to zero at large aspect ratios $(L/W \to \infty)$ due to the fact that the adiabatic wall was not divided into enough nodes to satisfy the uniform radiosity assumption along the wall.

Example 6-2

Reconsider the slot problem shown in Fig. 6-3 using the Monte Carlo method, which is effectively an infinite node solution, and solve for the dimensionless heat flux through the slot.

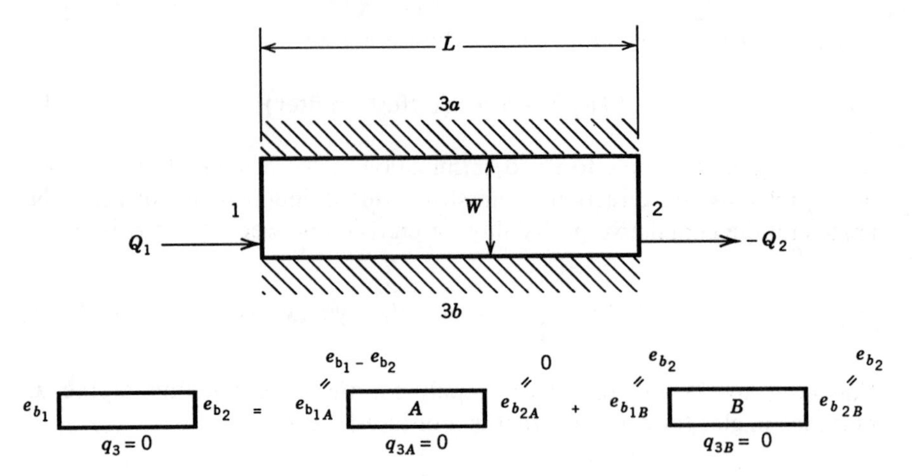

Figure 6-3 Two-dimensional refractory slot with black end walls (superposition formulation).

It is useful to make use of the superposition principle to simplify the formulation of this problem. Recall that if the surface properties are assumed to be independent of temperature, the governing equations from Chapter 3 are linear and therefore the powerful superposition principle can be used. Making use of this principle the general problem depicted in Fig. 6-3 can be formulated as the sum of two simpler problems, an A problem and a B problem. The A problem consists of a slot with a source blackbody hemispherical flux of $e_{b_1} - e_{b_2}$,

$$e_{b_{1A}} = e_{b_1} - e_{b_2} \tag{6-20}$$

and a 0 K blackbody sink.

$$e_{b_{2A}} = 0 \tag{6-21}$$

The B problem consists of the trivial problem of a slot with "source" and "sink" both having blackbody hemispherical flux of e_{b_2}.

$$e_{b_{1B}} = e_{b_{2B}} = e_{b_2} \tag{6-22}$$

By linear superposition, any of the quantities q_i^+, q_i^-, q_i or e_{b_i} in the general problem can be written as the sum of the corresponding quantities from the A and B problems. Thus the net radiative heat transfer is

$$Q_1 = Q_{1A} + Q_{1B} = Q_{1A} \tag{6-23}$$

since the heat transfer in the B problem is zero. It remains to determine the heat transfer for the A problem only. The net heat transfer rate for the A problem ($Q_{1A} = -Q_{2A}$) from Eq. (6-1) is

$$Q_{1A} = \varepsilon_1 e_{b_{1A}} A_1 - \alpha_1 q_{1A}^- A_1 \tag{6-24}$$

where the term representing the absorbed power can be replaced by the number of bundles absorbed at A_1 times the energy per bundle to give

$$Q_{1A} = \varepsilon_1 e_{b_{1A}} A_1 - w S_1 \tag{6-25}$$

The energy per bundle, from Eq. (6-4), is

$$w = \frac{\varepsilon_1 e_{b_{1A}} A_1}{N_1} \tag{6-26}$$

Substituting this expression into (6-25) and factoring the common term $\varepsilon_1 e_{b_{1A}} A_1$ gives

$$Q_{1A} = \varepsilon_1 e_{b_{1A}} A_1 \left(1 - \frac{S_1}{N_1}\right) \tag{6-27}$$

If it is assumed that the refractory surface is nonemitting and nonabsorbing ($\varepsilon_3 = 0, \rho_3 = 1$), the only surface from which bundles will be emitted is A_1. This assumption is justified since the emissivity of a diffuse, gray, adiabatic surface can be set to any value between 0 and 1 (see Chapter 3). Conservation of bundles thus requires that

$$N_1 = S_1 + S_2 \tag{6-28}$$

Therefore, Eq. (6-27) can be written, using (6-20), (6-23), (6-28) and the fact that $\varepsilon_1 = 1$, as

$$\frac{Q_1}{(e_{b_1} - e_{b_2}) A_1} = \frac{S_2}{N_1} \tag{6-29}$$

The nondimensional heat flux is the fraction of bundles emitted from surface 1, which are absorbed at surface 2. To solve the problem it is only necessary to emit bundles from 1 and count how many are absorbed at 2.

There are two geometrical configurations that must be analyzed to solve this problem. The first configuration corresponds to emission from A_1 and the second corresponds to reflection of bundles from A_3. In Fig. 6-4 the geometry is shown for a bundle or ray leaving A_1 going to A_{3b}. The relation between the incident angles θ_3, ϕ_3 and the emission angles θ_1, ϕ_1 can be

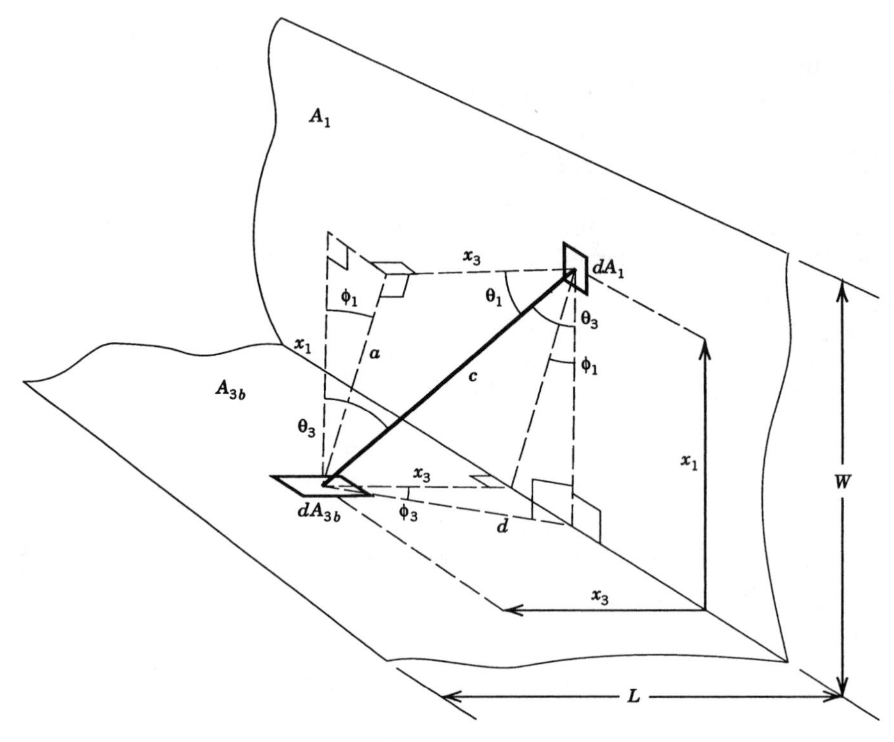

Figure 6-4 Geometry for emission from vertical wall to horizontal plane.

obtained from the identities

$$\frac{x_1}{c} = \frac{a}{c} \cdot \frac{x_1}{a} \to \cos\theta_3 = \sin\theta_1 |\cos\phi_1| \tag{6-30a}$$

$$\frac{x_3}{d} = \frac{x_3}{c} \cdot \frac{c}{d} \to \cos\phi_3 = \cos\theta_1 \frac{1}{\sin\theta_3} \tag{6-30b}$$

which follow from the right triangles pictured in Fig. 6-4. The absolute value is taken on $\cos\phi_1$ in Eq. (6-30a) because over the interval $\pi/2 < \phi_1 < 3\pi/2$, $\cos\phi_1 < 0$. From the triangles in Fig. 6-4 the tangent of θ_1 can be expressed as

$$\tan\theta_1 = \frac{x_1/\cos\phi_1}{x_3} \qquad (0 < \phi_1 < \pi/2 \quad \text{or} \quad 3\pi/2 < \phi_1 < 2\pi) \tag{6-31}$$

which can be solved for x_3 in terms of the initial position x_1 and the emission angles θ_1 and ϕ_1.

$$x_3 = \left| \frac{x_1}{\cos \phi_1 \tan \theta_1} \right| \tag{6-32}$$

The absolute value is taken in (6-32) because of the sign change associated with $\cos \phi_1$ over the interval $\pi/2 < \phi_1 < 3\pi/2$,

$$x_3 = \frac{x_1}{-\cos \phi_1 \tan \theta_1} \qquad (\pi/2 < \phi_1 < 3\pi/2) \tag{6-33}$$

By symmetry the interval for ϕ_1 can be restricted to $0 < \phi_1 < \pi$ without affecting the result of the calculation.

Next the configuration for bundles reflected at A_{3b} is considered. This geometry is shown in Fig. 6-5. The interval for θ_3 is $0 < \theta_3 < \pi/2$. Since the surface is a diffuse reflector, this angle is determined from the same equation that applies to a diffuse emitter, namely, Eq. (6-17).

$$\sin \theta_3 = \sqrt{R_{\theta_3}} \tag{6-34}$$

By symmetry the interval for ϕ_3 is limited to $0 < \phi_3 < \pi$ and this angle is

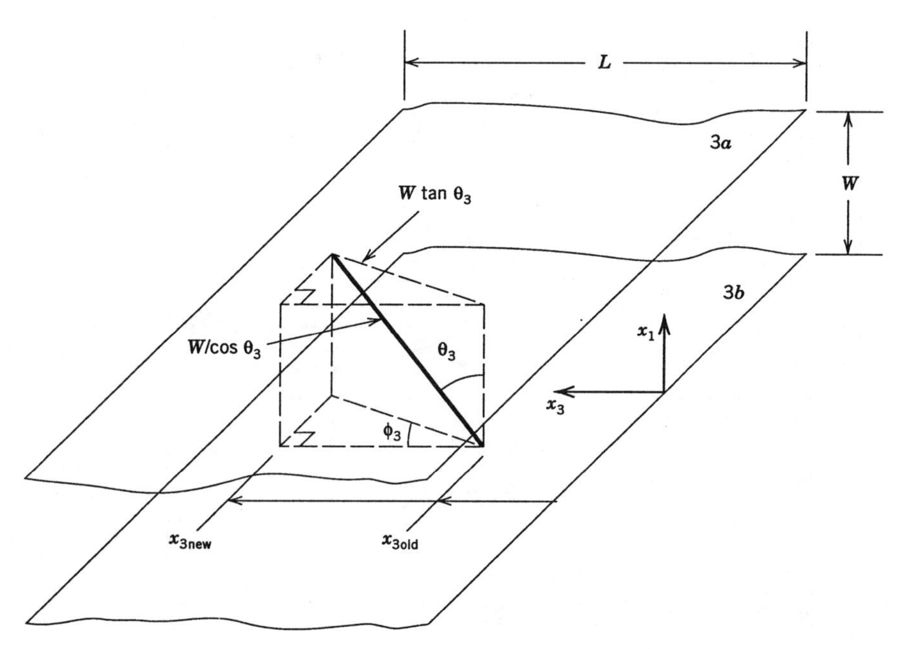

Figure 6-5 Geometry for reflection from $3b$ to $3a$ in Example 6-2.

determined from

$$\phi_3 = \pi R_{\phi_3} \tag{6-35}$$

The length of the ray from the point of reflection on 3*b* to the point of incidence on 3*a* is $W/\cos\theta_3$. The projection of this line segment into the plane of A_{3b} or A_{3a} is $W\tan\theta_3$. The projection into the direction of the x_3 coordinate, which is the difference between the new x_3 position and the old

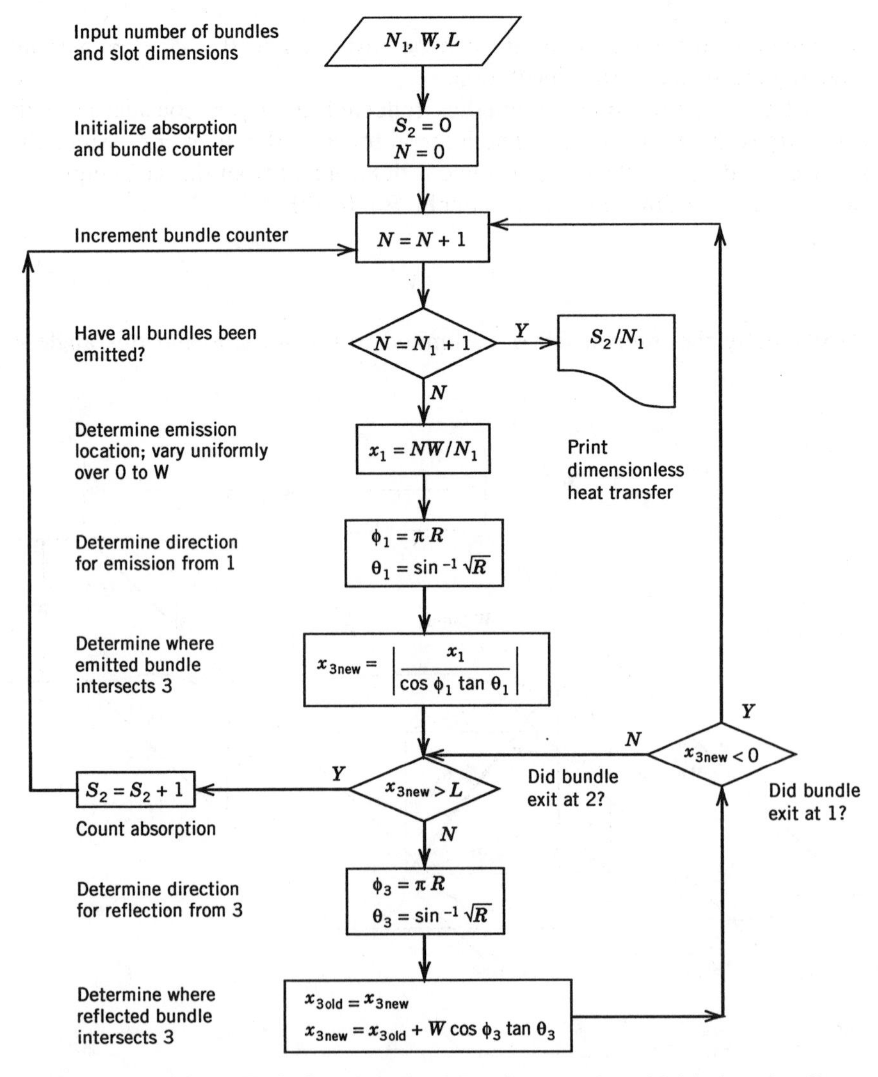

Figure 6-6 Monte Carlo flowchart for 2-D refractory slot with black end walls.

x_3 position is $W \tan \theta_3 \sin(\pi/2 - \phi_3)$.

$$x_{3_{\text{new}}} = x_{3_{\text{old}}} + W \tan \theta_3 \cos \phi_3 \tag{6-36}$$

This completes the formulation of the necessary geometrical relationships. The case of a bundle reflected at $3a$ toward $3b$ is the same as the case just considered and the same relationships can be used.

The algorithm for solving the Monte Carlo problem and tracking bundles is given in Fig. 6-6. The strategy of the algorithm is to keep track of the x_3 position of the bundles. The lateral position of the bundles does not matter because the slot is two dimensional. When the x_3 position of a bundle exceeds the length L, the bundle is absorbed by the black surface A_2 and the absorption counter S_2 is incremented. When x_3 is less than zero, the bundle is absorbed by the black surface A_1. After the prescribed number of bundles N_1 have all been emitted, the ratio S_2/N_1 is calculated to determine the dimensionless heat flux. The programming of this problem is left as an exercise.

The solution of this example is plotted in Figs. 6-7 and 6-8. Figure 6-7 compares the Monte Carlo solution, which is effectively an infinite node solution, with the three- and four-node approximate solutions from Chapter 3 and shows how the Monte Carlo solution gives the correct limiting behavior of heat flux going to zero as the aspect ratio $L/(2W)$ goes to infinity. Figure 6-8 shows the effect of the number of bundles on the accuracy of the Monte Carlo solution for an aspect ratio of 10. As the number of bundles increases, the error in the Monte Carlo solution diminishes.

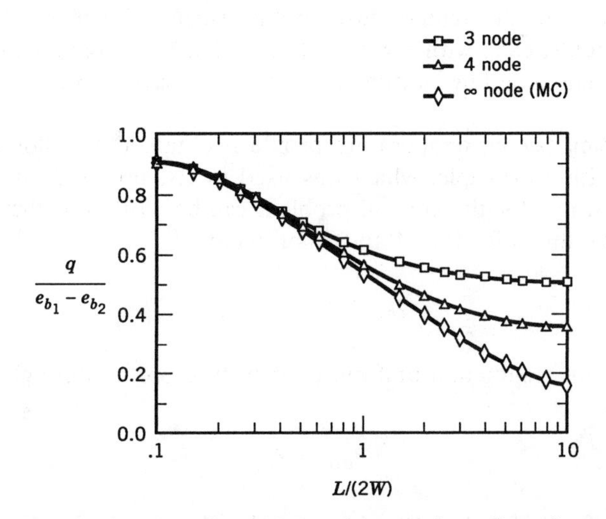

Figure 6-7 Dimensionless heat flux of 2-D refractory slot; effect of number of nodes.

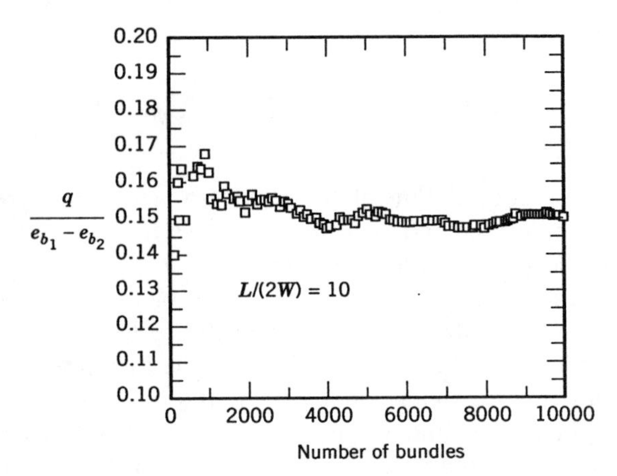

Figure 6-8 Dimensionless heat flux of 2-D refractory slot; effect of number of bundles on Monte Carlo accuracy.

Example 6-3

For the previous refractory slot problem, find the nondimensional temperature profile $(e_{b_3}(x_3) - e_{b_2})/(e_{b_1} - e_{b_2})$.

To determine the temperature profile of the refractory surface, it is necessary to change the assumption of the diffuse, gray refractory surface being a perfect reflector, because otherwise the temperature would be indeterminate, being decoupled from the radiative transfer. Instead, it will be assumed that the refractory surface is a perfect absorber-emitter.

To characterize the temperature profile, surface 3 will be divided into equal increments, each with a width of Δx_3. The adiabatic boundary condition will be maintained by emitting a bundle from surface 3 every time one is absorbed.

The nondimensional temperature profile is obtained as follows. Applying the superposition principle, which was used in Example 6-2, the blackbody hemispherical flux for the general problem can be written as the sum of the corresponding fluxes for the A and B problems.

$$e_{b_{3_i}} = e_{b_{3A_i}} + e_{b_{3B_i}} \tag{6-37}$$

The B problem results in a uniform temperature profile through the refractory wall.

$$e_{b_{3B_i}} = e_{b_2} \tag{6-38}$$

The profile of $e_{b_{3A_i}}$ for the A problem can be obtained by invoking Eq. (6-4), which defines the energy per bundle.

$$w = \frac{\varepsilon_3 e_{b_{3A_i}} A_{3_i}}{N_{3_i}} = \frac{\varepsilon_1 e_{b_{1A}} A_1}{N_1} \tag{6-39}$$

Solving for $e_{b_{3A_i}}$ and using $e_{b_{1A}} = e_{b_1} - e_{b_2}$ and $\varepsilon_1 = \varepsilon_3 = 1$ gives the nondimensional temperature profile as

$$\frac{e_{b_{3_i}} - e_{b_2}}{e_{b_1} - e_{b_2}} = \frac{N_{3_i}}{N_1} \frac{A_1}{A_{3_i}} \tag{6-40}$$

where $N_{3_i} = S_{3_i}$. The number of bundles emitted by each element of surface 3 is equal to the number of bundles absorbed at that element. This number can be determined by setting up a counter for each element and incrementing the counter when a bundle is absorbed at that element. The flowchart from Example 6-2 can also be used in this problem. The logic for representing the process of perfect, diffuse reflection (Example 6-2) is the same as that required for representing perfect, diffuse emission in this calculation. In fact, the heat transfer calculation of Example 6-2 would have been just the same if the refractory surface had been assumed to be a perfect emitter. The only significant change that is necessary in the flowchart of Fig. 6-6 to solve this problem is the inclusion of logic to find out which increment of surface 3 the bundles are absorbed at so that the counter representing N_{3_i} can be appropriately incremented. This programming is left as an exercise. The nondimensional temperature profile $[(e_{b_3}(x_3) - e_{b_2})/(e_{b_1} - e_{b_2})]$ is plotted for $\Delta x_3 = L/10$ in Fig. 6-9

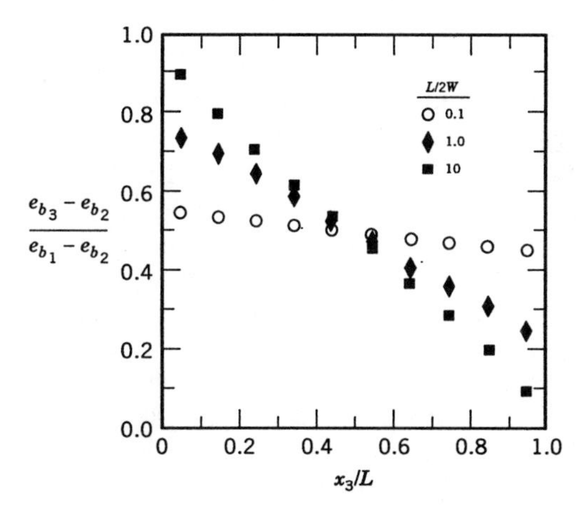

Figure 6-9 Monte Carlo temperature profile for diffuse, gray refractory slot.

ACCURACY OF MONTE CARLO METHOD

An important question with regard to the Monte Carlo method is the accuracy of the predicted solutions. Example 6-2 demonstrates that the Monte Carlo predictions fluctuate appreciably (see Fig. 6-8). The magnitude of the fluctuations depends on the number of bundles used. As the number of bundles increases the magnitude of the fluctuations decreases. One would like to know how to estimate the number of bundles necessary to guarantee a certain degree of accuracy. In general this is very difficult to do. However, some guidelines can be given.

The predicted Monte Carlo results can be regarded as a random variable x. In the context of Example 6-2 this random variable is nondimensional heat flux $x = S_2/N_1$. Different Monte Carlo experiments give different values x_i depending on the random numbers selected and the fate of the individual photon bundles. In the context of Example 6-2, if the photon bundle exits out surface 2, then $x_i = 1$. If the bundle exits out surface 1, then $x_i = 0$. If N experiments are run, the mean value $x_{m,N}$ based on N experiments can be calculated as

$$x_{m,N} = \frac{\sum_{i=1}^{N} x_i}{N} \tag{6-41}$$

As $N \to \infty$ the indicated mean approaches the true mean value, $x_{m,\infty}$. A measure of the deviation of the indicated mean $x_{m,N}$ away from the true mean $x_{m,\infty}$ is the standard deviation, which is defined as

$$s = \sqrt{\frac{\sum_{i=1}^{N} (x_i - x_{m,N})^2}{N - 1}} \tag{6-42}$$

The probability of the indicated mean lying within a given interval about the true mean is directly related to the standard deviation. For example, the probability of the indicated mean lying within $\pm s$ of the true mean is 68%. The probability of lying within $\pm 2s$ is 95%, etc. Equation (6-42) shows that for a large number of samples ($N \gg 1$) the standard deviation decreases as the inverse square root of the number of samples. Thus the accuracy of the method increases as the number of samples increases. These statistical results give an indication of how much improvement in accuracy can be expected for a given increase in the number of photon bundles emitted.

The other factor that influences accuracy, besides statistical uncertainty, is the uniformity of the distribution of random numbers used. Computer microprocessors do not generate true random numbers but rather they generate pseudo-random numbers using an algorithm. A common approach is to start with a seed number, perform some operation (such as taking a power or product), extract a portion of the digits of the result, and use this portion as the new random number. This new number then becomes the seed for the next random number. Such algorithms seeks to exploit the apparent randomness associated with certain mathematical operations. The term pseudo-random number is used to emphasize that the numbers are not really random but result from a deterministic process. The problem with such algorithms is that they are never truly random and eventually, over a large number of operations, patterns emerge. If these patterns correlate with the sequence of operations in the Monte Carlo algorithm the accuracy of the predicted results will be compromised. For example, suppose in the Monte Carlo algorithm that the step to decide whether a bundle is absorbed or reflected is followed by the decision of which direction to go if reflected (a common logical sequence). If there is any correlation between these two decisions then the direction of the reflected photons will be biased by the fact that this decision is made only when reflection occurs (i.e., when certain random numbers appear in the preceding step).

The best rule of thumb to observe to ensure maximum uniformity of computer-generated pseudo-random numbers and minimum execution time is to use a generator that was developed for a particular processor. The number of bits that represents a number in a processor influences the performance of some algorithms. Since the number of bits that represents a number varies from processor to processor, it is best to use a random number generator developed for a particular processor.

Accuracy of Monte Carlo predictions can also be improved by a number of shortcuts and strategies other than the one outlined in this chapter. One shortcut already mentioned is to use symmetry when possible to reduce the range of a variable. Another scheme is to allow the energy of the photon bundles to vary instead of forcing it to be constant. In this scheme, instead of following each bundle until it is completely absorbed, a portion of the energy in the bundle is removed by absorption at each interaction with a surface until a negligible amount of energy remains. Another scheme for reducing the number of bundles necessary to achieve a given degree of accuracy is called energy partitioning [Shamsundar, et al., 1973; Sparrow, et al., 1974]. In this scheme advantage is taken of the fact that some energy goes directly from one surface to another where it is absorbed without any intervening reflections. This energy can be accounted for by using the view factors discussed in Chapter 3. Thus the energy of a photon bundle is partitioned into that which is transferred directly and that which undergoes intermediate reflections. The savings in computer time that result from this scheme are demonstrated in Fig. 6-10, which is taken from [Shamsundar, et al., 1973]. Figure 6-10 shows the effective emissivity of an isothermal conical cavity with half apex angle of 20° for bulk emissivities of 0.5 and 0.9. The open symbols show the results for the conventional Monte Carlo method and the dark symbols show the results for the energy partitioning method. For a given

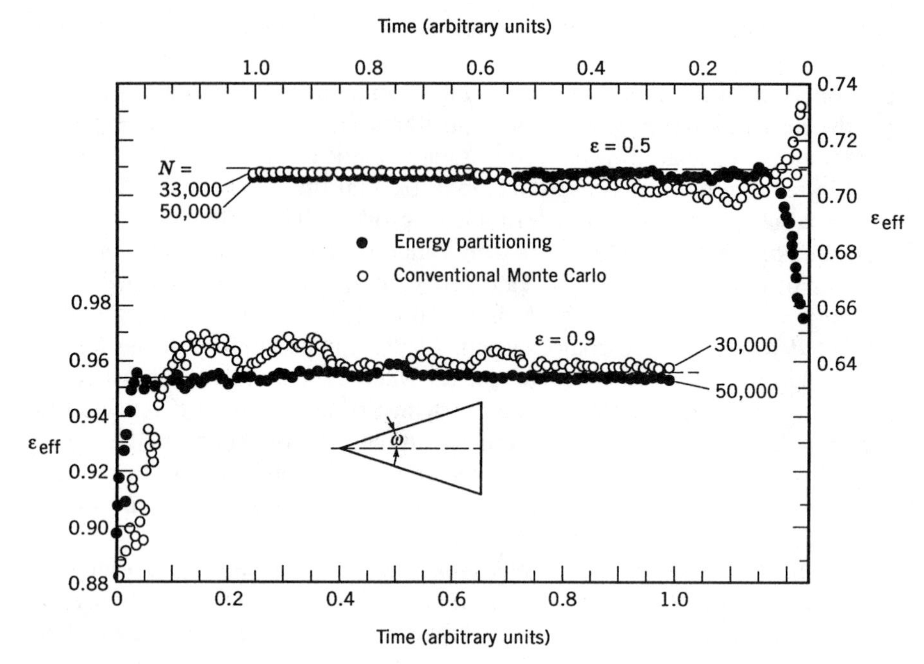

Figure 6-10 Effective emissivity for a conical cavity ($\omega = 20°$); effect of energy partitioning (from Shamsundar, et al., 1973).

CPU time (e.g., one unit), the energy partitioning method is able to process more bundles (50,000 compared with 30,000 for $\varepsilon = 0.9$) than the conventional method and thus achieve greater accuracy.

SUMMARY

The Monte Carlo method is a powerful tool for solving radiative transfer problems. Perhaps the greatest potential for this method is in the solution of nondiffuse, nongray transfer in complex geometries. The increase in complexity that occurs for nondiffuse, nongray surfaces is relatively minor as far as programming is concerned. Even polarization effects can be easily incorporated into the Monte Carlo formulation.

Some of the main drawbacks to the use of the Monte Carlo method are computer costs and speed, but these problems are rapidly becoming less of a factor and Monte Carlo calculations are finding wider acceptance in many disciplines, not just in radiative transfer calculations.

Another drawback of Monte Carlo radiative transfer calculations is the geometrical aspect. For geometries of significant complexity, analytical representation of the geometry is almost as formidable a task as determining many unknown view factors in the diffuse enclosure analysis. This obstacle will probably be overcome in the future with the development of multidimensional geometric modelers for computer-aided design applications.

REFERENCES

Edwards, D. K. and Tobin, R. D. (1967), "Effect of Polarization on Radiant Heat Transfer through Long Passages," *J. Heat Transfer*, Vol. 89, pp. 132–138.

Haji-Sheikh, A. (1988), "Monte Carlo Methods," *Handbook of Numerical Heat Transfer*, W. J. Minkowycz, E. M. Sparrow, R. H. Pletcher, and G. E. Schneider, Eds., Wiley, New York.

Haji-Sheikh, A. and Sparrow, E. M. (1969), "Probability Distributions and Error Estimates for Monte Carlo Solutions of Radiation Problems," *Prog. Heat Mass Transfer*, Vol. 2, pp. 1–22.

Howell, J. R. (1966), "Application of Monte Carlo to Heat Transfer Problems," *Advances in Heat Transfer*, Vol. 5, pp. 1–54.

Shamsundar, N., Sparrow, E. M., and Heinisch, R. P. (1973), "Monte Carlo Radiation Solutions—Effect of Energy Partitioning and Number of Rays," *Int. J. Heat Mass Transfer*, Vol. 16, pp. 690–694.

Sparrow, E. M., Heinisch, R. P., and Shamsundar, N. (1974), "Apparent Hemispherical Emittance of Baffled Cylindrical Cavities," *J. Heat Transfer*, Vol. 96, No. 1, pp. 112–114.

REFERENCES FOR FURTHER READING

Edwards, D. K., *Radiation Heat Transfer Notes*, Hemisphere Publishing Corporation, New York, 1981.

Siegel, R. and Howell, J. R., *Thermal Radiation Heat Transfer*, 2nd Ed., Taylor, Francis, Hemisphere, 1988.

Toor, J. S. and Viskanta, R., "A Numerical Experiment of Radiant Heat Exchange by the Monte Carlo Method," *Int. J. Heat and Mass Transfer*, Vol. 11, pp. 883–897, 1968.

PROBLEMS

1. Reconsider the two-dimensional rectangular crack in a furnace wall from Example 6-2. The wall thickness (crack length) is L and the crack width is W. The refractory wall material is diffuse and gray. The temperatures inside and outside the furnace are T_1 and T_2, respectively. Write a

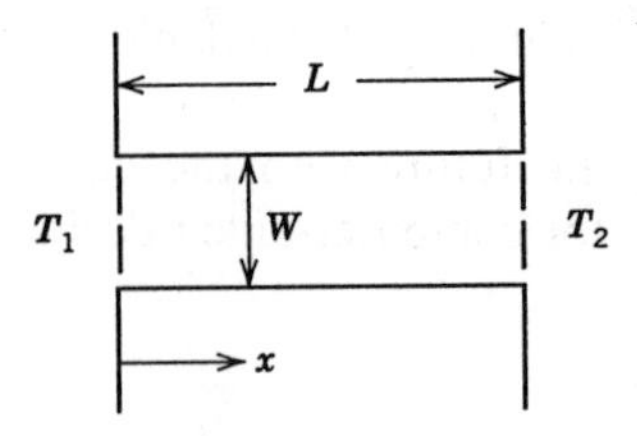

Figure P1

computer program to solve for the dimensionless heat transfer $q/(e_{b_1} - e_{b_2})$ and plot this as a function of $L/(2W)$ for $L/(2W) = 0.1$ to 10. Compare the results with Fig. 6-7

2. Write and execute a computer program to find the dimensionless temperature profile in Example 6-3 for $\Delta x = L/10$ and $L/(2W) = 0.1$, 1, and 10. Compare the predictions with the 3 and 4 node predictions from Examples 3-7 and 3-8 and with the results given in Fig. 6-9.

3. Reconsider the two-dimensional rectangular slot in an isothermal wall from Example 3-6 and Problem 3-3. The slot length is L and the width is W. Use the Monte Carlo method to calculate the effective emissivity/absorptivity of the slot for $L/W \to \infty$ and for wall emissivities of 0.25, 0.50, and 0.75. Compare these exact results with the approximate results from Example 3-6.

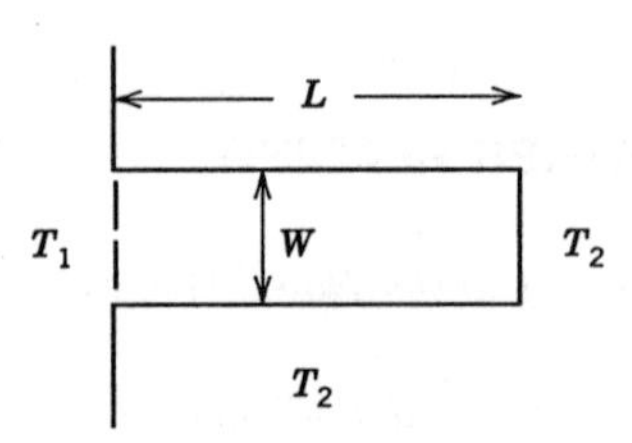

Figure P3

4. Using probability concepts, derive the integral expressions for the cumulative probability functions $R(\theta_r)$ and $R(\phi_r)$ that would be used in a Monte Carlo analysis to determine the reflectance angles θ_r and ϕ_r for a nondiffuse reflector. Verify that these expressions reduce to the appropriate results for a diffuse reflector.

5. A flat surface of length L has a temperature profile given by

$$T(x) = T_L\left(\frac{x}{L}\right)^{1/4}$$

Assuming the surface emissivity is independent of temperature (i.e., independent of x), derive the appropriate cumulative probability function $R(x)$ that could be used in a Monte Carlo analysis to determine the x location for emission of photon bundles.

6. A circular disc (radius r_0) with uniform temperature and emissivity is emitting thermal radiation. Derive an expression that could be used to determine the radial location for emission of photon bundles using a random number between zero and one.

7. Suppose that thermal emission from a surface was modeled by emitting photon bundles with a uniform distribution over both azimuthal *and* polar angles.

$$\phi = R_\phi 2\pi \qquad \theta = R_\theta \frac{\pi}{2}$$

What would be the corresponding variation in directional emissivity ε' with respect to polar angle θ for such a surface? Obtain an expression for ε' in terms of ε and θ.

8. Apply the Monte Carlo technique to Problem 2-17. Compare the Monte Carlo predictions for effective envelope transmissivity with the exact analytical result from Problem 2-17

$$\tau_{\text{eff}} = \frac{\pi I_2}{\pi I_1} = \frac{\tau}{\tau + \alpha}$$

where I_1 is the intensity emitted by the point source, I_2 is the intensity transmitted through the envelope, and τ and α are the intrinsic transmissivity and absorptivity of the envelope. Use $\alpha = 0.1$.

9. Use the Monte Carlo method to calculate the effective emissivity/absorptivity of an isothermal slot (see Problem 6-3) as a function of L/W for $L/W = 0$ to ∞ and for a wall emissivity of 0.25. Compare these exact results with the approximate results from Example 3-6.

CHAPTER 7

RADIATIVE TRANSFER EQUATION

Radiative transfer is usually classified as being either transfer between surfaces or transfer in participating media. However, all radiative transfer actually involves participating media. Even opaque surfaces are participating media in which the radiative emission, absorption and scattering are concentrated in a thin region near the surface of the material. Radiative transfer in participating media is governed by the transfer equation. In this chapter the radiative transfer equation and its integral along a single line of sight are discussed. The properties pertinent to radiative transfer in participating media are defined.

The transfer equation is an optical energy balance on a differential volume element along a single line of sight. Three types of processes can influence this optical energy balance: absorption, emission, and scattering. Absorption and emission are relatively easy to represent mathematically. Scattering is more difficult. The reason is that scattering is a mechanism that transfers energy from one line of sight to another. In accordance with the increased mathematical complexity in going from absorbing/emitting media to scattering media, the transfer equation is formulated in steps. First, a cold, nonemitting, nonscattering medium that only absorbs radiation is considered. Then the effects of emission, out-scattering, and finally in-scattering are included.

DERIVATION OF THE TRANSFER EQUATION

From the results of electromagnetic theory it was shown that the intensity of a plane wave propagating in a homogeneous absorbing medium with constant

n and k is attenuated as $\exp[-K_{a_\lambda} s]$ where K_{a_λ} is the spectral absorption coefficient for a homogeneous medium and s is the path length.

$$K_{a_\lambda} = \frac{4\pi k}{\lambda_0} \quad \text{(homogeneous medium)} \tag{7-1}$$

This exponential attenuation suggests that, in the absence of emission and scattering, intensity I_λ obeys the differential equation

$$\frac{dI_\lambda}{ds} = -K_{a_\lambda} I_\lambda \tag{7-2}$$

Equation (7-2) is the transfer equation for an absorbing, nonemitting, nonscattering medium. It is essentially an optical energy balance on a differential element along a single line of sight. The path length of the element is ds and the area is dA (Fig. 7-1). Equation (7-2) says that the change in intensity with distance is proportional to the incident intensity and that the proportionality constant is the absorption coefficient. The solution of (7-2) can be obtained by using an integrating factor. For the general case of variable absorption coefficient the solution is

$$I_\lambda(s) = I_{0_\lambda} \exp\left[-\int_0^s K_{a_\lambda}(s')\, ds'\right] \tag{7-3}$$

where I_{0_λ} is the incident intensity at $s = 0$. If K_{a_λ} is constant along s (homogeneous medium) the solution reduces to

$$I_\lambda = I_{0_\lambda} \exp\left[-K_{a_\lambda} s\right] \tag{7-4}$$

which agrees with the result of electromagnetic theory for a homogeneous medium.

Equation (7-2) also gives a new mathematical interpretation of the absorption coefficient, namely, the limit of the fraction of energy absorbed in a

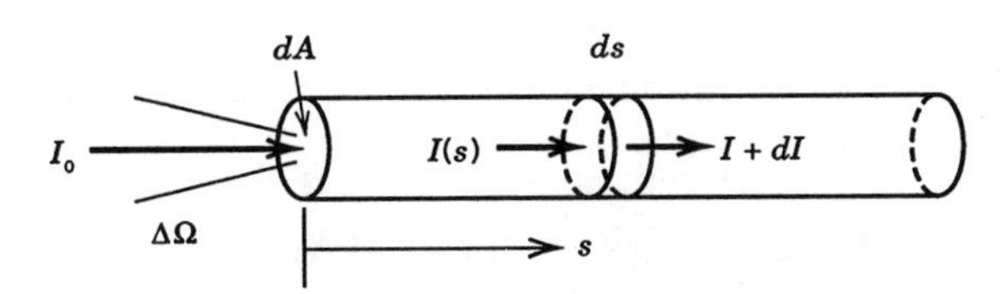

Figure 7-1 Optical energy balance along a line of sight.

small volume element divided by Δs as Δs approaches zero.

$$K_{a_\lambda} = \lim_{\Delta s \to 0} \frac{\text{energy absorbed in } \Delta V}{(\text{energy incident on } \Delta V)\Delta s} \tag{7-5}$$

The engineering radiative property definitions from Chapter 2 can now be applied to the column of matter between $s = 0$ and s in Fig. 7-1. Accordingly, the directional transmissivity and absorptivity of a homogeneous layer or column of participating medium are as follows:

$$\tau'_\lambda = \text{transmissivity} = \frac{I_\lambda \, \Delta\Omega \, dA}{I_{0_\lambda} \, \Delta\Omega \, dA} = \exp\left[-K_{a_\lambda} s\right] \tag{7-6}$$

$$\alpha'_\lambda = \text{absorptivity} = 1 - \tau'_\lambda = 1 - \exp\left[-K_{a_\lambda} s\right] \tag{7-7}$$

Although a nonemitting medium has been considered thus far, if the medium is hot enough to warrant consideration of emission, the emissivity is obtained by invoking Kirchhoff's law, assuming thermodynamic equilibrium.

$$\varepsilon'_\lambda = \text{emissivity} = \alpha'_\lambda = 1 - \exp\left[-K_{a_\lambda} s\right] \quad (\text{assuming } T = \text{constant}) \tag{7-8}$$

For a small path length Δs the argument of the exponential terms in (7-6) through (7-8) is small and the exponential can be linearized in a Taylor series to give

$$\tau'_\lambda = 1 - K_{a_\lambda} \Delta s + \cdots \quad \left(K_{a_\lambda} \Delta s \to 0\right) \tag{7-9}$$

$$\alpha'_\lambda = K_{a_\lambda} \Delta s + \cdots \quad \left(K_{a_\lambda} \Delta s \to 0\right) \tag{7-10}$$

$$\varepsilon'_\lambda = K_{a_\lambda} \Delta s + \cdots \quad \left(K_{a_\lambda} \Delta s \to 0\right) \tag{7-11}$$

Absorbing, Emitting, Nonscattering Medium

The transfer equation can be formally derived by making an optical energy balance on a differential element along a single line of sight as shown in Fig. 7-2.

Consider the differential element of an absorbing, emitting, nonscattering medium shown in Fig. 7-2. The projected area of the element normal to the direction of travel is ΔA and the path length is Δs. Radiant energy is incident on the element in the direction Ω with intensity $I_\lambda(s)$ and solid angle $\Delta\Omega$. The intensity leaving the element in the direction Ω is $I_\lambda(s + \Delta s)$. A balance of optical energy on the element in the direction Ω gives

$$\text{Energy in} + \text{energy emitted} = \text{energy out} + \text{energy absorbed}$$

$$I_\lambda(s)\,\Delta A\,\Delta\Omega + \varepsilon'_\lambda(\Delta s) I_{b_\lambda}(T)\,\Delta A\,\Delta\Omega = I_\lambda(s + \Delta s)\,\Delta A\,\Delta\Omega + \alpha'_\lambda(\Delta s) I_\lambda(s)\,\Delta A\,\Delta\Omega \tag{7-12}$$

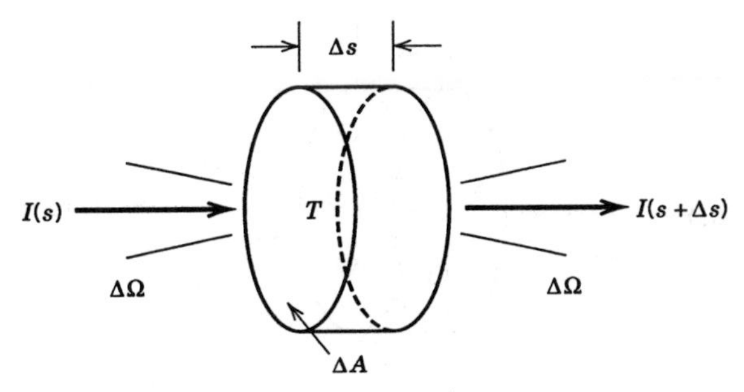

Figure 7-2 Optical energy balance on a differential, absorbing, emitting volume along a single line of sight.

where $\varepsilon_\lambda'(\Delta s)$ and $\alpha_\lambda'(\Delta s)$ denote the emissivity and absorptivity of the element with path length Δs. Substituting Eqs. (7-10) and (7-11) into (7-12), dividing by $\Delta A\, \Delta\Omega\, \Delta s$, and taking the limit as $\Delta s \to 0$ allows (7-12) to be written as

$$\lim_{\Delta s \to 0} \frac{I_\lambda(s + \Delta s) - I_\lambda(s)}{\Delta s} = -K_{a_\lambda} I_\lambda + K_{a_\lambda} I_{b_\lambda} \tag{7-13}$$

The left-hand side of (7-13) is the definition of differentiation of I_λ with respect to s.

$$\frac{dI_\lambda}{ds} = -K_{a_\lambda} I_\lambda + K_{a_\lambda} I_{b_\lambda} \tag{7-14}$$

Equation (7-14) is the transfer equation for an absorbing, emitting, nonscattering medium and the solution for a homogeneous, isothermal medium is obtained by using an integrating factor.

$$I_\lambda(s) = I_{0_\lambda} \exp\left[-K_{a_\lambda} s\right] + I_{b_\lambda}\left[1 - \exp(-K_{a_\lambda} s)\right] \tag{7-15}$$

Using the definitions of transmissivity and emissivity for an isothermal layer of matter, Eqs. (7-6) and (7-8), gives a result which shows that the intensity at any given point along a line-of-sight in a nonscattering medium is comprised of a transmitted contribution and an emitted contribution, in general.

$$I_\lambda(s) = \tau_\lambda' I_{0_\lambda} + \varepsilon_\lambda' I_{b_\lambda} \tag{7-16}$$

Scattering Medium, Out-scattering

Consider next a differential volume element ΔV containing scattering particles as shown in Fig. 7-3. An optical energy balance is made along a

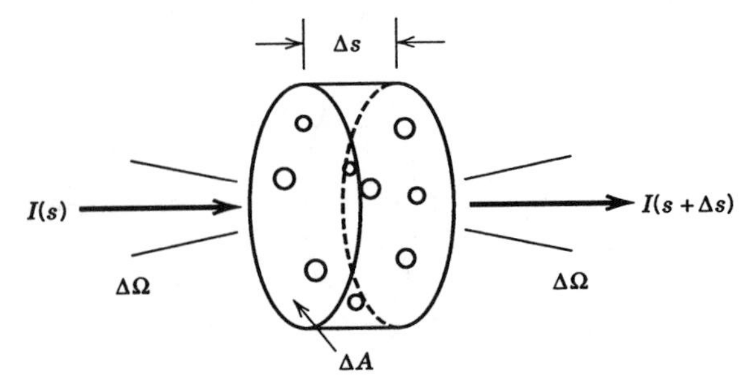

Figure 7-3 Optical energy balance on a differential, scattering volume along a single line of sight.

particular line of sight in the Ω direction. It is assumed the particles in ΔV are few enough and far enough apart that there is negligible multiple scattering and negligible shadowing of particles by one another.

Neglecting absorption and emission and including only scattering effects, the optical energy balance on ΔV is as follows:

$$\begin{aligned}&\text{Energy into } \Delta V \text{ along } \Omega + \text{energy scattered into } \Omega \text{ direction}\\ &\quad = \text{energy leaving } \Delta V \text{ along } \Omega + \text{energy scattered out of } \Omega \text{ direction}\end{aligned} \tag{7-17}$$

For the moment the in-scattering contribution on the left-hand side is neglected. Furthermore, all the particles in the volume are assumed to present the same cross-sectional area for scattering (such as for a monodispersion of identical particles). The balance thus becomes

$$I_\lambda(s)\,\Delta A\,\Delta\Omega + 0 = I_\lambda(s + \Delta s)\,\Delta A\,\Delta\Omega + I_\lambda(s)\,\Delta\Omega\,C_s N_0\,\Delta V \tag{7-18}$$

where C_s is the scattering cross section per particle, and N_0 is the number of particles per unit volume. Dividing (7-18) by $\Delta A\,\Delta\Omega\,\Delta s$ and taking the limit as $\Delta s \to 0$ gives the transfer equation for a scattering medium (neglecting in-scattering) as

$$\frac{dI_\lambda}{ds} = -K_{s_\lambda} I_\lambda \tag{7-19}$$

where the *scattering coefficient* is

$$K_{s_\lambda} = C_s N_0 \quad \text{(monodispersion)} \tag{7-20}$$

A more general definition of the scattering coefficient can be written analogous to Eq. (7-5) for the absorption coefficient as

$$K_{s_\lambda} = \lim_{\Delta s \to 0} \frac{\text{energy scattered out of } \Delta V}{(\text{energy incident on } \Delta V)\, \Delta s} \tag{7-21}$$

If the particles in ΔV absorb as well as scatter, then the absorption coefficient for the particles is the product of the absorption cross section per particle and the number density.

$$K_{a_\lambda} = C_a N_0 \quad (\text{monodispersion}) \tag{7-22}$$

where C_a is the absorption cross section per particle.

If absorbing particles are suspended in an absorbing continuous medium, then the absorption coefficient is a volume-weighted average of the particle and continuous absorption coefficients.

$$K_{a_\lambda} = C_a N_0 + \frac{4\pi k}{\lambda_0}(1 - f_v) \quad (f_v = \text{particle volume fraction}) \tag{7-23}$$

The absorption and scattering characteristics of particles are often expressed as efficiency factors. The absorption and scattering efficiencies $Q_{a,s}$ are defined as the ratio of the cross section to the geometric projected area of the particle, G.

$$Q_{a,s} = \frac{C_{a,s}}{G} \tag{7-24}$$

The sum of absorption and scattering effects gives the total extinction by the particles.

$$K_{e_\lambda} = K_{a_\lambda} + K_{s_\lambda} \tag{7-25a}$$

$$C_e = C_a + C_s \tag{7-25b}$$

$$Q_e = Q_a + Q_s \tag{7-25c}$$

The various scattering, absorption, and extinction parameters can be summarized as follows:

	Cross Section (m^2)	Coefficient (m^{-1})	Efficiency
Extinction	C_e	K_{e_λ}	Q_e
Absorption	C_a	K_{a_λ}	Q_a
Scattering	C_s	K_{s_λ}	Q_s

The scattering, absorption, and extinction coefficients for a *monodispersion of spherical particles* can be expressed in terms of the respective efficiencies, the particle volume fraction f_v, and the particle diameter d. From Eqs. (7-20) and (7-22) the coefficients are the product of the respective cross sections and the particle number density. The number density is related to the volume fraction as

$$N_0 = \frac{\text{particles}}{\text{cm}^3 \text{ total}} = \frac{\dfrac{\text{cm}^3 \text{ particle}}{\text{cm}^3 \text{ total}}}{\dfrac{\text{cm}^3 \text{ particle}}{\text{particle}}} = \frac{f_v}{\dfrac{\pi}{6} d^3} \quad \text{(spherical monodispersion)} \tag{7-26}$$

The cross sections can be expressed as the product of the efficiencies and the projected area where the projected area G is $\pi d^2/4$.

$$C_{s,a,e} = Q_{s,a,e} G = Q_{s,a,e} \frac{\pi}{4} d^2 \quad \text{(sphere)} \tag{7-27}$$

Multiplying (7-26) and (7-27) gives the scattering, absorption, and extinction coefficients for a spherical monodispersion.

$$K_{s,a,e_\lambda} = C_{s,a,e} N_0 = \frac{1.5 Q_{s,a,e} f_v}{d} \quad \text{(spherical monodispersion)} \tag{7-28}$$

For a *distribution of sizes (polydispersion)* the respective coefficients can be obtained by integrating the cross section with respect to the number density size distribution, $N(r)\,dr$ = number per unit volume with particle radius between r and $r + dr$.

$$K_{s,a,e_\lambda} = \int_0^\infty C_{s,a,e}(r) N(r)\, dr \quad \text{(polydispersion)} \tag{7-29}$$

For a spherical polydispersion (7-29) can also be written as

$$K_{s,a,e_\lambda} = \int_0^\infty \pi r^2 Q_{s,a,e}(r) N(r)\, dr \tag{7-30}$$

In-Scattering

The in-scattering term, which was neglected in the optical energy balance of Eq. (7-18), is now included. In order to determine the contribution of scattering into the Ω direction from other directions, it is first necessary to know the directional distribution of scattered energy. With reference to

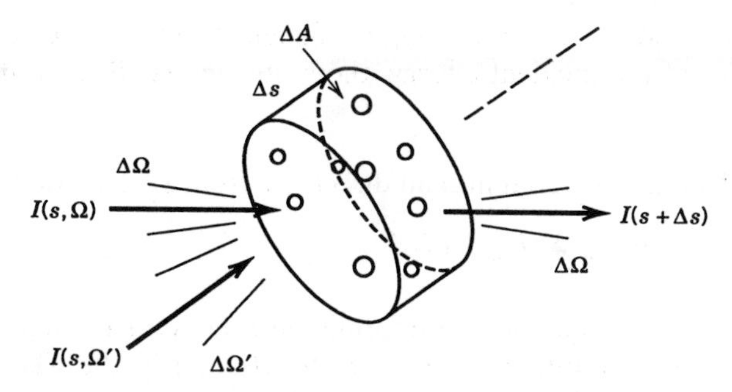

Figure 7-4 In-scattering contribution.

Fig. 7-4, the directional distribution of scattered energy is given by the scattering phase function $p(\Omega' \to \Omega)$, which is defined as follows.

Phase Function

$$p(\Omega' \to \Omega) \equiv \lim_{\Delta s \to 0} \frac{\text{energy scattered from } \Omega' \text{ into } \Omega \text{ direction}}{\text{energy scattered into } \Omega \text{ if scattering were isotropic}} \tag{7-31}$$

The phase function must satisfy the following statement of conservation of scattered energy:

$$\frac{1}{4\pi} \int_{4\pi} p(\Omega' \to \Omega)\, d\Omega' = 1 \tag{7-32}$$

A special case of scattering, which is analogous to diffuse reflection from surface transfer theory, is isotropic scattering. In this case the scattered energy is equally distributed in all directions.

$$p = 1 \quad \text{(isotropic scattering)} \tag{7-33}$$

Now consider the contribution of scattering into the Ω direction from other directions such as Ω'. This can be done most conveniently by allowing the differential volume ΔV to rotate so that it is always normal to the direction of incidence or in-scattering being considered. Allowing the volume element to rotate in this way does not affect the results since the same amount of energy is scattered with the same directional distribution. With reference to Fig. 7-4 the energy incident on ΔV from any direction (including the line of sight) is given as

$$\text{Energy incident from } \Omega' \text{ direction} = I_\lambda(\Omega')\, \Delta A\, \Delta\Omega' \tag{7-34}$$

Using the definition of the scattering coefficient (7-21), the scattered energy is the product of the incident energy, the scattering coefficient and the path length.

$$\text{Energy scattered into all directions from } \Omega' \text{ direction}$$
$$= K_{s_\lambda}\, \Delta s\, I_\lambda(\Omega')\, \Delta A\, \Delta\Omega' \tag{7-35}$$

By dividing the scattered energy uniformly into the 4π steradian solid angle, the energy scattered into a particular direction, say the Ω direction, if scattering were isotropic, is given as

$$\text{Energy scattered into } \Omega \text{ direction if isotropic} = \frac{\Delta\Omega}{4\pi} K_{s_\lambda}\, \Delta s\, I_\lambda(\Omega')\, \Delta A\, \Delta\Omega' \tag{7-36}$$

Then using the definition of phase function (7-31) the energy scattered into the Ω direction is

$$\text{Energy scattered into } \Omega \text{ direction from } \Omega' \text{ direction}$$
$$= \frac{\Delta\Omega}{4\pi} K_{s_\lambda}\, \Delta s\, I_\lambda(\Omega')\, \Delta A\, \Delta\Omega'\, p(\Omega' \to \Omega) \tag{7-37}$$

Integrating over all incoming directions gives

$$\text{Energy scattered into } \Omega \text{ direction from all directions}$$
$$= \frac{\Delta\Omega}{4\pi} K_{s_\lambda}\, \Delta s\, \Delta A \int_{4\pi} I_\lambda(\Omega') p(\Omega' \to \Omega)\, d\Omega' \tag{7-38}$$

Substituting this term back into the energy balance of Eq. (7-18) gives

$$I_\lambda(s)\, \Delta A\, \Delta\Omega + \frac{\Delta\Omega}{4\pi} K_{s_\lambda}\, \Delta s\, \Delta A \int_{4\pi} I_\lambda(\Omega') p(\Omega' \to \Omega)\, d\Omega'$$
$$= I_\lambda(s + \Delta s)\, \Delta A\, \Delta\Omega + K_{s_\lambda}\, \Delta s\, I_\lambda(s)\, \Delta\Omega\, \Delta A \tag{7-39}$$

which, upon dividing by $\Delta A\, \Delta\Omega\, \Delta s$ and taking the limit $\Delta s \to 0$, gives the transfer equation in a scattering medium, including in-scattering.

$$\frac{dI_\lambda(\Omega)}{ds} = -K_{s_\lambda} I_\lambda(\Omega) + \frac{K_{s_\lambda}}{4\pi} \int_{4\pi} I_\lambda(\Omega') p(\Omega' \to \Omega)\, d\Omega' \tag{7-40}$$

If the effects of absorption and emission are also included, the transfer equation becomes

$$\frac{dI_\lambda(\Omega)}{ds} = -(K_{s_\lambda} + K_{a_\lambda})I_\lambda(\Omega) + K_{a_\lambda}I_{b_\lambda} + \frac{K_{s_\lambda}}{4\pi}\int_{4\pi} I_\lambda(\Omega')p(\Omega' \to \Omega)\,d\Omega' \tag{7-41a}$$

which can also be written as

$$\mathbf{e}_\Omega \cdot \nabla I_\lambda(\Omega) = -K_{e_\lambda}I_\lambda(\Omega) + K_{a_\lambda}I_{b_\lambda} + \frac{K_{s_\lambda}}{4\pi}\int_{4\pi} I_\lambda(\Omega')p_\lambda(\Omega' \to \Omega)\,d\Omega' \tag{7-41b}$$

where $\mathbf{e}_\Omega$ is the unit vector in the Ω direction. Equation (7-41) is the radiative transfer equation for an absorbing, emitting, and scattering medium. It is an optical energy balance on a single-scattering volume element along a single line of sight in the Ω direction. Equation (7-41) is an integro-differential equation. In order to solve the transfer equation to obtain the intensity field in a particular geometry with multiple lines of sight (for example, in a planar slab) the transfer equation must first be written for all possible optical paths in the configuration of interest. The solution of the transfer equation in particular configurations with multiple lines of sight is discussed further in Chapter 12.

Radiative Heat Flux

Once the solution to the transfer equation has been obtained, the radiative flux vector can be obtained as

$$\mathbf{q}_r = \sum_{i=1}^{3} q_{r_i}\mathbf{e}_i \tag{7-42}$$

where q_{r_i} are the components of the radiative flux vector

$$q_{r_i} = \int_0^\infty \int_{4\pi} I_\lambda(\Omega)\mathbf{e}_\Omega \cdot \mathbf{e}_i\,d\Omega\,d\lambda \tag{7-43}$$

and $\mathbf{e}_i$ are the unit vectors in the coordinate system of interest.

Divergence of Radiative Heat Flux and Mean Coefficients

Another important quantity is the divergence of the radiative flux vector, which represents the net radiative energy per unit time and volume leaving a differential control volume at temperature T. This quantity can be obtained by integrating the transfer equation over all solid angles and wavelengths, giving

$$\nabla \cdot \mathbf{q}_r = 4K_{\mathrm{em}_p}(T)e_b(T) - \int_0^\infty K_{a_\lambda} \int_{4\pi} I_\lambda(\Omega)\, d\Omega\, d\lambda \tag{7-44a}$$

where

$$K_{\mathrm{em}_p}(T) = \frac{1}{I_b(T)} \int_0^\infty K_{a_\lambda}(T) I_{b_\lambda}(T)\, d\lambda = \int_0^1 K_{a_\lambda}(T)\, df(\lambda T) \tag{7-44b}$$

The coefficient $K_{\mathrm{em}_p}(T)$ is widely referred to as the *Planck mean absorption coefficient*. However, as can be seen from (7-44a), $K_{\mathrm{em}_p}(T)$ is more aptly referred to as a total *emission* coefficient rather than absorption coefficient since it represents total emission from a differential volume element. Assuming the intensity incident on the small volume element is blackbody radiation at a temperature of T_e, Eq. (7-44a) can be written as

$$\nabla \cdot \mathbf{q}_r = 4K_{\mathrm{em}_p}(T)e_b(T) - 4K_{a_p}(T, T_e)e_b(T_e) \quad \text{(blackbody irradiation)} \tag{7-44c}$$

where

$$K_{a_p}(T, T_e) = \frac{1}{I_b(T_e)} \int_0^\infty K_{a_\lambda}(T) I_{b_\lambda}(T_e)\, d\lambda = \int_0^1 K_{a_\lambda}(T)\, df(\lambda T_e) \tag{7-44d}$$

The coefficient $K_{a_p}(T, T_e)$ is the mean, total absorption coefficient for blackbody incident radiation. It should be noted that the scattering coefficient does not appear in Eq. (7-44a) since scattered energy is conserved. However, scattering does, in general, influence the overall energy transport since I_λ depends on scattering as determined by the solution of the transfer equation.

Single-Scattering Albedo

An important parameter in radiative transfer in participating media is the ratio of scattering to total extinction (absorption plus scattering). This ratio is called the *single-scattering albedo* ω_{0_λ}.

$$\omega_{0_\lambda} = \frac{K_{s_\lambda}}{K_{s_\lambda} + K_{a_\lambda}} \tag{7-45a}$$

$$1 - \omega_{0_\lambda} = \frac{K_{a_\lambda}}{K_{s_\lambda} + K_{a_\lambda}} \tag{7-45b}$$

The albedo varies between 0 and 1 and indicates the relative contributions of absorption and scattering to the total extinction. A value of 0 for the albedo indicates a nonscattering medium where extinction is due totally to absorption. A value of 1 indicates a pure scattering, nonabsorbing, nonemitting medium where the extinction is totally due to scattering.

When a participating medium consists of monodisperse particles in a nonparticipating continuous phase, the coefficients in (7-45) can be replaced with the corresponding particle efficiencies.

$$\omega_{0_\lambda} = \frac{Q_{s_\lambda}}{Q_{e_\lambda}} \quad \text{(single particle or monodispersion)} \tag{7-46a}$$

$$1 - \omega_{0_\lambda} = \frac{Q_{a_\lambda}}{Q_{e_\lambda}} \quad \text{(single particle or monodispersion)} \tag{7-46b}$$

In this case the albedo can be thought of as an effective particle reflectivity, and one minus the albedo is the effective particle absorptivity (and emissivity, if the particle is in thermodynamic equilibrium).

Optical Depth

Another important parameter in radiative transfer in participating media is the optical depth, also referred to as opacity or turbidity. The optical depth is the integral of the extinction coefficient along an optical line of sight.

$$t_\lambda = \int_0^s K_{e_\lambda}(s')\, ds' \tag{7-47}$$

or in differential form

$$dt_\lambda = K_{e_\lambda}\, ds \tag{7-48}$$

The optical depth is the appropriate nondimensional path length for radiative transfer in a participating medium, as can be seen from the transfer equation (7-41). For a uniform, participating medium, where the extinction coefficient is independent of position, the optical depth is simply the product of the extinction coefficient and path length.

$$t_\lambda = K_{e_\lambda} s = \left(K_{s_\lambda} + K_{a_\lambda}\right) s \quad \text{(uniform medium)} \tag{7-49}$$

Since the inverse of the extinction coefficient is the mean free photon path (MFP), the optical depth is the nondimensional path length in units of mean free photon paths.

$$t_\lambda = \frac{s}{\left(\dfrac{1}{K_{e_\lambda}}\right)} = \frac{s}{\text{MFP}} \tag{7-50}$$

If the path length exceeds the mean free path $s \gg$ MFP ($t_\lambda \gg 1$), the probability is high that a photon will be either absorbed or scattered before it traverses the entire path length. This condition is referred to as the *optically thick limit*. On the other hand if $s \ll$ MFP ($t_\lambda \ll 1$), the probability of absorption or scattering occurring along the path is small. This condition is referred to as the *optically thin limit*. The elemental volume considered in the derivation of the transfer equation is assumed to be optically thin and therefore non-self-absorbing. That is, the energy emitted by a differential volume is not absorbed by the same differential volume. These limits are summarized as follows.

$$t_\lambda \lesssim 0.1 \quad \text{(optically thin, non-self-absorbing)} \tag{7-51}$$

$$K_{s_\lambda} s = \omega_{0_\lambda} t_\lambda \lesssim 0.1 \quad \text{(single scattering)} \tag{7-52}$$

$$t_\lambda \gtrsim 2\text{-}3 \quad \text{(optically thick)} \tag{7-53}$$

Equation (7-52) describes a condition known as optically thin for scattering. In this condition the medium could possibly be optically thick for absorption but scattering is sufficiently unlikely that the probability of multiple scattering events occurring is small. Thus this condition is also referred to as single scattering. Substituting the definitions of optical depth and albedo into Eq. (7-41) gives the following form of the transfer equation:

$$\frac{dI_\lambda}{dt_\lambda} = -I_\lambda + (1 - \omega_{0_\lambda}) I_{b_\lambda} + \frac{\omega_{0_\lambda}}{4\pi} \int_{4\pi} I_\lambda p(\Omega' \to \Omega)\, d\Omega' \tag{7-54}$$

Example 7-1

When particles are present in high temperature industrial systems (such as fluidized beds or coal furnaces) there is a distinct possibility that radiative transfer by the particles may be important. If the particle properties can be estimated, then it is also possible to estimate the importance of radiative transfer using the principles outlined in this chapter.

(a) Estimate the mean free photon path in a packed or fluidized bed composed of 1-mm particles at 1000 K.

(b) Determine the relative importance of soot particle radiation and char particle radiation in a pulverized coal flame by estimating the extinction coefficient and mean free path associated with each type of particle.

(a) The extinction coefficient for particles can be estimated using Eq. (7-28). For particles much larger than the characteristic wavelength of radiation ($\lambda \sim 2900\ \mu\text{m K}/1000\ \text{K} \sim 3\ \mu\text{m} \ll 1000\ \mu\text{m}$) the extinction efficiency is 1. In a closely packed particle medium such as a packed or fluidized bed the particle volume fraction is approximately 0.6. Thus the extinction coefficient is approximately one over the particle diameter

$$K_e = \frac{1.5 Q_e f_v}{d} = \frac{1.5(1)(0.6)}{d} \sim \frac{1}{d}$$

and the photon mean free path is approximately the particle size

$$\text{MFP} = \frac{1}{K_e} \sim d$$

This means that in a closely packed particle medium a photon emitted or scattered by a particle will travel on the average a distance of one particle diameter before being absorbed or scattered by another particle.

(b) In a pulverized coal flame entrained coal particles quickly devolatilize and form char particles that burn out relatively slowly. These char particles are much larger than the characteristic wavelength of radiation ($d \sim 100\ \mu\text{m}$; $Q_e \sim 1$) and rather finely dispersed ($f_v \sim 10^{-4}$). Thus the extinction coefficient for char particle radiation is of the order of $1\ \text{m}^{-1}$

$$K_e \sim \frac{(1)(10^{-4})}{10^{-4}\ \text{m}} = 1\ \text{m}^{-1} \quad \text{(char particles)}$$

and the mean free path is 1 m. Soot particles formed during the devolatilization and combustion of the coal particles also participate radiatively in the flame. These particles are much smaller than the characteristic wavelength of radiation ($d \sim 10$ nm) and therefore their extinction efficiency is very small ($Q_e \sim 10^{-2}$). The volume fraction occupied by the soot particles is typically two orders of magnitude smaller than that of the coal/char particles [Altenkirch, et al., 1984]. Thus the extinction coefficient for soot particle radiation is also of the order of $1\ \text{m}^{-1}$

$$K_e \sim \frac{(10^{-2})(10^{-6})}{10^{-8}\ \text{m}} = 1\ \text{m}^{-1} \quad \text{(soot particles)}$$

and the mean free path is again 1 m. In comparing the properties of the char and soot particles, it can be seen that the smaller volume fraction and extinction efficiency of the soot particles are offset by their smaller size (i.e., greater specific surface area). As a result the two types of particles are predicted to be of comparable importance in determining the radiative properties of the flame. (A third type of particle that participates radiatively is fly ash; see Problem 7-12.)

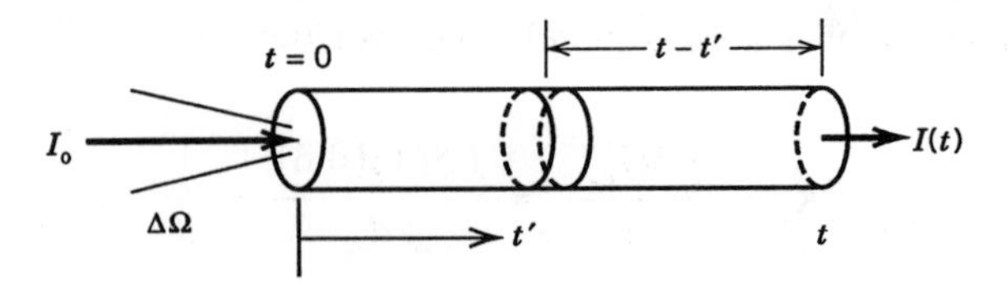

Figure 7-5 Transfer equation along a single line of sight.

INTEGRAL OF TRANSFER EQUATION ALONG A LINE OF SIGHT

It is possible to obtain a formal solution to the equation of transfer along a single line of sight by straightforward integration. In so doing, it is customary to define the *source function* as the emission plus in-scattering term

$$J_\lambda = \left(1 - \omega_{0_\lambda}\right) I_{b_\lambda} + \frac{\omega_{0_\lambda}}{4\pi} \int_{4\pi} I_\lambda p(\Omega' \to \Omega)\, d\Omega' \tag{7-55}$$

which gives the equation

$$\frac{dI_\lambda}{dt_\lambda} = -I_\lambda + J_\lambda \tag{7-56}$$

The integral of this equation, with an initial condition of $I_\lambda(t_\lambda = 0) = I_{0_\lambda}$ can be obtained by using an integrating factor to give

$$I_\lambda(t_\lambda) = I_{0_\lambda} e^{-t_\lambda} + \int_0^{t_\lambda} J_\lambda(t'_\lambda) e^{-(t_\lambda - t'_\lambda)}\, dt'_\lambda \tag{7-57}$$

With reference to Fig. 7-5 it can be seen from Eq. (7-57) that there are two contributions to the intensity in the Ω direction at the arbitrary path length location t_λ. The first comes from the incident intensity I_{0_λ}, which gets attenuated exponentially over the entire path length between 0 and t_λ. The second is the contribution from the source term, which is generated (either by emission or in-scattering) at the location t'_λ and gets attenuated exponentially over the remaining path length from t'_λ to t_λ.

Isotropic Intensity Approximation

The solution of the transfer equation is complicated by the unknown in-scattering term. Before the intensity can be obtained along a particular line of sight, it is necessary to know the intensity along all other lines of sight. That is, in the general problem, all lines of sight must be solved for simultaneously. A useful approximation to the full transfer equation can be made by assuming that the intensity distribution in the in-scattering term is isotropic or independent of direction.

$$\frac{dI_\lambda}{dt_\lambda} = -I_\lambda + (1 - \omega_{0_\lambda})I_{b_\lambda} + \frac{\omega_{0_\lambda}}{4\pi} I_\lambda \int_{4\pi} p(\Omega' \to \Omega)\, d\Omega' \qquad (7\text{-}58)$$

By virtue of the normalization condition for the phase function (7-32) the transfer equation simplifies to

$$\frac{dI_\lambda}{dt_\lambda} = -(1 - \omega_{0_\lambda})I_\lambda + (1 - \omega_{0_\lambda})I_{b_\lambda} \qquad (7\text{-}59)$$

or

$$\frac{dI_\lambda}{ds} = -K_{a_\lambda} I_\lambda + K_{a_\lambda} I_{b_\lambda} \qquad (7\text{-}60)$$

which is the same as Eq. (7-14) for a nonscattering medium. In essence, the scattering contribution to the transfer has fallen out of the problem. The isotropic intensity assumption is equivalent to assuming that the gain due to in-scattering is balanced by the loss due to out-scattering, therefore scattering can be ignored. This assumption is useful for obtaining approximate results in weakly scattering media ($\omega_{0_\lambda} t_\lambda < 0.1$). The accuracy of this approximation diminishes, however, as the degree of multiple scattering increases.

It is important to note that the isotropic intensity assumption does not apply to the scattered radiation field produced by highly directional (e.g., collimated laser or solar) incident radiation because the resulting intensity distribution is necessarily highly anisotropic. Rather, the assumption applies to the scattering of thermal radiation emitted by the medium. Mackowski, et al. [1983] and Altenkirch, et al. [1984] applied the isotropic intensity approximation to the thermal radiation emitted by pulverized coal flames and correctly indicated that the approximation did not apply to their collimated, spectral transmission measurements. Collimated transmission measurements generally measure the extinction coefficient and not the absorption coefficient.

Transmissivity Function

The integration of the transfer equation along a single line of sight can be facilitated, especially for nonhomogeneous paths, by introducing the transmissivity function. The transmissivity function is defined as the fraction of intensity transmitted between two points (s' and s) along a line of sight.

$$\tau_\lambda(s', s) = \exp\left[-\int_{s'}^{s} K_{e_\lambda}(s'')\, ds''\right] \qquad (7\text{-}61)$$

An equivalent definition could be made involving the optical depth by generalizing the definition of Eq. (7-47) to allow integration between arbitrary points (s' and s) along the line of sight.

$$\tau_\lambda(s', s) = \exp[-t_\lambda(s', s)] \tag{7-62}$$

where

$$t_\lambda(s', s) = \int_{s'}^{s} K_{e_\lambda}(s'')\, ds'' \tag{7-63}$$

Using this definition for transmissivity function, the integral of the transfer equation (7-57) can also be written as

$$I_\lambda(s) = I_{0_\lambda}\tau_\lambda(0, s) + \int_0^s J_\lambda(s')\tau_\lambda(s', s)K_{e_\lambda}(s')\, ds' \tag{7-64}$$

or, making use of the derivative properties of the transmissivity function, as

$$I_\lambda(s) = I_{0_\lambda}\tau_\lambda(0, s) + \int_0^s J_\lambda(s')\frac{\partial\tau_\lambda(s', s)}{\partial s'}\, ds' \tag{7-65}$$

The derivative of the transmissivity function can then be integrated along the line of sight to obtain the intensity at an arbitrary position s. By approximating a nonhomogeneous path as a series of homogeneous elements, the integral in Eq. (7-65) can be written as a summation involving the difference of the transmissivities from the endpoints of the elements (s_i and s_{i-1}) to the end of the path (s).

$$I_\lambda(s) = I_{0_\lambda}\tau_\lambda(0, s) + \sum_i J_\lambda(s_i - s_{i-1})[\tau_\lambda(s_i, s) - \tau_\lambda(s_{i-1}, s)] \tag{7-66}$$

For example, consider a nonscattering medium ($J_\lambda = I_{b_\lambda}$ and $K_{e_\lambda} = K_{a_\lambda}$) which is homogeneous over the path length from 0 to s. The intensity $I_\lambda(s)$ for this case is

$$I_\lambda(s) = I_{0_\lambda}\tau_\lambda(0, s) + I_{b_\lambda}[1 - \tau_\lambda(0, s)] \tag{7-67}$$

which agrees with Eq. (7-15).

Example 7-2

As an example of transfer in a nonhomogeneous medium, consider a plane, parallel, nonscattering, absorbing, emitting, gray medium ($n \approx 1$) that has a nonuniform temperature profile and nonuniform absorption coefficient as pictured in Fig. 7-6. The medium is 4 cm thick overall. The inner region (2 cm thick) is at $T_2 = 1500$ K and has a total absorption coefficient $K_{a_2} = 0.75\ \text{cm}^{-1}$. The outer edges (1 cm thick, each) are at $T_1 = 1000$ K and have a

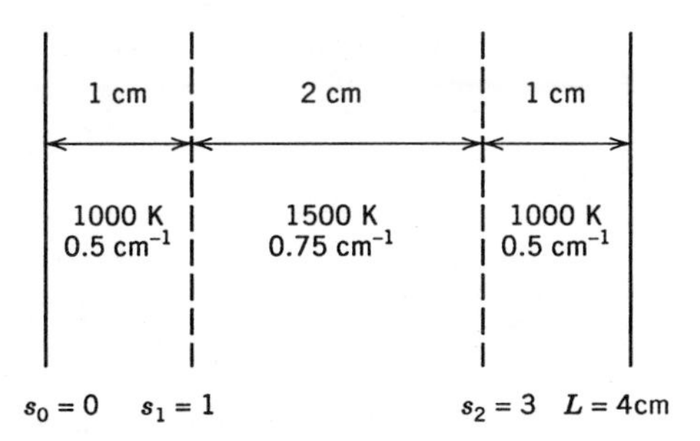

Figure 7-6 Radiative transfer along a nonhomogeneous line of sight.

total absorption coefficient of $K_{a_1} = 0.5\ \text{cm}^{-1}$. Calculate the total intensity emitted by the medium into the cold blackbody surroundings in the direction normal to the slab.

The solution can be found by integrating Eq. (7-66) over all wavelengths and applying the result to the three homogeneous elements of the slab.

$$I_e(L) = \frac{\sigma T_1^4}{\pi}[\tau(s_1, L) - \tau(s_0, L)] + \frac{\sigma T_2^4}{\pi}[\tau(s_2, L) - \tau(s_1, L)] + \frac{\sigma T_1^4}{\pi}[\tau(L, L) - \tau(s_2, L)]$$

This requires evaluation of the transmissivity functions

$$\begin{aligned}
\tau(s_0, L) &= \exp[-t(s_0, L)] \\
\tau(s_1, L) &= \exp[-t(s_1, L)] \\
\tau(s_2, L) &= \exp[-t(s_2, L)] \\
\tau(L, L) &= \exp[-t(L, L)]
\end{aligned}$$

which are obtained by evaluating the optical depth functions

$$\begin{aligned}
t(s_0, L) &= \int_{s_0}^{L} K_a(s'')\, ds'' = K_{a_1}(s_1 - s_0) + K_{a_2}(s_2 - s_1) + K_{a_1}(L - s_2) \\
&= 0.5 + 1.5 + 0.5 = 2.5 \\
t(s_1, L) &= \int_{s_1}^{L} K_a(s'')\, ds'' = K_{a_2}(s_2 - s_1) + K_{a_1}(L - s_2) = 1.5 + 0.5 = 2.0 \\
t(s_2, L) &= \int_{s_2}^{L} K_a(s'')\, ds'' = K_{a_1}(L - s_2) = 0.5 \\
t(L, L) &= \int_{L}^{L} K_a(s'')\, ds'' = 0
\end{aligned}$$

The resulting total emitted intensity is thus

$$I_e(L) = \frac{5.67}{\pi}\left[(1)^4(e^{-2.0} - e^{-2.5} + e^{-0} - e^{-0.5}) + (1.5)^4(e^{-0.5} - e^{-2.0})\right]$$

$$= 5.1\,\frac{\text{W}}{\text{cm}^2\ \text{sr}}$$

Radiation Pyrometry — An Application of the Transfer Equation

Using the solution of the transfer equation, it is possible to estimate the temperature of a weakly scattering, absorbing, emitting, isothermal medium from measurements of emitted and transmitted spectral intensity. The transfer equation for this situation is Eq. (7-60) whose solution is Eq. (7-15) or (7-67). This solution is first applied to a measurement of intensity made along a line of sight through the medium with no radiation incident upon the medium (see Fig. 7-7*a*). This intensity is the emitted intensity $I_{\lambda_e}(t_\lambda)$ and is given as

$$I_{\lambda_e}(t_\lambda) = I_{b_\lambda}(T)[1 - \exp(-t_\lambda)] \tag{7-68}$$

The solution of the transfer equation is then applied to a second measurement made along the same line of sight with a known intensity I_{0_λ} incident on the medium (see Fig. 7-7*b*). The transmitted plus emitted intensity measured in this case is given as

$$I_{\lambda_{t+e}}(t_\lambda) = I_{0_\lambda}\exp(-t_\lambda) + I_{b_\lambda}(T)[1 - \exp(-t_\lambda)] \tag{7-69}$$

With the two measured intensities $I_{\lambda_{t+e}}$ and I_{λ_e} as known input values, Eqs. (7-68) and (7-69) represent two equations that can be solved for the two unknowns T and t_λ. As a practical matter, to use this technique it is necessary to ensure that the transmitted and emitted intensities are the same

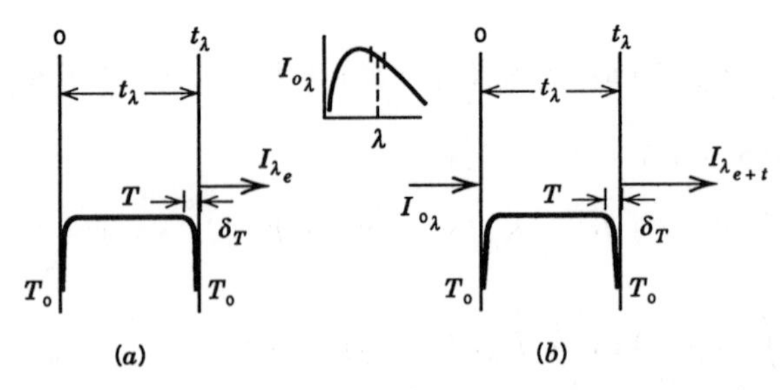

Figure 7-7 Measurement of spectral intensity from an isothermal medium; (*a*) emitted and (*b*) transmitted plus emitted.

order of magnitude, which is the same as saying that the flame emissivity must not approach either 0 or 1. Otherwise, whichever term were much smaller would be lost in the noise of the signal for the larger term in the right-hand side of (7-69).

Line Reversal Technique

The line reversal technique is a derivative of the technique just described, which is commonly used to measure the temperature of nonscattering flames. The idea of the line reversal technique is to seed the flame with a sodium containing compound (such as NaCl) to take advantage of the bright spectral lines emitted by sodium atoms at 589 nm. Using a continuous source for the incident intensity I_{0_λ} these spectral lines, when viewed through a spectrometer or some other dispersing instrument, will appear either in emission or absorption against the background continuum, depending on the relative magnitudes of I_{0_λ} and $I_{b_\lambda}(T)$ as shown in Fig. 7-8. The magnitude of I_{0_λ} is adjusted until the lines just change from emission to absorption, that is, they

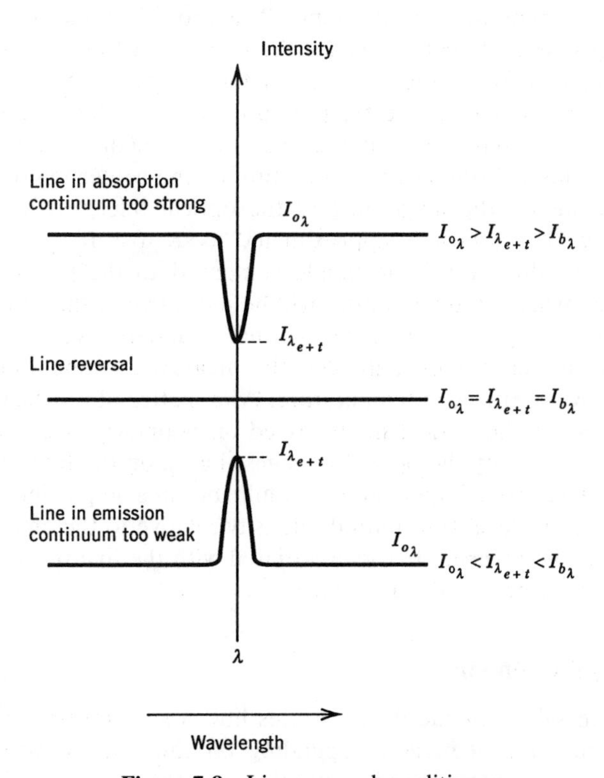

Figure 7-8 Line reversal condition.

disappear against the background continuum of the incident energy. This is the line reversal condition. At this point, assuming that the flame is nonparticipating in the spectral region just outside the lines, it can be said that the emitted plus transmitted intensity $I_{\lambda_{t+e}}$ is just equal to the incident intensity I_{0_λ}. Equation (7-69) then indicates that at the line reversal condition the transmitted plus emitted intensity, the incident intensity, and the blackbody intensity are all equal.

$$I_{b_\lambda}(T) = I_{0_\lambda} = I_{\lambda_{t+e}} \quad \text{(line reversal)} \tag{7-70}$$

The Planck function can be readily inverted to give the temperature of the flame if the incident intensity is known. This technique is relatively simple to use. It is only necessary to be able to adjust the strength of the incident intensity to achieve the line reversal condition. Also, the assumption that the flame is nonparticipating outside the spectral lines is not really necessary. Many flames are characterized by the presence of soot particles that emit and absorb as gray bodies over a limited spectral region (which is still wide with respect to the line width). In this case, since soot scatters only very weakly, the line reversal technique can still be shown to be valid, assuming that the soot and gas temperature are the same. This proof is left as an exercise.

Finally, it is important to consider the effect of a thin thermal boundary layer of thickness δ_T on the accuracy of the temperature predicted by the methods just presented. If the temperature in the center of the medium is greater than the temperature at the edge of the medium, is the measured temperature closer to the edge temperature or the center temperature? The answer depends on the magnitude of the optical thickness of the thermal boundary layer t_{δ_T} which is approximated as $K_{a_\lambda}\delta_T$. If $t_{\delta_T} \ll 1$ then the boundary layer thickness is negligible compared to the mean free photon path and the measured temperature will be indicative of the center temperature. However, if $t_{\delta_T} \gg 1$, then the thermal boundary layer is much thicker than the mean free photon path and the measured temperature would be more indicative of the edge temperature. As a matter of practical interest, in the latter case, if the optical depth based on boundary layer thickness was large, then of necessity the optical thickness based on the total medium path length would be even larger, and it would become impossible to measure accurately the low level transmitted intensity relative to the emitted intensity or to achieve line reversal. Errors associated with the line reversal technique are discussed further by Thomas [1968].

Self-Absorption Effects

The line reversal technique demonstrates how a spectral line can appear in either absorption or emission depending on the relative strength of the externally supplied background continuum and the line. A similar situation arises whenever significant temperature gradients are present in an emitting medium.

Consider an emitting medium composed of a "graybody" continuum emitter and a spectral line or band emitter. The continuum emitter might be soot particles in a hydrocarbon flame, refractory particles such as ash in a coal flame or aluminum oxide in a solid propellant flame, or a reticulated porous ceramic material inserted into a gas flame to increase luminosity. The spectral emitter might be an atomic line such as sodium or potassium lines, or a molecular band such as water vapor, carbon dioxide, or hydrogen chloride. Because of the presence of both continuum and spectral emitters, the optical depth in the region of the spectral line or band will be larger than that in the continuum. Neglecting temperature gradient effects for a moment, consider how the spectral variation in optical depth would affect the emitted spectral intensity. If the medium was isothermal, then the emitted intensity would be given by Eq. (7-67) or (7-15), which shows that the emitted intensity in the region of the spectral line or band would be greater than that in the continuum. Thus, the line or band would appear in emission. If the difference between the line/band optical depth and the continuum optical depth is large (i.e., if the line/band is optically thick and the continuum is optically thin), the line/band will appear very prominently in emission. If the difference is small (e.g., if both line/band and continuum are optically thick), the line/band will appear only very weakly, perhaps imperceptibly, in emission (but nevertheless in emission and not in absorption).

Now consider the effect of temperature gradients. Because of heat transfer from the medium, a significant temperature gradient would usually exist with temperature decreasing along a line of sight toward the observer. Such a temperature gradient might cause the line/band to appear in absorption if the line/band and continuum are both sufficiently optically thick. One way to understand this phenomenon of self-absorption is to think of an observer looking into the medium at different wavelengths. At the line/band wavelengths, one would not be able to see into the medium very far due to the large optical depth. Thus the observed photons would originate from the colder region nearer to the observer. On the other hand, looking into the medium at the continuum wavelengths, one would be able to see farther into the medium into the hotter interior region. Thus the line or band would appear in absorption against the continuum. This concept is demonstrated with a simple example.

Example 7-3

Consider a nonscattering, emitting, absorbing planar medium of thickness $L = 1$ cm that is composed of a solid, porous reticulated ceramic and gas (Fig. 7-9). The ceramic can be treated as a gray continuum emitter with uniform optical properties and overall optical thickness t_s. The gas can be treated as a nongray emitter with a band centered at $\eta_0 = 3700\ \text{cm}^{-1}$

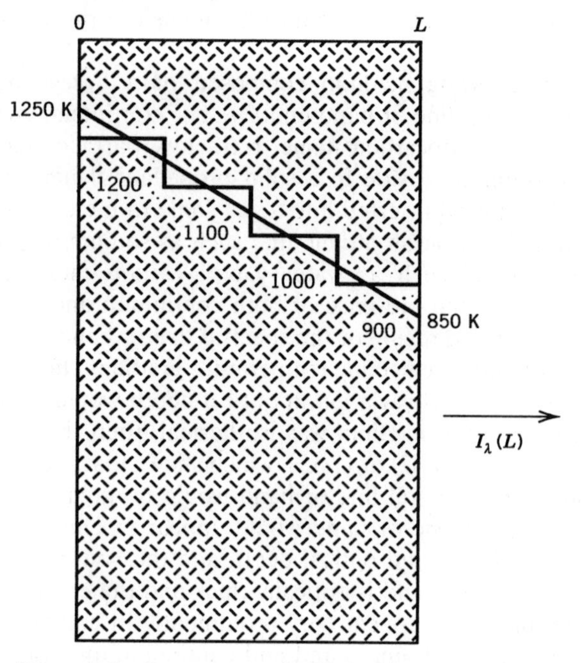

Figure 7-9 Porous ceramic filled with emitting gas.

($\lambda_0 = 2.7\ \mu$m). The absorption coefficient of the gas band K_{a_λ} decays exponentially with wavenumber away from the bandhead as given by

$$K_{a_\lambda} = K_{a_{\lambda 0}} \exp\left[\frac{-2|\eta_0 - \eta|}{\omega}\right]$$

with a bandwidth parameter of $\omega = 100\ \text{cm}^{-1}$ and $K_{a_{\lambda 0}} = 2\ \text{cm}^{-1}$. $K_{a_{\lambda 0}}$ is assumed to be uniform in space. The temperature decreases linearly from 1250 to 850 K through the medium. Divide the medium into four equal isothermal regions as shown in Fig. 7-9. Investigate the effect of the continuum optical depth t_s on the spectral intensity emitted from the cold side of the slab $I_\lambda(L)$ by calculating $I_\lambda(L)$ for various values of t_s. At what value of t_s does reversal of the band take place?

The optical depth for the slab is given by the sum of the optical depths for the continuum and gas band

$$t_\lambda(s', s) = t_{\lambda_s}(s', s) + t_{\lambda_g}(s', s)$$

where the solid optical depth is

$$t_{\lambda_s}(s', s) = t_s \frac{s - s'}{L}$$

and the gas optical depth is

$$t_{\lambda_g}(s', s) = t_0 \frac{s - s'}{L} \exp\left[\frac{-2|\eta_0 - \eta|}{\omega}\right], \qquad t_0 = K_{a_{\lambda 0}} L = 2$$

Using these expressions for optical depth in Eq. (7-62) allows the emitted intensity to be calculated from Eq. (7-66). Setting the incident intensity I_{0_λ} to zero and the source function J_λ to the Planck function in (7-66) gives

$$\begin{aligned} I_\lambda(L) = {} & I_{b_\lambda}(1200\ \text{K})\{\exp[-t_\lambda(s_1, L)] - \exp[-t_\lambda(s_0, L)]\} \\ & + I_{b_\lambda}(1100\ \text{K})\{\exp[-t_\lambda(s_2, L)] - \exp[-t_\lambda(s_1, L)]\} \\ & + I_{b_\lambda}(1000\ \text{K})\{\exp[-t_\lambda(s_3, L)] - \exp[-t_\lambda(s_2, L)]\} \\ & + I_{b_\lambda}(900\ \text{K})\{\exp[-t_\lambda(L, L)] - \exp[-t_\lambda(s_3, L)]\} \end{aligned}$$

where $s_0 = 0$, $s_1 = 0.25$, $s_2 = 0.50$, and $s_3 = 0.75$ cm. This expression is evaluated as a function of wavelength with the aid of a computer, and the results are plotted in Fig. 7-10.

When the continuum is optically thin, only the band emission appears. As the continuum optical thickness increases to 0.5, the band still appears totally

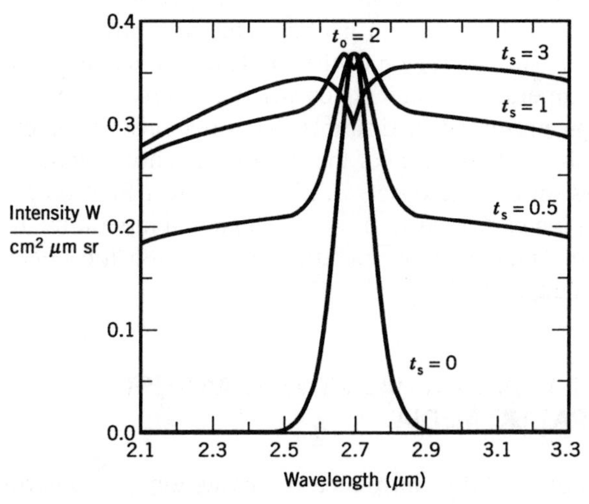

Figure 7-10 Emitted intensity from nonisothermal gray porous ceramic with optical thickness t_s and nongray gas band with optical thickness t_0.

in emission. However, at $t_s = 1$ the thickest center part of the band starts to appear in absorption (relative to the rest of the band) while the thinner wings of the band still appear in emission. At $t_s = 3$ the reversal of the band to full absorption is complete. Thus the reversal of the band cannot be described as occurring at a unique wavelength, but the process occurs over the range of continuum optical thicknesses from 1 to 3. The band profile assumed in this example is the exponential wide band model discussed in Chapter 8. In this example the absorption coefficient was assumed to be constant along the optical path. For a real gas it would be necessary to correct this assumption to account for the variation of $K_{a_{\lambda 0}}$ with temperature.

The preceding example considered self-absorption effects in terms of a continuum emitter and line emitter. Self-absorption effects can also appear in line or band emission alone without any background continuum. When a line is optically thick at the line center over a path length with substantial temperature variation such that photons emitted by the hotter gas are reabsorbed by the colder gas, self-absorption can occur at the center of the line. The observed spectrum of such a line appears with two peaks on either side of the line center instead of with a single peak at the line center, similar to the curve for $t_s = 1$ in Fig. 7-10. In spectroscopy this phenomenon sometimes results in an optically thick (self-absorbed) line being misinterpreted as a doublet which is a set of two distinct lines (e.g. sodium at 588.995 and 589.592 nm) (see Problem 7-8).

A well-known example of the phenomenon of self-absorption is the visible and infrared spectrum of the sun (the Fraunhofer spectrum). This spectrum shows the lines of several elements and ions (H, Na, Mg, Fe, Ca, and Si) appearing in absorption against a continuum that is produced by free-free and bound-free electronic transitions. The reason these lines appear in absorption is that the radiation in the visible and infrared region originates mainly from the surface region of the sun (the photosphere) where there is a decreasing temperature gradient toward the observer. The phenomenon of self-absorption presented in Example 7-3 is responsible for this observation. At altitudes above the surface of the sun (in the chromosphere) the temperature increases toward the observer. Since the ultraviolet portion of the solar spectrum originates primarily from this higher temperature region, most of the lines of the photospheric spectrum below 0.185 μm appear in emission [Goody and Yung, 1989].

WAYS OF CLASSIFYING RADIATIVE TRANSFER IN PARTICIPATING MEDIA

As a final note to this introductory discussion of radiative transfer in participating media, it is appropriate to mention the various ways of classifying radiative transfer in participating media. One way of classifying radiative

TABLE 7-1 Ways of Classifying Radiative Transfer in Participating Media

Nonscattering	versus	Scattering
Gas	versus	Particulate
Banded	versus	Continuum
Optically thin	versus	Optically thick
Decoupled (E) & (Tr) (prescribed *T* field)	versus	Coupled (E) & (Tr) (unknown *T* field)
Isothermal (mixing dominated)		Radiative equilibrium
Linear (conduction dominated)		Conduction/radiation
Boundary layer (convection dominated)		Convection/radiation

transfer is according to whether the medium is nonscattering or scattering. This distinction is important because the occurrence of scattering significantly increases the complexity of the solution. Without scattering it is possible to solve the transfer equation along each line of sight through the medium independently from the other lines of sight because there is no mechanism for energy to be transferred from one line of sight into another. Thus scattering problems are inherently more complex in that the whole intensity field must be solved at once.

Another way of classifying radiative transfer in a participating medium is according to whether the medium is composed of a gas or condensed phase (particulate). Classification can also be according to whether the spectral radiative properties of the medium are banded (nongray) or continuous (semigray). These categories are summarized in Table 7-1.

In general there is a correlation between the three types of classification just mentioned. Nonscattering media are usually gases which exhibit banded property behavior. Likewise, media where scattering is important are usually particulate media, which exhibit continuum behavior. However there are certain exceptions. For example, soot radiation in flames involves particles with continuum properties but is predominantly nonscattering in nature due to the small size of the particles and the large value of absorption index, as discussed in Chapter 9. Plasma radiation is another exception where the medium is gaseous but the properties have a strong continuum component in addition to the atomic line spectra.

Another way of classifying radiative transfer in participating media is according to whether the medium is optically thin or optically thick along some characteristic length scale of the system, such as the thermal boundary layer thickness or an enclosure dimension. It is important in this case to be clear about what length scale is involved when a medium is characterized as being optically thick. For example, a rocket motor containing a gas suspension of liquid Al_2O_3 particles may be optically thick based on some motor dimension, such as diameter, but optically thin over the thermal boundary layer near a cool surface, depending on the oxide particle concentration and boundary layer thickness.

Radiative transfer analysis can also be classified according to the coupling of the transfer and energy equations. Since temperature appears in the transfer equation via the Planck function, in general, the two equations are coupled. This means that in general radiative transfer both influences the temperature field and in turn is influenced by the temperature field. In some situations (such as stellar interiors and other plasmas) radiative transport represents the dominant term in the energy equation and therefore radiation alone establishes the temperature field. This condition is known as radiative equilibrium. In other situations, energy transport by conduction is comparable to the radiative transport, such as in porous, nonevacuated insulations. Or it is also possible that convective energy transport may be comparable to radiative transport. In all of these situations the energy and radiative transfer equations must be solved simultaneously. This coupling between the energy and transfer equations represents a formidable mathematical challenge in some cases.

In many situations the temperature profile may be assumed to be established by some energy transport mechanism other than radiation, such as conduction or convection. In these cases the radiative analysis is simplified significantly because the energy and transport equations are decoupled. The simplest such case is that of a flow dominated by strong recirculative backmixing (e.g., high intensity, turbulent, continuous flow combustors). In these cases the medium can be characterized as isothermal as long as it is optically thin through the thermal boundary layer. Another situation of interest is when conduction establishes the temperature profile. In that case the radiative transport could be estimated (for a steady, one-dimensional planar medium with constant conductivity) by assuming a linear temperature profile. If the temperature is established by a convective-diffusive balance, then a boundary layer profile could be used to calculate the radiative transport from the transfer equation. In all of these cases, the energy and transfer equations are decoupled and the temperature field is assumed known. This assumption represents a considerable simplification in the analysis of the problem. Table 7-1 summarizes the various ways of classifying radiative transfer in participating media.

SUMMARY

Radiative transfer in participating media is governed by the transfer equation. The important properties that appear in the transfer equation are the single-scattering phase function, albedo, and optical depth. The albedo and optical depth in turn depend on the extinction, scattering, and absorption coefficients of the medium. The extinction, scattering, and absorption coefficients represent the inverses of the mean free photon path lengths for these three respective processes.

The integral of the transfer equation along a single line of sight is quite readily obtained. However, this integral does not represent a solution to the full transfer problem since the in-scattering term still requires a knowledge of the intensity distribution for directions other than just the line of sight being considered. The solution of the transfer equation is further discussed in Chapters 10 through 12 where application is made to specific geometric configurations (such as the planar slab) and all possible optical paths through the medium are considered.

REFERENCES

Altenkirch, R. A., Mackowski, D. W., Peck, R. E., and Tong, T. W. (1984), "Effects of Soot on Pyrometer Measured Temperatures in Pulverized-Coal Flames," *Combustion Science and Technology*, Vol. 41, pp. 327–335.

Boothroyd, S. A., Jones, A. R., Nicholson, K. W., and Wood, R. (1987), "Light Scattering by Fly Ash and the Applicability of Mie Theory," *Combustion and Flame*, Vol. 69, pp. 235–241.

Goody, R. M. and Yung, Y. L. (1989), *Atmospheric Radiation Theoretical Basis*, 2nd ed., Oxford University Press, New York.

Mackowski, D. W., Altenkirch, R. E., Peck, R. E., and Tong, T. W. (1983), "A Method for Particle and Gas Temperature Measurement in Laboratory-Scale, Pulverized-Coal Flames," *Combustion Science and Technology*, Vol. 31, pp. 139–153.

Thomas, D. L. (1968), "Problems in Applying the Line Reversal Method of Temperature Measurement to Flames," *Combustion and Flame*, Vol. 12, No. 6, pp. 541–549.

PROBLEMS

1. The absorption coefficient along a path through a nonscattering participating medium varies trapezoidally,

$$K_a(s) = \begin{cases} K_{a_0} s/s_1 & 0 < s < s_1 \\ K_{a_0} & s_1 < s < s_2 \\ K_{a_0}(4 - s/s_1) & s_2 < s < L \end{cases}$$

where s is the coordinate along the path, $s_1 = L/4$ and $s_2 = 3L/4$ [i.e., $K_a(s)$ is constant along the middle half of the path and decreases linearly and symmetrically at the outer edges]. Obtain an expression for the absorptivity along the path $\alpha'(L)$ as a function of $t_L(= K_{a_0}L)$.

2. The total blackbody intensity $I_b(s)$ varies linearly along an optical path s through a gray, nonscattering medium, while the absorption coefficient K_{a_0} is assumed to be constant.

$$I_b(s) = I_{b_0} + \left(I_{b_L} - I_{b_0}\right)\frac{s}{L}$$

Obtain an expression for the emissivity $\varepsilon'(L)$ based on $I_{b_L}[= I_b(T_L)]$ along this optical path as a function of optical thickness $t_L = K_a L$ and the ratio $I_{b_0}/I_{b_L} = (T_0/T_L)^4$. Show that as $T_0 \to T_L$ the emissivity approaches the absorptivity.

3. A 5-W continuous wave (cw) laser is incident on a uniform medium containing scattering, nonabsorbing particles as shown in the diagram. The vertical entrance slit of a monochromator ($\Delta s = 1$ mm) is imaged 1:1 at 90° by an $f/5$ optical system onto the laser beam passing through the medium. The height of the slit is greater than the diameter of the laser beam so that the slit image covers the beam. The scattering particles are monodisperse spheres with a 1.5 μm diameter, scattering efficiency of 1, phase function at 90° of 1, and volume fraction of 10^{-6}. Estimate the scattered power incident at the entrance slit.

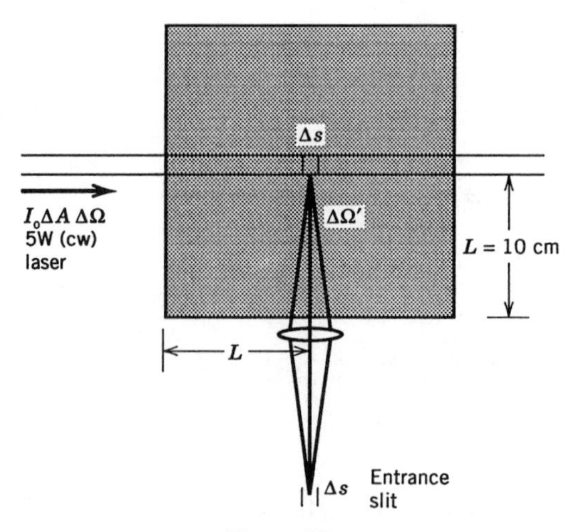

Figure P3

4. Show that the line reversal technique still holds (i.e., $I_{b_\lambda}(T) = I_{0_\lambda}$) for a sooty flame assuming that the soot temperature equals the gas temperature and that the soot is nonscattering. Note that the background continuum intensity is not just I_{0_λ} in this case. Assume that the soot is locally gray (soot spectral absorption coefficient independent of wavelength) in the spectral region near the line wavelength.

5. An otherwise nonparticipating medium is seeded with sodium atoms in order to determine the temperature of the medium by the sodium line reversal technique at $\lambda_0 = 0.589\ \mu m$. The temperature of the medium (T_e) is calculated assuming the medium is isothermal. It is then discovered that the temperature variation along the optical path $T(s)$ is such that the spectral blackbody intensity $I_{b_\lambda}[T(s)]$ at λ_0 actually varies linearly with distance as

$$I_{b_\lambda}(s) = I_{b_\lambda}(T_0) + \left[I_{b_\lambda}(T_L) - I_{b_\lambda}(T_0)\right]s/L$$

where T_0 is the temperature at the beginning of the path ($s = 0$) and T_L is the temperature at the end of the path ($s = L$). Assuming that the sodium vapor absorption coefficient is constant along s:

(a) Obtain an expression that could be solved for T_L assuming that T_0, T_e, and t_L (the optical depth at λ_0) are all known. (Hint: check the result by seeing if the limits $t_L \to 0$, $t_L \to \infty$, and $T_0 \to T_L$ give reasonable results.)

(b) Calculate T_L assuming $T_0 = 300$ K, $T_e = 977$ K, and $t_L = 1$.

6. Show how to obtain the divergence of the radiative flux vector Eq. (7-44a) from the transfer equation (7-41b).

7. Show that Eq. (7-65) can also be written as follows. Verify that using this form in Example 7-2 gives the same result as Eq. (7-65).

$$I_\lambda(s) = J_\lambda(s) + \tau_\lambda(0, s)\left[I_{0_\lambda} - J_\lambda(0)\right] - \int_0^s \tau_\lambda(s', s)\frac{\partial J_\lambda(s')}{\partial s'}\, ds'$$

8. The absorption coefficient of a collision-broadened gas radiation line has a Lorentz profile given by

$$K_{a_\lambda} = \frac{K_{a_{\lambda 0}}\gamma^2}{4(\lambda - \lambda_0)^2 + \gamma^2}$$

where $K_{a_{\lambda 0}}$ is the absorption coefficient at the center of the line, γ is the line width, and λ_0 is the line center wavelength. Self-absorption can occur when a line is optically thick at the line center over a path length with substantial temperature variation such that photons emitted by the hotter gas are reabsorbed by the colder gas. Consider a path length of 10 cm along which blackbody intensity is assumed to vary linearly as given in Problems 7-2 and 7-5. The temperature at $s = 0$ is $T_0 = 2500$ K and the temperature at $s = L$ is $T_L = 300$ K. For $\lambda_0 = 1000$ nm, $\gamma = 20$ nm, and $K_{a_{\lambda 0}} = 1\ \text{cm}^{-1}$ calculate and plot the spectral intensity emitted at $s = L$ as a function of wavelength from 800 to 1200 nm. Assume K_{a_λ} is constant along the path length.

9. Repeat Example 7-3 using the following model for the gas band:

$$K_{a_\lambda} = K_{a_{\lambda 0}} \frac{|\eta - \eta_0|}{\omega} \exp\left[\frac{-(\eta - \eta_0)^2}{\omega^2}\right]$$

(This is the rigid-rotator, harmonic oscillator molecular model, which is discussed in Chapter 8; see Fig. 8-13.) The bandwidth parameter ω has a value of 100 cm^{-1}. The center wavenumber is $\eta_0 = 3700\ \text{cm}^{-1}$ ($\lambda_0 = 2.7\ \mu\text{m}$).

10. Repeat Example 7-3 using the following model for the gas band:

$$K_{a_\lambda} = K_{a_{\lambda 0}} \exp\left[\frac{-2(\eta_0 - \eta)}{\omega}\right], \qquad \eta < \eta_0 \quad (\text{for } \eta > \eta_0, K_{a_\lambda} = 0)$$

(This is a model of a band that forms a head and degrades exponentially to the red such as the CO_2 4.3 μm band.) The bandwidth parameter is $\omega = 50\ \text{cm}^{-1}$ and the absorption coefficient at the bandhead is $K_{a_{\lambda 0}} = 2\ \text{cm}^{-1}$. The bandhead wavenumber is at $\eta_0 = 2410\ \text{cm}^{-1}$ ($\lambda_0 = 4.15\ \mu\text{m}$).

11. Combustion gases containing CO_2 and H_2O at 1700 K are flowing through a 3-cm-diameter porous ceramic tube as shown in the figure. The optical thickness of the gray, nonscattering ceramic, based on the 1 cm tube thickness, is $t_s = 3$. The gas has two prominent bands at $\lambda_{0_1} = 2.7$ μm (CO_2 and H_2O) and at $\lambda_{0_2} = 4.15\ \mu$m ($CO_2$). The 2.7-$\mu$m band can be modeled using the rigid-rotator, harmonic oscillator model of Problem 7-9 with $\omega_1 = 100\ \text{cm}^{-1}$ and $K_{a_{\lambda 0_1}} = 2\ \text{cm}^{-1}$ (independent of position). The 4.3-μm band can be modeled using the red-decaying exponential model of Problem 7-10 with $\omega_2 = 50\ \text{cm}^{-1}$ and $K_{a_{\lambda 0_2}} = 1\ \text{cm}^{-1}$ (independent of position). The gas fills the entire tube (hollow core and porous ceramic). Calculate and plot the spectral intensity emitted normal to the tube versus wavelength from 2 to 5 μm.

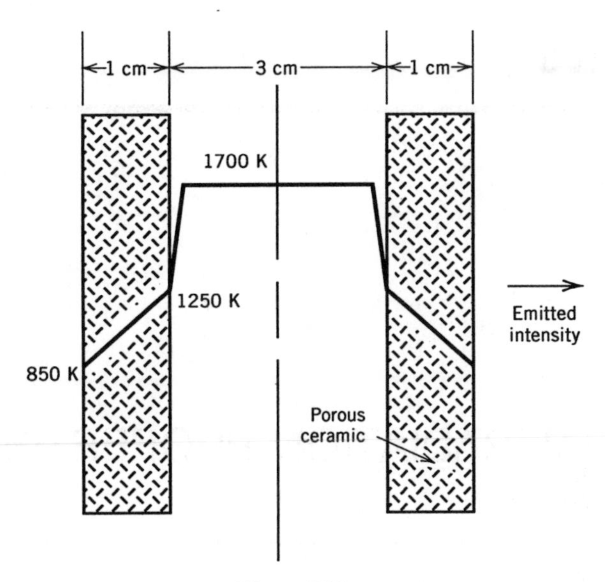

Figure P11

12. Estimate the importance of radiation by fly ash particles in pulverized coal flames by estimating the extinction coefficient and comparing with the values calculated for soot and char particles in Example 7-1. Assume that each char particle results in one ash particle upon burnout and that $N_{ash} \sim N_{char}$. Assume an average ash diameter of 10 μm [Boothroyd, et al., 1987], an average char particle size of 100 μm, char particle volume fraction of 10^{-4}, and an ash particle extinction efficiency of one.

13. Estimate the optical thickness of a 0.1-mm-thick paint layer containing pigment particles with a diameter of 0.3 μm, volume fraction of 0.3, and extinction efficiency of two. Would such a layer have good hiding power?

CHAPTER 8

THERMAL RADIATION PROPERTIES OF GASES

In this chapter the thermal radiative properties of gases are considered in more detail. This material is presented with two purposes in mind. One purpose is to provide a background of the important physics of gas radiation so that the reader may understand and appreciate the underlying principles that determine thermal radiative properties of gas mixtures. The other purpose is to allow the reader to become familiar with the use of band models for predicting gas radiation properties to be used in heat transfer analysis.

OVERVIEW OF GAS RADIATION PHYSICS

The internal energy state of a gas is characterized by four modes or types of energy states: (1) translational, (2) rotational, (3) vibrational, and (4) electronic [Penner, 1959; Goody and Yung, 1989]. The latter three modes are quantized,* that is, the energy states occur in discrete, quantum levels. When a gas molecule or atom undergoes a transition from one quantum energy state to another, it is possible for a photon to be emitted or absorbed if the energy states have an associated oscillating electric dipole. Such transitions are referred to as *radiative transitions*. Absorption is the type of radiative

*Translation is also quantized; but even at the lowest experimental temperatures, a quantum of translational energy is much smaller than the energy exchanged during collision and thus translational energy can be treated as effectively unquantized.

transition that occurs when a photon is absorbed by a radiating molecule causing the energy state of the molecule to increase from a lower level to a higher level. *Induced emission* is the opposite of absorption and occurs when a photon interacts with a radiating molecule causing the molecule to emit an identical photon. *Spontaneous emission* is a third type of radiative transition that occurs when a molecule in an excited state spontaneously emits a photon while dropping to a lower energy state.

When any of these types of radiative transitions occurs, the frequency of the emitted or absorbed photon corresponds to the energy difference between the two states divided by Planck's constant.

$$\nu = \frac{\Delta E}{h} \tag{8-1}$$

The relative magnitudes of ΔE for electronic, vibrational, and rotational state changes are $\Delta E_{\mathrm{elec}} > \Delta E_{\mathrm{vib}} > \Delta E_{\mathrm{rot}}$ as can be seen from the energy diagram for a diatomic molecule Fig. 8-1. Thus the frequency is in the visible and ultraviolet regions for electronic transitions, the infrared for vibrational transitions, and the far infrared and microwave regions for rotational transitions.

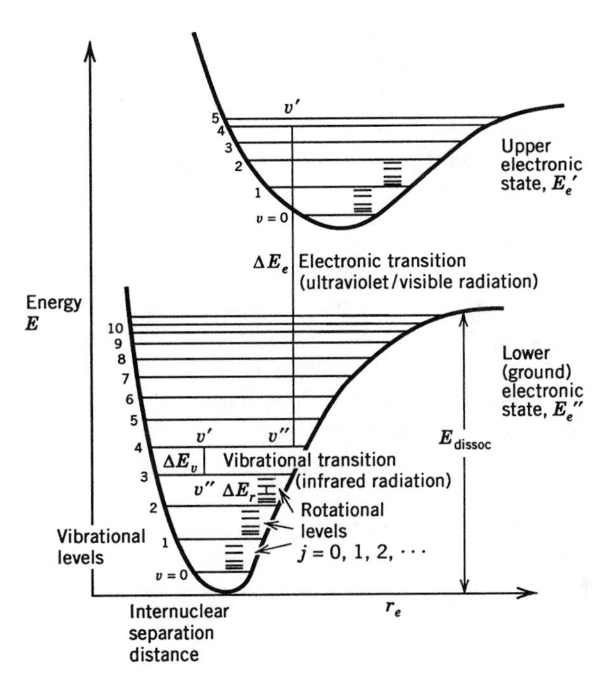

Figure 8-1 Energy diagram for diatomic molecule.

Radiative transitions can also be classified according to whether the beginning and ending energy states are "bound" or "free."

Bound-Bound Transitions. Bound-bound transitions are characterized by photon absorption and emission occurring at discrete frequencies associated with the energy difference between discrete quantum states. At moderate temperatures (below temperatures where ionization and dissociation occur) the majority of energy state transitions that occur in gases are bound-bound transitions. These include electronic transitions, which can occur in both atoms and molecules, rotational transitions, which can only occur in molecules, and vibrational transitions, which also can only occur in molecules. The electronic transitions that occur in monatomic gases, such as sodium and potassium vapor, give absorption and emission spectra consisting of discrete lines at the transition frequencies. The electronic transitions that occur in molecules are accompanied by simultaneous vibration and rotation transitions resulting in complex band systems consisting of many closely spaced rotational lines. Rotational and vibrational radiative transitions can only occur in molecules that have an oscillating dipole associated with the vibrational or rotational energy states. Such gases are referred to as infrared participating gases and include CO_2, H_2O, and CO. Gases such as N_2, O_2, and He are generally considered nonparticipating because they have no dipole moment associated with their rotation or vibration, and their electronic states are not readily excited. Usually vibrational and rotational transitions occur simultaneously, although rotational transitions can occur alone. The combination of rotational and vibrational transitions occurring together gives band spectra in the infrared region centered around the frequencies associated with the vibrational energy state transitions.

Bound-Free Transitions. The process of an atom being separated from a molecule as a result of photon absorption is referred to as photo-dissociation and occurs at temperatures of a few thousand degrees. Since the resulting free atom can occupy a continuum of translational energy states (depending on the velocity it acquires), the frequency necessary for this type of transition can be anything greater than that corresponding to the dissociation energy. The reverse process (photo-recombination) can also occur resulting in the emission of photons with energy equal to or greater than the dissociation energy. The resulting absorption and emission spectra are therefore semicontinuous in nature with a cutoff occurring at the dissociation energy, much like the energy band gap of semiconducting solids and liquids discussed in Chapter 4. Photoionization is also a bound-free process wherein an electron is removed from an atom by absorption of a photon with energy greater than the ionization energy. This process is important at high temperatures (> 6000 K) where the gas is really a weakly ionized plasma.

Free-Free Transitions. At temperatures on the order of 10,000 K and above free-free electronic transitions become important. The process of emission of photons by electrons undergoing free-free transitions is known as *Bremsstrahlung* and the absorption process is inverse *Bremsstrahlung*. Because both the initial and final electronic states are virtually continuous, the resulting emission and absorption spectra are continuous, much like the spectra of metals.

Of the various types of transitions discussed above, the most important transitions, as far as classical thermal radiative energy transfer is concerned, are the vibration-rotation transitions that involve photon energies in the infrared region. These transitions will be the primary subject of the remainder of this chapter.

Harmonic Oscillator Model

To illustrate the nature of vibrational and rotational transitions, the model of a rigid rotator, harmonic oscillator will be considered. From quantum theory the vibrational energy associated with discrete vibrational energy levels is given by

$$E_{\text{vib}} = h\nu_0\left(v + \tfrac{1}{2}\right) \tag{8-2}$$

where v is the vibrational quantum number, h is Planck's constant, and ν_0 is the oscillator frequency. From Eq. (8-1) the difference in energy between two vibrational states can be written in terms of an oscillator wavenumber η_0 as

$$E' - E'' = h\nu_0 = hc_0\eta_0 \tag{8-3}$$

The oscillator wavenumber can then be calculated from the vibrational quantum numbers, by combining (8-2) and (8-3).

$$\eta_0 = \frac{E' - E''}{hc_0} = \frac{\nu_0}{c_0}[v' - v''] \tag{8-4}$$

For the case of adjacent energy levels the oscillator wavenumber is simply

$$\eta_0 = \frac{\nu_0}{c_0} \qquad (v' = v'' + 1) \tag{8-5}$$

This wavenumber η_0 corresponds to the fundamental oscillation frequency. Figure 8-2 illustrates the vibrational energy level spacing for a harmonic oscillator. It should be noted that as a consequence of Eq. (8-2) the energy levels are equally spaced ($\Delta E = hc_0\eta_0$) and the photon wavenumber associated with transition between any two adjacent vibrational energy states is the

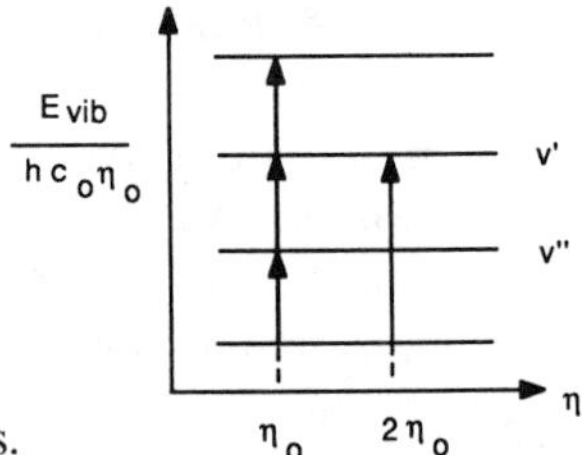

Figure 8-2 Harmonic oscillator vibrational energy levels.

same, namely, η_0. For the case of a transition between two energy levels skipping over an intermediate state ($v' = 2 + v''$) the wavenumber is simply $2\eta_0$. This wavenumber corresponds to the first overtone frequency. The strength of overtone bands is usually small compared with that of the fundamental frequency band.*

Rigid Rotator Model

Next consider the rotational energy states of a linear, rigid rotator. From quantum theory the energy associated with discrete rotational states of a rigid rotator is given by

$$E_{\text{rot}} = Bhc_0 j(j+1) \tag{8-6}$$

where

$$B = \frac{h}{8\pi^2 I c_0} \tag{8-7}$$

is the rotational constant, I is the angular moment of inertia,

$$I = r_e^2 \frac{m_1 m_2}{m_1 + m_2} \tag{8-8}$$

m_1 and m_2 are the atomic masses and r_e is the effective separation distance. The photon wavenumber η_j associated with rotational transition between states j'' and j' is then given by

$$\eta_j = \frac{E' - E''}{hc_0} = B[j'(j'+1) - j''(j''+1)] \tag{8-9}$$

*Selection rules of quantum theory prohibit a harmonic oscillator from (a) changing by increments of more than one in vibrational quantum number (overtone bands) and (b) changing more than one vibrational quantum number at a time (combination and difference bands). However, anharmonic effects in real molecules allow both of these possibilities.

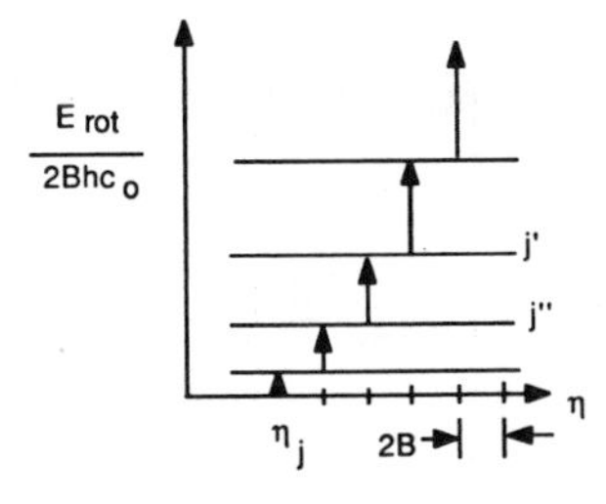

Figure 8-3 Rigid rotator rotational energy levels.

For adjacent rotational energy levels* the wavenumber reduces to

$$\eta_j = 2Bj' \qquad (j' = j'' + 1) \tag{8-10}$$

The frequency of a photon absorbed or emitted by a linear molecule undergoing a change from one rotational energy level to an adjacent energy level is proportional to the rotational quantum number. This situation is fundamentally different than that for vibrational energy states where the frequency was independent of quantum number. Figure 8-3 illustrates the energy level spacing for rotational energy states. A consequence of the spacing between rotational energy levels given by Eq. (8-6) is that absorption occurs at different frequencies, even for adjacent levels. Furthermore, the spectral spacing between rotational lines is a constant which is equal to $2B$ in units of wavenumber.

Rigid Rotator, Harmonic Oscillator Model

The rigid rotator and harmonic oscillator models can be combined to give an essentially complete model of the vibrational-rotational transitions of a linear molecule by simply adding the energies for vibration and rotation.

$$E = E_{\text{vib}} + E_{\text{rot}} \tag{8-11}$$

Substituting (8-2) and (8-6) into (8-11) and solving for the combined rotational and vibrational transition wavenumber gives

$$\eta_j = \frac{E' - E''}{hc_0} = \eta_0 \pm 2Bj' \quad \text{(rigid rotator)} \tag{8-12a}$$

for adjacent vibrational states ($v' = v'' + 1$) and adjacent rotational states ($j' = j'' \pm 1$). Therefore, Eq. (8-12a) is a complete representation of the spectral spacing of the combined rotation-vibration lines of a linear rigid rotator, harmonic oscillator (RRHO).

*Selection rules for rotation permit only $\Delta j = 0, \pm 1$.

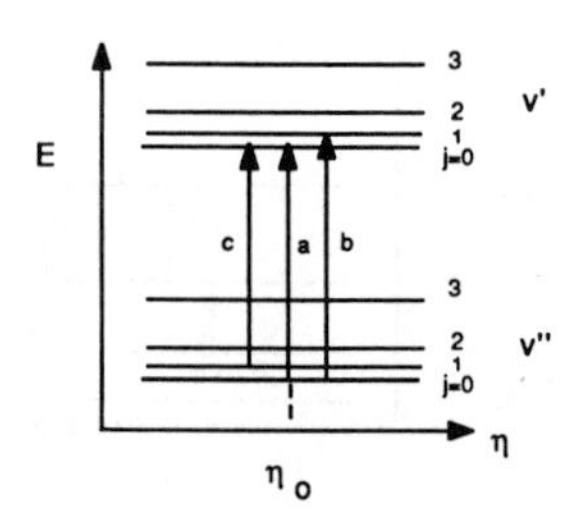

Figure 8-4 Rigid rotator harmonic oscillator vibrational-rotational energy levels.

The energy level diagram for a RRHO is illustrated in Fig. 8-4. It should be noted that the energy spacing between rotational levels is much less than that between vibrational levels, as indicated. In Fig. 8-4 three types of transitions are illustrated: (a) a transition from a lower to higher vibrational level with no change in rotational energy (a transition that rarely occurs due to the rotational energies being so much less than the vibrational energy), (b) a transition from a lower vibrational level to a higher vibrational level together with an increase in the rotational level, and (c) an increase in vibrational energy with a decrease in rotational energy. This combination of closely spaced spectral lines forms a *vibration-rotation band* about the vibrational frequency η_0.

Figure 8-5*a* illustrates what a vibration-rotation band would look like for a rigid rotator harmonic oscillator with the influence of line broadening neglected. The upper diagram of rotational quantum number j versus wavenumber is called a *Fortrat* diagram. It can be seen that there are two branches formed, an R branch and a P branch. The R, or positive, branch forms to the high frequency side of the band origin (η_0). It is formed by transitions in which both Δj and Δv have the same sign. The P, or negative, branch is formed by transitions in which Δj and Δv have opposite signs. Owing to the rigid rotator assumption both branches appear as straight lines since j is linear with η [see Eq. (8-12a)]. The lower diagram of Fig. 8-5*a* shows how the absorption coefficient or spectral emissivity varies for each of the lines. At the band origin the absorption coefficient is small for two reasons; very few molecules populate the ground rotational state ($j = 0$) and the probability of a vibrational transition occurring without a rotational transition is small, as noted earlier. The absorption coefficient reaches a maximum away from the center of the band origin because the most populated rotational energy levels are the intermediate ones. In the wings or tails of the band far from the origin, the absorption coefficient drops off because the population of the highest rotational energy levels also drops off.

Figure 8-1 shows that the assumption of a rigid rotator, harmonic oscillator is an approximation of the real behavior of a diatomic molecule. Generally, when a molecule undergoes a vibrational transition with $\Delta v > 0$, the internuclear separation distance r_e increases, and thus the rotational moment of inertia I increases. This causes the rotational constant B to be

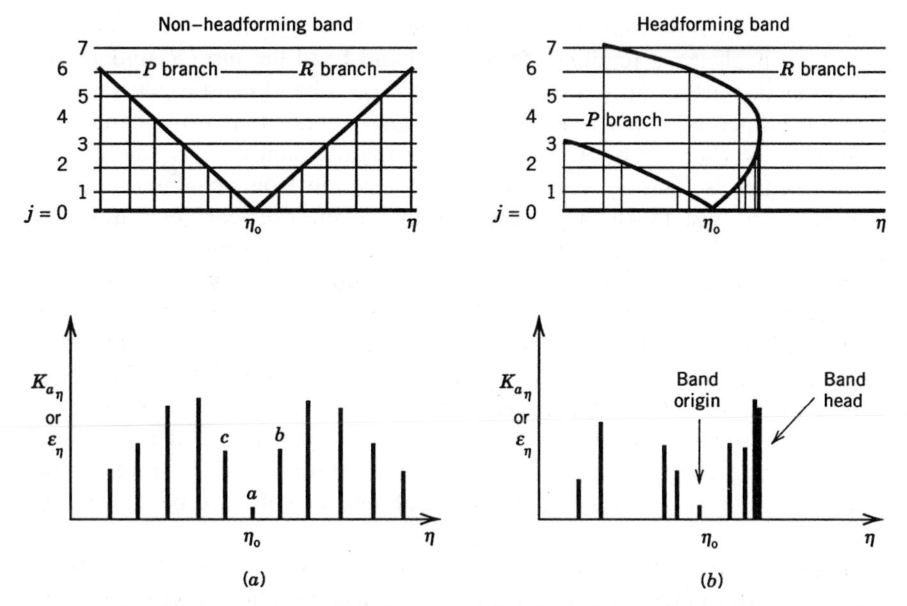

Figure 8-5 Vibration-rotation bands without line broadening; (*a*) non-head-forming band and (*b*) head-forming band ($B' < B''$).

different for the upper and lower states [see Eq. (8-7)]. The upper state has rotational constant B' and the lower state has rotational constant B''. For nonconstant B the expression for rotational transition frequency Eq. (8-12a) becomes Eq. (8-12b).

$$\eta_j = \eta_0 \pm (B' + B'')j' + (B' - B'')j'^2 \quad \text{(nonrigid rotator)} \quad (8\text{-}12b)$$

Equation (8-12b) is the equation for a parabola. The two branches of the Fortrat diagram now appear as parts of a parabola as shown in the upper diagram of Fig. 8-5*b* instead of as straight lines.

Figure 8-5*b* shows the case corresponding to an increase in moment of inertia in going from the lower vibrational state v'' to the upper vibrational state v' ($r_e' > r_e''$; $I' > I''$; $B' < B''$). This is the usual case for vibrational-rotational transitions as can be seen from Fig. 8-1. The value of j where the extremum in the parabola occurs can be estimated from Eq. (8-12b) as $j = B/\Delta B \gg 1$ assuming $B' - B'' = \Delta B \ll B''$, which is also the usual case for vibration-rotation bands. In the event that ΔB is large (such as the 4.3 μm CO_2 band) the extremum occurs at small values of j and the band forms a head as shown in Fig. 8-5*b*. Such bands are said to be red-degraded since the head forms at the high frequency side of the band origin and the band degrades to low frequencies (long wavelengths). The formation of a head is actually rare in vibration-rotation bands because usually $\Delta B \ll B$. Nevertheless there is always a tendency to form a band head because B' and B'' are never equal. Furthermore, the tendency to form a band head is always in the R branch since $B' < B''$. It is also interesting to note that while band head formation is the exception in vibration-rotation bands, it is the rule in electronic bands. This is because changes in the electronic configuration of a molecule produce much larger changes in the internuclear separation distance r_e than do vibrational transitions as can be seen in Fig. 8-1.

Nonrigid rotator effects can also be accounted for by introducing an equivalent rotational constant B_e [Ludwig, et al., 1973; Goody and Yung, 1989]. Anharmonic vibration effects can be adequately modeled by introducing a power series in vibrational quantum number into Eq. (8-12). To obtain good agreement between predicted and measured spectra, it is necessary to incorporate these nonideal effects into the oscillator model. For the purpose of illustrating the basic physics of vibration-rotation bands, however, the RRHO model is adequate, and nonideal effects will not be considered further.

The vibrational and rotational band properties of some important diatomic and triatomic molecules will next be considered in further detail. First a common diatomic molecule, carbon monoxide, will be considered, and then two triatomic molecules, carbon dioxide and water, will be discussed.

Diatomic Molecule: Carbon Monoxide

The fundamental oscillator wavenumber for stretching of the CO molecule is $\eta_0 = 2143\ \text{cm}^{-1}$ and the corresponding wavelength is $\lambda_0 = 4.67\ \mu\text{m}$. The first overtone occurs at $4286\ \text{cm}^{-1}$ or $2.34\ \mu\text{m}$. These bands are listed in Table 8-1 according to spectral location, incremental vibration quantum number ($\delta = v' - v''$), and band type.

The rotational constant for CO is $B = 1.94\ \text{cm}^{-1}$. Therefore the pure rotational transitions occur at wavenumbers of $\eta_j = 2Bj\ (j = 1, 2, \dots)$. These wavenumbers correspond to wavelengths of 2577, 1288, and 859 μm, ..., which are in the millimeter wave region and are relatively unimportant for thermal radiative transfer. The combined rotation-vibration transitions, however, occur at frequencies of $\eta_j = \eta_0 \pm 2Bj\ (j = 1, 2, \dots)$ or $2143 \pm 3.88\ \text{cm}^{-1}$, $2143 \pm 7.76\ \text{cm}^{-1}$, and $2143 \pm 11.64\ \text{cm}^{-1}$ etc. The corresponding

TABLE 8-1 CO Vibration-Rotation Bands

$\lambda_0\ (\mu\text{m})$	$\eta_0\ (\text{cm}^{-1})$	δ	Type
4.67	2143	1	Fundamental
2.34	4286	2	Overtone

wavelengths are

$$\lambda_j\ (\mu\text{m}) = \frac{10^4}{2143 \pm 3.88}, \frac{10^4}{2143 \pm 7.76}, \frac{10^4}{2143 \pm 11.64}, \ldots$$

which gives an array of spectral lines closely spaced around the fundamental vibrational wavelength of 4.67 μm.

Triatomic Molecules: CO_2 and H_2O

Triatomic vibration will be illustrated using two common molecules, a nonlinear molecule (H_2O) and a linear molecule (CO_2). As illustrated in Fig. 8-6 triatomic molecules are characterized by three modes of vibration: (1) symmetric stretching, (2) bending, and (3) asymmetric stretching.

The important bands for these two molecules are listed in Tables 8-2 and 8-3. The fundamental vibration wavenumbers are listed as η_1, η_2, and η_3 corresponding to each of the three fundamental modes of vibration. A statistical weight factor g_k is also listed indicating the number of ways that a particular vibration mode can occur. The statistical weight is equal to 1 for all of the modes except bending in the linear CO_2 molecule in which case it is 2 because a linear molecule can vibrate by bending with two degrees of freedom as illustrated in Fig. 8-6.

Tables 8-2 and 8-3 show that there are several more bands than just those corresponding to the fundamental and overtone bands. The reason is that now, in addition to the fundamental and overtone bands, there are also combination and difference bands, which can occur when transitions occur

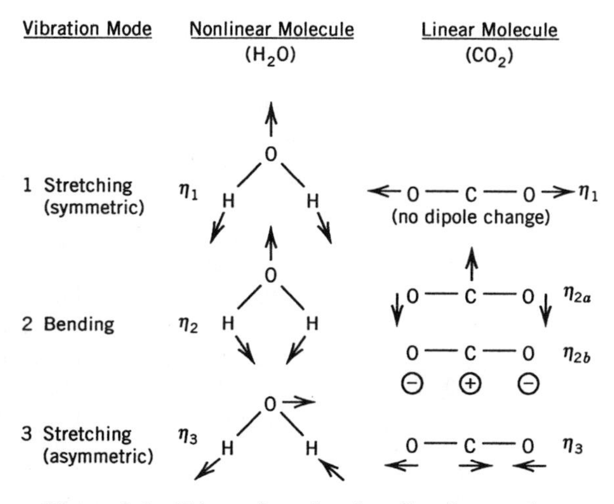

Figure 8-6 Triatomic molecular vibration modes.

TABLE 8-2 CO_2 Vibration-Rotation Bands

Fundamental Vibration Mode	Wavenumber (cm^{-1})	Statistical Degeneracy
Symmetric stretching	$\eta_1 = 1351$	$g_1 = 1$
Bending	$\eta_2 = 667$	$g_2 = 2$
Asymmetric stretching	$\eta_3 = 2396$	$g_3 = 1$

λ_0 (μm)	η_0 (cm^{-1})	$\delta_1\ \delta_2\ \delta_3$	Type
15	667	0, 1, 0	Fundamental η_2
10.4	960	−1, 0, 1	Difference
9.4	1060	0, −2, 1	Overtone/difference
4.3	2410 (bandhead)	0, 0, 1	Fundamental η_3, red-degraded
2.7	3660	1, 0, 1	Combination
2.0	5200	2, 0, 1	Overtone/combination

between two different vibrational modes simultaneously. When two vibrational modes increase in vibrational level simultaneously, the resulting band is a combination band. When one vibrational mode increases while another decreases, the resulting band is a difference band.

With this much background on the underlying physics of vibrational-rotational transitions and the spectral location of vibration-rotation bands, attention will now be turned to the issue of determining the magnitude or intensity of these lines. To do so it is necessary to consider the concepts of line shape and line broadening.

TABLE 8-3 H_2O Vibration-Rotation Bands

Fundamental Vibration Mode	Wavenumber (cm^{-1})	Statistical Degeneracy
Symmetric stretching	$\eta_1 = 3652$	$g_1 = 1$
Bending	$\eta_2 = 1596$	$g_2 = 1$
Asymmetric stretching	$\eta_3 = 3756$	$g_3 = 1$

λ_0 (μm)	η_0 (cm^{-1})	$\delta_1\ \delta_2\ \delta_3$	Type
6.3	1600	0, 1, 0	Fundamental η_2
4.7	2130	1, −1, 0	Difference
		0, −1, 1	Difference
2.7	3760	0, 2, 0	Overtone
		1, 0, 0	Fundamental η_1
		0, 0, 1	Fundamental η_3
1.87	5350	0, 1, 1	Combination
1.38	7250	1, 0, 1	Combination

Line Broadening

Pure spectral lines as illustrated in Fig. 8-5 cannot absorb or emit any energy because the probability of a photon having exactly the correct, required transition frequency approaches zero as the spectral width of the line becomes vanishingly small. There are, however, several mechanisms that act to broaden spectral lines, making it possible for a molecule undergoing a particular quantum transition of frequency ν_j to absorb or emit photons with frequencies $\nu_j \pm d\nu$, which are slightly higher or lower than the transition frequency. These mechanisms include collision (or pressure) broadening, Doppler broadening, and natural broadening.

Collision broadening is usually the most important of these mechanisms. The mechanism of collision broadening is that a molecule undergoing a collision with another molecule can still absorb (or emit) a photon with a frequency slightly different than the transition frequency ($\nu_j \pm d\nu$), and the energy deficit or surplus $h\,d\nu$ will be supplied or taken away by the collision.

$$h\nu_j = h(\nu_j \pm d\nu) \mp \text{collision energy} \tag{8-13}$$

In Doppler broadening the difference in frequency is made up by the difference in the velocity of the molecule relative to the photon, and in natural broadening the difference in frequency corresponds to the uncertainty in the momentum of the molecule associated with the Heisenberg uncertainty principle. However, both of these mechanisms are usually unimportant relative to collision broadening (except in atmospheric radiation at high altitudes where Doppler broadening is dominant).

Lorentz Profile

The broadening of spectral lines by collisions can be predicted from quantum theory by showing that collisions result in a broadening of energy states. However, this approach is equivalent to the damping mechanism that was introduced in the classical oscillator model of Chapter 4. Recalling the results of dispersion theory from Chapter 4, the optical constants for gases were given, in terms of circular frequency, as

$$n \approx 1 \tag{8-14}$$

$$k_j = \frac{1}{2}\frac{f_j N_j e^2}{m_j \varepsilon_0} \cdot \frac{\gamma_j \omega}{4(\omega - \omega_j)^2 \omega^2 + \gamma_j^2 \omega^2} \tag{8-15}$$

and the associated volume absorption coefficient was

$$K_{a_{\eta j}} = \frac{4\pi k_j}{\lambda} = \frac{2\omega k_j}{c_0} = \frac{f_j N_j e^2}{m_j \varepsilon_0 c_0} \cdot \frac{\gamma_j}{4(\omega - \omega_j)^2 + \gamma_j^2} \tag{8-16}$$

which is the Lorentz profile. A new factor has been added to Eqs. (8-15) and (8-16), which is the *oscillator strength* f_j. This parameter represents the fraction of oscillators that participate in the radiative transition.

Mass Absorption Coefficient

Dividing the volume absorption coefficient by the partial density of the absorbing species ρ_a and converting from circular frequency to wavenumber using

$$\omega = 2\pi\nu = \frac{2\pi c_0}{\lambda} = 2\pi c_0 \eta \tag{8-17}$$

gives the *mass absorption coefficient* κ_j as

$$\kappa_j = \frac{K_{a_{\eta j}}}{\rho_a} = \frac{f_j N_j e^2}{\rho_a m_j \varepsilon_0 c_0^2 4\pi} \frac{\Delta_j}{(\eta - \eta_j)^2 + \Delta_j^2} \tag{8-18}$$

where

$$\Delta_j = \frac{\gamma_j}{4\pi c_0} \tag{8-19}$$

is the line half-width in units of wavenumber (cm^{-1}). The mass absorption coefficient is often used in the field of gas radiation properties, as opposed to the volume absorption coefficient, to remove the primary pressure dependence associated with the mass of material in the line of sight. (The secondary pressure dependence, which cannot be removed, is the influence of pressure on broadening.)

Integrated Line Intensity

The area under the curve of κ_j is called the *integrated line intensity* S_j and is a measure of the strength of the line.

$$S_j = \int_{-\infty}^{\infty} \kappa_j \, d(\eta - \eta_j) = \frac{f_j N_j e^2}{4\rho_a m_j \varepsilon_0 c_0^2} = \frac{f_j e^2}{4 m_j^2 \varepsilon_0 c_0^2} \tag{8-20}$$

In Eq. (8-20) the relation $m_j = \rho_a / N_j$ relating the mass of an oscillator to the partial density and number density has been used. With Eq. (8-20) the mass absorption coefficient (8-18) can be rewritten as

$$\kappa_j = \frac{S_j}{\pi} \frac{\Delta_j}{\Delta_j^2 + (\eta - \eta_j)^2} \tag{8-21}$$

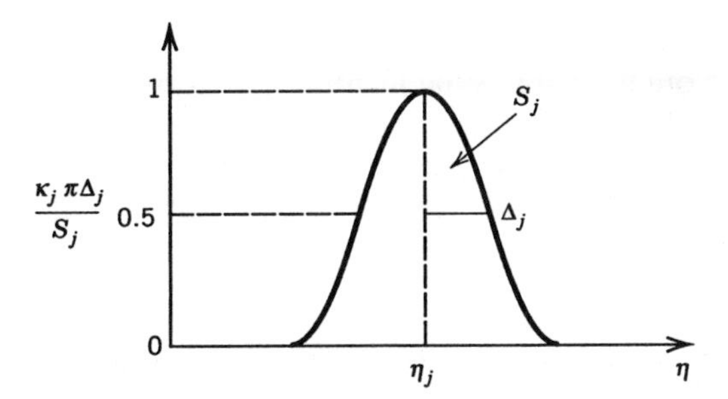

Figure 8-7 Lorentz line profile.

Figure 8-7 shows the shape of the collision-broadened, Lorentz line profile. The three parameters in the profile are the line wavenumber η_j, the line half-width Δ_j, and the integrated line intensity S_j. For a gas at atmospheric pressure and temperature, the line width is on the order of 10^{-1} cm^{-1} while the spacing between rotation lines is on the order of 1–5 cm^{-1}. The kinetic theory of gases predicts that the line width should vary as the first power of pressure and inverse square root of temperature. Thus the line width at conditions other than atmospheric can be estimated by scaling using a factor of $p/\sqrt{T}$. The line intensity S_j is independent of pressure and its temperature dependence can be estimated using vibrational and rotational partition functions as discussed by McLatchey, et al. [1973].

Monochromatic Emissivity

Consider the pure monochromatic emissivity, denoted by ε_{η_j}, which is the emissivity over a spectral region much narrower than the line width. Defining the *mass path length* X as the integral of partial density of the absorbing species along path length s,

$$X = \int_0^s \rho_a \, ds' \tag{8-22a}$$

$$X = \rho_a s \quad \text{(homogeneous path)} \tag{8-22b}$$

the monochromatic emissivity along an isothermal, homogeneous line of sight can be written using Eqs. (7-8) and (8-18) as

$$\varepsilon_{\eta_j} = 1 - \exp\left[-\kappa_j X\right] \tag{8-23}$$

As can be seen from (8-23) the monochromatic emissivity obeys Beer's law. However, it is not a very useful quantity because it varies so rapidly with wavenumber, as evidenced by Eq. (8-21), as to make line-by-line integration unfeasible. A typical heat transfer calculation might involve 10^5 to 10^6 such lines. Given the other complicating factors involved with the transfer problem (i.e., spatial and directional integration), such a detailed spectral integration is not a justifiable use of computational resources. A more useful kind of emissivity, at least as far as a single line is concerned, would be one that was integrated with respect to wavenumber.

Integrated Line Emissivity and Effective Line Width

Consider the emissivity that is obtained when the monochromatic emissivity is integrated with respect to wavenumber over a single line:

$$\varepsilon_{\Delta\eta_j} = \frac{\int_0^\infty \varepsilon_{\eta_j} I_{b_\eta}\, d\eta}{\int_0^\infty I_{b_\eta}\, d\eta} \tag{8-24}$$

This emissivity can be referred to as the *integrated line emissivity*. Over the narrow spectral region where ε_{η_j} is nonzero, the Planck function can be assumed to be constant and taken out of the spectral integration, leaving

$$\varepsilon_{\Delta\eta_j} = \frac{\pi}{\sigma T^4} I_{b_{\eta_j}} A_j \tag{8-25}$$

where the *effective line width* A_j is defined as

$$A_j = \int_{-\infty}^{\infty} \varepsilon_{\eta_j}\, d(\eta - \eta_j) \tag{8-26}$$

and the limits in (8-26) mean integration over one single broadened line. The interpretation of the effective line width is the width of an equivalent black rectangle with the same area, as shown in Fig. 8-8. The integral of Eq. (8-26)

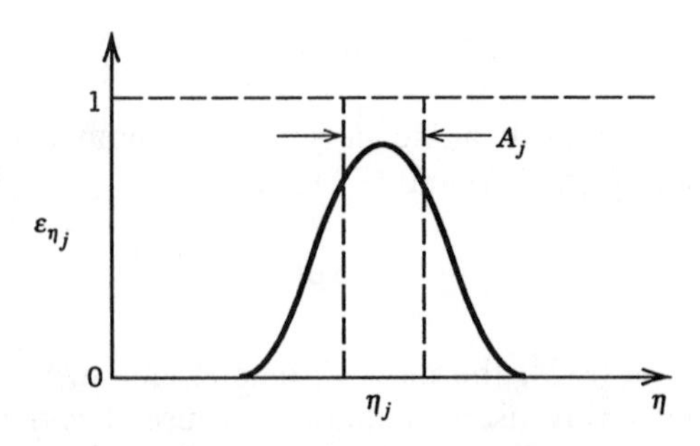

Figure 8-8 Effective line width.

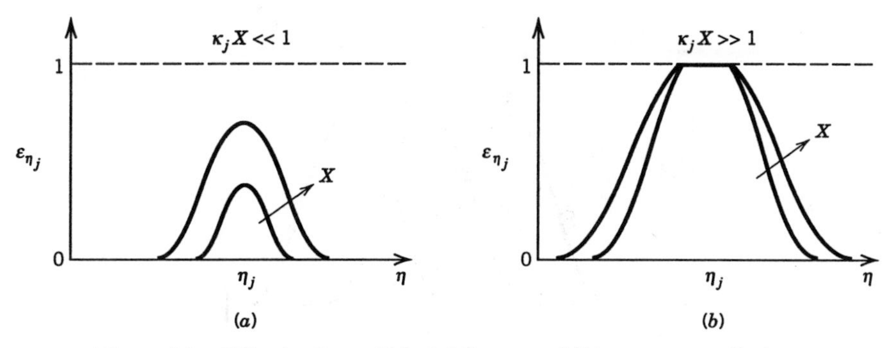

Figure 8-9 Effective line width; (*a*) linear and (*b*) square root limits.

can be expressed, for a Lorentz line profile, in terms of Bessel's functions as discussed by Ludwig, et al. [1973]. However, more important than the exact expression are the two limiting expressions for optically thin (weak) and optically thick (strong) lines.

$$A_j = XS_j \; (\kappa_j X \ll 1, \text{weak line, independent of shape}) \tag{8-27a}$$

$$A_j = 2\sqrt{\Delta_j X S_j} \; (\kappa_j X \gg 1, \text{strong line, Lorentz}) \tag{8-27b}$$

The effective line width and integrated line emissivity vary linearly with mass path length for weak lines, and as the square root of mass path length for strong lines. Thus it can be seen that the integrated line emissivity does not obey Beer's law. Figure 8-9 shows the linear and square root limits for integrated line emissivity. The linear limit is simply a consequence of the linearization of the exponential function for small arguments (i.e., truncated Taylor series). For small mass path lengths, the area under the curve of ε_{η_j} increase proportionally with X (Fig. 8-9*a*) while for large mass path lengths the increase in area with X reduces to a square root dependence because the center of the line has become saturated and only the "wings" or outer edges of the line are contributing additional area. The relation between the effective line width A_j and path length is called the *curve of growth*.

Line Overlap

When closely spaced vibration-rotation lines are sufficiently broadened by collisions there is the possibility that adjacent lines and possibly even more remotely spaced lines can overlap each other as shown in Fig. 8-10.

The overlap parameter β_j is defined to characterize the extent of line overlapping.

$$\beta_j = \frac{4\Delta_j}{d_j} \tag{8-28}$$

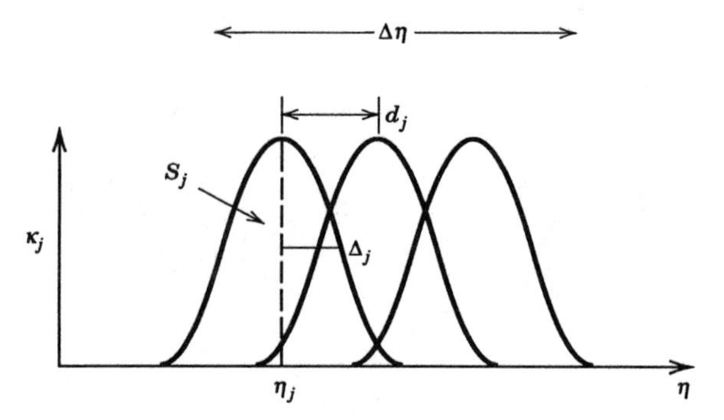

Figure 8-10 Overlapping lines.

When the line half width Δ_j is small relative to the line spacing d_j ($\beta < 1$), then the overlapping is weak and when the opposite is true ($\beta > 1$) the overlapping is strong.

When overlapping occurs, the question arises of how to handle the influence of overlapping lines, which are now "competing" to absorb the same photons. The correct method for treating the effect of overlapping lines is to sum the spectral volumetric absorption coefficients. Therefore, the spectral volumetric absorption coefficient is the sum of the coefficients of all the lines that contribute to absorption.

$$K_{a_\eta} = \sum_j K_{a_{\eta j}} \tag{8-29}$$

For cases where neighboring lines have the same partial density ρ_a, as in the case of neighboring lines in a vibration-rotation band, the summation can be distributed over partial density, and the spectral mass absorption coefficients are also additive (this would not be true for neighboring lines of different gases).

$$\kappa_\eta = \sum_j \kappa_j \tag{8-30}$$

Now consider what effect overlapping of lines has on the spectral emissivity. Using Eqs. (8-23) and (8-30) the spectral emissivity is

$$\varepsilon_\eta = 1 - \exp\left[-X \sum_j \kappa_j\right] = 1 - \exp\left[-X\kappa_\eta\right] \tag{8-31}$$

Therefore the spectral transmissivity is

$$\tau_\eta = \exp\left[-X\sum_j \kappa_j\right] = \prod_j \exp[-X\kappa_j] = \prod_j \tau_{\eta_j} \tag{8-32}$$

which indicates that the spectral transmissivities of neighboring lines are multiplicative to give the net spectral transmissivity. The net spectral emissivity is then

$$\varepsilon_\eta = 1 - \tau_\eta = 1 - \prod_j \left(1 - \varepsilon_{\eta_j}\right) \tag{8-33a}$$

For example, for two neighboring lines with spectral emissivities of ε_{η_1} and ε_{η_2}, the net spectral emissivity in the neighborhood of either line ε_η is

$$\varepsilon_\eta = 1 - \left(1 - \varepsilon_{\eta_1}\right)\left(1 - \varepsilon_{\eta_2}\right) = \varepsilon_{\eta_1} + \varepsilon_{\eta_2} - \varepsilon_{\eta_1}\varepsilon_{\eta_2} \tag{8-33b}$$

which shows that spectral emissivities are additive minus a multiplicative overlap term. Integrating (8-33b) with respect to wavenumber over both lines gives

$$A_{12} = A_1 + A_2 - \int_0^\infty \varepsilon_{\eta_1}\varepsilon_{\eta_2}\, d\eta \tag{8-34}$$

which shows that effective line widths are also additive minus an integrated multiplicative overlap term, which is a function of β. When both overlapping lines are weak ($\varepsilon_{\eta_{1,2}} \ll 1$), the overlap term will be negligible and the effective line width A_{12} will be linear in X just like a single weak line [Eq. (8-27a)].

BAND MODELS

From the previous discussion, it can be seen that the spectral emissivity ε_{η_j} is a very rapidly varying function of wavenumber superposed on a gradually varying function. The rapid variation is associated with the line structure as indicated in Fig. 8-10 and occurs on the order of inverse centimeters. The gradual variation is associated with the line intensity $S_j(\eta)$ and occurs on the order of tens and hundreds of inverse centimeters. Thus the radiative properties of gases can be described at various levels of complexity, depending on the scale of the spectral resolution involved.

At the limit of theoretically perfect spectral resolution is the line-by-line description. This description accounts for detailed line shape in terms of the pure monochromatic emissivity ε_{η_j} (Fig. 8-11*a*). Line-by-line parameters for many gases (H_2O, CO_2, O_3, N_2O, CO, CH_4, and many more) are available

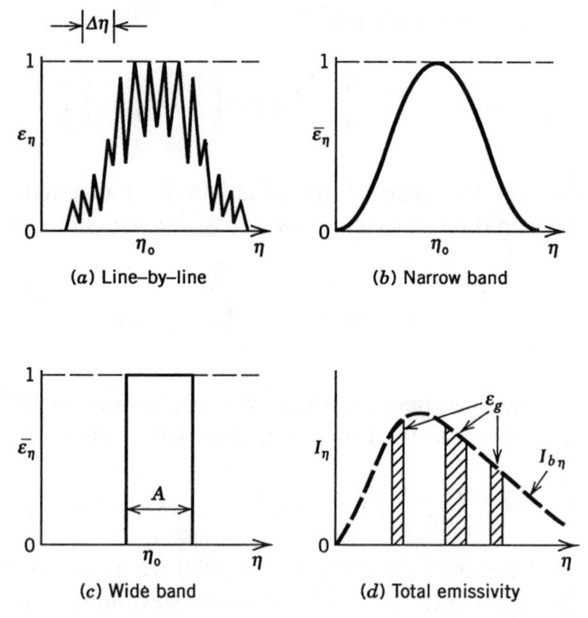

Figure 8-11 Various descriptions of gas emissivity; (*a*) line by line, (*b*) narrow band, (*c*) wide band, (*d*) total emissivity.

on the AFGL tape (formerly AFCRL), which can be obtained from the National Climatic Data Center. Documentation for the line parameter data is given by McLatchey, et al. [1973], Park, et al. [1981], and Rothman [1983]. The line-by-line description, however, is usually too detailed and time-consuming for heat transfer calculations and will not be discussed further here.

At the next level of spectral detail are *narrow band models*, which integrate out the rapid spectral variations associated with line structure and characterize the gas properties in terms of a *smoothed spectral emissivity* $\bar{\varepsilon}_\eta$ (Fig. 8-11*b*). At a coarser level are *wide band models*, which integrate out the gradual spectral variation associated with line intensity $S_j(\eta)$ and characterize an entire band in terms of an *effective bandwidth* A (Fig 8-11*c*). At a still coarser level is the total emissivity approach, which integrates over all bands and characterizes the gas in terms of a total emissivity or absorptivity (Fig. 8-11*d*).

A comparison of the predictions of a wide band model and a narrow band model are shown in Fig. 8-12. In this illustration the spectral intensity emitted by the 6.3-μm band of water at 500 K along a path length of 1 m is calculated by both models. It can be seen how the narrow band model description is able to represent the spectral structure of the band rather well, whereas the wide band model gives an equivalent black area representation of the band.

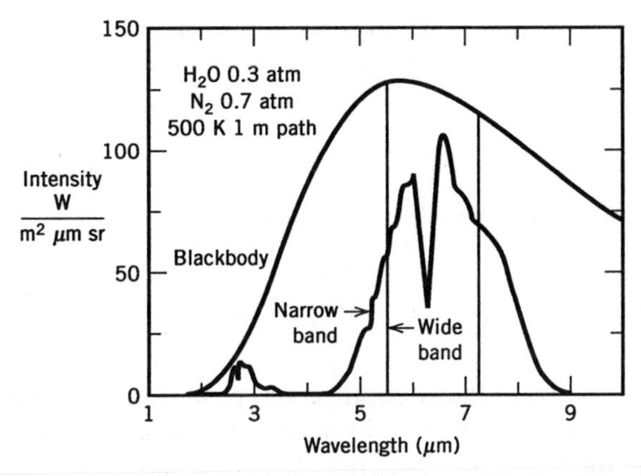

Figure 8-12 Comparison of narrow and wide band models for 6.3 μm water band.

Two narrow band models, the regular (Elsasser) and the random (Goody) models, and two wide band models, the rigid rotator, harmonic oscillator (RRHO) and the Edwards exponential models, are discussed in more detail as follows. Further discussion of band models can also be found in Tien [1968], Cess and Tiwari [1972], and Tiwari [1978].

Narrow Band Models

Narrow band models are used to describe the essential features of spectral line structure with spectral variations on the order of inverse centimeters. These models utilize local integration or spectral smoothing over a spectral region $\Delta\eta$ (see Fig. 8-11), which is narrow with respect to the entire band width but wide with respect to a line width. The essence of a narrow band model is to specify the functional dependence of spectral absorption coefficient as a function of wavenumber.

$$\kappa_j = f\left(\beta_j, \frac{S_j}{d_j}, \frac{\eta - \eta_j}{d_j}\right) \tag{8-35}$$

Given a functional relationship as indicated by (8-35), the spectral emissivity can be integrated over a small interval $\Delta\eta$ to give the smoothed spectral emissivity.

$$\bar{\varepsilon}_\eta = \frac{1}{\Delta\eta}\int_{\Delta\eta} \varepsilon_\eta \, d\eta = \frac{1}{\Delta\eta}\int_{\Delta\eta} \left[1 - \exp(-\kappa_\eta X)\right] d\eta \tag{8-36}$$

The smoothed spectral emissivity is a function of two parameters, u_η and β

where β is the mean line width-to-spacing parameter

$$\beta = 4\frac{\bar{\Delta}}{d} \tag{8-37}$$

and u_η is the smoothed spectral optical depth

$$u_\eta = X\frac{\bar{S}}{d} = X\bar{\kappa}_\eta \tag{8-38}$$

The smoothed spectral optical depth is the product of the mass path length X and the smoothed spectral absorption coefficient, which can also be interpreted as the mean line intensity-to-spacing ratio.

$$\bar{\kappa}_\eta = \frac{\bar{S}}{d} = \frac{1}{\Delta\eta}\int_{\Delta\eta} \kappa_\eta \, d(\eta - \eta_j) \tag{8-39}$$

Elsasser Narrow Band Model

The Elsasser narrow band model [Elsasser, 1943] assumes that all lines are identical, that is, have the same width and intensity, and are equally spaced. The spectral absorption coefficient is then obtained by using Eqs. (8-21) and (8-30) to sum over an infinite array of identical lines giving

$$\kappa_\eta = \frac{S}{\pi}\sum_{n=-\infty}^{\infty}\frac{\Delta}{\Delta^2 + (\eta - \eta_j - nd)^2} = \frac{S}{d}\sum_{n=-\infty}^{\infty}\frac{\beta}{\beta^2 + \pi^2\left(\dfrac{\eta - \eta_j}{d} - n\right)^2} \tag{8-40}$$

This expression can be summed over all terms giving

$$\kappa_\eta = \frac{\bar{S}}{d}\frac{\sinh(\pi\beta/2)}{\cosh(\pi\beta/2) - \cos(\pi z/2)} \tag{8-41}$$

where

$$\frac{\bar{S}}{d} = \frac{S}{d} = \text{constant} \tag{8-42}$$

$$\beta = \frac{4\Delta}{d} \tag{8-43}$$

$$z = \frac{4(\eta - \eta_j)}{d} \tag{8-44}$$

In the limit of strongly overlapping lines, the line structure becomes smeared out giving

$$\kappa_\eta = \frac{S}{d} \qquad (\beta \gg 1) \tag{8-45}$$

Substituting Eq. (8-41) into Eq. (8-36) gives

$$\bar{\varepsilon}_\eta = 1 - \frac{1}{2}\int_0^2 \exp\left[\frac{-u_\eta \sinh(\pi\beta/2)}{\cosh(\pi\beta/2) - \cos(\pi z/2)}\right] dz \tag{8-46}$$

where the interval of integration has been taken as $-d/2$ to $d/2$ in η due to the identical nature of the lines. This integral has no closed-form solution but some limiting expressions are

$$\bar{\varepsilon}_\eta = u_\eta \qquad (u_\eta \ll 1) \tag{8-47}$$

$$\bar{\varepsilon}_\eta \to 1 \qquad (u_\eta \gg 1) \tag{8-48}$$

$$\bar{\varepsilon}_\eta = 1 - \exp[-u_\eta] \qquad (\beta \gg 1) \tag{8-49}$$

Thus, in general, the smoothed spectral emissivity does not obey Beer's law, except in the case of strongly overlapping lines where line structure has been removed.

Due to the assumption of equal spaced lines, the Elsasser model is most applicable to linear molecules such as CO, CO_2, etc. It may be recalled that the RRHO model for a linear molecule predicts equally spaced lines.

Goody Random Narrow Band Model

The Goody narrow band model [Goody, 1952] assumes that vibration-rotation lines are randomly spaced. Random line spacing is a reasonable representation for the spectra of most nonlinear molecules, especially asymmetric top molecules such as water vapor. The resulting smoothed spectral emissivity for equal intensity, randomly spaced Lorentz lines is given by an approximate algebraic expression as [Ludwig, et al., 1973]

$$\bar{\varepsilon}_\eta = 1 - \exp\left[\frac{-u_\eta}{\sqrt{1 + \dfrac{u_\eta}{\beta}}}\right] \tag{8-50}$$

Equation (8-50) also satisfies the same limits as the Elsasser model given in Eqs. (8-47) through (8-49). Again, the smoothed spectral emissivity does not, in general, obey Beer's law unless $\beta \gg 1$ and line structure has been smeared out.

A further discussion of narrow band models is given by Ludwig, et al. [1973]. A good discussion of how to apply narrow band calculations to predict radiative heat transfer for various combustion gas mixtures has also been given by Grosshandler [1980].

Wide Band Models

Wide band models are used to describe the variation of mean line intensity $\bar{S}/d$ or mean absorption coefficient $\bar{\kappa}_\eta$ with wavenumber, over the width of the band. More specifically, a wide band model provides the functional relationship between mean line intensity-to-spacing $\bar{S}/d$, dimensionless wavenumber $(\eta - \eta_0)/\omega$, and effective mass absorption coefficient α/ω

$$\frac{\bar{S}}{d} = f\left(\frac{\eta - \eta_0}{\omega}, \frac{\alpha}{\omega}\right) \tag{8-51}$$

where

$$\alpha = \int_{-\infty}^{\infty} \frac{\bar{S}}{d}\, d(\eta - \eta_0) = \text{integrated band intensity} \tag{8-52}$$

$$\omega = \text{bandwidth parameter (depends on wide band model)} \tag{8-53}$$

The integrated band intensity is the area under the curve of $\bar{S}/d$ or $\bar{\kappa}_\eta$ versus wavenumber and is analogous by its definition to the integrated line intensity S_j [see Eq. (8-20)] except that in (8-52) the integration extends over the entire vibration-rotation band while in (8-20) the integration is only over a single line. The bandwidth parameter is a characteristic of the particular band shape chosen and is indicative of the spectral width of the band. This parameter is not to be confused with the *effective bandwidth* A, which is the width of an equivalent black rectangle.

$$A = \int_{-\infty}^{\infty} \bar{\varepsilon}_\eta\, d(\eta - \eta_0) \tag{8-54}$$

More is said about the bandwidth parameter ω in connection with the specific wide band models in the following sections.

Once the functional relationship given by Eq. (8-51) is specified, Eq. (8-54) can be integrated to give the nondimensional effective bandwidth $A^*(u, \beta)$

$$A^* = \frac{A}{\omega} = A^*(u, \beta) \tag{8-55}$$

where

$$u = \frac{1}{\omega} \int_{-\infty}^{\infty} u_\eta \, d(\eta - \eta_0) = \frac{X}{\omega} \int_{-\infty}^{\infty} \frac{\bar{S}}{d} \, d(\eta - \eta_0) = \frac{X\alpha}{\omega} \qquad (8\text{-}56)$$

is the integrated optical depth parameter for the band and

$$\beta = \gamma(T) P_e(P) \qquad (8\text{-}57)$$

is the mean overlap parameter for the band. In Eq. (8-57) the overlap parameter has been expressed as the product of a temperature-dependent function $\gamma(T)$ and a pressure-dependent function $P_e(P)$, the reduced pressure.

Rigid Rotator, Harmonic Oscillator Wide Band Model

From statistical mechanics [Goody and Yung, 1989] the integrated line intensity for the RRHO model is given by

$$S_j = \frac{\alpha h c_0 B_j}{k_B T} \exp\left[-\frac{h c_0 B j^2}{k_B T} \right] \qquad (8\text{-}58)$$

assuming a large number of lines ($j \gg 1$). Since the lines are equally spaced

$$\eta - \eta_0 = \pm 2Bj \quad \text{and} \quad j = \frac{|\eta - \eta_0|}{2B} \qquad (8\text{-}59)$$

Thus the mean line intensity-to-spacing is given by

$$\frac{\bar{S}}{d} = \frac{\alpha}{\omega} \frac{|\eta - \eta_0|}{\omega} \exp\left[\frac{-(\eta - \eta_0)^2}{\omega^2} \right] \qquad (8\text{-}60)$$

where the bandwidth parameter has been defined as

$$\omega = \sqrt{\frac{4 k_B T B}{h c_0}} \qquad (8\text{-}61)$$

Figure 8-13 shows the spectral distribution for the RRHO model and the interpretation of the bandwidth parameter ω. The fact that the line intensity goes to zero at the center of the band is a result of the large spacing between vibrational energy levels relative to rotational levels. The probability of a pure vibrational transition occurring without a simultaneous rotational transition is very small.

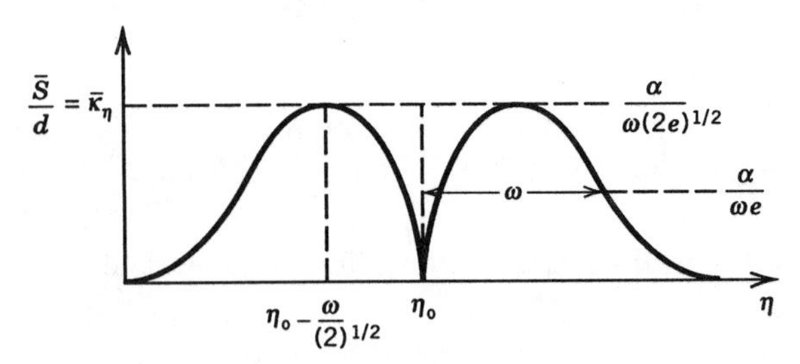

Figure 8-13 Rigid rotator harmonic oscillator wide band model.

Edwards Exponential Wide Band Model

The basis of the exponential wide band model [Edwards and Menard, 1964; Edwards and Balakrishnan, 1973; Edwards, 1981] is to assume a redistribution of lines such that an exponential profile of mean intensity-to-spacing versus wavenumber results.

$$\frac{\bar{S}}{d} = \frac{\alpha}{\omega} \exp\left[\frac{-2|\eta - \eta_0|}{\omega}\right] \tag{8-62}$$

The exponential band profile is shown in Fig. 8-14, including the interpretation of the bandwidth parameter ω. Combining the exponential wide band model (8-62) with the random narrow band model (8-50) gives

$$A = \int_{-\infty}^{\infty}\left[1 - \exp\left\{\frac{-u \exp\left[\dfrac{-2|\eta - \eta_0|}{\omega}\right]}{\sqrt{1 + \dfrac{u}{\beta}\exp\left[\dfrac{-2|\eta - \eta_0|}{\omega}\right]}}\right\}\right] d(\eta - \eta_0) \tag{8-63}$$

It should be noted that at the narrow band level β is actually a spectral

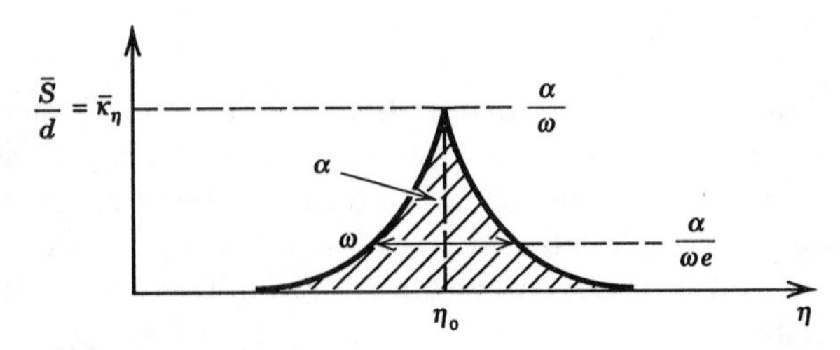

Figure 8-14 Edwards exponential wide band model.

variable, although its variation with frequency is much less severe than that of u_η. Thus, Eq. (8-63) also defines a spectral averaging process for β, although this dependence is not explicitly noted. Equation (8-63) can be written using the variable substitution $x' = 2(\eta - \eta_0)/\omega$ as

$$A^* = \int_0^\infty \left[1 - \exp\left\{ \frac{-u \exp[-x']}{\sqrt{1 + \frac{u}{\beta} \exp[-x']}} \right\} \right] dx' \tag{8-64}$$

The integral in Eq. (8-64), which has no closed-form solution, gives the nondimensional effective bandwidth $A^*(u, \beta)$. Edwards and co-workers [1964; 1973; 1976; 1981] have recommended the following curve fit for A^*.

$$\beta < 1 \qquad A^* = \begin{cases} u & 0 < u < \beta \\ 2\sqrt{\beta u} - \beta & \beta < u < 1/\beta \\ \ln(\beta u) + 2 - \beta & 1/\beta < u \end{cases} \tag{8-65}$$

$$\beta > 1 \qquad A^* = \begin{cases} u & u < 1 \\ \ln(u) + 1 & u > 1 \end{cases}$$

where

$$u = \frac{X\alpha}{\omega} \tag{8-66}$$

$$X = \rho_a s \tag{8-67}$$

$$\beta = \gamma(T) P_e \tag{8-68}$$

$$P_e = \left[\frac{P}{P_0} + \frac{P_a}{P_0}(b - 1) \right]^n \qquad (P_0 = 1 \text{ atm}) \tag{8-69}$$

The total pressure is P and the partial pressure of the absorbing gas is P_a. The parameters γ, α, and ω have been correlated, using statistical mechanical considerations, to give

$$\gamma(T) = \gamma_0 \sqrt{\frac{T_0}{T}} \frac{\Phi(T)}{\Phi(T_0)} \qquad (T_0 = 100 \text{ K}) \tag{8-70}$$

$$\omega(T) = \omega_0 \sqrt{\frac{T}{T_0}} \tag{8-71}$$

$$\alpha(T) = \alpha_0 \frac{1 - \exp\left(-\sum_{k=1}^{m} u_k \delta_k \right)}{1 - \exp\left(-\sum_{k=1}^{m} u_{0k} \delta_k \right)} \frac{\Psi(T)}{\Psi(T_0)} \tag{8-72}$$

where

$$u_k = \frac{hc_0\eta_k}{k_B T} \tag{8-73}$$

$$u_{0k} = \frac{hc_0\eta_k}{k_B T_0} \tag{8-74}$$

The Φ and Ψ functions are given by

$$\Psi(T) = \frac{\prod_{k=1}^{m} \sum_{v_k=v_{0k}}^{\infty} [(v_k + g_k + |\delta_k| - 1)!/(g_k - 1)!v_k!]e^{-u_k v_k}}{\prod_{k=1}^{m} \sum_{v_k=0}^{\infty} [(v_k + g_k - 1)!/(g_k - 1)!v_k!]e^{-u_k v_k}} \tag{8-75}$$

$$\Phi(T) = \frac{\left(\prod_{k=1}^{m} \sum_{v_k=v_{0k}}^{\infty} \{[(v_k + g_k + |\delta_k| - 1)!/(g_k - 1)!v_k!]e^{-u_k v_k}\}^{1/2}\right)^2}{\prod_{k=1}^{m} \sum_{v_k=v_{0k}}^{\infty} [(v_k + g_k + |\delta_k| - 1)!/(g_k - 1)!v_k!]e^{-u_k v_k}} \tag{8-76}$$

$$v_{0k} = \begin{cases} 0 & \delta_k > 0 \\ |\delta_k| & \delta_k < 0 \end{cases} \tag{8-77}$$

The parameters η_k, g_k, δ_k, b, n, α_0, ω_0, and γ_0 are listed in Table 8-4 for several important gases. An example program for evaluating the exponential wide band parameters for the H_2O 2.7-μm band is given in Appendix F.

The correlation given in Eq. (8-65) has two regions, a region of nonoverlapping lines ($\beta < 1$), where line structure influences the effective bandwidth, and a region of overlapping lines ($\beta > 1$), where the line structure does not influence the effective bandwidth. Both regions are fit by a linear relation with optical depth for optically thin band conditions and a logarithmic relation for optically thick conditions. In addition, for nonoverlapping lines there is an intermediate square root region [recall Eq. (8-27b)] where the spectral emissivity at the center of the lines has become saturated at a value of one and the additional increase in effective bandwidth with optical depth is coming from the edges of the lines. For fundamental bands the band intensity parameter α is to a first approximation independent of temperature. The variation of u with temperature at constant partial pressure of participating gas P_a can be estimated from Eq. (8-66). Since X varies as ρ_a, which varies as T^{-1}, and ω varies as $T^{1/2}$, the integrated optical depth parameter u varies as $T^{-3/2}$.

Other wide band correlations similar to that of Edwards and co-workers have also been proposed [Tien and Lowder, 1966; Cess and Ramanathan, 1972; Felski and Tien, 1974]. However, some of these correlations do not satisfy the square root limit for strong, nonoverlapping Lorentz lines ($\beta \ll 1$, $u/\beta \gg 1$, $u\beta \ll 1$) in which the nondimensional bandwidth should approach $2\sqrt{u\beta}$ (see Problem 8-2). A continuous correlation that does satisfy the square root limit as well as the optically thin and thick limits is that of Cess and Ramanathan [1972].

$$A^* = 2\ln\left[1 + \frac{u}{2 + \sqrt{u(1 + 1/\beta)}}\right]$$

Wide Band Representation of Spectral Emissivity

Wide band models, such as the exponential wide band model, express gas band properties in terms of an effective bandwidth A. If one wishes to specify the value of the spectral emissivity for a gas band using wide band results, there is some arbitrariness involved because A is defined as the integral of spectral emissivity over a given band, according to Eq. (8-54). Two representations of spectral emissivity are given in Fig. 8-15.

The simplest representation of spectral emissivity is the black band model, which assumes the spectral emissivity is constant at a value of one over a wavenumber interval of A. A slightly more complicated representation is the gray band model, which assumes the spectral emissivity is constant at a value of $\varepsilon_\eta = A/\Delta\eta$ over a wavenumber interval of $\Delta\eta$. The gray band model is able to give a somewhat better representation of the actual spectral emissivity than the black band model (which is important particularly when bands overlap), but it also requires knowing the additional parameter $\Delta\eta$. Edwards [1981] has recommended using

$$\frac{A}{\Delta\eta} = 1 - \frac{u}{A^*}\frac{dA^*}{du} \tag{8-78}$$

to estimate the value of $\Delta\eta$ if the calculated value of $A/\Delta\eta$ according to (8-78) is greater than 0.1 and 0.1 if it is less. Using (8-78) with (8-65) gives the following results for spectral emissivity for the exponential wide band model:

$$\varepsilon_\eta = \frac{A}{\Delta\eta} = \begin{cases} 0.1 & \text{linear} \\ 1 - \dfrac{\sqrt{u\beta}}{A^*} & \text{square root} \\ 1 - \dfrac{1}{A^*} & \text{log} \end{cases} \tag{8-79}$$

TABLE 8-4 Exponential Wide Band Parameters

Gas		Band,	Band center		Pressure Parameters ($T_0 = 100$ K)		α_0,		ω_0,
m, η (cm^{-1}), g		μm	η_0, cm^{-1}	$\delta_1 \cdots \delta_m$	b	n	cm^{-1}/(g m^{-2})	γ_0	cm^{-1}
	CO_2[a]	15	667	0, 1, 0	1.3	0.7	19.0	0.06157	12.7
$m = 3$	$\eta_1 = 1351,\ g_1 = 1$	10.4	960	-1, 0, 1	1.3	0.8	2.47×10^{-9}	0.04017	13.4
	$\eta_2 = 667,\ g_2 = 2$	9.4	1060	0, -2, 1[b]	1.3	0.8	2.48×10^{-9}	0.11888	10.1
	$\eta_3 = 2396,\ g_3 = 1$	4.3	2410[c]	0, 0, 1	1.3	0.8	110.0	0.24723	11.2
		2.7	3660	1, 0, 1	1.3	0.65	4.0	0.13341	23.5
		2.0	5200	2, 0, 1	1.3	0.65	0.066	0.39305	34.5
	CH_4	7.66	1310	0, 0, 0, 1	1.3	0.8	28.0	0.08698	21.0
$m = 4$	$\eta_1 = 2914,\ g_1 = 1$	3.31	3020	0, 0, 1, 0	1.3	0.8	46.0	0.06973	56.0
	$\eta_2 = 1526,\ g_2 = 2$	2.37	4220	1, 0, 0, 1	1.3	0.8	2.9	0.35429	60.0
	$\eta_3 = 3020,\ g_3 = 3$	1.71	5861	1, 1, 0, 1	1.3	0.8	0.42	0.68598	45.0
	$\eta_4 = 1306,\ g_4 = 3$								
	H_2O	Rotational	140	0, 0, 0	$8.6(T_0/T)^{1/2} + 0.5$	1	10,400.0[e]	0.14311[e]	57.1
$m = 3$	$\eta_1 = 3652,\ g_1 = 1$	6.3	1600	0, 1, 0	$8.6(T_0/T)^{1/2} + 0.5$	1	41.2	0.09427	56.4
	$\eta_2 = 1595,\ g_2 = 1$	2.7[d]	3760	0, 2, 0	$8.6(T_0/T)^{1/2} + 0.5$	1	0.19	0.13219	60.0
	$\eta_3 = 3756,\ g_3 = 1$			1, 0, 0			2.30		
				0, 0, 1			22.40		
		1.87	5350	0, 1, 1	$8.6(T_0/T)^{1/2} + 0.5$	1	3.0	0.08169	43.1
		1.38	7250	1, 0, 1	$8.6(T_0/T)^{1/2} + 0.5$	1	2.5	0.11628	32.0

	CO	4.7	2143	1	1.1	0.8	20.9	0.07506	25.5
$m = 1$	$\eta_1 = 2143,\ g_1 = 1$	2.35	4260	2	1.0	0.8	0.14	0.16758	20.0
	NO	5.34	1876	1	0.65	1.0	9.0	0.18050	20.0
$m = 1$	$\eta_1 = 1876,\ g_1 = 1$								
	SO_2	19.27	519	0, 1, 0	0.7	1.28	4.22	0.05291	33.08
$m = 3$	$\eta_1 = 1151,\ g_1 = 1$	8.68	1151	1, 0, 0	0.7	1.28	3.674	0.05952	24.83
	$\eta_2 = 519,\ g_2 = 1$	7.35	1361	0, 0, 1	0.65	1.28	29.97	0.49299	8.78
	$\eta_3 = 1361,\ g_3 = 1$	4.34	2350	2, 0, 0	0.6	1.28	0.423	0.47513	16.45
		4.0	2512	1, 0, 1	0.6	1.28	0.346	0.58937	10.91

[a] The 1, 0, 0 band of the linear CO_2 molecule appears only at high pressures when a dipole is induced by collisions.

[b] Because of Fermi resonance between the η_1 and $2\eta_2$ levels, the Ψ and Φ functions for the 1060 cm^{-1} band are to be those of the 960 cm^{-1} band; i.e., use the set of δ's for the 960 cm^{-1} band to get Ψ and Φ for either band.

[c] Upper band limit.

[d] $\alpha_{2.7} = \Sigma_{j=1}^{3}\alpha_j$ and $\gamma_{2.7} = (1/\alpha_{2.7})\{\Sigma_{j=1}^{3}\sqrt{\alpha_j\gamma_j}\}^2$.

[e] For the rotational band of H_2O $\alpha(T) = \alpha_0 \exp(-5.0\sqrt{T_0/T})$ and $\gamma(T) = \gamma_0\sqrt{T_0/T}$. Otherwise α_0, γ_0, and ω_0, apply to Eqs. (8-70) to (8-72). If calculated bandwidth A_{calc} gives $\eta_0 - A_{calc}/2 < 0$ for lower band limit of rotational band, use zero for lower limit, $\eta_0 + A_{calc}/2$ as upper limit and calculate the bandwidth as $A = \eta_0 + A_{calc}/2$. For details on rotational water band see Modak [1979].

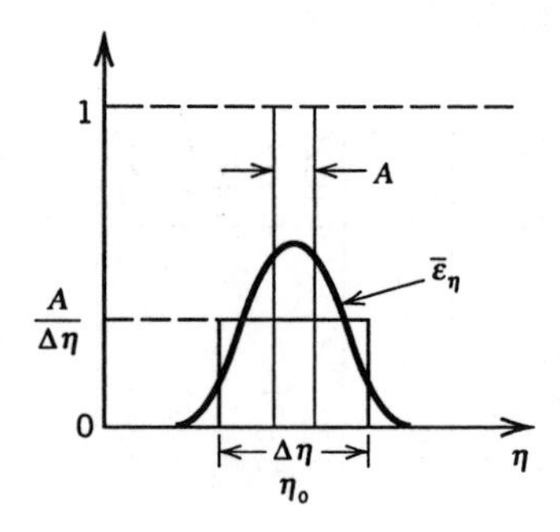

Figure 8-15 Two wide band representations of spectral emissivity; black band ($A \times 1$) and gray band ($A/\Delta\eta \times \Delta\eta$).

Equation (8-79) can be used to evaluate both the gray bandwidth parameter $\Delta\eta$ and the corresponding spectral emissivity (or equivalent transmissivity, $\tau_\eta = 1 - \varepsilon_\eta$).

Total Intensity — Isothermal, Homogeneous Path

Consider a homogeneous column of gas at temperature T_g subject to incident intensity I_{0_η}. The solution of the transfer equation for this case, on a spectral basis, was given by Eq. (7-67) or (7-15).

$$I_\eta(s) = \varepsilon_\eta I_{b_\eta}(T_g) + \tau_\eta I_{0_\eta} \tag{8-80}$$

In Eq. (8-80) τ_η denotes the spectral transmissivity $\tau_\eta(0, s)$ along the homogeneous path from the origin to an arbitrary location s, and ε_η is the corresponding emissivity (or absorptivity). The total intensity emitted and transmitted by the gas along path length s is obtained by integrating (8-80) over wavenumber.

$$I(s) = \int_0^\infty \varepsilon_\eta I_{b_\eta}(T_g)\, d\eta + \int_0^\infty \tau_\eta I_{0_\eta}\, d\eta \tag{8-81}$$

There are two approaches for evaluating the integrals in (8-81). One approach is to assume that the spectral Planck function and incident intensity are constant over the width of each band and equal to their respective values at the center of the bands ($I_{b_{\eta 0}}$ and $I_{0_{\eta 0}}$). This assumption gives

$$I(s) = \sum_{k \text{ bands}} I_{b_{\eta 0 k}}(T_g) A_k(T_g) + I_0 - \sum_{k \text{ bands}} I_{0_{\eta 0 k}} A_k(T_g) \tag{8-82}$$

where the first term on the right-hand side of (8-82) accounts for emission by the gas and the second and third terms account for transmission of incident radiation. Equation (8-82) can also be written in terms of total properties as

$$I(s) = \varepsilon_g I_b(T_g) + \tau_g I_0 \tag{8-83}$$

where the total properties are defined as

$$\varepsilon_g = \frac{1}{\sigma T_g^4} \sum_k e_{b_{\eta 0k}}(T_g) A_k(T_g) \tag{8-84}$$

$$\tau_g = 1 - \alpha_g = 1 - \frac{1}{I_0} \sum_k I_{0_{\eta 0k}} A_k(T_g) \tag{8-85}$$

If the incident intensity is proportional to blackbody radiation at T_w, the total absorptivity becomes

$$\alpha_g = \frac{1}{\sigma T_w^4} \sum_k e_{b_{\eta 0k}}(T_w) A_k(T_g), \qquad I_{0_\eta} \sim I_{b_\eta}(T_w) \tag{8-86}$$

This approach is known as the *band approximation* and is accurate for vibration-rotation bands, which are narrow relative to the spectral distribution of the Planck function. The band approximation is not recommended when wide, rotational bands are under consideration because the assumption that the Planck function is relatively constant over the width of the band is not satisfied.

The other approach to integrating Eq. (8-81) is to assume that the spectral emissivity is constant over the band and remove it from under the integral. This approach is known as the *block approximation*. Assuming the spectral emissivity is constant at a value of $A/\Delta\eta$ over a bandwidth of $\Delta\eta$ (the gray band model; see Fig. 8-15) gives Eq. (8-83) with the total emissivity and absorptivity defined as

$$\varepsilon_g = \sum_k \frac{A_k(T_g)}{\Delta\eta_k} \left[f\left(\frac{T_g}{\eta_{0k} - \frac{\Delta\eta_k}{2}} \right) - f\left(\frac{T_g}{\eta_{0k} + \frac{\Delta\eta_k}{2}} \right) \right] \tag{8-87}$$

$$\alpha_g = \sum_k \frac{A_k(T_g)}{\Delta\eta_k} \left[f\left(\frac{T_w}{\eta_{0k} - \frac{\Delta\eta_k}{2}} \right) - f\left(\frac{T_w}{\eta_{0k} + \frac{\Delta\eta_k}{2}} \right) \right], \qquad I_{0_\eta} \sim I_{b_\eta}(T_w) \tag{8-88}$$

In (8-87) and (8-88) f is the blackbody fractional function. The block approximation is valid with both wide, rotational bands and narrow, vibration-rotation bands. For this reason the block approximation is preferable over the band approximation for computer solutions.

Example 8-1

What is the total intensity emitted by a mixture of CO_2 (0.5 atm) and N_2 (0.5 atm) at $T_g = 1000$ K along a 0.5-m path length for the 2.7-μm band?

The mass path length is calculated from Eq. (8-67) using the ideal gas equation.

$$X = \frac{P_a M}{R T_g} s = \frac{(0.5 \text{ atm})(44 \text{ g/mole})(1.01 \times 10^5 \text{ J/m}^3 \text{ atm})(0.5 \text{ m})}{(8.314 \text{ J/mole K})\ (1000 \text{ K})}$$

$$= 134 \text{ g/m}^2$$

The effective pressure is calculated from Eq. (8-69) as

$$P_e = [1 + 0.5(0.3)]^{0.65} = 1.10$$

Using the exponential wide band model, the following band parameters result:

$$\alpha = 4.80 \frac{\text{m}^2}{\text{g cm}} \qquad \omega = 74.3 \text{ cm}^{-1} \qquad \gamma = 1.12$$

The line overlap parameter can then be determined from Eq. (8-68)

$$\beta = (1.12)(1.10) = 1.23$$

and the band optical depth parameter from Eq. (8-66)

$$u = \frac{(134 \text{ g/m}^2)(4.80 \text{ m}^2/\text{g cm})}{74.3 \text{ cm}^{-1}} = 8.66$$

Since $\beta > 1$ and $u > 1$, the nondimensional effective bandwidth from (8-65) is

$$A^* = \ln(8.66) + 1 = 3.16$$

Using the definition of A^*, Eq. (8-55), the effective bandwidth is thus

$$A = A^* \omega = (3.16)(74.3 \text{ cm}^{-1}) = 235 \text{ cm}^{-1}$$

Using the band approximation (8-82) the total intensity emitted by the 2.7-μm band is

$$I = \frac{\varepsilon_g \sigma T_g^4}{\pi} = I_{b_{\eta 0}} A = \left(2.95 \frac{\text{W}}{\text{m}^2\ \text{cm}^{-1}\ \text{sr}}\right)(235\ \text{cm}^{-1}) = 692 \frac{\text{W}}{\text{m}^2\ \text{sr}}$$

Overlapping Bands

In extending the process illustrated in Example 8-1 to include all bands, situations arise where adjacent bands of the same gas (e.g., 9.4- and 10.4-μm CO_2 bands) or bands of different gases (e.g., 2.7-μm CO_2 and H_2O bands) overlap spectrally. In the case of overlapping bands the combined effective bandwidth can be calculated using the additivity of absorption coefficients (or multiplicativity of transmissivities) as discussed earlier. Equation (8-34), which was developed for two overlapping lines, can also be applied to two overlapping bands. Using the same type of analysis leading to Eq. (8-34) the band overlap term can be evaluated using the gray band model, assuming no statistical correlation between the lines of the two bands. The resulting effective bandwidth for two overlapping bands (A and B) can be shown to be

$$A_{A+B} = A_A + A_B - \frac{A_A}{\Delta\eta_A}\frac{A_B}{\Delta\eta_B}\Delta\eta_{AB} \tag{8-89}$$

where, assuming $\eta_{0A} < \eta_{0B}$, the overlap region is

$$\Delta\eta_{AB} = 0 \quad \text{if } \eta_{LB} > \eta_{UA} \quad \text{(no overlap)} \tag{8-90a}$$

$$\Delta\eta_{AB} = \eta_{UA} - \eta_{LB} \quad \text{if } \eta_{UB} > \eta_{UA} > \eta_{LB} > \eta_{LA} \quad \text{(partial overlap)} \tag{8-90b}$$

$$\Delta\eta_{AB} = \Delta\eta_A \quad \text{if } \eta_{UB} > \eta_{UA} \quad \text{and} \quad \eta_{LB} < \eta_{LA} \quad \text{(complete overlap; } A \text{ inside } B\text{)} \tag{8-90c}$$

$$\Delta\eta_{AB} = \Delta\eta_B \quad \text{if } \eta_{UB} < \eta_{UA} \quad \text{and} \quad \eta_{LB} > \eta_{LA} \quad \text{(complete overlap; } B \text{ inside } A\text{)} \tag{8-90d}$$

and the upper (U) and lower (L) wavenumbers for the A and B bands are calculated from

$$\eta_U = \eta_0 + \frac{\Delta\eta}{2} \quad \text{and} \quad \eta_L = \eta_0 - \frac{\Delta\eta}{2} \tag{8-91}$$

The derivation of these results is left as an exercise (see Problem 8-1).

Example 8-2

What is the total intensity emitted by a mixture of CO_2 (0.5 atm) and H_2O (0.5 atm) at 1000 K along a 0.5-m path length for the 2.7-μm band?

Since both CO_2 and H_2O emit at 2.7 μm, the individual bandwidths must be determined and the overlap considered. The band center wavenumber for CO_2 at 2.7 μm (3660 cm^{-1}) from Table 8-4 is lower than that for H_2O (3760 cm^{-1}). Therefore, the CO_2 band will be referred to as the A band and the H_2O band will be referred to as the B band.

$$A = CO_2 \qquad B = H_2O$$

The effective bandwidth of the A band was calculated in Example 8-1. Since overlap is to be considered, a gray band analysis will be performed. This requires that the effective gray emissivity and gray bandwidth be determined. Using Eq. (8-79) and (8-65) the effective gray band emissivity for the A band is

$$\frac{A_A}{\Delta\eta_A} = 1 - \frac{1}{A_A^*} = 0.68$$

since the curve of growth is in the log region. The corresponding effective gray bandwidth for the A band is

$$\Delta\eta_A = \frac{235 \text{ cm}^{-1}}{0.68} = 344 \text{ cm}^{-1}$$

The exponential wide band results for the B band can be calculated in a manner similar to that demonstrated in Example 8-1. These results are as follows:

$$X = 54.7 \frac{\text{g}}{\text{m}^2} \qquad P_e = 2.11$$

$$\alpha = 24.9 \frac{\text{m}^2}{\text{g cm}} \qquad \omega = 190 \text{ cm}^{-1} \qquad \gamma = 0.210$$

$$\beta = 0.442 \qquad u = 7.18\,(> 1/\beta) \qquad A_B^* = 2.71 \qquad A_B = 515 \text{ cm}^{-1}$$

$$\frac{A_B}{\Delta\eta_B} = 1 - \frac{1}{A_B^*} = 0.63 \qquad \Delta\eta_B = 816 \text{ cm}^{-1}$$

$$\eta_{UB} = 3760 + \frac{816}{2} = 4168 > \eta_{UA} = 3660 + \frac{344}{2} = 3832$$

$$\eta_{LB} = 3760 - \frac{816}{2} = 3352 < \eta_{LA} = 3660 - \frac{344}{2} = 3488$$

According to Eq. (8-90c) the bands are entirely overlapped (A inside B) and the band overlap is $\Delta\eta_A$. From Eq. (8-89) the effective combined bandwidth is thus

$$A_{A+B} = 235 + 515 - (0.68)(0.63)(344) = 603 \text{ cm}^{-1}$$

and the total intensity emitted by the overlapping 2.7-μm bands is

$$I = I_{b_{\eta 0}} A_{A+B} = \left(2.95 \frac{\text{W}}{\text{m}^2 \text{ cm}^{-1} \text{ sr}}\right)(603 \text{ cm}^{-1}) = 1780 \frac{\text{W}}{\text{m}^2 \text{ sr}}$$

TRANSFER IN NONHOMOGENEOUS GASES

In the previous sections it has been assumed that the gas is homogeneous (constant temperature, pressure, and density) over the path length under consideration. In many practical situations the temperature and composition of the gas vary considerably along path length. The appropriate solution to the transfer equation along a nonhomogeneous path is that given by Eqs. (7-61) through (7-66). Applied to a nonscattering medium, Eq. (7-65) becomes

$$I_\eta(s) = I_{0_\eta} \tau_\eta(0, s) + \int_0^s I_{b_\eta}[T(s')] \frac{\partial \tau_\eta(s', s)}{\partial s'} \, ds' \tag{8-92}$$

This form of the transfer equation is the pure spectral form and is appropriate for doing line-by-line calculations. The transmissivity function is a spectral quantity that would be evaluated from

$$\tau_\eta(s', s) = \exp\left[-t_\eta(s', s)\right] \tag{8-93}$$

and

$$t_\eta(s', s) = \int_{s'}^s K_{a_\eta}(s'') \, ds'' = \int_{s'}^s \rho_a(s'') \kappa_\eta(s'') \, ds'' \tag{8-94}$$

This type of analysis requires extensive spectral data and considerable computational effort.

Narrow Band Scaling

Nonhomogeneous effects can also be included within the narrow band description. Spectrally smoothing Eq. (8-92) gives

$$\bar{I}_\eta(s) = I_{0_\eta} \bar{\tau}_\eta(0, s) + \int_0^s I_{b_\eta}[T(s')] \frac{\partial \bar{\tau}_\eta(s', s)}{\partial s'} \, ds' \tag{8-95}$$

for the smoothed spectral intensity. The function $\bar{\tau}_\eta(s', s)$ is the spectrally smoothed, narrow band transmissivity function, which, for the random model can be written as

$$\bar{\tau}_\eta(s', s) = \exp\left[\frac{-u_{e\eta}(s', s)}{\sqrt{1 + u_{e\eta}(s', s)/\beta_{e_{nb}}}}\right] \tag{8-96}$$

Formal evaluation of this function is not feasible for most engineering calculations. Therefore, an approximation is usually employed. The so-called Curtis-Godson approximation is the most popular approach used and was proposed independently by van de Hulst, Curtis, and Godson [Goody and Yung, 1989]. The essence of this approach is to determine effective values for the spectrally smoothed optical depth $u_{e\eta}$ and the overlap parameter β_e, which satisfy the weak and strong line limits and give a reasonable estimate of the correct results for intermediate cases. The appropriate definitions for these effective parameters that satisfy the aforementioned constraints are

$$u_{e\eta}(s', s) = \int_{s'}^{s} \rho_a \bar{\kappa}_\eta \, ds'' \tag{8-97}$$

$$\beta_{e_{nb}}(s', s) = \frac{\int_{s'}^{s} \rho_a \bar{\kappa}_\eta \beta \, ds''}{\int_{s'}^{s} \rho_a \bar{\kappa}_\eta \, ds''} \tag{8-98}$$

Using the definitions (8-97) and (8-98), the effective, nonhomogeneous spectrally smoothed optical depth and line overlap parameters can be evaluated and used in (8-96) to evaluate the transmissivity function. The transmissivity function, in turn, can be integrated according to (8-95) to give the spectrally smoothed intensity.

Wide Band Scaling

The concept of nonhomogeneous scaling can also be extended to wide band models. The total intensity transmitted and emitted by a nonhomogeneous column of gas is obtained by integrating Eq. (8-92) over wavenumber.

$$I(s) = \int_0^\infty I_{0_\eta}\{1 - [1 - \tau_\eta(0, s)]\} \, d\eta + \int_0^\infty \int_0^s I_{b_\eta}[T(s')] \frac{\partial}{\partial s'}[\tau_\eta(s', s)] \, ds' \, d\eta \tag{8-99}$$

Using the band approximation on (8-99) and approximating the nonhomogeneous path as a series of homogeneous elements gives

$$I(s) = I_0 - \sum_k I_{0_{\eta 0k}}[1 - \tau_k(0, s)](\eta_{U_k} - \eta_{L_k})$$
$$+ \sum_i \sum_k I_{b_{\eta 0k}}[T(s_i)][\tau_k(s_i, s) - \tau_k(s_{i-1}, s)](\eta_{U_k} - \eta_{L_k}) \quad (8\text{-}100)$$

where

$$\tau_k(s_i, s) = \prod_j \tau_{kj}(s_i, s) = \prod_j \left[1 - \frac{A_j(s_i, s)}{\Delta\eta_j(s_i, s)}\right] \quad (8\text{-}101)$$

i = homogeneous path element index
j = species or band index
k = wavenumber interval index

In Eq. (8-100) η_{U_k} and η_{L_k} represent the upper and lower wavenumbers, respectively, of a given wavenumber interval and $\Delta\eta_j(s_i, s)$ represents the gray bandwidth of band j. The effective bandwidth function $A_j(s_i, s)$ for a nonhomogeneous path can be calculated from Eq. (8-65) using the scaled exponential wide band parameters recommended by Edwards and Morizumi [1970]

$$\alpha_e(s', s) = \frac{\int_{s'}^{s} \rho_a \alpha \, ds''}{\int_{s'}^{s} \rho_a \, ds''} = \frac{\sum_i \alpha_i X_i}{\sum_i X_i} \quad (8\text{-}102)$$

$$\omega_e(s', s) = \frac{\int_{s'}^{s} \rho_a \alpha\omega \, ds''}{\int_{s'}^{s} \rho_a \alpha \, ds''} = \frac{\sum_i \alpha_i \omega_i X_i}{\sum_i \alpha_i X_i} \quad (8\text{-}103)$$

$$\beta_{e_{wb}}(s', s) = \frac{\int_{s'}^{s} \rho_a \alpha\omega\beta \, ds''}{\int_{s'}^{s} \rho_a \alpha\omega \, ds''} = \frac{\sum_i \alpha_i \omega_i \beta_i X_i}{\sum_i \alpha_i \omega_i X_i} \quad (8\text{-}104)$$

where

$$u_e(s', s) = \frac{X(s', s)\alpha_e(s', s)}{\omega_e(s', s)} \quad (8\text{-}105)$$

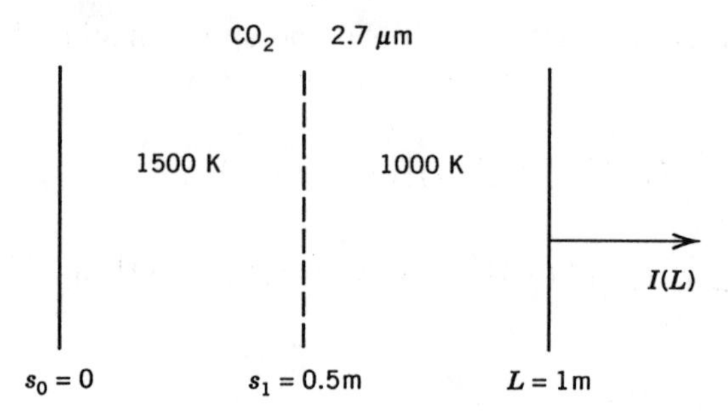

Figure 8-16 Nonhomogeneous emitting, absorbing, nonscattering medium.

and $X(s', s)$ is given by

$$X(s', s) = \int_{s'}^{s} \rho_a \, ds'' = \sum_i X_i \tag{8-106}$$

Example 8-3

What is the total intensity emitted by a mixture of CO_2 (0.5 atm) and N_2 (0.5 atm) that is at $T_1 = 1500$ K along a 0.5-m path length and $T_2 = 1000$ K along another 0.5-m path length, as pictured in Fig. 8-16 for the 2.7-μm band?

The relation needed for evaluating the total intensity at L is Eq. (8-100). However, since band overlap is not present in this example, a simpler version of (8-100) may be used. If band overlap is neglected, the wavenumber intervals in (8-100) $(\eta_{U_k} - \eta_{L_k})$ are the same as the bandwidths in (8-101) $\Delta\eta_k$ and (8-100) reduces to

$$I(s) = I_0 - \sum_k I_{0_{\eta 0k}} A_k(0, s)$$
$$+ \sum_i \sum_k I_{b_{\eta 0k}}[T(s_i)][A_k(s_{i-1}, s) - A_k(s_i, s)]$$

For the medium of Fig. 8-16, which is composed of two homogeneous elements ($i = 1, 2$) with no incident intensity ($I_0 = 0$) and only one band under consideration ($k = 1$), this relation becomes

$$I(L) = I_{b_{\eta 0}}(T_1)[A(s_0, L) - A(s_1, L)] + I_{b_{\eta 0}}(T_2)[A(s_1, L) - A(L, L)]$$

The bandwidth $A(L, L)$ is zero. The bandwidth $A(s_1, L)$ covers a homogeneous path and has already been evaluated in Example 8-1. The band parameters over the path s_1 to L are, from Example 8-1:

$$X(s_1, L) = 134 \frac{\text{g}}{\text{m}^2} \qquad \alpha(s_1, L) = 4.80 \frac{\text{m}^2}{\text{g cm}} \qquad \omega(s_1, L) = 74.3 \text{ cm}^{-1}$$

$$\beta(s_1, L) = 1.23 \qquad u(s_1, L) = 8.66 \qquad A(s_1, L) = 235 \text{ cm}^{-1}$$

The bandwidth $A(s_0, L)$ covers a nonhomogeneous path. The parameters needed to evaluate $A(s_0, L)$ can be obtained from Eqs. (8-102) through (8-106) along the nonhomogeneous path from s_0 to L as

$$X(s_0, L) = X(s_0, s_1) + X(s_1, L) = 89.1 + 134 = 223 \frac{\text{g}}{\text{m}^2}$$

$$\alpha_e(s_0, L) = \frac{\alpha(s_0, s_1) X(s_0, s_1) + \alpha(s_1, L) X(s_1, L)}{X(s_0, L)}$$

$$= \frac{(5.95)(89.1) + (4.80)(134)}{223} = 5.26 \frac{\text{m}^2}{\text{g cm}}$$

$$\omega_e(s_0, L) = \frac{\alpha(s_0, s_1)\omega(s_0, s_1) X(s_0, s_1) + \alpha(s_1, L)\omega(s_1, L) X(s_1, L)}{\alpha_e(s_0, L) X(s_0, L)}$$

$$= \frac{(5.95)(91)(89.1) + (4.80)(74.3)(134)}{(5.26)(223)} = 81.9 \text{ cm}^{-1}$$

$$\beta_e(s_0, L) = \frac{\begin{array}{c}\alpha(s_0, s_1)\omega(s_0, s_1)\beta(s_0, s_1) X(s_0, s_1) \\ +\alpha(s_1, L)\omega(s_1, L)\beta(s_1, L) X(s_1, L)\end{array}}{\alpha_e(s_0, L)\omega_e(s_0, L) X(s_0, L)}$$

$$= \frac{(5.95)(91)(3.04)(89.1) + (4.80)(74.3)(1.23)(134)}{(5.26)(81.9)(223)} = 2.14$$

$$u_e(s_0, L) = \frac{X(s_0, L)\alpha_e(s_0, L)}{\omega_e(s_0, L)} = \frac{(223)(5.26)}{81.9} = 14.3$$

$$A(s_0, L) = \omega_e(s_0, L)\{\ln[u_e(s_0, L)] + 1\} = 300 \text{ cm}^{-1}$$

The total emitted intensity can then be calculated by substituting the parameters back into the transfer equation giving

$$I(L) = 17.8[300 - 235] + 2.95[235] = 1870 \ \frac{\text{W}}{\text{m}^2\ \text{sr}}$$

Example 8-4

What is the total intensity emitted by the mixture of Example 8-3 for the 2.7 μm band if the N_2 is replaced by H_2O?

In this case both the CO_2 and H_2O have bands at 2.7 μm and the mixture spectral transmissivity must be calculated as the product of the component transmissivities according to Eq. (8-101). Furthermore, because there is overlap of two bands, the gray bandwidths $\Delta\eta_k$ do not correspond exactly to the wavenumber intervals $\eta_{U_k} - \eta_{L_k}$ as in Example 8-3.

For two homogeneous elements, with no incident radiation, Eq. (8-100) becomes

$$I(s) = I_{b_{\eta 0}}(T_1) \sum_k [\tau_k(s_1, L) - \tau_k(s_0, L)](\eta_{U_k} - \eta_{L_k})$$
$$+ I_{b_{\eta 0}}(T_2) \sum_k [1 - \tau_k(s_1, L)](\eta_{U_k} - \eta_{L_k})$$

where the $I_{b_{\eta 0}}$ term has been taken out of the summation over k and evaluated at the band center ($\eta_0 = 3760$ cm^{-1}) since it is relatively constant over the spectral region under consideration. The transmissivities are the product of the transmissivities for the CO_2 band and the H_2O band (let $A = CO_2$ and $B = H_2O$ as in the earlier examples). Consider first the transmissivity from s_1 to L.

$$\tau_k(s_1, L) = \tau_{k_A}(s_1, L)\tau_{k_B}(s_1, L)$$

The component transmissivities $\tau_{k_A}(s_1, L)$ and $\tau_{k_B}(s_1, L)$ and their corresponding band limits were previously obtained in Example 8-2. These component transmissivities and their product $\tau_k(s_1, L)$ are illustrated in Fig. 8-17*a*. Consider next the transmissivity from s_0 to L.

$$\tau_k(s_0, L) = \tau_{k_A}(s_0, L)\tau_{k_B}(s_0, L)$$

The transmissivity for the A (CO_2) component can be obtained from the results of Example 8-3 as

$$\tau_{k_A}(s_0, L) = \frac{1}{A_A^*(s_0, L)} = \frac{\omega_e(s_0, L)}{A_A(s_0, L)} = \frac{81.9}{300} = 0.27$$

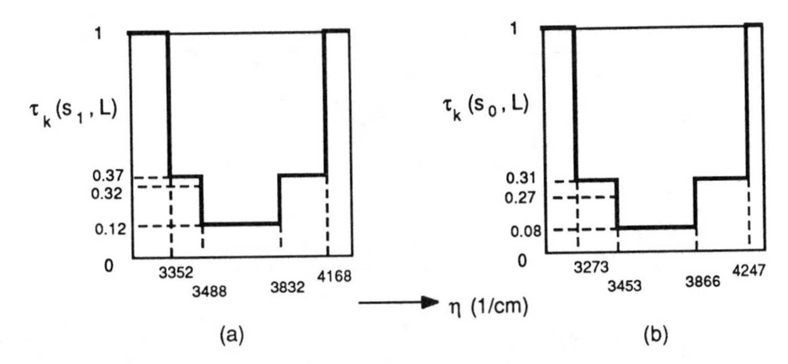

Figure 8-17 Spectral transmissivity along (*a*) homogeneous path s_1 to L and (*b*) nonhomogeneous path s_0 to L with overlapping bands.

inside the A band (and one outside the A band), where the corresponding gray bandwidth limits for the A band are defined by

$$\Delta\eta_A(s_0, L) = \frac{A_A(s_0, L)}{1 - \tau_{k_A}(s_0, L)} = \frac{300}{1 - 0.27} = 413 \text{ cm}^{-1}$$

The lower and upper limits for the A band from s_0 to L are thus $3660 - (0.5)(413) = 3453 \text{ cm}^{-1}$ and $3660 + (0.5)(413) = 3866 \text{ cm}^{-1}$ as shown in Fig. 8-17*b*.

The transmissivity for the B (H_2O) component from s_0 to L can be obtained by carrying out an analysis for H_2O similar to that which was performed for CO_2 in Example 8-3. The results are given below:

$$X(s_1, L) = 54.7 \frac{\text{g}}{\text{m}^2} \qquad \alpha(s_1, L) = 24.9 \frac{\text{m}^2}{\text{g cm}}$$

$$\omega(s_1, L) = 190 \text{ cm}^{-1} \qquad \beta(s_1, L) = 0.442$$

$$X(s_0, s_1) = 36.4 \frac{\text{g}}{\text{m}^2} \qquad \alpha(s_0, s_1) = 25.0 \frac{\text{m}^2}{\text{g cm}}$$

$$\omega(s_0, s_1) = 232 \text{ cm}^{-1} \qquad \beta(s_0, s_1) = 0.743$$

$$X(s_0, L) = X(s_0, s_1) + X(s_1, L) = 36.4 + 54.7 = 91.1 \frac{\text{g}}{\text{m}^2}$$

$$\alpha_e(s_0, L) = \frac{\alpha(s_0, s_1)X(s_0, s_1) + \alpha(s_1, L)X(s_1, L)}{X(s_0, L)}$$

$$= \frac{(24.9)(54.7) + (25.0)(36.4)}{91.1} = 24.9 \frac{\text{m}^2}{\text{g cm}}$$

$$\omega_e(s_0, L) = \frac{\alpha(s_0, s_1)\omega(s_0, s_1)X(s_0, s_1) + \alpha(s_1, L)\omega(s_1, L)X(s_1, L)}{\alpha_e(s_0, L)X(s_0, L)}$$

$$= \frac{(24.9)(190)(54.7) + (25.0)(232)(36.4)}{(24.9)(91.1)} = 207 \text{ cm}^{-1}$$

$$\beta_e(s_0, L) = \frac{\alpha(s_0, s_1)\omega(s_0, s_1)\beta(s_0, s_1)X(s_0, s_1) + \alpha(s_1, L)\omega(s_1, L)\beta(s_1, L)X(s_1, L)}{\alpha_e(s_0, L)\omega_e(s_0, L)X(s_0, L)}$$

$$= \frac{(24.9)(190)(0.442)(54.7) + (25.0)(232)(0.743)(36.4)}{(24.9)(207)(91.1)} = 0.578$$

$$u_e(s_0, L) = \frac{X(s_0, L)\alpha_e(s_0, L)}{\omega_e(s_0, L)} = \frac{(91.1)(24.9)}{207} = 10.96$$

$$A_B(s_0, L) = \omega_e(s_0, L)\{\ln[u_e(s_0, L)] + 1\} = 676 \text{ cm}^{-1}$$

$$\tau_{k_B}(s_0, L) = \frac{1}{A_B^*(s_0, L)} = \frac{\omega_e(s_0, L)}{A_B(s_0, L)} = \frac{207}{676} = 0.31$$

$$\Delta\eta_B(s_0, L) = \frac{A_B(s_0, L)}{1 - \tau_{k_B}(s_0, L)} = \frac{676}{1 - 0.31} = 974 \text{ cm}^{-1}$$

The lower and upper limits for the B band from s_0 to L are thus 3760 − (0.5)(974) = 3273 cm^{-1} and 3760 + (0.5)(974) = 4247 cm^{-1} as shown in Fig. 8-17*b*. From Fig. 8-17 a table can be constructed listing all the transmissivities for the various bandwidth intervals. Summing over the last two columns in the Table 8-5 gives the required terms for evaluating Eq. (8-100).

$$I(L) = 17.8[168] + 2.95[600] = 4760 \frac{\text{W}}{\text{m}^2 \text{ sr}}$$

TABLE 8-5 Bandwidth Intervals and Corresponding Transmissivities for Example 8-4

η_{L_k} (cm^{-1})	η_{U_k}	$\tau_k(s_1, L)$	$\tau_k(s_0, L)$	$[\tau_k(s_1, L) - \tau_k(s_0, L)] \cdot (\eta_{U_k} - \eta_{L_k})$	$[1 - \tau_k(s_1, L)] \cdot (\eta_{U_k} - \eta_{L_k})$
0	3273	1	1	0	0
3273	3352	1	0.31	55	0
3352	3453	0.37	0.31	6	64
3453	3488	0.37	0.08	10	22
3488	3832	0.12	0.08	14	303
3832	3866	0.37	0.08	10	21
3866	4168	0.37	0.31	18	190
4168	4247	1	0.31	55	0
4247	∞	1	1	0	0
Sum				168	600

The corresponding result from narrow band calculations [Grosshandler, 1980] is 5180 W/m^2 sr.

TOTAL GAS EMISSIVITY AND ABSORPTIVITY

In addition to line-by-line and band models, another way to present gas radiative properties is in the form of total emissivity and absorptivity data. Since only total emissivity and absorptivity are involved, this description of gas properties represents the lowest level of spectral resolution. Indeed there is no spectral resolution (see Fig. 8-11*d*). The most widely used total gas emissivity/absorptivity data are a set of charts for CO_2 and H_2O prepared by Hottel [1954]. A more recent tabulation of properties for these two gases has been given by Edwards and Matavosian [1984, 1987], which improves on the pressure scaling and absorptivity determination of Hottel's charts.

Tables 8-6 and 8-7 give the total emissivity for H_2O and CO_2, respectively, at an equivalent broadening pressure of $P_e = 1$. The data in Tables 8-8 and 8-9 can be used to scale to other values of P_e using the following relations:

$$\varepsilon_g(T_g, P_aL, P_e) = \varepsilon_g(T_g, P_aL', P_e = 1) \tag{8-107}$$

$$P_aL' = P_aLP_e^m \tag{8-108}$$

TABLE 8-6 Total Emissivity ε_g of H_2O at $P_e = 1$

	T_g				
P_aL (atm-m)	400 K	800 K	1200 K	1600 K	2000 K
0.001	0.0200	0.0073	0.0033	0.0017	0.0010
0.01	0.0795	0.0421	0.0275	0.0165	0.0098
0.1	0.2031	0.1414	0.1140	0.0949	0.0750
1	0.3811	0.3505	0.3203	0.2947	0.2634
10	0.5754	0.5780	0.5427	0.0522	0.4871
100	0.7782	0.7271	0.7363	0.7319	0.6572

TABLE 8-7 Total Emissivity ε_g of CO_2 at $P_e = 1$

	T_g				
P_aL (atm-m)	400 K	800 K	1200 K	1600 K	2000 K
0.0001	0.0028	0.0026	0.0014	0.0007	0.0004
0.001	0.0141	0.0211	0.0137	0.0070	0.0039
0.01	0.0444	0.0578	0.0462	0.0311	0.0207
0.1	0.0945	0.1156	0.1074	0.0827	0.0611
1	0.1510	0.1916	0.1900	0.1576	0.1242
10	0.2257	0.2913	0.2960	0.2365	0.2227

TABLE 8-8 Pressure Scaling Exponent m for H_2O

		$P_a L$ (atm-m)			
T_g (K)	P_e	0.01	0.1	1	10
800	0.1	0.70	0.91	0.96	1.00
	0.3	0.70	0.85	0.92	0.97
	3	0.63	0.90	0.83	0.80
	10	0.46	0.72	0.70	0.60
1200	0.1	0.47	0.80	0.92	0.94
	0.3	0.45	0.87	0.95	0.87
	3	0.20	0.72	0.76	0.63
	10	0.17	0.51	0.57	0.52
1600	0.1	0.26	0.60	0.87	0.89
	0.3	0.21	0.58	0.85	0.87
	3	0.18	0.48	0.54	0.54
	10	0.15	0.27	0.32	0.26
2000	0.1	0.14	0.46	0.77	0.82
	0.3	0.14	0.41	0.68	0.70
	3	0.14	0.25	0.40	0.34
	10	0.13	0.15	0.20	0.15

The total absorptivity can also be determined from the emissivity tables when the incident intensity is proportional to blackbody radiation at a temperature T_w. The total absorptivity is determined by using a scaled equivalent broadening pressure P_e' and a scaled partial pressure–path length product $P_a L_\alpha'$ as shown in the following relations:

$$\alpha_g(T_w, T_g, P_a L, P_e) = \sqrt{\frac{T_g}{T_w}}\, \varepsilon_g(T_w, P_a L_\alpha', P_e') \tag{8-109}$$

$$P_a L_\alpha' = P_a L \left(\frac{T_w}{T_g}\right)^r \qquad H_2O\colon r = 1.5 \quad CO_2\colon r = 1.0 \tag{8-110}$$

$$P_e' = P_e \left(\frac{T_g}{T_w}\right)^s \qquad H_2O\colon s = 1.6 \quad CO_2\colon s = 2.4 \tag{8-111}$$

The temperature of the incident intensity T_w is substituted for the gas temperature T_g in looking up ε_g as indicated in Eq. (8-109). The rationale for this is that α_g is a function of both T_w and T_g as indicated by Eq. (8-86). The temperature dependence of $e_{b_\eta}(T_w)$ is stronger than that of $A_k(T_g)$ so ε_g is evaluated at T_w. To correct for the fact that A_k has been evaluated at the

TABLE 8-9 Pressure Scaling Exponent m for CO_2

		P_aL (atm-m)			
T_g (K)	P_e	0.01	0.1	1	10
400	0.1	0.47	0.64	0.65	0.65
	0.3	0.45	0.66	0.62	0.60
	3	0.41	0.61	0.45	0.45
	10	0.35	0.49	0.38	0.38
800	0.1	0.30	0.40	0.40	0.40
	0.3	0.21	0.41	0.30	0.30
	3	0	0.09	0.11	0.11
	10	0	0.05	0.06	0.06
1200	0.1	0.08	0.10	0.15	0.18
	0.3	0	0	0	0
	3	0	0	0	0
	10	0	0	0	0
1600	0.1	0	0.02	0.02	0.03
	0.3	0	0	0	0
	3	0	0	0	0
	10	0	0	0	0
2000	0.1	0	0	0	0
	0.3	0	0	0	0
	3	0	0	0	0
	10	0	0	0	0

wrong temperature (T_w instead of T_g) the value of ε_g is multiplied by $(T_g/T_w)^{1/2}$ since bandwidth increases as $T^{1/2}$ [see Eq. (8-71)].

The formula for calculating the equivalent broadening pressure P_e is given by Eq. (8-69). The values of b and n are given in Table 8-4. For CO_2 $n = 0.8$. This formula accounts for the fact that collisions between absorbing molecules (e.g., CO_2 or H_2O) and nonabsorbing molecules (e.g., N_2) are less effective in broadening the absorption lines than collisions between two absorbing molecules.

Tables 8-6 and 8-7 show that for small values of P_aL the total emissivity grows linearly with P_aL since each additional molecule adds equally to the total emission (optically thin limit). At larger values of P_aL the total emissivity grows as $\sqrt{P_aL}$ since the centers of the lines are saturated and additional growth occurs only in the wings of the lines [see Eq. (8-27) and Fig. 8-9]. At still larger values of P_aL the total emissivity grows as $\log(P_aL)$ since the region between lines is saturated and additional growth occurs only in the wings of the band. Figure 8-18 illustrates the typical curve of growth for a gaseous emitter.

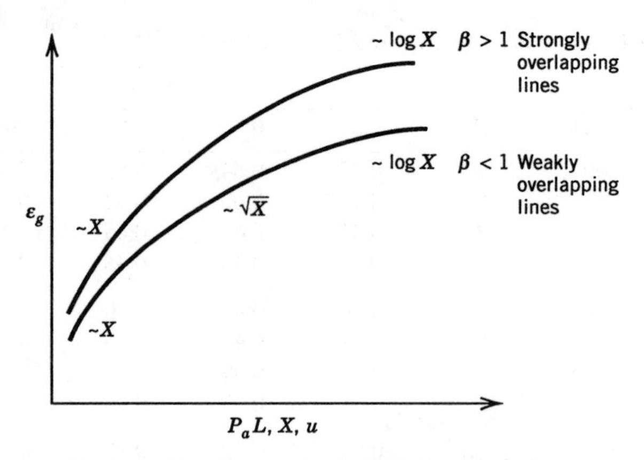

Figure 8-18 Curve of growth for gas emission.

Example 8-5

Determine the total emitted intensity (all bands) for the CO_2 and N_2 mixture of Example 8-1 using the total emissivity tables.

From Example 8-1 the total pressure is $P = 1$ atm, the partial pressure is $P_a = 0.5$ atm, the temperature is $T_g = 1000$ K, and the path length is $L = 0.5$ m. Thus the partial pressure–path length product is

$$P_a L = (0.5 \text{ atm})(0.5 \text{ m}) = 0.25 \text{ atm-m}$$

The equivalent broadening pressure from Eq. (8-69) is

$$P_e = [1 + 0.5(1.3 - 1)]^{0.8} = 1.12$$

Interpolation from Table 8-9 gives a pressure scaling exponent of $m \sim 0.1$. Thus the scaled partial pressure–path length is essentially unchanged from the actual value.

$$P_a L' = (0.25)(1.12)^{0.1} \sim 0.25$$

Logarithmic interpolation from Table 8-7 gives

$$n = \frac{\log[\varepsilon_g(P_a L_2)/\varepsilon_g(P_a L_1)]}{\log[P_a L_2/P_a L_1]} = \frac{\log[0.191/0.111]}{\log[1/0.1]} = 0.236$$

$$\varepsilon_g(P_a L) = \varepsilon_g(P_a L_2)\left(\frac{P_a L}{P_a L_2}\right)^n = 0.191\left(\frac{0.25}{1}\right)^{0.236} = 0.138$$

The total emitted intensity is thus

$$I_e = \varepsilon_g \sigma T_g^4/\pi = (0.138)(5.67 \times 10^{-8})(1000)^4/\pi = 2470 \text{ W/m}^2 \text{ sr}$$

The gas emissivity data given in Tables 8-6 through 8-9 are convenient for evaluating gas emission and absorption when only a single participating gas is present (either CO_2 or H_2O). For mixtures of CO_2 and H_2O it is necessary to account for band overlap. In this case the spectral block method of Edwards [1976] using the exponential wide band model is recommended.

SUMMARY

Gases emit and absorb thermal radiation by various physical mechanisms, including electronic, vibrational, and rotational radiative transitions. For heat transfer purposes vibration-rotation transitions, which involve infrared radiation, are the most important transitions. Vibration-rotation transitions result in line spectra that are broadened by collisions between radiating molecules and both similar and dissimilar species. An important aspect of the physics of gas radiation is the relatively small energy difference between rotational levels compared with vibrational energy levels. This difference results in vibration-rotation bands that are centered around the vibrational transition frequency and are relatively narrow compared to the spectral distribution of the Planck function but relatively wide compared with the spectral width of the individual lines.

Four different types of models have been developed to describe gas radiation properties, depending on the level of spectral resolution required. Line-by-line calculations involve spectral resolution on the order of the line structure but are seldom warranted for heat transfer purposes. Narrow band models utilize spectrally smoothed parameters, with spectral resolution on the order of a region spanning several lines. Wide band models characterize gas properties with parameters having spectral resolution of the order of a bandwidth. Total emissivity and absorptivity can also be obtained from tabulated data for a reference pressure and scaled to other pressures. The bulk of this chapter has been devoted to the description of several well-known band models, both wide and narrow.

REFERENCES

Cess, R. D. and Ramanathan, V. (1972), " Radiative Transfer in the Atmosphere of Mars and that of Venus Above the Cloud Deck," *J. Quant. Spectr. Radiative Transfer*, Vol. 12, p. 933.

Cess, R. D. and Tiwari, S. N. (1972), "Infrared Radiative Energy Transfer in Gases," *Advances in Heat Transfer*, T. F. Irvine, Jr. and J. P. Hartnett, Eds. Vol. 8, p. 229. Academic Press, New York.

Edwards, D. K. (1976), "Molecular Gas Radiation," *Advances in Heat Transfer*, T. F. Irvine, Jr. and J. P. Hartnett, Eds., Vol. 12, pp. 115–193. Academic Press, New York.

Edwards, D. K. (1981), *Radiation Heat Transfer Notes*, Hemisphere Publishing Company, New York.

Edwards, D. K. and Balakrishnan, A. (1973), "Thermal Radiation by Combustion Gases," *Int. J. Heat and Mass Trans.*, Vol. 16, pp. 25–40.

Edwards, D. K. and Matavosian, R. (1984), "Scaling Rules for Total Absorptivity and Emissivity of Gases," *J. Heat Transfer*, Vol. 106, Nov. pp. 684–689.

Edwards, D. K. and Matavosian, R. (1987), "Emissivity Data for Gases," Section 5.5.5, *Heat Exchanger Design Handbook*, E. U. Schlunder, Ed., Hemisphere, New York.

Edwards, D. K. and Menard, W. A. (1964), "Comparison of Models for Correlation of Total Band Absorption," *Applied Optics*, Vol. 3, p. 621.

Edwards, D. K. and Morizumi, D. J. (1970), "Scaling of Vibration-Rotation Band Parameters for Nonhomogeneous Gas Radiation," *J. Quant. Spect. Rad. Trans.*, Vol. 10, pp. 175–188.

Elsasser, W. M. (1943), "Heat Transfer by Infrared Radiation in the Atmosphere," *Harvard Meteorological Studies*, No. 6, Harvard University Press, Cambridge, MA.

Felske, J. D. and Tien, C-L. (1974), "A Theoretical Closed-Form Expression for the Total Band Absorptance of Infrared Radiating Gases," *Int. J. Heat. Mass Transfer*, Vol. 17, pp. 155–158.

Goody, R. M. (1952), "A Statistical Model for Water Vapor Absorption," *Quart. J. R. Meteorol. Soc.*, Vol. 78, p. 165.

Goody, R. M. and Yung, Y. L. (1989) *Atmospheric Radiation Theoretical Basis*, 2nd ed., Oxford University Press, New York.

Grosshandler, W. L. (1980), "Radiative Heat Transfer in Non-Homogeneous Gases: A Simplified Approach," *Int. J. Heat and Mass Trans.*, Vol. 23, pp. 1447–1459.

Hottel, H. C. (1954), "Radiant Heat Transmission," Chap. 4 in *Heat Transmission*, W. H. McAdams, Ed., McGraw-Hill, New York.

Ludwig, C. B., Malkmus, W., Reardon, J. E., and Thomson, J. A. L. (1973), *Handbook of Infrared Radiation from Combustion Gases*, NASA SP-3080, R. Goulard and J. A. L. Thomson, Eds., available from NTIS, publ. no. N73-27807.

McLatchey, R. A., Benedict, W. S., Clough, S. A., Burch, D. E., Calfee, R. F., Fox, K., Rothman, L. S., and Garing J. S. (1973), "AFCRL Atmospheric Absorption Line Parameters Compilation," AFCRL-TR-73-0096, January.

Modak, A. T., (1979), "Exponential Wide Band Parameters for the Pure Rotational Band of Water Vapor," *J. Quant. Spect. Rad. Trans.*, Vol. 21, pp. 131–142.

Park, J. H., Rothman, L. S., Rinsland, C. P., Smith, M. A. H., Richardson, D. J., and Larsen, J. C. (1981), "Atlas of Absorption Lines from 0 to 17,900 cm^{-1}," NASA Ref. Publ. 1084, available from NTIS.

Penner, S. S. (1959), *Quantitative Molecular Spectroscopy and Gas Emissivities*, Addison-Wesley, Reading, MA.

Rothman, L. S. (1983), "AFGL Trace Gas Compilation: 1982 Version," *Appl. Optics*, Vol. 22, pp. 1616, 2247.

Tien, C. L. (1968), "Thermal Radiation Properties of Gases," in *Advances in Heat Transfer*, T. F. Irvine, Jr. and J. P. Hartnett, Eds., Vol. 5, pp. 253–324.

Tien, C. L., and Lowder, J. E. (1966), "A Correlation for Total Band Absorption of Radiating Gases," *Int. J. Heat Mass Transfer*, Vol. 9, p. 698.

Tiwari, S. N. (1978), "Models for Infrared Atmospheric Radiation," *Adv. Geophys.*, Vol. 20, p. 1.

PROBLEMS

1. Consider two partially overlapping gas bands A and B with band centers at η_{0A} and η_{0B}, bandwidths of A_A and B_B, and effective gray bandwidths of $\Delta\eta_A$ ($> A_A$) and $\Delta\eta_B$ ($> A_B$). Using the gray, block approximation, derive the expression for the combined effective bandwidth in Eq. (8-89) A_{A+B} and the overlap correction term $\Delta\eta_{AB}$ as given in Eq. (8-90b).

2. For the condition of strong, nonoverlapping lines ($\beta \ll 1$, $u/\beta \gg 1$, $u\beta \ll 1$) show that the exponential wide band, random narrow band model gives the limiting expression for nondimensional bandwidth of $A^* = 2(u\beta)^{1/2}$.

3. Using the Elsasser narrow band, rigid rotator harmonic oscillator wide band model obtain an integral expression for the nondimensional bandwidth A^* in terms of the optical depth parameter u and the overlap parameter β.

4. Using the Elsasser narrow band model, verify the general result that the spectrally smoothed absorption coefficient $\bar{\kappa}_\eta$ is independent of the line overlap parameter β.

5. Verify that for the rigid rotator harmonic oscillator model the spectral spacing between the band center and the location of maximum smoothed absorption coefficient is $\omega/\sqrt{2}$ in wavenumbers.

6. Show that the Curtis-Godson approximation exactly satisfies the weak line limit for the spectrally smoothed narrow band emissivity of a nonhomogeneous gas $\bar{\varepsilon}_\eta = u_{e\eta}$.

7. Using the exponential wide band model, find the total intensity emitted by a mixture of CO_2 (0.5 atm) and N_2 (0.5 atm) at 500 K along a 1-m path length between 8 and 12 μm (i.e., include the 9.4- and 10.4-μm bands).

8. Using the exponential wide band model, find the total intensity emitted by a mixture of H_2O (0.3 atm) and N_2 (0.7 atm) at 500 K along a 1-m path length between 4 and 10 μm (i.e., include the 6.3-μm band).

9. Using the exponential wide band model, find the total intensity emitted by the nonhomogeneous mixture of Example 8-3 (2.7 μm band), at s_0 in the direction normal to and away from the slab.

10. Using the exponential wide band model, find the total intensity emitted by the nonhomogeneous mixture of Example 8-4 (2.7 μm band) at s_0 in the direction normal to and away from the slab.

11. Calculate the total intensity emitted along a 1.43-m path by a mixture of water vapor (0.7 atm) and nitrogen at a temperature of 2000 K and total pressure of 2 atm. Use two methods:
    (a) total emissivity tables
    (b) exponential wide band model

CHAPTER 9

RADIATIVE PROPERTIES OF PARTICLES

Absorption, emission, and scattering of radiation by particles play an important role in many engineering and environmental systems. Soot, coal, char, and ash particles significantly enhance radiative heat transfer in industrial flames [Tien and Lee, 1982; Boothroyd and Jones, 1986]. Burning metal and metal oxide particles play a dominant role in determining the radiative characteristics of solid rocket motors and plumes [Brewster, 1989, 1991]. Through scattering of solar radiation, clouds and other aerosols play a role in the earth's energy balance that is as important as that of infrared absorbing gases (but much less predictable) [Cess, 1978]. Other examples of systems in which particle radiation is important include particles in fixed and moving beds, pigment particles in paint layers, microsphere cryogenic insulations, and fibrous insulations. To predict radiative transfer in these systems or to interpret laser diagnostic information in such systems, it is necessary to know how the particles individually interact with radiation incident upon them.

The problem of radiation interacting with a particle can be solved by considering a plane, monochromatic wave incident upon a particle of a given shape, size, and optical constants and applying Maxwell's equations. This has been done for spheres (both homogeneous and layered), cylinders, and spheroids. Solutions to these various problems may be found in standard texts on electromagnetic scattering theory [Stratton, 1941; Kerker, 1969; van de Hulst, 1981; Bohren and Huffman, 1983]. These solutions provide the fundamental radiative properties for particle emission, absorption, and scattering, which are the extinction efficiency Q_e, albedo ω_0, and phase function $p(\theta)$ as defined in Chapter 7. The most widely used solution is the Mie theory for homogeneous spheres. This solution is applicable even for nonspherical

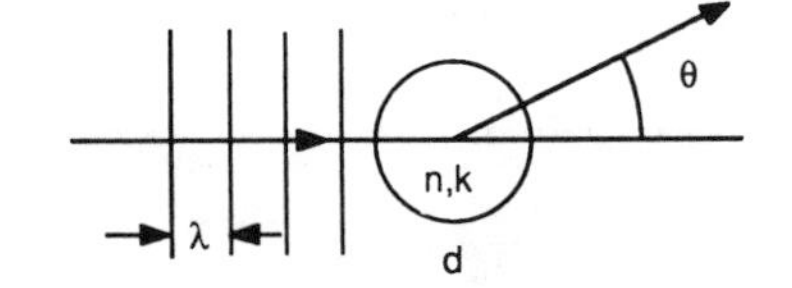

Figure 9-1 Mie scattering by a spherical particle.

particles if there are many particles that are randomly oriented. This chapter is devoted to consideration of the radiative properties of spherical particles.

SINGLE-PARTICLE PROPERTIES

In 1908 Gustav Mie developed a theory based on Maxwell's equations for a plane, monochromatic wave incident upon a homogeneous sphere in a nonabsorbing medium in order to explain the colors associated with light scattering by small colloidal gold particles suspended in water (Fig. 9-1). Solutions to the equations were later found that provided expressions for the extinction efficiency, albedo, and phase function as functions of only the optical constants ($\tilde{n} = n - ik$) relative to the surrounding medium and the particle size parameter, x,

$$x = \frac{\pi d}{\lambda} \tag{9-1}$$

where d is the particle diameter and λ is the wavelength in the surrounding medium.

The extinction efficiency Q_e is the ratio of the extinction cross section for the particle C_e to the geometric cross section G. The albedo is the ratio of the scattering efficiency (or cross section) of the particle to the extinction efficiency (or cross section). The phase function gives the normalized angular distribution of the intensity scattered from the particle. These fundamental properties are defined in Chapter 7. Except for the angular dependence of the phase function, these key properties are functions of only three parameters, n, k, and x.

$$Q_e(n, k, x) = \frac{C_e}{G} \tag{9-2}$$

$$\omega_0(n, k, x) = \frac{Q_s}{Q_e} \tag{9-3}$$

$$p = p(n, k, x, \theta) \tag{9-4}$$

In Eq. (9-4) it has been assumed that the incident radiation is unpolarized or circularly polarized. Hence the phase function is independent of azimuthal angle ϕ. For nonspherical particles or for polarized incident radiation on spherical particles, the phase function is also dependent on ϕ.

When characterizing the scattering behavior of particles, the full angular distribution $p(\theta)$ is often too much detailed information. Rather than deal with the full phase function, it is appropriate and useful to consider some moment of the distribution that describes the relative forward-to-backward ratio. For this purpose the asymmetry factor $\langle p \rangle$ is defined as follows.

$$\langle p \rangle = \frac{1}{4\pi} \int_{4\pi} p \cos\theta \, d\Omega \tag{9-5}$$

The value of $\langle p \rangle$ ranges from -1 (for maximum backward scattering) to 1 (for maximum forward scattering). Since p is independent of ϕ for spheres with unpolarized incident radiation, Eq. (9-5) can also be written as

$$\langle p \rangle = \frac{1}{2} \int_{-1}^{1} p \cos\theta \, d(\cos\theta) \tag{9-6}$$

The asymmetry factor is a function of only n, k, and x.

$$\langle p \rangle = \langle p \rangle(n, k, x) \tag{9-7}$$

The optical constants (n and k) appearing in the Mie scattering relations are those of the particle relative to the surrounding matrix. If the surrounding matrix refractive index n_∞ is greater than 1 (as is the case for particles or bubbles in an optically dense matrix), then the actual n and k of the particle are to be divided by n_∞. If n_∞ is complex, then the question comes up whether to divide by the complex value, real part, or magnitude (since Mie theory is based on a pure real n_∞). The practical answer is that this situation will almost never arise. If k_∞ is big enough to influence the magnitude of $n_\infty - ik_\infty$, then it is almost assuredly big enough that scattering by a particle in such a matrix will never be an issue. The reason is that the matrix would be so strongly absorbing that negligible radiative energy would be propagating through the medium in the first place.

LIMITING SOLUTIONS

Before considering the full Mie solution it is useful to examine some limiting regions where simpler analytic results apply. These limiting regions are defined by the magnitude of the particle size parameter x and the optical constants $\tilde{n} = n - ik$. When the particle is much smaller than the wavelength ($x \ll 1$) and the refractive index is moderate, i.e., not extremely large

($x|\tilde{n} - 1| \ll 1$), the limiting solution of Mie theory is known as Rayleigh scattering. When the particle is much larger than the wavelength ($x \gg 1$) and the refractive index is not extremely small ($x|\tilde{n} - 1| \gg 1$), geometric optics (ray tracing) and diffraction theory can be used to predict the scattering behavior. When $n \to 1$ and $k \ll 1$ such that the condition $x|\tilde{n} - 1| \ll 1$ holds for arbitrary x, the limiting solution referred to as Rayleigh-Gans scattering applies. And finally, when the particle is very large ($x \gg 1$) but the refractive index change with respect to the surroundings is very small $|\tilde{n} - 1| \ll 1$, the limiting behavior referred to as anomalous diffraction by van de Hulst is obtained. Each of these limiting regions is described by expressions that are simpler than those for the full Mie theory.

$x \ll 1$ and $x|\tilde{n} - 1| \ll 1$ (Rayleigh scattering)

$x \gg 1$ and $x|\tilde{n} - 1| \gg 1$ (Geometric optics and diffraction theory)

$|\tilde{n} - 1| \ll 1$ and $x|\tilde{n} - 1| \ll 1$ (Rayleigh-Gans scattering)

$x \gg 1$ and $|\tilde{n} - 1| \ll 1$ (Anomalous diffraction)

Arbitrary x and $\tilde{n}$ (Mie theory)

Rayleigh Scattering, $x \ll 1$ and $x|\tilde{n} - 1| \ll 1$

When a particle (spherical or otherwise) is much smaller than the wavelength of the incident radiation, and the particle refractive index is moderate such that $x \ll 1$ and $x|\tilde{n} - 1| \ll 1$, the approximation may be made that the electromagnetic field inside the particle is uniform. The particle thus behaves like an oscillating dipole, with the charged particles (electrons and ions) in the particle set into synchronous motion by the incident electromagnetic field. As far as the scattering properties are concerned, the shape of the particle is irrelevant and only its volume V matters. The scattering and absorption cross sections for a Rayleigh scatterer are given as

$$C_a = \frac{36\pi V}{\lambda} \frac{nk}{(n^2 - k^2 + 2)^2 + 4n^2k^2} \tag{9-8}$$

$$C_s = \frac{24\pi^3 V^2}{\lambda^4} \left| \frac{\tilde{n}^2 - 1}{\tilde{n}^2 + 2} \right|^2 \tag{9-9}$$

Since C_a goes like V^1 and C_s goes like V^2, for small particles ($V \to 0$) absorption, if present at all, is usually the dominant mechanism of extinction. Another observation that can be made from Eqs. (9-8) and (9-9) is that for nonabsorbing Rayleigh scatterers since C_s goes like λ^{-4}, light of shorter wavelengths will be preferentially scattered compared with light of longer wavelengths. This spectral selectivity accounts for the blue color of the sky, whereby sunlight is selectively scattered by atmospheric gas molecules.

If a Rayleigh particle is assumed to be a sphere of diameter d, the efficiencies can be written from (9-8) and (9-9) as

$$Q_a = x f_1(n, k) \tag{9-10}$$

$$Q_s = x^4 f_2(n, k) \tag{9-11}$$

where

$$f_1 = \frac{24nk}{(n^2 - k^2 + 2)^2 + 4n^2k^2} \tag{9-12}$$

$$f_2 = \frac{8}{3} \frac{\left[(n^2 - k^2 - 1)(n^2 - k^2 + 2) + 4n^2k^2\right]^2 + 36n^2k^2}{\left[(n^2 - k^2 + 2)^2 + 4n^2k^2\right]^2} \tag{9-13}$$

The albedo is obtained as the ratio

$$\omega_0 = \frac{Q_s}{Q_s + Q_a} = \frac{x^4 f_2(n, k)}{x^4 f_2(n, k) + x f_1(n, k)} \tag{9-14}$$

For the case of a lossy dielectric particle ($k \ll n$) Eq. (9-14) can be simplified to give

$$\omega_0 = \frac{1}{1 + \dfrac{9nk}{x^3(n^2 - 1)^2}}, \qquad xn < 0.5, \qquad k \ll n \tag{9-15}$$

which indicates that, for lossy dielectrics, a rapid transition between absorption dominated extinction ($\omega_0 \to 0$) and scattering dominated extinction ($\omega_0 \to 1$) occurs in the Rayleigh region as size parameter varies. The critical value of x where this transition occurs is $[9nk/(n^2 - 1)^2]^{1/3}$.

$$x \gg \left[\frac{9nk}{(n^2 - 1)^2}\right]^{1/3}: \qquad \omega_0 \to 1 \tag{9-16}$$

$$x \ll \left[\frac{9nk}{(n^2 - 1)^2}\right]^{1/3}: \qquad \omega_0 \to 0 \tag{9-17}$$

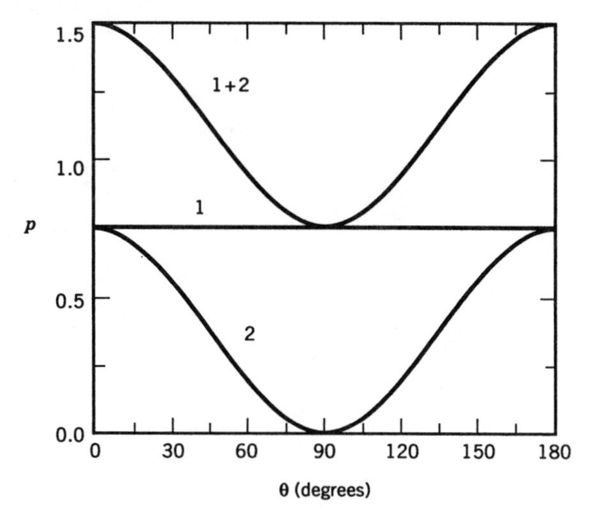

Figure 9-2 Rayleigh scattering phase function for unpolarized incident radiation (1 = perpendicular, 2 = parallel component).

Thus the magnitude of k alone is not an indication of the importance of absorption relative to scattering. A Rayleigh particle with $k \ll n$ can still be absorption dominated if x is small enough.

The phase function for Rayleigh scattering, assuming unpolarized incident radiation (Fig. 9-2), is

$$p(\theta) = \tfrac{3}{4}(1 + \cos^2\theta) \tag{9-18}$$

$$\langle p \rangle = 0 \tag{9-19}$$

This type of scattering is equally balanced in the forward and backward hemispheres. At 90° the scattered energy is polarized perpendicular to the plane of scattering (the plane containing the incident and scattering directions).

Geometric Optics Limit, $x \gg 1$ and $x|\tilde{n} - 1| \gg 1$

When a particle is much larger than the characteristic wavelength of radiation ($x \gg 1$), the series expansions used to evaluate the expressions in the full Mie theory converge very slowly. The number of terms in the series required for convergence is approximately x. Thus exact computations according to the Mie equations are rather time-consuming when $x \gg 1$. However, since a large particle size parameter is equivalent to the limiting case $\lambda \to 0$ in electromagnetic theory, the simpler laws of geometric optics hold even in the neighborhood of the particle.

Limitations of Geometric Optics

Before proceeding to outline how geometric optics may be used to calculate the various scattering properties, it is beneficial to discuss when this approximation is valid. The minimum value of x for which geometric optics applies can be estimated by comparing the magnitudes of the particle diameter and the minimum width of a pencil of radiation incident on the particle. If the diameter is sufficiently large compared to the width of an incident pencil ($x \gg 1$), the concept of a ray will adequately describe the behavior of the electromagnetic field and geometric optics will be relevant. If not ($x \approx 1$), then Mie theory must be employed.

The question is raised at this point as to what determines the "width" of a ray. A geometric ray, of course, has zero width. But a pencil of light or other radiation, which behaves like a ray, i.e., propagates rectilinearly over some given path length, has a certain minimum width associated with it. A formula for the magnitude of that width was given in 1818 when Fresnel formulated Huygen's principle quantitatively [van de Hulst, 1981].

Fresnel demonstrated that the effective area of a rectilinearly propagating plane wave front, which contributes to the field at a distance L from the wave front, is $L\lambda$. Thus a pencil of light (or ray) of length L can exist only if its width is large compared to $(L\lambda)^{1/2}$. Stated another way, a pencil of width $p\lambda$ can lead an independent existence over a length of the order of $p^2\lambda$.

Now Fresnel's result can be applied to a spherical particle to determine the relation between x and the width of an incident pencil of light. Consider first a sphere with $x \approx 30$ or $d \approx 10\lambda$. By requiring that the pencil of light maintain its identity over a length equal to the sphere diameter, 10λ, it is apparent that the "width" of the pencil is 3λ, or approximately one-third of the particle diameter.

For a smaller particle, e.g., $x \approx 5$, the corresponding ray width would be on the same order as the particle itself. Obviously geometric optics would be a poor model for scattering by particles with a very small size parameter. Just how large x needs to be for reasonable accuracy depends on the specific application, but for opaque particles $x = 5$ is about the lowest value for which the phase function predicted by geometric optics plus diffraction theory reasonably approximates the exact phase function of Mie theory.

Radiative Properties from Geometric Optics

The expressions for Q_e, Q_s and $p(\theta)$ pertaining to geometric particles can be derived either by direct application of the laws of geometric optics or as a limiting case of the complete Mie theory ($x \to \infty$). Van de Hulst [1981] demonstrates the equivalence of both approaches in detail. In the following only the results are presented.

Efficiency Factors

The expressions for the scattering, extinction, and absorption efficiencies are rather simple for large particles. For all large particles, irrespective of composition and shape, the expression for the extinction efficiency is

$$Q_e = 2 \quad (1 \text{ if diffraction is neglected}) \tag{9-20}$$

A contribution to Q_e of 1, corresponding to the geometric shadow of the particle, is due to the radiation either reflected, refracted, or absorbed by the particle. The remaining contribution of 1 arises from the diffracted radiation passing near the edges of the particle, known in the far-field as Fraunhofer diffraction.

The scattering and absorption efficiencies depend on the nature of the particle [van de Hulst, 1981; Siegel and Howell, 1988].

$$Q_a = 1 - \rho_\lambda - \tau_\lambda = \alpha_\lambda \tag{9-21}$$

$$Q_s = 1 + \rho_\lambda + \tau_\lambda \quad (\rho_\lambda + \tau_\lambda \text{ if diffraction is neglected}) \tag{9-22}$$

In these equations ρ_λ represents the fraction of incident energy reflected at the front surface of the particle (the surface facing the collimated incident radiation), and τ_λ represents the fraction of incident energy that is transmitted into the particle, undergoes possible internal reflection, and leaves the particle without being absorbed.

The reflectivity for a spherical particle subject to collimated irradiation can be shown to be equivalent to the hemispherical reflectivity of a flat surface subject to diffuse irradiation (see Problem 2-18). Thus ρ_λ can be obtained from Eq. (2-28) as

$$\rho_\lambda = \frac{1}{\pi} \int_{2\pi} \rho'_\lambda \cos\beta \, d\Omega = 2 \int_0^{\pi/2} \rho'_\lambda \cos\beta \sin\beta \, d\beta$$

where β has been used to indicate the angle between the incident rays and the local normal to the surface of the sphere and ρ'_λ is the directional reflectivity of the surface of the particle. This change in angle notation is necessary because θ, which was used for local polar angle in Chapter 2, now stands for the particle scattering angle (see Fig. 9-1).

The particle transmissivity depends on the opacity of the particle. The opacity of the particle is determined by the magnitude of the optical depth based on particle diameter, $4\pi kd/\lambda = 4xk$. If $4xk$ is large, the particle is opaque and the transmissivity is zero. If $4xk$ is small, the particle is nonabsorbing.

Large nonabsorbing particle:	$4xk \ll 1 \quad (\rho_\lambda + \tau_\lambda = 1)$
Large partially absorbing particle:	$4xk \sim 1$
Large opaque particle:	$4xk \gg 1 \quad (\tau_\lambda = 0)$

In the case of partially absorbing particles it is necessary to carry out ray tracing to simulate the internal reflection of energy within the particle to determine the transmissivity as outlined by van de Hulst [1981].

Phase Function

The angular distribution, or phase function, of the scattered energy must be considered in two parts. The part that is due to diffraction is given for a sphere in terms of a Bessel function of the first kind (Fig. 9-3).

$$p(\theta) = x^2\left[\frac{2J_1(x \sin\theta)}{x \sin\theta}\right]^2 \tag{9-23}$$

This contribution to the scattering phase function is independent of the optical constants of the particle and is exclusively in the forward direction ($\theta \sim 0$ to 6° depending on x). The other part of the phase function, arising from reflected radiation, depends on the optical constants of the particle but is independent of size.

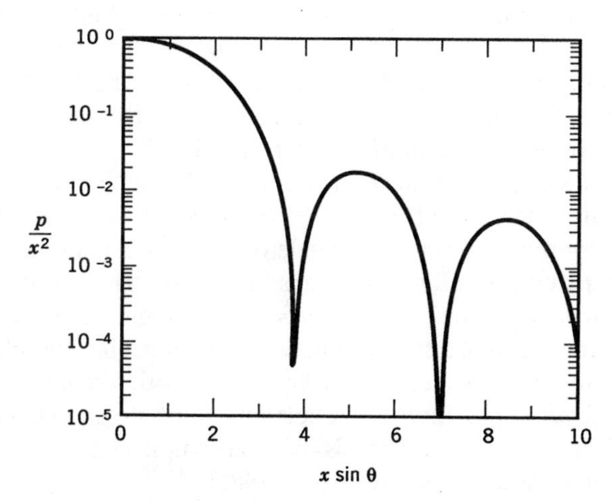

Figure 9-3 Phase function for diffraction by large sphere.

For opaque, specularly reflecting spheres [$\beta = (\pi - \theta)/2$] the phase function corresponding to reflection is [Siegel and Howell, 1988]

$$p(\theta) = \frac{\rho'_\lambda\left[\frac{1}{2}(\pi - \theta)\right]}{\rho_\lambda} \tag{9-24}$$

where ρ'_λ is the directional reflectivity, θ is the single scattering angle, and $\beta = (\pi - \theta)/2$ is the local incident angle relative to the particle surface. Recalling the angular dependence of ρ'_λ from Chapter 4 (Fig. 4-9), Eq. (9-24) indicates that large metallic specular spheres are nearly isotropic scatterers while large, opaque, specular, dielectric spheres have a strong forward scattering tendency.

The phase function corresponding to reflection from large opaque diffusely reflecting, diffusely absorbing spheres is [van de Hulst, 1981; Siegel and Howell, 1988]

$$p(\theta) = \frac{8}{3\pi}(\sin\theta - \theta\cos\theta) \tag{9-25}$$

which is predominantly in the backward hemisphere ($\langle p \rangle = -0.14$).

Next the case of large transparent spheres will be considered. These are spheres for which k is small enough that $4xk \ll 1$ even though $x \gg 1$. For large, transparent, nonabsorbing particles the nondiffractive phase function is dominated by refracted rays that pass through the droplet unreflected. Most of these refracted rays propagate in the forward direction with only a slight variation from the direction of incidence. Thus the phase function for transparent, spherical particles is strongly forward biased. However, the contribution to $p(\theta)$ near $\theta = 0$ is usually greater for diffraction than for refraction. There are also several spikes in $p(\theta)$ in the backward direction that arise from stationary points of reflection for rays undergoing multiple internal reflections. For example water droplets ($n = 1.33$ in the visible region) have a stationary reflection point for the set of rays undergoing a single internal reflection at $\theta = 138°$. This spike is responsible for the primary rainbow that appears at an angle of 42° between a line connecting an observer and a cloud of falling raindrops ($x \simeq 5000$) and the direction of the sun's rays. For a refractive index of $n = 1.6$, a single rainbow occurs at an angle of $\theta = 165°$, as illustrated in Fig. 9-4. To determine the phase function for a large transparent sphere using geometric optics requires that rays be traced through multiple internal reflections until negligible energy is left inside the droplet. This exercise is best accomplished with the aid of a computed as illustrated by van de Hulst [1981].

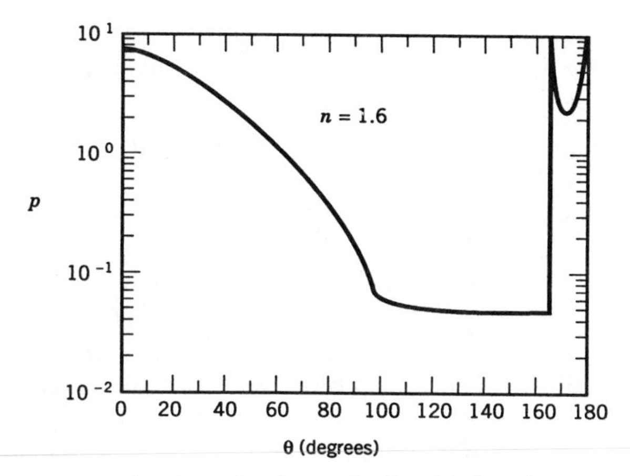

Figure 9-4 Phase function for reflection and refraction by a large transparent sphere from ray tracing.

Neglect of Diffraction

For most situations involving multiple scattering and particularly for heat transfer applications, the diffraction contribution can be ignored and a total extinction efficiency of one used. This change is not only convenient but desirable. The extremely large values of the exact phase function in the forward direction ($\theta \approx 0°$), which give rise to the diffraction pattern as $x \to \infty$, cause erroneous results in the numerical solution of the transfer equation unless extremely fine angular resolution is employed. Accordingly, the exact phase function should be modified by reducing the extremely large values in the neighborhood $\theta \approx 0°$ to 5°. This effectively removes diffraction from consideration as a source of scattering. The exact amount of reduction necessary is dictated by energy conservation. The relation

$$\frac{1}{4\pi}\int_0^{2\pi}\int_0^{\pi} p(\theta)\sin\theta \, d\theta \, d\phi = 1 \tag{9-26}$$

must be satisfied according to conservation of scattered energy. Reducing the forward values of $p(\theta)$ until the integral is one-half corresponds to subtracting one from both the scattering and extinction efficiencies. The modification is successfully completed by doubling the remaining values of $p(\theta)$, so that Eq. (9-26) is once more satisfied. The phase functions given in Eqs. (9-24) and (9-25) have already been normalized with the diffraction contribution removed.

Rayleigh-Gans Scattering, $2x|\tilde{n}-1| \ll 1$ and $|\tilde{n}-1| \ll 1$ (x Arbitrary)

When $n \to 1$ and $k \ll 1$ such that the condition $2x|\tilde{n}-1| \ll 1$ holds, the limiting solution referred to by van de Hulst as Rayleigh-Gans scattering applies. The distinction between this case and Rayleigh scattering is that in this case the limiting condition is being imposed by the optical constants and the value of x is arbitrary. Compared with Rayleigh scattering, Rayleigh-Gans scattering is more restrictive in the allowable values of n and k and less restrictive in the allowable values of x. Since the parameter $2x|\tilde{n}-1|$ corresponds to the change of phase of a light ray passing along the diameter of a sphere, this limiting region corresponds to scattering with small phase shift (also called Born's approximation). As discussed by van de Hulst, a necessary consequence of the assumptions employed in Rayleigh-Gans theory is that $Q_e \ll 1$. The important scattering results for the Rayleigh-Gans case are as follows:

$$Q_a = \tfrac{8}{3}nkx \tag{9-27}$$

$$Q_s = |\tilde{n}-1|^2\varphi(x) \tag{9-28}$$

$$\varphi(x) = \tfrac{4}{9}x^4\int_0^{\pi} g^2\left(2x\sin\frac{\theta}{2}\right)(1+\cos^2\theta)\sin\theta\,d\theta \tag{9-29}$$

$$g(u) = \frac{3}{u^3}(\sin u - u\cos u) \tag{9-30}$$

$$\langle p\rangle = \frac{4x^4}{9\varphi(x)}\int_0^{\pi} g^2\left(2x\sin\frac{\theta}{2}\right)(1+\cos^2\theta)\cos\theta\sin\theta\,d\theta \tag{9-31}$$

In Eqs. (9-29) and (9-31), the argument of the function g is $2x\sin(\theta/2)$.

Of the Rayleigh-Gans results the asymmetry factor is particularly important because it displays the essential features of the phase function for more general situations. At small x the conditions of Rayleigh-Gans scattering match those of Rayleigh scattering, and the phase function becomes identical to that of Rayleigh scattering with $\langle p\rangle \to 0$. As x increases, maxima and minima begin to appear in the phase function and the distribution becomes more biased in the forward direction. For large values of x, a pattern resembling Fraunhofer diffraction develops in the region near $\theta = 0$ and $\langle p\rangle \to 1$. This type of behavior is representative of that for other values of $\tilde{n}$, which do not satisfy the Rayleigh-Gans criteria. Thus, the Rayleigh-Gans result for scattering asymmetry, which is only a function of size parameter, can be used to estimate the asymmetry for more general situations. Edwards and Babikian [1989] have provided a curve fit that closely matches the exact integral results for $\langle p\rangle$.

$$\langle p\rangle = 0.07729x^2 + 0.13334x^3 - 0.05322x^4, \qquad x < 2 \tag{9-32}$$

$$\langle p\rangle = 1 - 2.21919\exp(-0.78857x), \qquad x > 2 \tag{9-33}$$

Anomalous Diffraction, $|\tilde{n} - 1| \ll 1$ and $x \gg 1$

Another limiting region of some practical interest is the region known as anomalous diffraction. This region is defined by the limiting values $|\tilde{n} - 1| \ll 1$ and $x \gg 1$ such that the product $x|\tilde{n} - 1|$ remains finite. The pertinent results for this region are as follows:

$$Q_e = 2 + 4u^2\cos(2\beta) - 4\exp(-\rho\tan\beta)\left[u\sin(\rho - \beta) + u^2\cos(\rho - 2\beta)\right] \tag{9-34}$$

$$Q_a = 1 + (2/v)e^{-v} - (2/v^2)(1 - e^{-v}) \tag{9-35}$$

where

$$\rho = 2x(n - 1) \tag{9-36}$$

$$\tan\beta = \frac{k}{n - 1} \tag{9-37}$$

$$u = \frac{\cos\beta}{\rho} \tag{9-38}$$

$$v = 4xk = 2\rho\tan\beta \tag{9-39}$$

Because of the large size parameter throughout this domain, the contribution to the phase function from diffraction is concentrated in the forward direction. Likewise, because of the small value of $n - 1$ in this domain, any contribution to the phase function from reflection and transmission is also concentrated in the forward direction.

The extinction efficiency is plotted in Fig. 9-5. The extinction efficiency is small for $\rho < 1$ and oscillates around the geometric optics limiting value of 2 for $\rho > 2$. These oscillations are due to alternate constructive and destructive interference between the rays transmitted through the particle and those passing by the edge, which are diffracted. The oscillations are damped out as β increases due to the stronger absorption of transmitted rays as k increases and the weaker refraction of transmitted rays as $n - 1$ decreases.

The albedo is plotted in Fig. 9-6. As $\rho \to 0$ the albedo approaches zero and the extinction is dominated by absorption. As ρ increases the albedo increases due to the increasing fraction of extinction by scattering. For small but nonzero values of β the albedo oscillates and approaches 0.5 as $\rho \to \infty$. This limit corresponds to blackbody behavior, even though the nonzero value of ω_0 seems to indicate otherwise. The contribution of $\frac{1}{2}$ to the albedo comes entirely from diffraction. All of the radiation "incident" upon the particle is absorbed ($Q_a \to 1$). It may be recalled from Chapter 4 [Eq. (4-117)] that the conditions necessary for a material to be a perfect absorber-emitter are exactly those of anomalous diffraction with $\rho \to \infty$ and nonzero β.

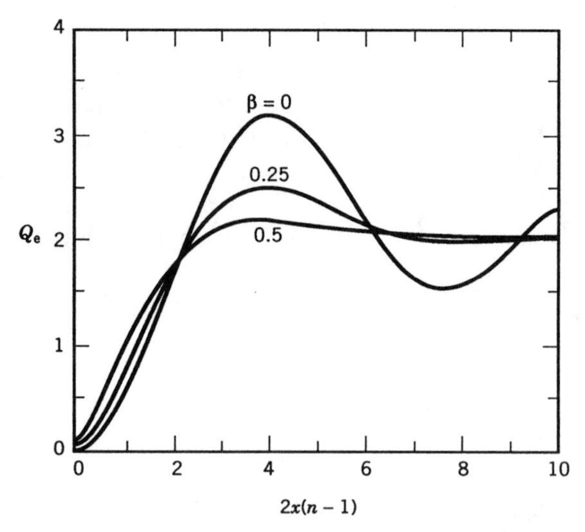

Figure 9-5 Extinction efficiency for anomalous diffraction.

The restrictions on the values of n and k, which have been assumed in the limiting condition known as anomalous diffraction, are $n \sim 1$ and $k \ll 1$. Since the range of optical constants $1 < n < 2$ and $k \ll 1$ applies to a wide variety of lossy dielectric materials in the visible and infrared regions, these results are quite useful in their own right for estimating particle efficiencies (particularly Q_a) when $x > 1$.

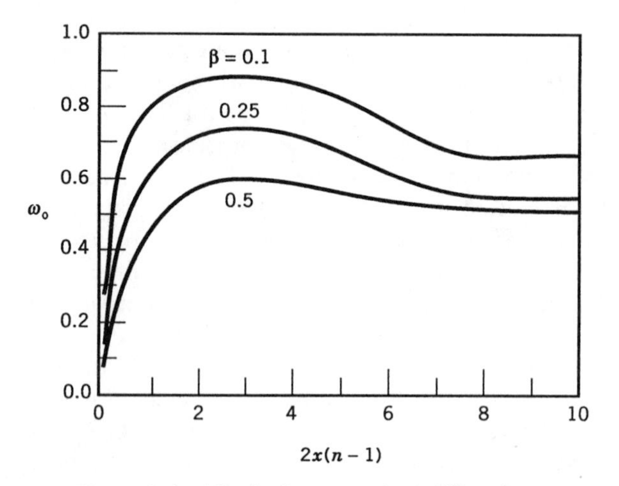

Figure 9-6 Albedo for anomalous diffraction.

MIE SCATTERING

For a rigorous treatment of the scattering of radiation by a sphere of arbitrary size and optical constants, it is necessary to solve Maxwell's equations. Details of the solution can be found in Stratton [1941], Kerker [1969], van de Hulst [1981] and Bohren and Huffman [1983]. The resulting expressions for extinction and scattering efficiencies are

$$Q_e = \frac{2}{x^2} \sum_{n=1}^{\infty} (2n+1) \mathrm{Re}\{a_n + b_n\} \tag{9-40}$$

$$Q_s = \frac{2}{x^2} \sum_{n=1}^{\infty} (2n+1) \left(|a_n|^2 + |b_n|^2 \right) \tag{9-41}$$

where

$$a_n = \frac{\psi_n'(\tilde{n}x)\psi_n(x) - \tilde{n}\psi_n(\tilde{n}x)\psi_n'(x)}{\psi_n'(\tilde{n}x)\xi_n(x) - \tilde{n}\psi_n(\tilde{n}x)\xi_n'(x)} \tag{9-42}$$

$$b_n = \frac{\tilde{n}\psi_n'(\tilde{n}x)\psi_n(x) - \psi_n(\tilde{n}x)\psi_n'(x)}{\tilde{n}\psi_n'(\tilde{n}x)\xi_n(x) - \psi_n(\tilde{n}x)\xi_n'(x)} \tag{9-43}$$

The functions ψ_n and ξ_n are Ricatti-Bessel functions, which obey the recursion relations

$$\psi_{n+1}(x) = \frac{2n+1}{x}\psi_n(x) - \psi_{n-1}(x) \tag{9-44}$$

$$\chi_{n+1}(x) = \frac{2n+1}{x}\chi_n(x) - \chi_{n-1}(x) \tag{9-45}$$

where

$$\xi_n = \psi_n - i\chi_n \tag{9-46}$$

$$\psi_{-1}(x) = \cos x \qquad \psi_0(x) = \sin x \tag{9-47}$$

$$\chi_{-1}(x) = -\sin x \qquad \chi_0(x) = \cos x \tag{9-48}$$

The phase function for unpolarized incident radiation is

$$p(\theta) = \frac{2\left(|S_1|^2 + |S_2|^2\right)}{Q_s x^2} \tag{9-49}$$

where

$$S_1(\theta) = \sum_{n=1}^{\infty} \frac{2n+1}{n(n+1)} (a_n \pi_n + b_n \tau_n) \tag{9-50}$$

$$S_2(\theta) = \sum_{n=1}^{\infty} \frac{2n+1}{n(n+1)} (a_n \tau_n + b_n \pi_n) \tag{9-51}$$

$$\pi_n(\cos\theta) = \frac{1}{\sin\theta} P_n^1(\cos\theta) \tag{9-52}$$

$$\tau_n(\cos\theta) = \frac{d}{d\theta} P_n^1(\cos\theta) \tag{9-53}$$

and P_n^1 are associated Legendre polynomials. The asymmetry factor, which is independent of polarization of the incident radiation, is given by

$$\langle p \rangle = \frac{4}{Q_s x^2} \sum_{n=1}^{\infty} \frac{n(n+2)}{n+1} \operatorname{Re}\{a_n a_{n+1}^* + b_n b_{n+1}^*\} + \frac{2n+1}{n(n+1)} \operatorname{Re}\{a_n b_n^*\} \tag{9-54}$$

Many computer codes are available for computing Mie scattering parameters. One of the earliest and most widely utilized is that of Dave [1968]. Bohren and Huffman [1983] also provide a version of the Mie code that incorporates some improvements in computational efficiency made by Wiscombe [1980] as well as some of their own modifications. A few specific results for Mie scattering are given in the remainder of this section.

The results for Q_e, ω_0, and $\langle p \rangle$ for a sphere of arbitrary size, $n = 1.8$ and $k = 0$ to 10 are given in Figs. 9-7 through 9-9. The results for Q_e are given in Figs. 9-7*a* and 9-7*b*. The value of Q_e varies from very small for Rayleigh scattering ($x \ll 1$) to 2 for the geometric optics limit (including diffraction).

In the Rayleigh region ($x < 0.5$) for $k \ll n$ there are two subdivisions. There is a region of pure absorption ($\omega_0 \to 0$), as discussed in the previous section on Rayleigh scattering, and a region of pure scattering ($\omega_0 \to 1$). The transition between the two regions occurs where the slope changes from 1 to 4 on a log-log plot (Fig. 9-7*b*) and depends on the magnitude of k as indicated by Eq. (9-15). In the pure absorption region Eq. (9-10) for Q_a is also the expression for Q_e. Since the extinction coefficient K_e for an assembly of particles is proportional to $Q_e f_v / d$, Eq. (9-10) indicates that the extinction coefficient (and hence optical depth) approaches a constant in the limit as $d \to 0$ for constant particle volume fraction f_v. Thus, lines of constant extinction coefficient appear as straight lines with slope of 1 on Fig. 9-7*b*, parallel to the nonscattering Rayleigh solutions.

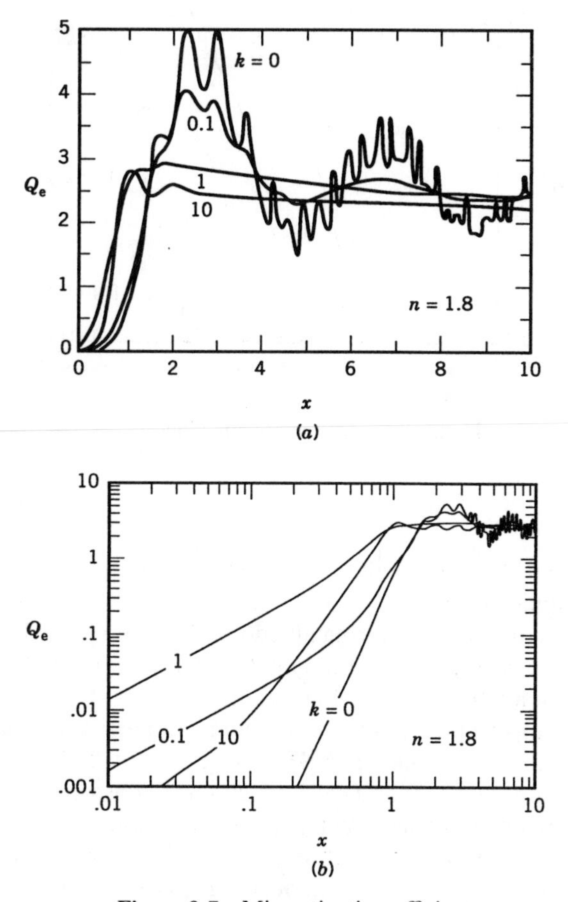

Figure 9-7 Mie extinction efficiency.

In the geometric optics region ($x \gtrsim 5$) Q_e approaches a constant value of 2 with a contribution of 1 corresponding to the geometric shadow and the remaining contribution of 1 corresponding to diffraction. For $x > 1$ the values of Q_e oscillate with x due to interference effects. The cause of the interference oscillations is the same as that described for anomalous diffraction, namely, the interference between internally transmitted (refracted) rays and externally diffracted rays. As the value of k increases, these oscillations are diminished. In addition to the interference oscillations, there is also a high frequency ripple component that appears for $k < 0.1$, which is ascribed to resonance effects. As the magnitude of k increases, both the interference and ripple oscillations are smoothed out. The structure of the extinction curves for $x > 1$, including magnitude and location of maxima and

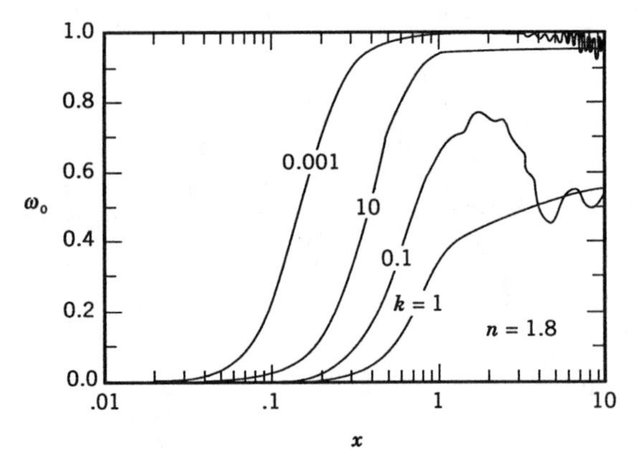

Figure 9-8 Mie scattering albedo.

minima, is reasonably well predicted by the anomalous diffraction results, except for the ripple structure.

Figure 9-8 gives the Mie scattering albedo results for a particle of arbitrary size. In accordance with Eq. (9-15) the albedo changes rapidly from 0 to 1 in the Rayleigh region when $k \ll n$. For example, for $k = 10^{-3}$ and $n = 1.8$, the value of ω_0 approaches 1 for $x > 0.5$ and the particles behave as pure scatterers due to the small value of k. However, for $x < 0.05$ (and $k = 10^{-3}$) the value of ω_0 approaches 0 and the particles are nonscattering.

Figure 9-9 shows how the asymmetry factor $\langle p \rangle$ varies with x for a sphere of arbitrary size. It can be seen that as $x \to 0$, $\langle p \rangle$ approaches 0 (Rayleigh scattering) and as $x \to \infty$, $\langle p \rangle$ approaches a constant on the order of 1

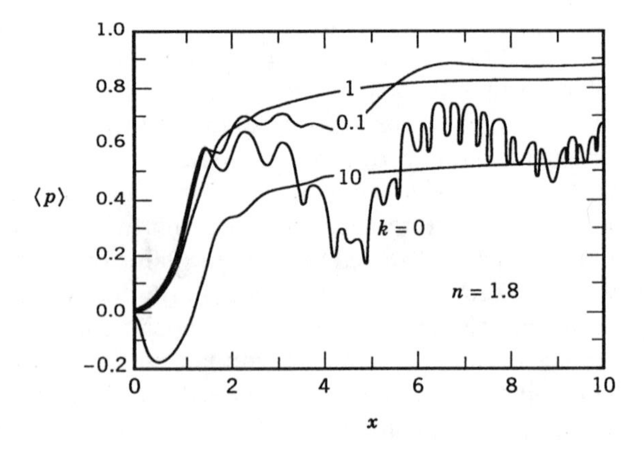

Figure 9-9 Mie scattering asymmetry parameter.

(diffraction dominated scattering). Also, for $x > 1$ and $k \ll n$, $\langle p \rangle$ exhibits interference oscillations and ripple. Although Fig. 9-9 clearly indicates that $\langle p \rangle$ is not only a function of x but is also a function of optical constants, comparison of the Rayleigh-Gans result for $\langle p \rangle$ indicates that this approximation gives a reasonable estimate of the scattering asymmetry, particularly in the critical transition region $1 < x < 2$.

Summarizing the important trends in Q_e, ω_0, and $\langle p \rangle$, it can be said that as particle size parameter increases (e.g., as particle size increases at a fixed wavelength), the extinction efficiency increases, passes through a maximum, and approaches 2. For finite k the albedo begins at values near zero for very small particles and increases as size increases. For $k \ll n \sim 1$ the albedo approaches a value of 0.5 for very large (i.e., opaque, blackbody) particles. The scattering distribution shifts from quasi-isotropic (Rayleigh scattering) to forward diffraction dominated scattering as particle size increases.

This completes the consideration of the radiative properties of single properties. Attention is now turned to radiative properties of assemblies of particles.

DISPERSION PROPERTIES

The particle radiative properties defined in Chapter 7 that are required for integration of the transfer equation are the extinction coefficient, albedo, and phase function. Two of these properties, albedo and phase function, are similar in name and definition to the single-particle properties defined in this chapter, but there is an important distinction. The properties of Chapter 7 were derived for an optically thin, differential volume of space occupied by many particles of possibly different sizes and properties, whereas the particle properties obtained from Mie theory apply to single particles. The volumetric properties can easily be obtained from the Mie scattering properties by adding up the effect of all the particles of different sizes and properties in a small volume [Deirmendjian, 1969].

$$K_{e,s} = \sum_i C_{e,s_i} N_i \tag{9-55a}$$

$$\overline{\omega}_0 = \frac{K_s}{K_e} \tag{9-56a}$$

$$\bar{p}(\theta, \phi) = \frac{1}{K_s} \sum_i C_{s_i} p_i(\theta, \phi) N_i \tag{9-57a}$$

The bar over the albedo and phase function indicates that these are average properties that include the effects of many particles with different properties. There is no bar used on the extinction and scattering coefficients because these properties are not defined for a single particle but only for dispersions consisting of many particles. The properties defined in Eqs. (9-55a) through (9-57a) are referred to as single-scattering properties as are the corresponding single-particle properties defined in Eqs. (9-2) through (9-4). This might seem contradictory since by definition the dispersion properties involve multiple particles. It must be remembered, however, that the volumetric properties defined by Eqs. (9-55a) through (9-57a) apply to a small volume element that by definition is optically thin with negligible shadowing and negligible multiple scattering (see discussion associated with Fig. 7-3). Such a volume element is by definition characterized by single scattering and hence these volumetric properties are also referred to as single-scattering properties.

Naturally occurring particle dispersions usually consist of particles that have variations in both their optical properties (n and k) and size. In such cases the summations in Eqs. (9-55a) and (9-57a) refer to summation over both size and optical properties. Sometimes it is reasonable to characterize a dispersion of particles as a *monodispersion* in which case all particles are assumed to have the same size but possibly different optical properties. The following example considers such a case.

Example 9-1

Many studies of the optical properties of atmospheric aerosols have been based on the amount of energy absorbed when light is transmitted through a homogeneous layer obtained from a collection of aerosol particles [Toon, et al., 1976]. In these studies the transmission is assumed to be given by

$$\tau = \exp[-K_a h] \qquad K_a = \frac{4\pi k}{\lambda}$$

where h is the layer thickness and reflection losses have been neglected or otherwise accounted for. Suppose a value of $k = 0.05$ were obtained from such an experiment. Determine the aerosol single-scattering albedo for two different assumed compositions: (a) all particles have the same composition, and (b) 99% (by volume) of the particles are actually nonabsorbing ($k = 0$) and only 1% of the particles are absorbing. Assume all particles have the same size and $n = 1.5$. Use a particle size of $d = 0.18$ μm and wavelength of $\lambda = 0.5$ μm.

(a) Under the assumption that the monodispersion is homogeneous, the single-scattering albedo and single-particle albedo are equivalent and can be calculated directly from the Mie scattering properties as follows:

$$n = 1.5 \qquad k = 0.05 \qquad x = \frac{\pi(0.18)}{0.5} = 1.13$$

$$\overline{\omega}_0 = \omega_0(n, k, x) = \frac{Q_s}{Q_e} = \frac{0.31}{0.48} = 0.65$$

(b) For the nonhomogeneous case it is necessary to note that the absorption coefficient of the homogeneous layer is a volume average of the absorption coefficients of the constituents

$$K_a = \frac{4\pi}{\lambda} \sum_i k_i f_{v_i}$$

Using $i = 1$ to denote the nonabsorbing particles ($k_1 = 0$) and $i = 2$ to denote the absorbing particles, the absorption index of the absorbing particles k_2 can be calculated from the apparent homogeneous absorption index $k = 0.05$ as

$$k = f_{v_1}k_1 + f_{v_2}k_2 \rightarrow 0.05 = (0.99)(0) + (0.01)k_2 \rightarrow k_2 = 5$$

The single-scattering albedo is calculated using Eqs. (9-55a) and (9-56a). For spherical particles Eqs. (9-55a) and (9-56a) can be written as

$$K_{e,s} = 1.5 \sum_i \frac{Q_{e,s_i} f_{v_i}}{d_i} \qquad \overline{\omega}_0 = \frac{\sum_i \dfrac{Q_{s_i} f_{v_i}}{d_i}}{\sum_i \dfrac{Q_{e_i} f_{v_i}}{d_i}}$$

The Mie scattering properties can be calculated from the optical constants and size parameter for each type of particle as follows:

$$n_1 = 1.5 \qquad k_1 = 0 \qquad x_1 = 1.13 \rightarrow Q_{s_1} = Q_{e_1} = 0.33$$

$$n_2 = 1.5 \qquad k_2 = 5 \qquad x_2 = 1.13 \rightarrow Q_{s_2} = 2.75 \qquad Q_{e_2} = 3.31$$

Noting that $d_1 = d_2$ gives

$$\overline{\omega}_0 = \frac{Q_{s_1} f_{v_1} + Q_{s_2} f_{v_2}}{Q_{e_1} f_{v_1} + Q_{e_2} f_{v_2}} = \frac{(0.33)(0.99) + (2.75)(0.01)}{(0.33)(0.99) + (3.31)(0.01)} = 0.98$$

This change in albedo (from 0.65 to 0.98) in going from the homogeneous dispersion to the nonhomogeneous dispersion is enough to change a radiative transfer climate model prediction from global warming to a prediction of global cooling [Toon, et al., 1976]. This example points out not only the importance of composition on dispersion radiative properties but also the difficulty in determining optical properties from collected samples of particles.

SIZE DISTRIBUTION EFFECTS

A dispersion consisting of particles with the same composition but different sizes is referred to as a homogeneous *polydispersion*. In the case of a homogeneous polydispersion the optical constants are the same for all the particles, and the summations in Eqs. (9-55a) and (9-57a) refer to summation over particle size only. Assuming the size distribution is a continuous function $N(r)$, Eqs. (9-55a) and (9-57a) can be written as integrals.

$$K_{e,s} = \int_0^\infty \pi r^2 Q_{e,s}(r) N(r)\, dr \tag{9-55b}$$

$$\bar{p}(\theta,\phi) = \frac{1}{K_s}\int_0^\infty \pi r^2 Q_s(r) p(\theta,\phi,r) N(r)\, dr \tag{9-57b}$$

$N(r)$ is the number density distribution and $N(r)\,dr$ is interpreted as the number of particles per unit volume with radius in the interval between r and $r + dr$. The total number of particles per unit volume N_0 is

$$N_0 = \int_0^\infty N(r)\, dr \tag{9-58}$$

and the particle volume fraction is

$$f_v = \frac{4}{3}\pi \int_0^\infty r^3 N(r)\, dr \tag{9-59}$$

Multiplying and dividing Eq. (9-55b) by f_v and using Eq. (9-59) gives the following form for K_e

$$K_e = \frac{3}{4}\overline{\left(\frac{Q_e}{r}\right)} f_v \tag{9-60}$$

where the average extinction efficiency to radius ratio is defined as

$$\overline{\left(\frac{Q_e}{r}\right)} = \frac{\int_0^\infty r^2 Q_e(r) N(r)\, dr}{\int_0^\infty r^3 N(r)\, dr} \tag{9-61}$$

The motivation for doing this is that Eq. (9-60) has the same form as the monodisperse case. Various moments of size and extinction efficiency may conceivably be used to evaluate the mean extinction efficiency to radius parameter. One particular choice that has special significance for particles in the size range $x > 1$ is the 32 moment.

$$\overline{\left(\frac{Q_e}{r}\right)} = \frac{\overline{Q}_e}{r_{32}} \tag{9-62}$$

where

$$\overline{Q}_e = \frac{\int_0^\infty r^2 Q_e(r) N(r)\, dr}{\int_0^\infty r^2 N(r)\, dr} \tag{9-63}$$

and

$$r_{32} = \frac{\int_0^\infty r^3 N(r)\, dr}{\int_0^\infty r^2 N(r)\, dr} \tag{9-64}$$

Gamma Distribution Function

To further explore the implications of polydispersity on particle radiative properties, it is expedient at this point to introduce a representative size distribution function. The simplest description of size distribution, which is often adequate, is to assume a monomodal distribution described by two parameters. Examples of such functions are the gamma, log-normal, and Rosin-Rammler distribution functions. For purposes of illustration the gamma function distribution that has parameters a and b will be used here as suggested by Hansen and Travis [1974].

$$N(r) = \frac{N_0 b^{a+1}}{\Gamma(a+1)} r^a e^{-br} \tag{9-65}$$

Figure 9-10 shows three sample gamma distributions with a 32 radius of 5 (in arbitrary units) where the 32 radius from Eqs. (9-64) and (9-65) is

$$r_{32} = \frac{3+a}{b} \tag{9-66}$$

and the most probable radius is the radius where $dN/dr = 0$ and is given by

$$r_{\mathrm{mp}} = \frac{a}{b} \tag{9-67}$$

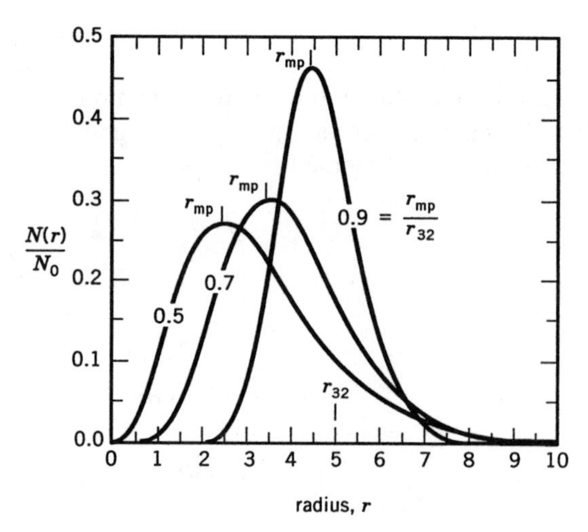

Figure 9-10 Gamma size distribution function for $r_{32} = 5$ and $r_{mp}/r_{32} = 0.5$, 0.7, and 0.9.

The ratio r_{mp}/r_{32} (or x_{mp}/x_{32}) can be thought of as a width parameter for the size distribution. Figure 9-10 shows three distributions, for $r_{mp}/r_{32} = 0.5$, 0.7, and 0.9. A monodispersion would be given by $r_{mp}/r_{32} = 1$. The widest distribution possible for a given r_{32} would correspond to $a = 0$, and $r_{mp}/r_{32} = 0$. This case ($a = 0$) results in a simple exponential distribution and has the advantage of simplicity. It is a function of only a single parameter, r_{32}. It has the disadvantage that it predicts $r_{mp} = 0$ which is physically unrealistic. Even particles formed by homogeneous nucleation have a minimum size, below which particles are not thermodynamically stable. Fortunately, this minimum size is often so small that the corresponding cross sections are negligible in comparison with the cross sections of the larger particles in the distribution and this discrepency is of no consequence. Therefore the exponential distribution is useful for representing many naturally occurring monomodal polydispersions such as rocket exhaust particles [Lyons, et al., 1982].

For any two parameter, monomodal distribution function (such as the gamma function), the polydisperse Mie scattering properties can be represented functionally by the following relations:

$$\overline{Q}_e = \overline{Q}_e(x_{32}, x_{mp}, n, k) \tag{9-68}$$

$$\overline{\omega}_0 = \overline{\omega}_0(x_{32}, x_{mp}, n, k) \tag{9-69}$$

$$\langle \overline{p} \rangle = \langle \overline{p} \rangle(x_{32}, x_{mp}, n, k) \tag{9-70}$$

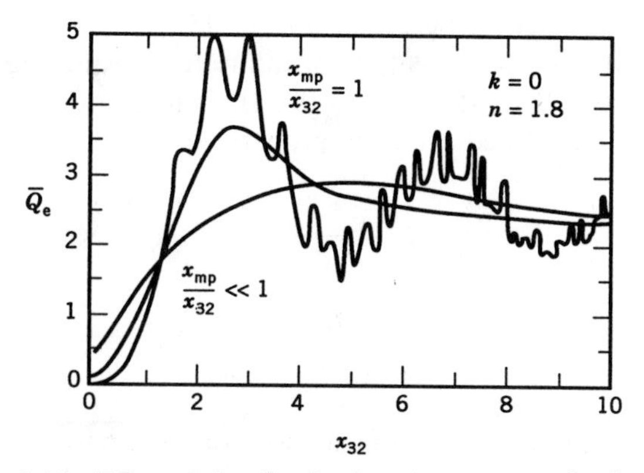

Figure 9-11 Effect of size distribution on average extinction efficiency.

where $x_{32,\,mp}$ means particle size parameter based on either d_{32} or d_{mp}. Equations (9-68) through (9-70) can be verified by substituting (9-65) into (9-63) and (9-57). The important distinction from the monodisperse case is that an extra particle size parameter has been introduced. Now there are two size parameters, x_{mp} and x_{32}, instead of just one, x.

The most noticeable effect of polydispersity on the size-averaged radiative properties $\overline{Q}_e$, $\overline{\omega}_0$, and $\langle \overline{p} \rangle$ is that the oscillations in Q_e, ω_0, and $\langle p \rangle$ that occur for $k \ll 1$ and $x > 1$ are integrated out due to the distribution of sizes. Otherwise, the same general trends are followed by $\overline{Q}_e$, $\overline{\omega}_0$, and $\langle \overline{p} \rangle$ as were displayed for the monodisperse case by Q_e, ω_0, and $\langle p \rangle$. Figure 9-11 shows the effect of size distribution on average extinction efficiency. As the width of the size distribution increases (corresponding to a decreasing value of x_{mp}/x_{32}), the oscillations in extinction efficiency are smoothed out to a greater extent.

Limiting Cases of Polydispersion Properties

In general the radiative properties for a polydispersion must be obtained by integration of Eqs. (9-55), (9-56), and (9-57). This requires a knowledge of the full particle size distribution $N(r)$. However, for two limiting cases, $x \gg 1$ and $x \ll 1$, Eqs. (9-55), (9-56), and (9-57) simplify considerably, and the radiative properties can be expressed in terms of the simple monodisperse relations based on an equivalent particle size.

Consider first the large, opaque particle limit, $x \gg 1$ and $4xk \gg 1$, for which the following limiting results hold for a single particle of a given radius r (neglecting diffraction):

$$Q_e \to 1 \tag{9-71}$$

$$Q_s \to \rho_\lambda \tag{9-72}$$

The phase function is given by either Eq. (9-24) or (9-25) depending on whether the particle reflects specularly or diffusely. Substituting Eqs. (9-71), (9-72) and either (9-24) or (9-25) into (9-55), (9-56), and (9-57) gives the following results for the sized-averaged properties:

$$K_e = \frac{1.5 f_v}{d_{32}} \tag{9-73}$$

$$\bar{\omega}_0 = \rho_\lambda \tag{9-74}$$

$$\bar{p} = p \tag{9-75}$$

Thus r_{32} (or d_{32}) is an appropriate mean size for large particles.

For the small particle limit, $x \ll 1$, the scattering and absorption efficiencies for a single particle of a given radius r vary as

$$Q_s \sim r^4 \tag{9-76}$$

$$Q_a \sim r^1 \tag{9-77}$$

from Eqs. (9-10) and (9-11), and the phase function is the Rayleigh phase function, which is independent of r. Since absorption efficiency Q_a varies as r^1, the absorption coefficient

$$K_a = 0.75 f_v \frac{\int_0^\infty Q_a r^2 N\, dr}{\int_0^\infty r^3 N\, dr} = 0.75 f_v \left(\frac{Q_a}{r} \right) \Bigg|_{\text{any } r} \tag{9-78}$$

can be determined by evaluating Q_a/r at any moment of r (e.g., r_{32} or otherwise). Therefore scattering considerations will be allowed to determine the appropriate moment. Since $Q_s \sim r^4$, it can be seen that the scattering coefficient can be written as

$$K_s = 0.75 f_v \frac{\int_0^\infty Q_s r^2 N\, dr}{\int_0^\infty r^3 N\, dr} = 0.75 f_v \left(\frac{Q_s}{r^4} \right) \Bigg|_{\text{any } r} r_{63}^3 \tag{9-79}$$

where the 63 moment of radius is defined as

$$r_{63}^3 = \frac{\int_0^\infty r^6 N\, dr}{\int_0^\infty r^3 N\, dr} \tag{9-80}$$

and Q_s/r^4 can be evaluated at any radius that is convenient. Choosing the 63 moment to evaluate at gives

$$K_s = 0.75 f_v \frac{Q_s(x_{63})}{r_{63}} \tag{9-81}$$

Thus the d_{63} moment is also adopted for absorption and hence extinction.

$$K_{a,s,e} = \frac{1.5 f_v Q_{a,s,e}(x_{63})}{d_{63}} \tag{9-82}$$

The size-averaged albedo from (9-56a) and (9-82) would be

$$\bar{\omega}_0 = \frac{Q_s(x_{63})}{Q_e(x_{63})} \tag{9-83}$$

where $Q_s(x_{63})$ and $Q_a(x_{63})$ are given by Eqs. (9-10) and (9-11). For a lossy dielectric ($k \ll n$) Eq. (9-15) holds with x replaced by x_{63}. The size-averaged phase function would be simply the Rayleigh phase function. Thus d_{63} is an appropriate mean size for particles that are small with respect to wavelength. From the properties of the gamma function it can be shown that the 63 moment is related to the 23 moment by the equation

$$\frac{r_{63}}{r_{32}} = \frac{[(6-3\zeta)(5-2\zeta)(4-\zeta)]^{1/3}}{3} \tag{9-84}$$

where

$$\zeta = \frac{r_{\mathrm{mp}}}{r_{32}} \tag{9-85}$$

which indicates that r_{63} is bounded by r_{32} (for a monodispersion $\zeta = 1$) and $1.64 r_{32}$ (for an "infinitely wide" distribution $\zeta = 0$).

$$r_{32} < r_{63} < 1.64 r_{32} \tag{9-86}$$

Application to Coal Polydispersions

The preceding considerations show that in the large and small particle limits, the radiative properties, extinction coefficients, albedo, and phase function become independent of the size distribution and can be calculated using appropriate mean sizes (d_{32} for large particles, d_{63} for small nonabsorbing particles, and any d for small absorbing particles). In the intermediate size region ($x \sim 1$) the radiative properties depend on the size distribution.

These considerations have been applied to coal particles to obtain useful correlations for polydispersion radiative properties in terms of appropriately scaled variables. Figure 9-12, which is adapted from Buckius and Hwang [1980], shows that the normalized extinction efficiency $\overline{Q}_e/[x_{32}f_1(n,k)]$ for coal particles can be represented over the entire size parameter range as a function of the single variable $x_{32}f_1(n,k)$ where $f_1(n,k)$ is defined in Eq. (9-12). Over 300 calculations were used to generate Fig. 9-12 with optical constants between $2.3 - i1.1$ and $1.5 - i0.5$ and various size distributions based on Eq. (9-65). The region indicated by broken lines in the intermediate size region reflects the variation due to different size distributions and optical constants. In the large particle limit the results converge to a single curve with a slope of -1, which is indicative of the geometric optics limit where extinction efficiency approaches 2. In the small particle limit the results converge to a single curve with a slope of zero, which is indicative of the nonscattering Rayleigh limit. (Recall that in the nonscattering Rayleigh limit the extinction efficiency is proportional to x [Eq. (9-10)] and can thus be based on any size— including d_{32}—as discussed in connection with Eq. (9-78).) The fact that the normalized extinction efficiency can be reasonably scaled as a function of a single variable in the case of coal particles is a result of the absorption index being large enough that the nonscattering Rayleigh region (which requires an x_{63} scaling) is not experienced. Later studies of coal optical properties [Brewster and Kunitomo, 1984] showed that the infrared imaginary index of coal was probably an order of magnitude less than that used to generate Fig. 9-12. However, the trends discussed above are unchanged by this finding.

The trend exhibited in Fig. 9-12 is important for understanding the dependence of extinction coefficient on particle size. As far as particle size

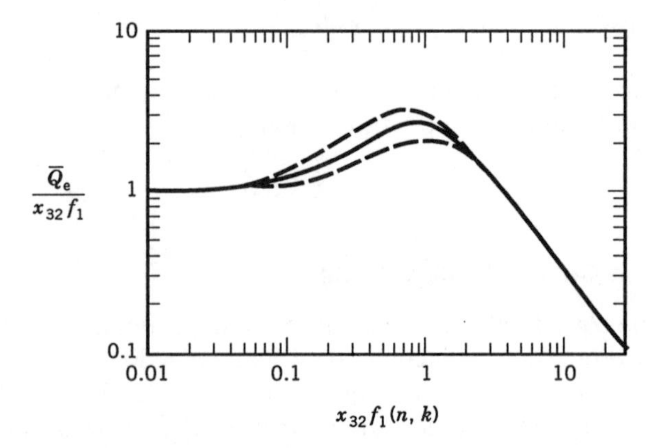

Figure 9-12 Normalized extinction efficiency of coal particles with $n = 1.5$ to 2.3 and $k = 0.5$ to 1.1 and various size distributions.

dependence is concerned, $\overline{Q}_e/(x_{32}f_1)$ is proportional to extinction coefficient via Eqs. (9-60) and (9-62).

$$K_e = \frac{\overline{Q}_e}{x_{32}f_1} \frac{1.5\pi f_1 f_v}{\lambda}$$

Thus for f_1 of order 1 (indeed $f_1 = 1$ for $n = 1.5$ and $k = 0.5$), the maximum extinction coefficient occurs at particle sizes with $x_{32} \sim 1$. For large values of x_{32} the extinction coefficient falls off as $1/x_{32}$ and is independent of n and k. For small values of x_{32} the extinction coefficient initially falls and then levels out at a value that depends on the optical constants through the value of the function f_1. For the values of k considered in Fig. 9-12 ($k \sim 1$), f_1 is also of order 1 and the extinction coefficient does not fall very far in the small particle limit before leveling out. Thus small strongly absorbing particles are almost as effective at extinguishing radiation as intermediate size particles, especially considering that intermediate particles scatter strongly in the forward directions while small particles scatter symmetrically in the forward and backward directions.

The underlying reason for the variation of extinction coefficient with size as noted above is a trade-off between particle surface area and volume. In small particles radiation effects penetrate throughout the entire volume of each particle. Therefore in the small particle limit the extinction coefficient is proportional to total particle volume. Dividing a given mass (or volume) of particles into finer particles does not change the total particle volume and thus the extinction coefficient is constant in the small particle limit. In large particles radiation effects are confined to the surface of the particles. Therefore in the large particle limit the extinction coefficient is proportional to total particle surface area. In this limit, dividing a given mass of particles into finer particles (i.e., decreasing particle size) increases the surface area and thus the extinction coefficient varies as the surface area to volume ratio, which varies as $1/d$.

$$\text{Small particle limit } (x \ll 1): \quad K_e \sim \frac{\text{particle volume}}{\text{total volume}} \sim d^0$$

$$\text{Large particle limit } (x \gg 1): \quad K_e \sim \frac{\text{particle surface area}}{\text{total volume}} \sim d^{-1}$$

Finally, since the foregoing discussion has concentrated on strongly absorbing particles, it is important to consider the situation of weak absorbers. For k values approaching zero, f_1 also approaches zero and the scaling of Fig. 9-12 is no longer valid in the small particle limit. In this case the extinction coefficient decreases as x^4 through the pure scattering Rayleigh regime before leveling out at a constant value defined by f_1 (see Fig. 12-9).

Thus small nonabsorbing particles (unlike small absorbing particles) are relatively inefficient at extinguishing radiation.

Example 9-2

In particle-laden flames the radiant emission is usually dominated by particle optical properties. One of the most important particle properties that determines the radiative properties of the dispersion is particle size. Another is particle loading. Figure 9-12 indicates that there is a particular particle size that will result in maximum extinction for any given wavelength at a fixed particle loading. What is the mean size (d_{32}) for which a dispersion of coal particles with $f_v = 10^{-4}$, $n = 1.5$ and $k = 0.5$ will have a maximum extinction coefficient at 2 μm wavelength? What is the corresponding extinction coefficient?

The maximum value of K_e corresponds to the maximum value of normalized extinction efficiency for a given wavelength and a given set of optical constants. For $n = 1.5$ and $k = 0.5$, Eq. (9-12) gives $f_1 = 1$. The maximum value of normalized extinction efficiency from Fig. 9-12 occurs at $x_{32} f_1 \sim 0.9$ and is approximately two. Thus the particle size for maximum extinction is

$$d_{32} = \frac{x_{32}\lambda}{\pi} = \frac{(0.9)(2\ \mu\text{m})}{\pi} = 0.6\ \mu\text{m}$$

and the corresponding extinction coefficient and mean free photon path (MFP) are

$$K_e = (2)\frac{1.5\pi 10^{-4}}{(2 \times 10^{-4}\ \text{cm})} = 5\ \text{cm}^{-1} \qquad \text{MFP} = \frac{1}{K_e} = 0.2\ \text{cm}$$

As the particle size decreases below the optimum value of 0.6 μm, the normalized extinction efficiency decreases by a factor of 2. Thus the value of K_e would only drop by a factor of 2 and remain constant as size continued to decrease. As the particle size increases above the optimum value of 0.6 μm, the normalized extinction efficiency, and thus the extinction coefficient, would decrease as the inverse of size. In typical pulverized coal flames the particle size is between 2 and 3 orders of magnitude larger than 0.6 μm, and thus the extinction coefficient would be 2 to 3 orders of magnitude smaller than that determined in this example (see Example 7-1). These calculations illustrate the importance of particle size on the radiative properties of particle dispersions. Other important particle properties are the particle loading (f_v) and, in the small particle limit, the particle optical constants.

Example 9-3

When determination of particle sizes is made difficult because collection of the particles alters their sizes, optical methods are often used to measure particle size in situ. The simplest optical measurement to make is the transmission (or extinction) measurement. The results of these measurements are sometimes ambiguous, however. Measuring extinction coefficient is equivalent to measuring the normalized extinction efficiency. As Fig. 9-12 shows, there may be two possible solutions for the particle size parameter for certain values of extinction coefficient.

Consider the plume of an aluminized rocket motor that contains molten aluminum oxide particles ($f_v = 10^{-5}$). A Nd-YAG laser ($\lambda = 1.06\ \mu$m) is used to measure transmission through the plume and an extinction coefficient of 0.14 cm^{-1} is obtained (Fig. 9-13). At this wavelength the optical constants are $n = 1.6$ and $k = 10^{-3}$. Assume the detector field of view can be made small enough that no appreciable amount of diffracted energy is measured (a difficult task when refractive beam steering is severe). What are the possible values of mean alumina particle size?

The large particle solution is first investigated by assuming the average particle size is large enough that the geometric optics approximation holds

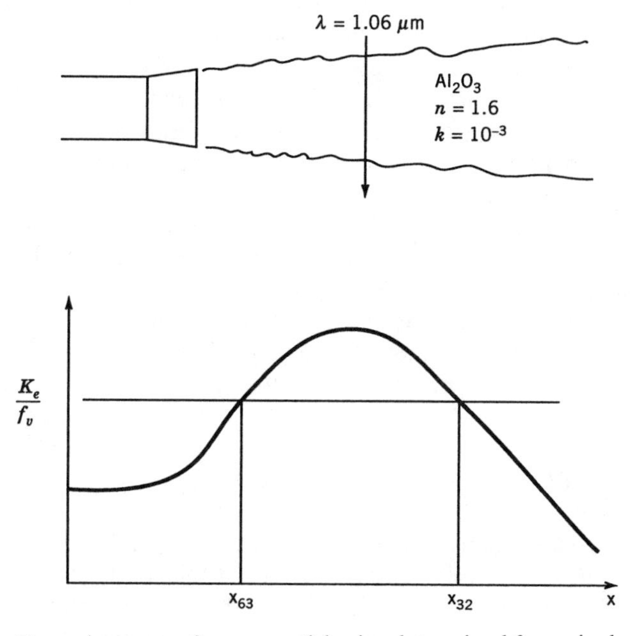

Figure 9-13 Nonuniqueness of mean particle size determined from single wavelength transmission measurements.

($Q_e = 2$). Thus Eq. (9-55b) gives

$$K_e = \frac{1.5 f_v \overline{Q}_e}{d_{32}} = \frac{1.5(10^{-5})(2)}{d_{32}} = 1.4 \times 10^{-5}\ \mu\text{m}^{-1} \rightarrow d_{32} = 2.2\ \mu\text{m}$$

Since the corresponding size parameter is $x_{32} = 6.5$, the assumption of geometric optics was justified. The small particle solution is investigated by assuming the particle size is small enough that the Rayleigh scattering approximation holds. Thus Eqs. (9-82), (9-10), and (9-11) give

$$K_e = \frac{1.5 f_v Q_e(x_{63})}{\dfrac{\lambda x_{63}}{\pi}} = \frac{1.5\pi f_v [f_1 + x_{63}^3 f_2]}{\lambda}$$

solving for x_{63} and noting that $f_1 \ll f_2$ gives

$$x_{63} = \left[\frac{K_e \lambda}{1.5\pi f_v f_2}\right]^{1/3} = \left[\frac{(1.4 \times 10^{-5}\ \mu\text{m}^{-1})(1.06\ \mu\text{m})}{1.5\pi(10^{-5})(0.312)}\right]^{1/3} = 1.0$$

$$d_{63} = 0.34\ \mu\text{m}$$

In spite of the seemingly large value of x_{63} an implicit solution using a Mie scattering program to calculate Q_e instead of Eqs. (9-10) and (9-11) gives the same result, thus justifying the assumption of Rayleigh scattering. From these results it can be seen that there are indeed two valid solutions for the average particle size as shown in Fig. 9-13. There is a small particle solution, which represents the 63 moment of the unknown size distribution ($d_{63} = 0.34\ \mu$m), and a large particle solution, which represents the 32 moment of the unknown size distribution ($d_{32} = 2.2\ \mu$m). To determine which is the correct solution requires more information. This information could be obtained by making another transmission measurement at a different wavelength, for example, near 0.5 or 3 μm. This would give a measurement at a different value of size parameter. The correct solution could then be determined according to the change in the extinction coefficient (assuming the change in optical constants is known; see Problem 9-13).

DEPENDENT SCATTERING

Throughout this chapter independent scattering by individual particles has been assumed. Independent scattering means that particles are assumed to interact with the radiation incident upon them as if uninfluenced by the presence of neighboring particles. Under this assumption the single-particle properties Q_e, ω_0, and $p(\theta)$ can be determined using Mie theory and then the corresponding properties of an optically thin, single-scattering volume element K_e, $\overline{\omega}_0$, and $\overline{p}(\Omega' \rightarrow \Omega)$ can be determined by summing over the single-particle properties using Eqs. (9-55), (9-56), and (9-57). Multiple scattering between particles is accounted for in the integration of the transfer equation.

In certain circumstances the assumption of independent scattering is invalid and a phenomenon known as dependent scattering (and absorption) occurs. Under dependent scattering, the interaction of a given particle with the radiation incident upon it is affected by the presence of neighboring particles. This mutual dependence between neighboring particles refers to more than just the effect of multiple scattering between particles, which is represented by the transfer equation. Dependent scattering refers to the fact that a given particle will generally scatter less radiation and with a different directional distribution if other particles are nearby due to mutual interaction with the same incident wave and interference effects.

Experiments have shown that the key parameter in determining the onset of dependent scattering is the interparticle clearance to wavelength ratio c/λ [Brewster and Tien, 1982; Yamada, et al., 1986]. If c/λ is decreased below about 0.5, particles will effectively scatter less and absorb more radiation per particle than if they were not surrounded by neighboring particles. The occurrence of dependent scattering can be monitored by evaluating c/λ using the approximate formula for interparticle spacing in a rhombohedral array

$$\frac{\delta}{d} = \frac{0.9}{f_v^{1/3}} \tag{9-87}$$

where δ is the center-to-center particle spacing

$$\delta = c + d \tag{9-88}$$

$$\frac{c}{\lambda} = \left(\frac{0.9}{f_v^{1/3}} - 1\right)\frac{d}{\lambda} > 0.5 \quad \text{(independent scattering)} \tag{9-89}$$

When the inequality of (9-89) is satisfied, experimental evidence has indicated that independent scattering prevails. The onset of dependent scattering occurs when the inequality in Eq. (9-89) is not satisfied. Thus it is quite possible for scattering to be independent even for relatively high particle volume fractions in systems of large particles ($d/\lambda \gg 1$). A map of the independent and dependent scattering regimes is shown in Fig. 9-14. Thermal systems subject to dependent scattering are those with small c/λ ratios such as cryogenic microsphere insulations.

Dependent scattering is often used to allow selective transmission of short wavelength radiation through a screen while longer wavelength radiation is absorbed or reflected by the screen. Infrared cameras mounted in stealth

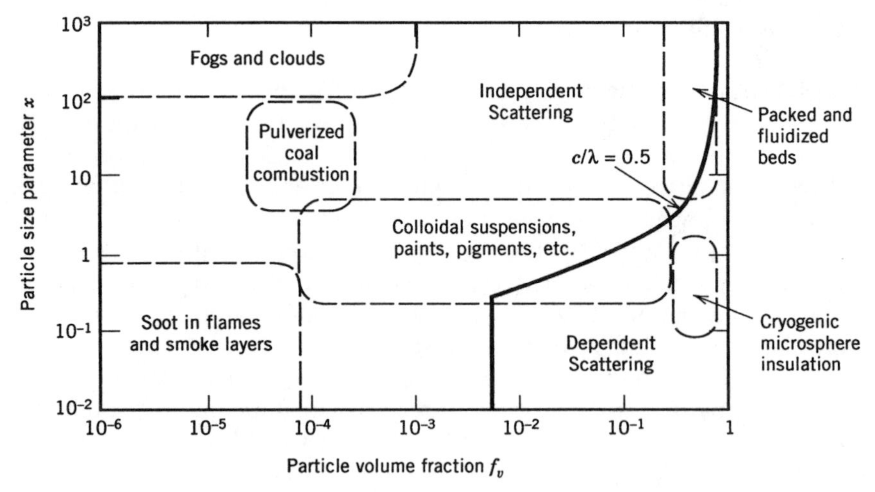

Figure 9-14 Map of dependent and independent scattering regimes.

aircraft are shielded behind a wire mesh coated with radar absorbing material. The mesh spacing (c) is designed to be much smaller than the smallest anticipated radar wavelength ($c/\lambda_{\text{radar}} \ll 1$) and yet much larger than the infrared camera wavelength ($c/\lambda_{\text{ir}} \gg 1$) to allow clear infrared imaging by the camera while preventing a strong radar return from the camera. Similarly microwave ovens utilize a screen mounted in the window which has holes (diameter $= c$) that are large enough to allow visual observation of the cooking process from the outside ($c/\lambda_{\text{vis}} \gg 1$) but small enough to prevent the escape of microwave energy ($c/\lambda_{\mu\text{wave}} \ll 1$). In each of these examples short wavelength radiation is able to penetrate through gaps in the material that are much larger than the wavelength. Further discussion of the dependent scattering problem, including some attempts at analyzing dependent scattering using Maxwell's equations, can be found in Cartigny, et al. [1986], Drolen and Tien [1987], Kumar and Tien [1990], Ma, et al. [1990] and Tien [1988].

Example 9-4

One interesting practical problem involving dependent scattering is the problem of maximizing the hiding power or opacity of a paint that uses scattering pigment particles such as TiO_2, SiO_2, or $CaCO_3$ in a resin vehicle. Determine the optimum pigment particle size for maximum hiding power of a TiO_2 pigment ($n = 2.6$ and $k = 0$) in an alkyd resin vehicle ($n = 1.45$) in the visible region. Also determine the maximum pigment particle volume fraction to avoid dependent scattering.

This problem amounts to one of trying to maximize the extinction coefficient, which can also be expressed as

$$K_e = \frac{\pi}{\lambda} 1.5 f_v \frac{\overline{Q}_e}{x_{32}} \tag{9-90}$$

The maximum in $\overline{Q}_e/x_{32}$ for a relative index of $2.6/1.45 = 1.8$ can be estimated from the Mie calculations presented in Fig. 9-7*b*. Figure 9-7*b* shows that lines of constant $\overline{Q}_e/x_{32}$ are straight lines with a slope of 1. A maximum in $\overline{Q}_e/x_{32}$ for $k = 0$ occurs at $x_{32} \sim 2$. Thus paint pigment particles are generally most efficient in the visible region ($\lambda \approx 0.5\ \mu\text{m}$) when their mean optical size d_{32} is chosen to be

$$d_{32} = \frac{x_{32}\lambda}{\pi} = \frac{(2)(0.5\ \mu\text{m})}{\pi} = 0.3\ \mu\text{m}$$

For $d = 0.3\ \mu\text{m}$ and $\lambda = 0.5\ \mu\text{m}$, Eq. (9-89) gives

$$\left(\frac{0.9}{f_v^{1/3}} - 1\right)\frac{0.3}{0.5} > 0.5 \rightarrow f_v < 0.22$$

which indicates that the pigment particle volume fraction f_v should be 0.22 or less to avoid a decrease in hiding power due to dependent scattering effects.

SUMMARY

Particles play an important role in the radiative transfer in many engineering and environmental systems, and it is important to be able to at least estimate the particulate contribution to the overall radiative transfer in such systems. The workhorse theory for estimating particle radiative properties is the Mie theory for scattering and absorption by a sphere of arbitrary size and arbitrary optical constants. The simpler relationships of Rayleigh scattering, geometric optics, diffraction theory, Rayleigh-Gans scattering, and anomalous diffraction also play a useful role in their respective regions of limiting values for particle size and optical constants.

Nowadays Mie codes are widely available, computers are faster and less expensive, and compilers are more powerful than ever before. Thus it is quite feasible to include full Mie calculations in simulations of even complex systems. This advanced capability is a mixed blessing, however. Because it is so easy to include a Mie subroutine and calculate efficiencies and phase functions to almost any number of significant figures, it is easy to forget that the optical constants of the particles may not be known to better than one figure of accuracy or that the particle size distribution may be variable and uncertain. It is also easy to overlook the fact that the solution of the transfer equation may not be able to resolve the highly anisotropic scattered intensity distribution of large particles. One of the most common errors made in heat transfer analysis is to include full Mie calculations of efficiencies (i.e., $Q_e \to 2$ for $x \gg 1$) and ignore the implications of highly anisotropic forward scattered diffraction ($Q_s = 1$) by assuming isotropic scattering or using a low resolution multiflux scheme to directionally integrate the transfer equation. The validity of applying Mie calculations to nonspherical particle systems is also an issue that deserves case-by-case consideration. In general, heat transfer calculations are less sensitive to nonspherical effects. However, polarized, laser diagnostic scattering measurements is clearly one application where particle shape would play a crucial role.

In short, the application of Mie theory to predict particle radiative properties is fraught with numerous pitfalls. The best that can be said is that a careful sensitivity analysis should always be conducted, keeping in mind the uncertainty in the particle optical constants and size distribution and the resolution of the transfer equation solution. In many cases, an educated guess or an approximate representation, guided by a knowledge of the limiting behavior of the Mie solution, will be as accurate or better than extensive Mie calculations.

REFERENCES

Bohren, C. F. and Huffman, D. R. (1983), *Absorption and Scattering of Light by Small Particles*, John Wiley & Sons, New York.

Boothroyd, S. A. and Jones, A. R. (1986), "Comparison of Radiative Characteristics for Fly Ash and Coal," *Int. J. Heat Mass Transfer*, Vol. 29, No. 11, pp. 1649–1654.

Brewster, M. Q. (1989), "Radiation-Stagnation Flow Model of Aluminized Solid Rocket Motor Internal Insulator Heat Transfer," *J. Thermophysics and Heat Transfer*, Vol. 3, No. 2, pp. 132–139.

Brewster, M. Q. (1991), "Heat Transfer in Heterogeneous Propellant Combustion Systems," Chap. 6, in *Annual Review of Heat Transfer*, Vol. IV, C-L. Tien, Ed. Hemisphere, New York.

Brewster, M. Q. and Kunitomo, T. (1984), "The Optical Constants of Coal, Char and Limestone," *J. Heat Transfer*, Vol. 106, pp. 678–683.

Brewster, M. Q. and Tien, C-L. (1982), "Radiative Transfer in Packed and Fluidized Beds: Dependent vs. Independent Scattering," *J. Heat Transfer*, Vol. 104, pp. 573–579.

Buckius, R. O. and Hwang, D. C. (1980), "Radiation Properties for Polydispersions: Application to Coal," *J. Heat Transfer*, Vol. 102, No. 1, pp. 99–103.

Cartigny, J. D., Yamada, Y., and Tien, C.-L. (1986), "Radiative Transfer with Dependent Scattering by Particles: Part I-Theoretical Investigation," *J. Heat Transfer*, Vol. 108, pp. 608–613.

Cess, R. D. (1978), "Biosphere-Albedo Feedback and Climate Modelling," *J. Atmos. Sci.*, Vol. 35, pp. 1765–1768.

Dave, J. V. (1968), "Subroutines for Computing the Parameters of the Electromagnetic Radiation Scattered by a Sphere," IBM order no. 360D-17.4.002.

Deirmendjian, D. (1969), *Electromagnetic Scattering on Spherical Polydispersions*, Elsevier, New York.

Drolen, B. L. and Tien, C.-L. (1987), "Independent and Dependent Scattering in Packed Sphere Systems," *J. Thermophysics and Heat Transfer*, Vol. 1, pp. 63–68.

Edwards, D. K. and Babikian, D. S. (1989), "Radiation from a Nongray Scattering Emitting and Absorbing SRM Plume," AIAA-89-1721, 24th Thermophysics Conference, Buffalo, NY, June 12–14.

Hansen, J. E. and Travis, L. D. (1974), "Light Scattering in Planetary Atmospheres," *Space Science Reviews*, Vol. 16, pp. 527–610.

Kerker, M. (1969), *The Scattering of Light and Other Electromagnetic Radiation*, Academic Press, New York.

Kumar, S. and Tien, C.-L. (1990), "Dependent Absorption and Extinction of Radiation by Small Particles," *J. Heat Transfer*, Vol. 112, pp. 178–185.

Lyons, R. B., Wormhoudt, J., and Kolb, C. E. (1982), "Calculation of Visible Radiation from Missile Plumes," in *Spacecraft Radiative Transfer and Temperature Control*, *Prog. In Astronautics and Aeronautics Series*, T. E. Horton, Ed., AIAA, New York, Vol. 83, pp. 128–148.

Ma, Y., Varadan, V. K., and Varadan, V. V. (1990), "Enhanced Absorption due to Dependent Scattering," *J. Heat Transfer*, Vol. 112, pp. 402–407.

Siegel, R. and Howell, J. R. (1988), *Thermal Radiation Heat Transfer*, 2nd ed., Taylor, Francis, Hemisphere, New York.

Stratton, J. A. (1941), *Electromagnetic Theory*, McGraw-Hill Book Co., New York.

Tien, C.-L. (1988), "Thermal Radiation in Packed and Fluidized Beds," *J. Heat Transfer*, Vol. 110, Nov., pp. 1230–1242.

Tien, C-L. and Lee, S. C. (1982), "Flame Radiation," *Prog. Energy Comb. Sci.*, Vol. 8, pp. 41–59.

Toon, O. B., Pollack, J. B., and Khare, B. N. (1976), "The Optical Constants of Several Atmospheric Aerosol Species: Ammonium Sulfate, Aluminum Oxide, and Sodium Chloride," *J. Geophys. Res.*, Vol. 81, No. 33, pp. 5733–5748.

van de Hulst, H. C. (1981), *Light Scattering by Small Particles*, Dover Publications, Inc., New York.

Wiscombe, W. J. (1980), "Improved Mie Scattering Algorithms," *Appl. Opt.*, Vol. 19, pp. 1505–1509.

Yamada, Y., Cartigny J. D., and Tien, C.-L. (1986), "Radiative Transfer with Dependent Scattering by Particles: Part 2—Experimental Investigation," *J. Heat Transfer*, Vol. 108, pp. 614–618.

PROBLEMS

1. Derive the expression for the phase function and scattering efficiency of a specularly reflecting, opaque sphere that is much larger than the incident wavelength, Eq. (9-24).

2. Consider a monodispersion of Rayleigh spherical particles with optical constants n and k, diameter d, and volume fraction f_v in a nonattenuating medium. For a given wavelength λ:
   (a) Obtain expressions for the absorption and scattering coefficients in terms of n, k, d, λ, and f_v for the two limiting cases:
   (i) f_v, n, k, and λ nonzero and fixed; $d \to 0$
   (ii) f_v, n, λ, and d nonzero and fixed; $k \to 0$
   (b) Calculate the single-scattering albedo for the following cases:
   (i) Soot; $\lambda = 2.5\ \mu m$, $n = 2.04$, $k = 1.15$, $d = 0.01\ \mu m$
   (ii) Alumina (liq.); $\lambda = 1.06\ \mu m$, $n = 1.62$, $k = 10^{-3}$, $d = 0.1\ \mu m$

3. Write a computer program to determine the scattering efficiency and scattering phase function for opaque, specular, geometric optical spheres given n, k, and x as input, including diffraction. Assume unpolarized incident radiation.
   For a specular aluminum droplet with $d = 2\ \mu m$, $\lambda = 1.06\ \mu m$, $n = 1.2$, and $k = 10$ determine:
   (a) Scattering efficiency
   (b) Albedo, and
   (c) Phase function vs. scattering angle for 0° to 180°

4. A satellite is in geosynchronous orbit (GSO) about the earth (see figure). The satellite is a flat plate with area A_1 (on one side). Assume the earth is a diffuse reflector. Note that $h^2 \gg r_2^2$ for GSO.

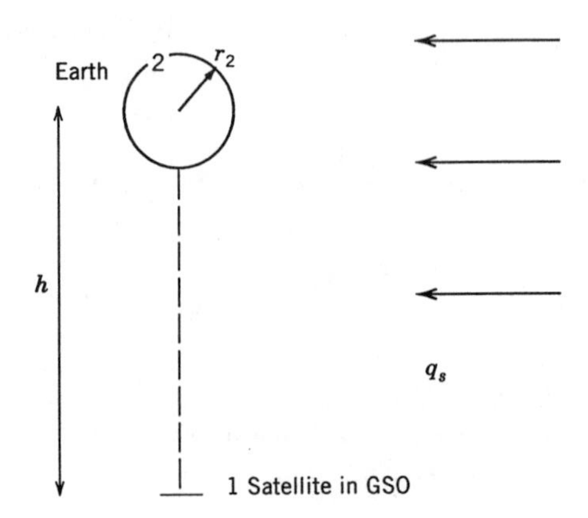

Figure P4

For the orientation shown (satellite 90° from direction of collimated solar incident flux, q_s) and by treating the earth as a diffusely reflecting spherical particle, obtain an expression for the reflected solar flux that is incident on the satellite q_1^- in terms of r_2, h, ρ_2, and q_s. Compare the result with that of Example 2-3.

5. Obtain an expression for the scattering efficiency Q_s (ignoring diffraction) of a large ($2\pi r/\lambda \gg 1$), opaque, circular cylinder in terms of the cylinder's directional reflectivity $\rho'_\lambda(\beta)$. The axis of the cylinder is oriented perpendicular to the incident radiation and β is the polar angle of incidence ($0 < \beta < \pi/2$).

6. Using the definition of phase function and Eq. (9-49), derive a relationship for flux $q_s(\theta)$ scattered from a spherical particle (radius r) in terms of $S_{1,2}$, the flux incident on the particle q_0, and the distance from the particle R (for $R \gg r$), which shows that $q_s(\theta)$ is proportional to q_0 and inversely proportional to R^2.

7. The optical extinction theorem states that the extinction efficiency of a spherical particle can be obtained by evaluating the intensity function $S_{1,2}$ in the forward direction as

$$Q_e = \frac{4}{x^2} \operatorname{Re}\{S_{1,2}(\theta = 0)\}$$

Using the optical extinction theorem, Eq. (9-50), and the properties of associated Legendre functions, derive Eq. (9-40).

8. Consider a large, semitransparent sphere in the limit of anomalous diffraction ($x \gg 1$, $n \sim 1$, and $k \ll n$). Derive Eq. (9-35) for the absorption efficiency. Show that for $v \ll 1$ a result compatible with that of Rayleigh-Gans scattering is obtained.

9. Calculate Q_e, ω_0, and $\langle p \rangle$ for a 1-μm-diameter particle of liquid aluminum oxide at 1 μm wavelength using Mie theory. The optical constants are $n = 1.62$ and $k = 0.001$. Compare Q_e and ω_0 with the predictions of anomalous diffraction theory and $\langle p \rangle$ with Eq. (9-33) for Rayleigh-Gans scattering. Are the restrictions for these limiting theories satisfied in this situation?

10. Show that the albedo of a mixture of large ($x \gg 1$) equal sized spherical particles with different absorption indices is equal to the volume-weighted average of the albedos of the constituent particles. Does this result apply generally to other size regimes ($x \sim 1$ and $x \ll 1$)? Explain.

11. Consider two clouds of opaque ($4xk \gg 1$), monodisperse, geometric optics ($x \gg 1$) particles. One cloud is homogeneous with particle optical constants $n = 1.5$ and $k = 0.05$. The other cloud is nonhomogeneous such that 10% (by volume) of the particles have $n = 1.5$ and $k = 0.5$ and the rest are nonabsorbing with $n = 1.5$ and $k = 0$. Determine the single-scattering albedo for each cloud for two cases:

(a) Including diffraction

(b) Neglecting diffraction

For computing the particle albedo assume that the hemispherical reflectivity is 1.05 times the normal reflectivity.

12. A white pigment (TiO_2) and a black pigment (carbon) are to be mixed in an alkyd resin to achieve a gray paint with a single-scattering albedo of 0.5 at a vacuum wavelength of 0.725 μm. The resin has $n_\infty = 1.45$ and $k_\infty = 0$. The white pigment has $n_1 = 2.6$, $k_1 = 0$, and $d_1 = 0.2$ μm. The black pigment has $n_2 = 1.5$, $k_2 = 0.5$, and $d_2 = 0.05$ μm. The total pigment particle volume fraction is to be 0.3 (the rest is resin vehicle). Assuming independent Mie scattering, determine the needed volume fraction of each of the pigments.

13. Reconsider the single wavelength exhaust plume transmission measurement of Example 9-3, which produced two possible solutions for average particle size. A second transmission measurement is conducted at 0.6328 μm wavelength where $n = 1.65$ and $k = 10^{-2}$. Again care is taken to exclude diffracted energy from the detector field of view. The resulting extinction coefficient is 1 cm^{-1}. Based on this information, which is the correct solution obtained in Example 9-3, the large or small particle solution?

14. Many solid propellants contain small aluminum particles that often dominate the optical properties of the propellant. Since lasers and other sources of visible radiation are often used to ignite solid propellants and since microwaves are sometimes used as a diagnostic tool in solid propellant combustion studies, it is important to estimate the extinction characteristics of aluminized solid propellants at various frequencies.

Consider a dielectric, homogeneous propellant containing 15% (by volume) of 30 μm aluminum particles. Assuming independent scattering,

(a) Estimate the mean free photon path and

(b) Indicate whether the extinction is dominated by absorption or scattering (i.e., calculate single-scattering albedo) for two frequences:

(i) 16 GHz (KU-band) microwave radiation, and

(ii) 0.5 μm visible radiation.

The dc resistivity of aluminum is 2.91×10^{-6} Ω-cm. The dc (and microwave) dielectric constant of the propellant (excluding aluminum) is $5 - i0$ and the visible refractive index of the propellant (excluding aluminum) is $1.5 - i0$. Neglect diffraction if geometric optics is the governing optical regime.

(c) Evaluate the assumption of independent scattering in parts (a) and (b) and discuss any impact on the calculated results.

CHAPTER 10

RADIATIVE TRANSFER IN NONSCATTERING HOMOGENEOUS MEDIA

In Chapter 7 the integral of the transfer equation along a single line of sight was discussed. This integral is sufficient if all the information that is needed is intensity along a single line of sight. Most radiative transfer problems in participating media, however, involve many lines of sight, each with a different path length through the medium. This chapter begins to address the problem of radiative transfer in participating media when multiple lines of sight are involved.

The simplest problem of radiative transfer in a participating medium is that involving a homogeneous, nonscattering medium. Many participating media can be adequately described as nonscattering and homogeneous. High intensity continuous flow combustors are often characterized by a high degree of recirculative backmixing that creates a nearly isothermal condition in the enclosure with a thin nonisothermal boundary layer near the walls. Assuming that the optical depth based on thermal boundary layer thickness is small, such a medium can be treated as isothermal with uniform properties.

Scattering will be neglected in the present treatment. The complication introduced by including scattering is that radiative transfer along a given line of sight can influence the transfer along another line of sight by in-scattering. When scattering is included in the transfer equation solution, all lines of sight must be solved simultaneously. However, by neglecting scattering each line of sight can be solved independently. For many heat transfer calculations, in particular when the scattering coefficient is not much larger than the absorption coefficient, the neglect of scattering in the solution of the transfer equation is a justifiable assumption [Goody, 1989; Altenkirch, et al., 1984].

NONPARTICIPATING MEDIUM IN DIFFUSE ENCLOSURE

The problem of radiative transfer in an enclosure filled with a homogeneous, nonscattering medium is very similar to that developed in Chapter 3 for diffuse transfer in an enclosure without a participating medium. Without a participating medium, the transfer is formulated by a system of algebraic equations involving the radiosity $q_{\lambda i}^{+}$, irradiation $q_{\lambda i}^{-}$ and net flux $q_{\lambda i}$ at each surface A_i (see Fig. 10-1).

The irradiation at surface A_i is given by the expression

$$q_{\lambda i}^{-} A_i = \sum_j \int_{A_i} \int_{A_j} \frac{I_{\lambda i}^{-} \cos\theta_j \cos\theta_i}{s^2} \, dA_j \, dA_i \tag{10-1}$$

which is obtained from the definitions of intensity and solid angle. The solution of the transfer equation for a nonparticipating medium $(dI/ds = 0)$ gives the result that the intensity incident at A_i is equal to the intensity leaving A_j; $I_{\lambda i}^{-} = I_{\lambda j}^{+}$. Assuming that $I_{\lambda j}^{+}$ is *uniform* and *diffuse* allows it to be brought out of both integrals in (10-1) giving

$$q_{\lambda i}^{-} = \sum_j \pi I_{\lambda j}^{+} \frac{1}{\pi A_i} \int_{A_i} \int_{A_j} \frac{\cos\theta_i \cos\theta_j}{s^2} \, dA_j \, dA_i . \tag{10-2}$$

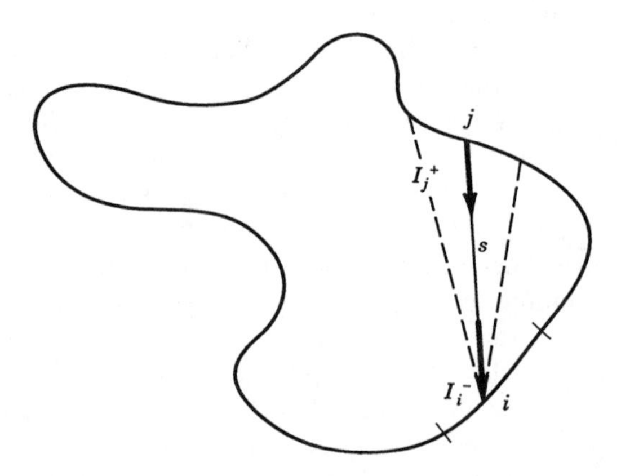

Figure 10-1 Radiative transfer in a diffuse enclosure.

By defining the view factor F_{ij} as

$$F_{ij} = \frac{1}{\pi A_i} \int_{A_i} \int_{A_j} \frac{\cos\theta_i \cos\theta_j}{s^2} \, dA_j \, dA_i \tag{10-3}$$

and noting that for diffuse radiosity

$$q^+_{\lambda j} = \pi I^+_{\lambda j} \tag{10-4}$$

the system of equations for radiosity, irradiation, and net flux becomes

$$q^+_{\lambda i} = \varepsilon_{\lambda i} e_{b_{\lambda i}} + \rho_{\lambda i} q^-_{\lambda i} \tag{10-5}$$

$$q^-_{\lambda i} = \sum_j q^+_{\lambda j} F_{ij} \tag{10-6}$$

$$q_{\lambda i} = q^+_{\lambda i} - q^-_{\lambda i} \tag{10-7}$$

If the preceding equations and ideas are not familiar it would be a good idea to review Chapter 3 before proceeding with this chapter.

ISOTHERMAL NONSCATTERING PARTICIPATING MEDIUM IN DIFFUSE ENCLOSURE

Now consider the case where the enclosure in Fig. 10-1 is filled with an isothermal nonscattering participating medium at temperature T_g. The transfer equation along an arbitrary path is

$$\frac{dI_\lambda}{ds} = -K_{a_\lambda} I_\lambda + K_{a_\lambda} I_{b_\lambda}(T_g) \tag{10-8}$$

The solution of (10-8), as presented in Eq. (7-15), is

$$I^-_{\lambda i} = \tau_\lambda(s) I^+_{\lambda j} + \varepsilon_\lambda(s) I_{b_\lambda}(T_g) \tag{10-9}$$

where the emissivity and transmissivity along the path s are

$$\tau_\lambda(s) = \exp\left[-K_{a_\lambda} s\right] \tag{10-10}$$

$$\varepsilon_\lambda(s) = 1 - \exp\left[-K_{a_\lambda} s\right] \tag{10-11}$$

If (10-9) is substituted into (10-1) and $I^+_{\lambda j}$ is again assumed to be *uniform* and *diffuse*, the irradiation at A_i becomes

$$q_{\lambda i}^{-} = \sum_j \left[q_{\lambda j}^{+} \bar{\tau}_{\lambda_{ij}} + e_{b_\lambda}(T_g) \bar{\varepsilon}_{\lambda_{ij}} \right] F_{ij} \tag{10-12}$$

where the mean transmissivity and emissivity factors are defined as

$$\bar{\varepsilon}_{\lambda_{ij}} = \frac{1}{\pi A_i F_{ij}} \int_{A_i} \int_{A_j} \frac{\varepsilon_\lambda(s) \cos\theta_j \cos\theta_i}{s^2} \, dA_j \, dA_i \tag{10-13}$$

$$\bar{\tau}_{\lambda_{ij}} = 1 - \bar{\varepsilon}_{\lambda_{ij}} \tag{10-14}$$

The integration defined by (10-13) is a geometric averaging process that takes into account the various path lengths across the isothermal medium between the two surfaces A_i and A_j. Depending on the orientation of A_i and A_j, some of the paths may be very short (perhaps optically thin) and others very long (perhaps optically thick).

As a matter of convenience a single equivalent path length, called the *mean beam length*, $L_{m_{\lambda_{ij}}}$, is defined that allows Eq. (10-13) to be evaluated using the simple exponential formula of Eq. (10-11).

$$\bar{\varepsilon}_{\lambda_{ij}} = \varepsilon_\lambda\left(L_{m_{\lambda_{ij}}}\right) = 1 - \exp\left[-K_{a_\lambda} L_{m_{\lambda_{ij}}}\right] \tag{10-15}$$

The mean beam length, which is defined by (10-15), is greater than the shortest path but less than the longest path between A_i and A_j. Its value is such that the various path lengths with differing optical depths are all accounted for to give the correct mean emissivity factor.

It can now be seen that the system of equations for an isothermal, nonscattering medium is equivalent to that for an enclosure without a participating medium. If the radiosity term $q_{\lambda j}^{+}$ appearing in Eq. (10-6) is replaced by $q_{\lambda j}^{+} \bar{\tau}_{\lambda_{ij}} + e_{b_\lambda}(T_g) \bar{\varepsilon}_{\lambda_{ij}}$ the system of equations is essentially unchanged. This system of equations can be solved by the methods of Chapter 3. One distinction that should be pointed out is that the mean emissivity factor for gases, unlike the emissivity of solids, is highly temperature dependent. Its product with $e_b(T_g)$ introduces a nonlinear term as an unknown into the equations in the case when the specified gas condition is heat transfer rate and the unknown condition is temperature.

NETWORK FORMULATION FOR ISOTHERMAL GAS

One of the methods for formulating the solution of the system of enclosure equations is the network representation. This method was first applied to an enclosure containing an isothermal absorbing medium by Bevans and Dunkle

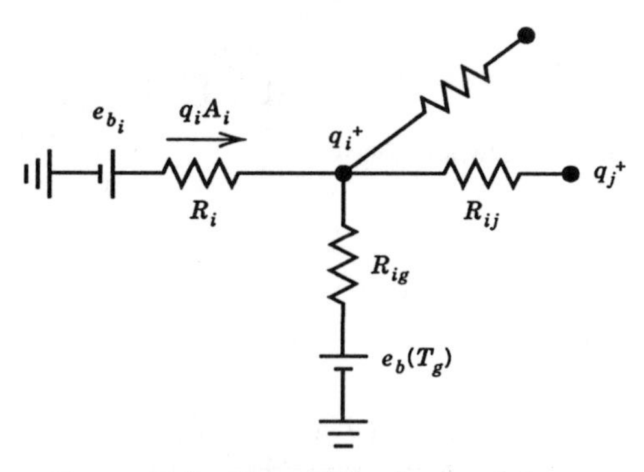

Figure 10-2 Network for isothermal gas.

[1960]. It is presented here because of its usefulness for solving small systems of equations (two or three nodes).

The equivalent network for an enclosure containing a homogeneous, nonscattering participating medium is shown in Fig. 10-2. Upon comparing the network of Fig. 10-2 with the nonparticipating network of Chapter 3 (Fig. 3-9), it can be seen that an extra node appears, which accounts for the gas.

The resistances in the network are modified to account for the presence of the gas. The resistance between surfaces [Eq. (10-17)] is increased by a factor of $1/\bar{\tau}_{\lambda_{ij}}$ to account for the transmission loss through the gas along the various paths between A_i and A_j.

$$R_i = \frac{1 - \varepsilon_{\lambda_i}}{\varepsilon_{\lambda_i} A_i} \tag{10-16}$$

$$R_{ij} = \frac{1}{A_i F_{ij} \bar{\tau}_{\lambda_{ij}}} \tag{10-17}$$

$$R_{ig} = \frac{1}{\sum_j A_i F_{ij} \bar{\varepsilon}_{\lambda_{ij}}} \tag{10-18}$$

In the gas resistance term (10-18) it is important not to forget the $i = j$ term in the summation.

Two Nodes: Source (T_g), Sink (T_w)

As an example, consider transfer from a hot, isothermal gas (T_g) to a colder, isothermal wall (T_w). The network for this system is shown in Fig. 10-3. The

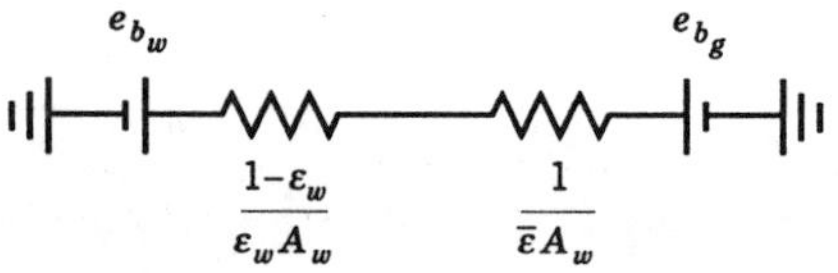

e_{b_w} e_{b_g}

$\frac{1-\varepsilon_w}{\varepsilon_w A_w}$ $\frac{1}{\bar{\varepsilon} A_w}$

Figure 10-3 Two-node network (gas-wall).

spectral heat flux is given by

$$q_\eta = \frac{e_{b_\eta}(T_g) - e_{b_\eta}(T_w)}{1/\varepsilon_{w_\eta} + 1/\bar{\varepsilon}_\eta - 1} \tag{10-19}$$

Equation (10-19) must be integrated over wavenumber or wavelength to obtain the total heat flux.

$$q = \int_0^\infty q_\eta \, d\eta \tag{10-20}$$

This integration will be carried out for two special cases.

Black Wall, Nongray Gas

If the wall of Fig. 10-3 is black, $\varepsilon_{w_\eta} = 1$, Eq. (10-20) can be integrated such that the total heat flux can be written in terms of a total gas emissivity and total absorptivity,

$$q = \bar{\varepsilon}_g \sigma T_g^4 - \bar{\alpha}_g \sigma T_w^4 \tag{10-21}$$

where the total gas emissivity and absorptivity are defined as

$$\bar{\varepsilon}_g = \frac{\int_0^\infty \bar{\varepsilon}_\eta(T_g) e_{b_\eta}(T_g) \, d\eta}{\int_0^\infty e_{b_\eta}(T_g) \, d\eta} \tag{10-22}$$

$$\bar{\alpha}_g = \frac{\int_0^\infty \bar{\varepsilon}_\eta(T_g) e_{b_\eta}(T_w) \, d\eta}{\int_0^\infty e_{b_\eta}(T_w) \, d\eta} \tag{10-23}$$

If continuum emission (such as soot participation) is negligible, the total gas emissivity and absorptivity can be written, using the band approximation (see

Chapter 8) as

$$\bar{\varepsilon}_g = \frac{1}{\sigma T_g^4} \sum_{k \text{ band}} e_{b_{\eta 0k}}(T_g) \bar{A}_k(T_g) \tag{10-24}$$

$$\bar{\alpha}_g = \frac{1}{\sigma T_w^4} \sum_{k \text{ band}} e_{b_{\eta 0k}}(T_w) \bar{A}_k(T_g) \tag{10-25}$$

The mean effective bandwidth for the kth band $\bar{A}_k$ is defined by integrating Eq. (10-13) over wavenumber.

$$\bar{A} = \int_{-\infty}^{\infty} \bar{\varepsilon}_\eta d(\eta - \eta_0) \tag{10-26}$$

Gray Wall, Nongray Gas

If the wall is gray, $\varepsilon_{w_\eta} = \varepsilon_w$ = constant, and continuum emission is negligible, Eq. (10-19) can still be integrated using the band approximation and q expressed in terms of $\bar{\varepsilon}_g$ and $\bar{\alpha}_g$, as

$$q = \varepsilon_w \left[\bar{\varepsilon}_g \sigma T_g^4 - \bar{\alpha}_g \sigma T_w^4 \right] \tag{10-27}$$

The total mean gas emissivity and absorptivity are still given by (10-24) and (10-25) for this case.

The concepts of total gas emissivity and absorptivity that have emerged in these simple cases are widely used in engineering radiative transfer. However, it is important to realize that these quantities are defined for only a few specific cases (i.e., black isothermal wall or gray, isothermal wall without continuum gas emission). For multiple wall systems, nongray walls, or gray walls with soot participation, the integration of the spectral governing equations does not allow for simple definitions such as (10-22) and (10-23). For these more complicated systems, integration of the spectral governing equations must be done on a case-by-case basis.

EVALUATION OF MEAN BANDWIDTH FOR GASES

The preceding examples demonstrate that integration of the spectral governing transfer equations results in a mean bandwidth that must be evaluated. The definition of the nondimensional mean bandwidth is given by (10-28).

$$\bar{A}_{ij}^* = \frac{1}{\pi A_i F_{ij}} \int_{A_i} \int_{A_j} \frac{A^*(s) \cos\theta_j \cos\theta_i}{s^2} dA_j \, dA_i \tag{10-28}$$

Equation (10-28) is obtained by integrating (10-13) with respect to wavenumber, interchanging the order of integration, and using the definition of effective bandwidth (10-26). In Eq. (10-28) $A^*(s)$ is the effective bandwidth along an arbitrary path through the gas and $\bar{A}^*_{ij}$ is the geometrically averaged effective bandwidth, which accounts for short (possibly optically thin) path lengths between A_i and A_j as well as longer (possibly optically thick) path lengths. The bandwidth $A^*(s)$ can be obtained from a wide band model such as Edwards' exponential wide band model.

For relatively simple geometries Eq. (10-28) can be evaluated analytically. However, for more complicated geometries integration of (10-28) can be prohibitive. In either case it would be convenient to have an accurate method for evaluating $\bar{A}^*$ that did not require complicated integration. Fortunately, a simple method is available for evaluating $\bar{A}^*$ based on the mean beam length concept introduced in Eq. (10-15).

The basis for rapid and simple evaluation of $\bar{A}^*$ is to evaluate A^* at an equivalent path length that gives the correct value for $\bar{A}^*$. The equivalent path length is the mean beam length $L_{m_{ij}}$, defined for gas band radiation by the following relation.

$$\bar{A}^*_{ij} = A^*\left(L_{m_{ij}}\right) \tag{10-29}$$

Equation (10-29) indicates that $\bar{A}^*$ can be obtained simply by evaluating the wide band model at the mean beam length $L_{m_{ij}}$. Since $L_{m_{ij}}$ depends on both geometry and gas optical properties, a procedure for evaluating $L_{m_{ij}}$ is to first obtain a limiting value of $L_{m_{ij}}$, which is only a function of geometry, and then correct that value to account for the optical properties of the gas. Such a limiting value is called the geometric mean beam length.

Geometric (Optically Thin) Mean Beam Length

The geometric (or optically thin) mean beam length $L_{m_{0_{ij}}}$ is obtained by considering the limit of all slant paths being optically thin, $s \to 0$ or equivalently $u = \alpha\rho_a s/\omega \to 0$. In this limit, A^* increases linearly with path length

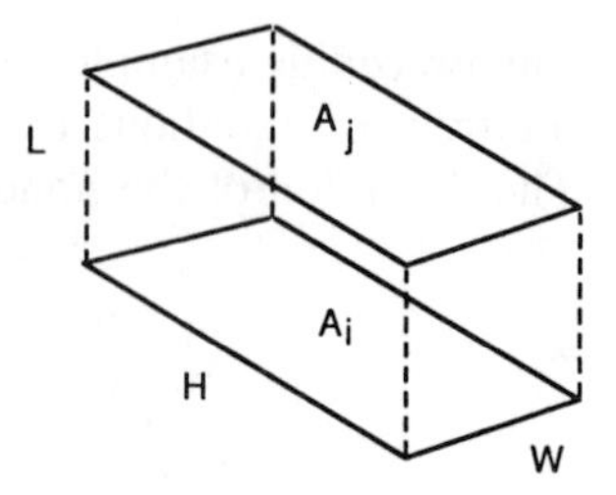

Figure 10-4 Slab of gas between opposite parallel rectangles.

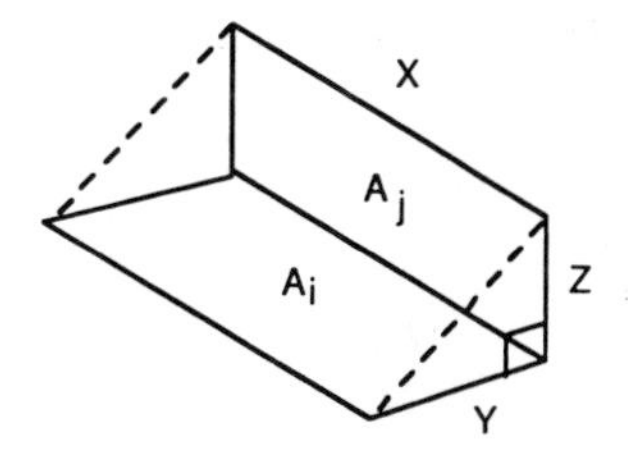

Figure 10-5 Wedge of gas between adjacent perpendicular rectangles.

$A^* \sim s$ (see Chapter 8) and (10-28) gives

$$L_{m_{0_{ij}}} = \frac{1}{\pi A_i F_{ij}} \int_{A_i} \int_{A_j} \frac{\cos\theta_j \cos\theta_i}{s} \, dA_j \, dA_i \tag{10-30}$$

which is only a function of geometry. For a one-wall system ($A_i = A_j$, $F_{ij} = 1$) it can be shown that (10-30) reduces to

$$L_{m_0} = \frac{4V}{A_w} \quad \text{(one wall)} \tag{10-31}$$

where V is the volume of the gas and A_w is the surface area of the gas (and wall). For multiple wall systems the integrations in Eq. (10-30) must be evaluated. This has been done for a slab between opposite rectangles (Fig. 10-4) and a wedge between adjacent rectangles (Fig. 10-5) by Oppenheim and Bevans [1960] and Dunkle [1964]. The annulus between two concentric infinite cylinders has been evaluated by Andersen and Hadvig [1989]. The geometric mean beam lengths for several one-wall and multiple wall configurations are tabulated in Table 10-1.

Once a value of $L_{m_{0_{ij}}}$ has been obtained, it is corrected for finite optical depth using a multiplying factor C.

$$L_{m_{ij}} = C L_{m_{0_{ij}}} \tag{10-32}$$

A value of $C = 0.9$ works well in many situations and is recommended as a general rule. (When no value for C is given in Table 10-1 the recommended value is 0.9.) However, no single value of C can be valid for every situation. The appropriate value of C depends on the geometric configuration of the enclosure as well as the optical depth along the various slant paths through the enclosure. Two simple geometries, a sphere and slab, will next be considered to illustrate how the best choice of the constant C can be made.

TABLE 10-1 Geometric Mean Beam Length and Optically Thick Correction Factor for Various Configurations

Configuration	$L_{m_{0ij}}$	C
One-wall systems		
Sphere, diameter D	$2D/3$	0.9
Infinite slab, thickness L	$2L$	0.83
Infinite circular cylinder, diameter D	D	0.9
Infinite circular annulus, radii R_i and R_0	$2(R_0 - R_i)$	
Two-wall systems		
Infinite slab, thickness L, one wall to the other	$2L$	0.83
Finite circular cylinder of diameter D to base		
Semi-infinite height	$0.81D$	0.80
Height $= 2D$	$0.73D$	0.82
Height $= D$	$0.60D$	0.86
Height $= D/2$	$0.48D$	0.90
Infinite circular annulus, radii R_i and R_0, inner wall to outer (or vice versa)		
$R_i/R_0 = 0$	$4R_0/\pi$	
$R_i/R_0 = 0.5$	$0.72R_0$	
$R_i/R_0 \to 1$	$2(R_0 - R_i)$	
Finite slab of thickness L between opposite parallel rectangles of area WH (Fig. 10-4)		
$W/L = 0.1$ $H/L = 0.1$	$1.00L$	
$W/L = 0.1$ $H/L = 1$	$1.06L$	
$W/L = 1$ $H/L = 1$	$1.11L$	
$W/L = 1$ $H/L = 10$	$1.30L$	
$W/L = 10$ $H/L = 10$	$1.62L$	
Wedge of thickness X between adjacent rectangles of area XY and XZ (Fig. 10-5)		
$Y/X = 0.1$ $Z/X = 0.1$	$0.072X$	
$Y/X = 0.1$ $Z/X = 1$	$0.15X$	
$Y/X = 1$ $Z/X = 1$	$0.56X$	
$Y/X = 1$ $Z/X = 10$	$0.89X$	
$Y/X = 10$ $Z/X = 10$	$3.20X$	

Isothermal Sphere of Nongray Gas

Consider an isothermal sphere of gas with overlapping lines as shown in Fig. 10-6. Due to symmetry and the fact that this is a one-wall system, the mean bandwidth for the entire sphere to itself is the same as for the entire surface A_i to a differential area on the surface dA_j.

$$\bar{A}^* = \bar{A}^*_{ij} = \frac{1}{\pi}\int_{A_i} \frac{A^*(s)\cos\theta_i\cos\theta_j}{s^2}\,dA_i \tag{10-33}$$

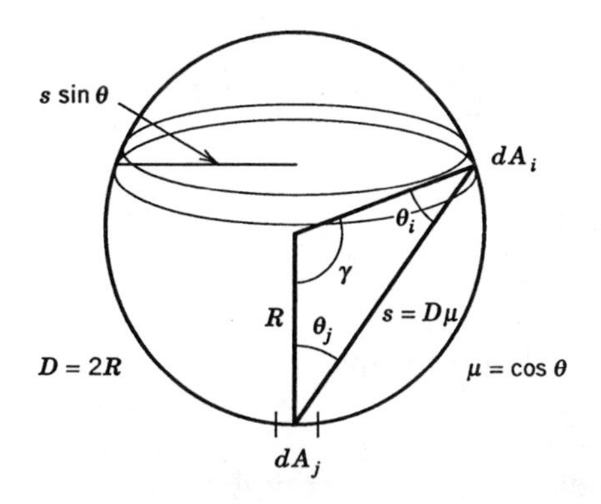

Figure 10-6 Geometry for radiative transfer in sphere of isothermal gas.

From Fig. 10-6 the following geometric relations hold:

$$\theta = \theta_i = \theta_j \tag{10-34}$$

$$\mu = \cos\theta \tag{10-35}$$

$$s = D\mu \tag{10-36}$$

$$\gamma = \pi - 2\theta \tag{10-37}$$

$$R\,d\gamma = -2R\,d\theta = \frac{ds}{\sin\theta} = \frac{D\,d\mu}{\sin\theta} \tag{10-38}$$

$$dA_i = \left(\frac{D\,d\mu}{\sin\theta}\right)2\pi s\sin\theta = 2\pi D^2\mu\,d\mu \tag{10-39}$$

Substituting (10-35), (10-36), and (10-39) into (10-33) gives

$$\overline{A}^* = 2\int_0^1 A^*(D\mu)\mu\,d\mu \tag{10-40}$$

The effective bandwidth A^* along a single line of sight of length $s = D\mu$ is obtained from the exponential wide band model Eq. (8-65) for overlapping lines ($\beta > 1$).

$$A^*(D\mu) = \begin{cases} \dfrac{\alpha\rho_a D\mu}{\omega} & \mu < \mu_c \quad (10\text{-}41a) \\[2ex] \ln\left[\dfrac{\alpha\rho_a D\mu}{\omega}\right] + 1 & \mu > \mu_c \quad (10\text{-}41b) \end{cases}$$

where the critical direction cosine μ_c is defined as the direction where the slant path optical depth is equal to 1.

$$\mu_c = \begin{cases} 1 & u_D < 1 \quad (10\text{-}42a) \\ \dfrac{1}{u_D} & u_D > 1 \quad (10\text{-}42b) \end{cases}$$

When the sphere is optically thin across the diameter ($u_D = \alpha \rho_a D/\omega \leq 1$), all paths will be optically thin and $\mu_c = 1$. When the sphere is optically thick on the diameter ($u_D > 1$), there will be a slant path at some direction μ ($0 < \mu < 1$) where a transition occurs from optically thick conditions to optically thin conditions. At that direction ($\mu = \mu_c$), the optical depth on the slant path is 1 ($\alpha \rho_a D \mu_c/\omega = 1$), which defines the value of μ_c for that case to be $1/u_D$.

Substituting (10-41) into (10-40) and performing the integration gives

$$\bar{A}^* = \begin{cases} \frac{2}{3} u_D & u_D < 1 \quad (10\text{-}43a) \\ \dfrac{1}{6u_D^2} + \ln u_D + \frac{1}{2} & u_D > 1 \quad (10\text{-}43b) \end{cases}$$

For the case of all paths optically thin ($u_D < 1$), Eq. (10-43a) corresponds exactly to the result predicted using the geometric mean beam length $L_{m_0} = 4V/A_w = 2D/3$ according to (10-29) and (10-41a). For the limiting case of all paths optically thick ($u_D \gg 1$), Eq. (10-43b) simplifies to give

$$\bar{A}^* = \ln u_D + \frac{1}{2} = \ln\left[\frac{u_D}{\sqrt{e}}\right] + 1 = A^*(0.606\text{D}) \qquad (u_D \gg 1) \quad (10\text{-}44)$$

This corresponds almost exactly to the result predicted using the mean beam length $L_m = 0.9 L_{m_0} = 0.60D$ according to (10-32), (10-29), and (10-41b) with a correction factor of $C = 0.9$. A more detailed comparison between the exact results $\bar{A}^*$ and the approximate prediction A^* is given in Table 10-2. This comparison is made for intermediate values of u_D and indicates that the approximate method based on mean beam length works well for the spherical geometry when a correction factor of $C = 0.9$ is used for all values of $u_D > 1$. For $u_D < 1$ the geometric mean beam length ($C = 1$) should be used.

Isothermal Slab of Nongray Gas

Another relatively simple geometry for which $\bar{A}^*$ can be evaluated analytically is the slab of thickness L. The steps are very similar to those for a sphere and are left as an exercise (Problem 10-6). The slant path is L/μ as shown in Fig. 10-7. It can be shown that Eq. (10-40) still applies with $D\mu$

TABLE 10-2 Comparison between Exact and Approximate Evaluation of $\bar{A}^*$ for Sphere Using Exponential Band Model with Overlapping Lines

u_D	$\bar{A}^*$	$A^*(L_{m_0} = 0.67D)$ $C = 1$	$A^*(L_m = 0.60D)$ $C = 0.9$	$\bar{A}^*/A^*$
0.1	0.0667	0.0667	—	1.00
0.2	0.133	0.133	—	1.00
0.5	0.333	0.333	—	1.00
1.0	0.667	0.667	—	1.00
2.0	1.23	—	1.18	1.04
5.0	2.12	—	2.10	1.01
10.0	2.80	—	2.79	1.004

replaced by L/μ. The result for $\bar{A}^*$ for a slab of thickness L of a gas band with overlapping lines ($\beta > 1$) according to the exponential band model is thus

$$\bar{A}^* = \begin{cases} 2u_L - \frac{1}{2}u_L^2 & u_L < 1 \qquad (10\text{-}45a) \\ \ln(u_L) + \frac{3}{2} & u_L > 1 \qquad (10\text{-}45b) \end{cases}$$

For the limiting case of all slant paths optically thin ($u_L \ll 1$), (10-45a) simplifies to

$$\bar{A}^* = 2u_L \qquad (u_L \ll 1) \tag{10-46}$$

which agrees with (10-29) using the geometric mean beam length $L_{m_0} = 4V/A_w = 2L$. For the case of all slant paths optically thick ($u_L > 1$), (10-45b)

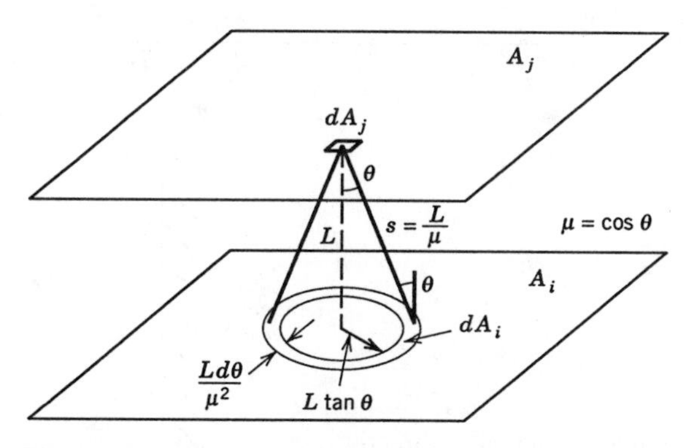

Figure 10-7 Geometry for radiative transfer in a slab.

TABLE 10-3 Comparison between Exact and Approximate Evaluation of $\bar{A}^*$ for Slab Using Exponential Band Model with Overlapping Lines

u_L	$\bar{A}^*$	$A^*(L_{m_0} = 2L)$ $C = 1$	$A^*(L_m = 1.66L)$ $C = 0.83$	$\bar{A}^*/A^*$
0.1	0.195	0.20		0.975
0.2	0.380	0.40		0.950
0.5	0.875	1.0		0.875
1.0	1.50	—	1.50	1.0
2.0	2.19	—	2.19	1.0
5.0	3.11	—	3.11	1.0
10.0	3.80	—	3.80	1.0

can be rewritten as

$$\bar{A}^* = \ln\left[\sqrt{e}\, u_L\right] + 1 = A^*(1.65L), \qquad (u_L > 1) \tag{10-47}$$

which indicates that a correction factor of $C = 0.83$ ($= 1.65/2$) would be better suited in (10-32) for the slab geometry when all or most of the slant paths are optically thick. A comparison between $\bar{A}^*$ and $A^*(L_m)$ for the slab geometry at some intermediate optical thickness values u_L is given in Table 10-3.

Example 10-1

A simple model of a direct-fired furnace is two parallel walls, one a load and the other a refractory (both assumed to be black), separated by an isothermal gas (Fig. 10-8). Calculate the radiant heat flux to the flat plate load at $T_1 = 500$ K. The gas is a slab of thickness $L = 0.3$ m composed of CO_2 (0.5 atm) and H_2O (0.5 atm) at 1 atm total pressure and $T_g = 1000$ K. Consider only the 2.7-μm bands.

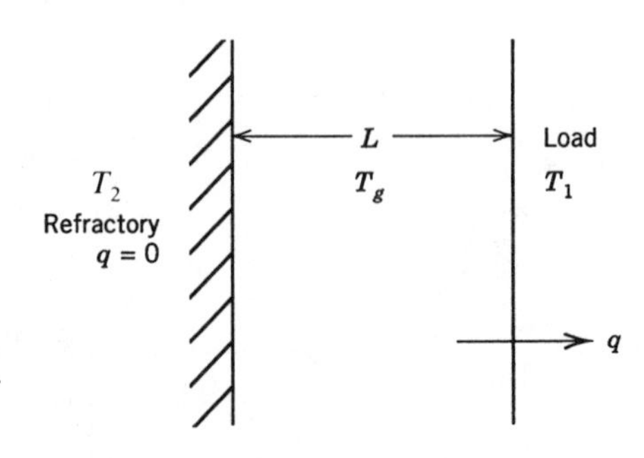

Figure 10-8 Simple model of a direct-fired furnace.

This is a three-node problem with two walls and a gas. The nongray gas at T_g is the source, the load at T_1 is the sink and the other wall at T_2 is an adiabatic surface. The solution to this type of problem can be formulated in terms of the enclosure equations (10-5, 12, 7) with $\varepsilon_{\eta_i} = 1$ and $\rho_{\eta_i} = 0$ as

$$q_{\eta_1} = \bar{\varepsilon}_{\eta_{12}}\left[e_{b_\eta}(T_1) - e_{b_\eta}(T_g)\right] + \left(1 - \bar{\varepsilon}_{\eta_{12}}\right)\left[e_{b_\eta}(T_1) - e_{b_\eta}(T_2)\right] \quad \text{(10-48a)}$$

$$q_{\eta_2} = \bar{\varepsilon}_{\eta_{12}}\left[e_{b_\eta}(T_2) - e_{b_\eta}(T_g)\right] + \left(1 - \bar{\varepsilon}_{\eta_{12}}\right)\left[e_{b_\eta}(T_2) - e_{b_\eta}(T_1)\right] \quad \text{(10-48b)}$$

Equations (10-48) can be integrated over wavenumber by assuming that $\bar{\varepsilon}_{\eta_{12}}$ is zero outside the bands and one inside the bands to give

$$\begin{aligned} q_1 = {} & \sum_k \bar{A}_{k_{12}}\left[e_{b_{\eta_{0k}}}(T_1) - e_{b_{\eta_{0k}}}(T_g)\right] + \sigma\left(T_1^4 - T_2^4\right) \\ & - \sum_k \bar{A}_{k_{12}}\left[e_{b_{\eta_{0k}}}(T_1) - e_{b_{\eta_{0k}}}(T_2)\right] \end{aligned} \quad \text{(10-49)}$$

$$\begin{aligned} 0 = q_2 = {} & \sum_k \bar{A}_{k_{12}}\left[e_{b_{\eta_{0k}}}(T_2) - e_{b_{\eta_{0k}}}(T_g)\right] + \sigma\left(T_2^4 - T_1^4\right) \\ & - \sum_k \bar{A}_{k_{12}}\left[e_{b_{\eta_{0k}}}(T_2) - e_{b_{\eta_{0k}}}(T_1)\right] \end{aligned} \quad \text{(10-50)}$$

where the bandwidth for the kth band from surface 1 to 2 would be the same as that given in (10-45) if the lines were all overlapping [$\beta > 1$ was a condition of (10-45)]. However, in this example $\beta < 1$ for the water vapor. To proceed with the solution of the problem there are two possible approaches. One would be to obtain a relation corresponding to (10-45) for $\beta < 1$. The other approach would be to use the mean beam length concept. The latter approach is adopted here. Assuming that most of the slant paths are optically thick (this can be checked later) gives $C = 0.83$ and $L_m = (0.83)(2)(0.3 \text{ m}) = 0.5$ m. This is the same path length that was used in Examples 8-1 and 8-2. The appropriate bandwidth is the same as that determined in Example 8-2. Since only the 2.7-μm bands are being considered, there is only a single term in the summations of Eqs. (10-49, 50). The temperature T_2 can thus be determined from Eq. (10-50) and the net radiant heat flux to the load $(-q_1)$ from (10-49) as

$$T_2 = \left[500^4 + 603(8.94 - 0.04)/5.67 \times 10^{-8}\right]^{1/4} = 630 \text{ K}$$

$$\begin{aligned} -q_1 &= 603(8.94 - 0.04) + 5.67(6.3^4 - 5^4) - 603(0.37 - 0.04) \\ &= 5370 + 5190 = 10{,}560 \text{ W/m}^2 \end{aligned}$$

The heat fluxes to the load inside the band (from the gas) and outside the band (from the refractory surface) are 5370 and 5190 W/m^2, respectively. Thus it should be noted that while the refractory surface is adiabatic on a total basis, it is not so on a spectral basis.

SOOT RADIATION

During combustion of hydrocarbon fuels, when fuel-air ratio and mixing conditions are such that the fuel is heated under locally oxygen-deficient conditions, there is a tendency for solid phase soot particles to form that are on the order of tens and hundreds of angstroms in size. Because soot appears as a condensed phase, it emits in a continuum spectrally and can be responsible for emitting a radiant heat flux contribution several times larger than the gas band contribution [Siddall and McGrath, 1963; Kunitomo and Sato, 1970; Felske and Tien, 1973; Buckius and Tien, 1977; Yuen and Tien, 1977; Tien and Lee, 1982; Grosshandler and Vantelon, 1985; Mackowski, et. al., 1989]. Because of the continuum emission, there is also emission in the visible region, and the flame often appears yellow due to the retina being most sensitive to yellow light.

The absorption index of soot is typically on the order of 1 in the near infrared. The size of soot particles is on the order of 10^{-2} μm, resulting in size parameters on the order of 10^{-2}. Due to the combination of small size and relatively large absorption index, soot behaves as a nonscattering Rayleigh emitter [see Eq. (9-15)]. From Eqs. (9-10), (9-12), and (9-78) the absorption coefficient for soot can be written as

$$\frac{K_{a_\lambda}}{f_v} = \frac{f(n,k)}{\lambda} \tag{10-51}$$

where

$$f(n,k) = \frac{36\pi nk}{\left(n^2 - k^2 + 2\right)^2 + 4n^2k^2} \tag{10-52}$$

The optical constants of soot have been measured by various investigators [Stull and Plass, 1960; Howarth, et al., 1966; Foster and Howarth, 1968; Dalzell and Sarofim, 1969; Lee and Tien, 1980]. These results are plotted in Fig. 10-9. It can be seen that there are minor differences between the results of different investigators. These differences may be a result of variation in optical constants between different soot samples or the different techniques used. In the study of Dalzell and Sarofim [1969] n and k were determined from Fresnel reflectance measurements obtained from compressed layers of soot particles collected from acetylene and propane flames. Tien and Lee [1982] argued that these measurements were erroneous based on Janzen's observation that the conditions necessary to invoke the Fresnel relations are seldom satisfied even in visibly smooth, packed carbon surfaces [Janzen, 1979]. Their results [Lee and Tien, 1980] were derived from in situ flame transmission measurements. The difference between these two sets of data, however, is not especially large.

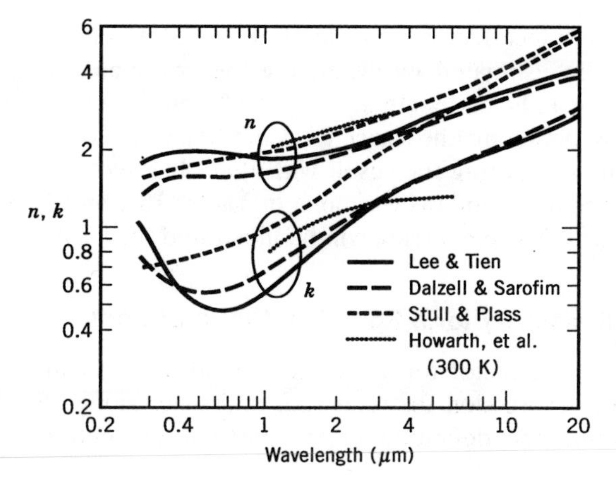

Figure 10-9 Optical constants of soot.

The spectral emissivity due to soot can be calculated from the optical constants with the aid of Eq. (10-51) and (10-52) as

$$\varepsilon_\lambda = 1 - \exp(-K_{a_\lambda} s) \tag{10-53a}$$

and the total emissivity and absorptivity are thus given by

$$\varepsilon, \alpha = \int_0^1 \left[1 - \exp(-K_{a_\lambda} s)\right] df(\lambda T_{g,w}) \tag{10-53b}$$

Although it can be seen from Fig. 10-9 that n and k generally increase with wavelength, the function $f(n, k)$ in Eq. (10-51) can be taken as a first approximation to be constant. With this assumption and using Wien's law as an approximation of the Planck function, Eq. (10-53b) gives

$$\varepsilon, \alpha = 1 - \frac{1}{(1+a)^4} \tag{10-54}$$

where

$$a = \frac{f(n,k) f_v s T_{g,w}}{C_2} \tag{10-55}$$

is the total optical depth parameter based on path length s. The value of $f(n, k)$ has been found experimentally to vary between 4 and 6. The value of f_v is typically 10^{-7} to 10^{-5} but depends strongly on the flame conditions [Kunugi and Jinno, 1966].

The total emittance factor for a cloud of soot in an enclosure of arbitrary geometry can be estimated by evaluating the parameter a in (10-54) and (10-55) at the mean beam length $s = L_m$. The accuracy of this procedure can be checked by obtaining the exact total emittance factor for a few simple geometries and comparing the result with that obtained by the approximate mean beam length method (as was done in Tables 10-2 and 10-3 for pure gas band radiation). This comparison will be illustrated for a spherical enclosure.

Mean Total Emissivity for a Spherical Cloud of Soot

The total emittance factor for a spherical cloud of soot can be obtained by substituting the expression for total soot emissivity along an arbitrary slant path (10-54) into the definition of the mean total emittance factor for a sphere

$$\bar{\varepsilon} = 2\int_0^1 \varepsilon(s)\mu\, d\mu \tag{10-56}$$

with $s = D\mu$. Integrating (10-56) gives

$$\bar{\varepsilon} = 1 - \frac{3 + a_D}{3(1 + a_D)^3} \tag{10-57}$$

where

$$a_D = \frac{f(n,k) f_v D T_g}{C_2} \tag{10-58}$$

is the total optical depth parameter based on sphere diameter. A comparison between the mean total emissivity and the mean beam length approximation is given in Table 10-4. This comparison shows that when $a_D > 1$ a geometric

TABLE 10-4 Comparison of Mean Emissivity with Mean Beam Length Approximation for Isothermal Sphere of Rayleigh Continuum Emitters (e.g., Soot)

a_D	$\bar{\varepsilon}$	$\varepsilon(L_{m_0} = 0.67D)$ $C = 1$	$\varepsilon(L_m = 0.60D)$ $C = 0.9$	$\bar{\varepsilon}/\varepsilon$
0.1	0.223	0.228	—	0.978
0.2	0.383	0.395	—	0.970
0.5	0.654	0.685	—	0.955
1.0	0.833	—	0.847	0.983
2.0	0.938	—	0.957	0.980
5.0	0.988	—	0.996	0.992
10.0	0.997	—	0.999	0.997

mean beam length correction factor of $C = 0.9$ again works well in the approximate method.

Example 10-2

A long tube, 4.5 cm in diameter, carries a mixture of combustion products at 1 atm and 1500 K. The inside wall is at 1000 K and is black. The gas composition is CO_2 (0.06 atm), H_2O (0.18 atm), N_2 (0.76 atm), and soot ($f_v = 10^{-6}$). Assuming $f(n, k)$ for soot [see Eq. (10-52)] is a constant equal to 5, calculate the total radiant heat flux to the tube wall.

The appropriate expression for the total radiant heat flux for this one-wall system can be obtained from Eqs. (10-19) and (10-20).

$$q = \int_0^{\infty} \bar{\varepsilon}_{\eta}\left[e_{b_\eta}(T_g) - e_{b_\eta}(T_w)\right] d\eta \tag{10-59}$$

The emittance factor $\bar{\varepsilon}_{\eta}$ is approximated using the mean beam length approach and using the additivity of absorption coefficients.

$$\bar{\varepsilon}_{\eta} = 1 - \exp\left[-K_{a_\lambda} L_m\right] = 1 - \exp\left[-\left(K_{a_{\lambda_s}} + K_{a_{\lambda_H}} + K_{a_{\lambda_C}}\right) L_m\right] \tag{10-60}$$

The mean beam length is calculated using Eqs. (10-31) and (10-32). Anticipating that most of the slant paths through the cylinder will be optically thin ($C = 1$), the resulting mean beam length is $L_m = L_{m_0} = D = 0.045$ m. Expanding Eq. (10-60) and rearranging terms the emittance factor can also be written as

$$\bar{\varepsilon}_{\eta} = \bar{\varepsilon}_{\eta_s} + \bar{\varepsilon}_{\eta_H} + \bar{\varepsilon}_{\eta_C} - \bar{\varepsilon}_{\eta_s}\bar{\varepsilon}_{\eta_H} - \bar{\varepsilon}_{\eta_s}\bar{\varepsilon}_{\eta_C} - \bar{\varepsilon}_{\eta_H}\bar{\varepsilon}_{\eta_C} + \bar{\varepsilon}_{\eta_s}\bar{\varepsilon}_{\eta_H}\bar{\varepsilon}_{\eta_C} \tag{10-61}$$

where s = soot, $C = CO_2$ and $H = H_2O$. Equation (10-61) shows that in this system it is possible for three species to overlap simultaneously, CO_2, H_2O, and soot. Substituting (10-61) into (10-59) and utilizing the band approximation for the gas bands gives

$$q = \bar{\varepsilon}_s \sigma T_g^4 - \bar{\alpha}_s \sigma T_w^4 + \sum_{k \text{ band}} \left[e_{b_{\eta 0k}}(T_g) - e_{b_{\eta 0k}}(T_w)\right] \bar{A}_k(T_g) \tag{10-62}$$

where

$$\bar{A}_k = \bar{A}_H + \bar{A}_C - \frac{\bar{A}_H}{\Delta\eta_H}\bar{\varepsilon}_{\eta 0k_s}\Delta\eta_H - \frac{\bar{A}_C}{\Delta\eta_C}\bar{\varepsilon}_{\eta 0k_s}\Delta\eta_C - \frac{\bar{A}_H}{\Delta\eta_H}\frac{\bar{A}_C}{\Delta\eta_C}\Delta\eta_{HC}$$
$$+ \frac{\bar{A}_H}{\Delta\eta_H}\frac{\bar{A}_C}{\Delta\eta_C}\bar{\varepsilon}_{\eta 0k_s}\Delta\eta_{HC} \tag{10-63}$$

TABLE 10-5 Exponential Wide Band Parameters for Example 10-2

	λ_0 (μm)	η_0 (cm^{-1})	u	$\bar{A}^*$	$\bar{A}$	$\dfrac{\bar{A}}{\Delta\eta}$	$\Delta\eta$	η_L	η_U	$\bar{\varepsilon}_{\eta_{0_s}}$
H_2O	1.38	7250	0.025	0.025	3.13	0.1	31	7234	7266	0.150
	1.87	5350	0.028	0.028	4.62	0.1	46	5327	5373	0.113
	2.7	3760	0.127	0.127	29.5	0.1	295	3612	3908	0.081
	6.3	1600	0.223	0.223	48.6	0.1	486	1357	1843	0.035
CO_2	2.0	5200	0.001	0.001	0.133	0.1	2	5199	5201	0.106
	2.7	3660	0.063	0.063	5.73	0.1	57	3631	3689	0.081
	4.3	2323	2.44	1.89	82.1	0.47	174	2236	2410	0.051
	9.4	1060	0.004	0.004	0.15	0.1	2	1059	1061	0.024
	10.4	960	0.003	0.003	0.15	0.1	2	959	961	0.021
	15	667	0.364	0.364	17.9	0.1	179	577	757	0.015

The bandwidths obtained from the exponential wide band model are given in Table 10-5. The spectral soot emittance factors in Table 10-5 are calculated using Eqs. (10-51) and (10-64).

$$\bar{\varepsilon}_{\eta_s} = 1 - \exp\left[-K_{a_{\lambda_s}} L_m\right] \tag{10-64}$$

The only region where the gas bands overlap is the 2.7-μm region where the H_2O band completely overlaps the CO_2 band. In this region (2.7 μm) the full expression of Eq. (10-63) is utilized because soot, CO_2, and H_2O are all overlapping. In other spectral regions where either CO_2 or H_2O bands occur, only the terms corresponding to one gas band [either C or H in Eq. (10-63)] are utilized, reducing the number of terms needed on the right-hand side of Eq. (10-63) from six to two. The bandwidth terms $\bar{A}_k$ for only the bands that contribute significantly to the total radiant heat flux (2.7, 4.3, and 6.3 μm) are

$$\begin{aligned}\bar{A}_{2.7} &= 29.5 + 5.73 - (0.1)(0.081)(295) - (0.1)(0.081)(57) \\ &\quad - (0.1)(0.1)(57) + (0.1)(0.1)(0.081)(57) = 31.8 \text{ cm}^{-1}\end{aligned}$$

$$\bar{A}_{4.3} = 82.1 - (82.1)(0.051) = 77.9 \text{ cm}^{-1}$$

$$\bar{A}_{6.3} = 48.6 - (48.6)(0.035) = 46.9 \text{ cm}^{-1}$$

The total soot emittance and absorptance terms are calculated from Eq. (10-54) as

$$\bar{\varepsilon}_s = 0.089 \quad \text{and} \quad \bar{\alpha}_s = 0.060$$

Thus the total radiant heat flux from (10-62) is

$$q = (5.67)\left[(0.089)(1.5)^4 - (0.060)(1)^4\right] + (0.00555 - 0.00089)(31.8)$$
$$+ (0.00566 - 0.00172)(77.9) + (0.00421 - 0.00170)(46.9)$$
$$= 2.79 \text{ W/cm}^2$$

SUMMARY

This chapter describes a method for solving radiative transfer in an enclosure containing an isothermal, nonscattering medium. The method is based on the concept of a mean beam length. It applies to gas bands as well as continuum emission. An exact method for determining the mean beam length is demonstrated for simple geometries (sphere and slab). An approximate method for estimating the mean beam length in more complex geometries is also demonstrated. Comparisons between the rigorous and approximate mean beam length methods for the slab and sphere give some guidance for estimating an appropriate correction factor to be used in more complex geometries, which have not been validated. The mean beam length concept has also been extended to scattering media. [Tong and Tien, 1980; Cartigny, 1986; Yuen, 1990].

There are two common heat transfer situations that can be described by the methods of this chapter. The first is a well-mixed combustion chamber containing gases such as CO_2, H_2O, CO, etc. and perhaps soot, where heat is transferred from the gas to the walls or some other load. This case has no restriction on the optical depth of the isothermal bulk of the gas. (Of course, the nonisothermal boundary layer must be optically thin.) If most slant paths through the bulk of the gas are optically thin, the geometric mean beam length applies. If most slant paths through the bulk of the gas are optically thick, a correction factor of approximately 0.9 should be applied to the geometric mean beam length. In the optically thin case the heat transfer rate to the walls is proportional to the volume of the gas. In the optically thick case the heat transfer rate to the walls is proportional to the surface area of the gas.

A second heat transfer situation is one in which heat is transferred from one surface (a source) to another surface (a sink) with an intervening participating gas. In this case the gas is usually adiabatic and acts only as a barrier to the heat transfer between the surfaces. If most of the slant paths through the gas are optically thin, the gas will remain nearly isothermal and the methods of this chapter can be applied to determine the heat transfer rate and gas temperature. On the other hand, if most of the slant paths through the gas are optically thick and the gas is not well mixed, the gas will not remain isothermal. To describe this situation it is necessary to address nonisothermal effects—the subject of Chapter 11.

REFERENCES

Altenkirch, R. A., Mackowski, D. W., Peck, R. E., and Tong, T. W. (1984), "Effects of Soot on Pyrometer Measured Temperatures in Pulverized-Coal Flames," *Combustion Science and Technology*, Vol. 41, pp. 327–335.

Andersen, K. M. and Hadvig, S. (1989), "Geometric Mean Beam Length for the Space Between Two Coaxial Cylinders," *J. Heat Transfer*, Vol. 111, pp. 811–813.

Bevans, J. T. and Dunkle, R. V. (1960), "Radiative Exchange within an Enclosure," *J. Heat Transfer*, Vol. 82, pp. 1–19.

Buckius, R. O. and Tien, C-L. (1977), "Infrared Flame Radiation," *Int. J. Heat Mass Transfer*, Vol. 20, No. 2, pp. 93–106.

Cartigny, J. D. (1986), "Mean Beam Length for a Scattering Medium," *Heat Transfer–1986*, 8th International Heat Transfer Conference Proceedings, Hemisphere, C-L. Tien, V. P. Carey, and J. K. Farrell, Eds., Vol. 2, pp. 769–772.

Dalzell, W. H. and Sarofim, A. F. (1969), "Optical Constants of Soot and Their Application to Heat-flux Calculations," *J. Heat Transfer*, Vol. 91, No. 1, pp. 100–104.

Dunkle, R. V. (1964), "Geometric Mean Beam Lengths for Radiant Heat Transfer Calculations," *J. Heat Transfer*, Vol. 86, pp. 75–80.

Felske, J. D. and Tien, C-L. (1973), "Calculation of the Emissivity of Luminous Flames," *Comb. Sci. Tech.*, Vol. 7, No. 1, pp. 25–31.

Foster, P. J. and Howarth, C. R. (1968), "Optical Constants of Carbons and Coals in the Infrared," *Carbon*, Vol. 6, pp. 719–729.

Goody, R. M. and Yung, Y. L. (1989), *Atmospheric Radiation Theoretical Basis*, 2nd ed., Oxford University Press, New York, p. 216.

Grosshandler, W. L. and Vantelon, J. P. (1985), "Predicting Soot Radiation in Laminar Diffusion Flames," *Combustion Science and Technology*, Vol. 44, pp. 125–141.

Howarth, C. R., Foster, P. J., and Thring, M. W. (1966), "The Effect of Temperature on the Extinction of Radiation by Soot Particles," *Third International Heat Transfer Conference AIChE*, Vol. 5, pp. 122–128.

Janzen, J. (1979), "The Refractive Index of Colloidal Carbon," *J. Coll. Int. Science*, Vol. 69, No. 3, pp. 436–447.

Kunitomo, T. and Sato, T. (1970), "Experimental and Theoretical Study on the Infrared Emission of Soot Particles in Luminous Flames," *Fourth International Heat Transfer Conference*, Elsevier, Amsterdam.

Kunugi, M. and Jinno, H. (1966), "Determination of Size and Concentration of Soot Particles in Diffusion Flames by a Light Scattering Technique," *Eleventh Symposium (International) on Combustion*, The Combustion Institute, Pittsburgh, pp. 257–266.

Lee, S. C. and Tien, C-L. (1980), "Optical Constants of Soot in Hydrocarbon Flames," *Eighteenth Symposium (International) on Combustion*, The Combustion Institute, Pittsburgh, pp. 1159–1166.

Mackowski, D. W., Altenkirch, R. A., and Menguc, M. P. (1989), "Comparison of Electromagnetic Wave and Radiative Transfer Equation Analyses of a Coal

Particle Surrounded by a Soot Cloud," *Combustion and Flame*, Vol. 76, pp. 415–420.

Oppenheim, A. K. and Bevans, J. T. (1960), "Geometric Factors for Radiant Heat Transfer Through an Absorbing Medium in Cartesian Coordinates," *J. Heat Transfer*, Vol. 82, pp. 360–368.

Siddall, R. G. and McGrath, I. A. (1963), "The Emissivity of Luminous Flames," *Ninth Symposium (International) on Combustion*, The Combustion Institute, Academic Press, New York, pp. 102–110.

Stull, V. R. and Plass, G. N. (1960), "Emissivity of Dispersed Carbon Particles," *J. Opt. Soc. Am.*, Vol. 50, No. 2, pp. 121–129.

Tien, C-L. and Lee, S. C. (1982), "Flame Radiation," in *Prog. Energy Comb. Sci.*, Vol. 8, pp. 41–59.

Tong, T. W. and Tien, C-L. (1980), "Resistance Network Representation of Radiative Heat Transfer with Particulate Scattering," *J. Quant. Spect. Rad. Transfer*, Vol. 24, pp. 491–503.

Yuen, W. W. (1990), "Development of a Network Analogy and Evaluation of Mean Beam Lengths for Multidimensional Absorbing/Isotropically Scattering Media," *J. Heat Transfer*, Vol. 112, pp. 408–414.

Yuen, W. W. and Tien, C-L. (1977), "A Simple Calculation Scheme for the Luminous Flame Emissivity," *Sixteenth Symposium (International) on Combustion*, The Combustion Institute, Pittsburgh, pp. 1481–1487.

REFERENCES FOR FURTHER READING

Grosshandler, W. L. and J. P. Vantelon, "Predicting Soot Radiation in Laminar Diffusion Flames," *Combustion Science and Technology*, Vol. 44, 1985, pp. 125–141.

Mackowski, D. W., R. A. Altenkirch, and M. P. Menguc, "Comparison of Electromagnetic Wave and Radiative Transfer Equation Analyses of a Coal Particle Surrounded by a Soot Cloud," *Combustion and Flame*, Vol. 76, 1989, pp. 415–420.

PROBLEMS

1. A long pipe, 10 cm in diameter, carries steam at 10 atm pressure and 1200 K. The inside wall is at 800 K, diffuse, gray, and has an emissivity of 0.9. Including the 2.7 and 6.3 μm bands, use the exponential wide band model to estimate the net radiant heat flux to the pipe.

2. Consider a cylinder of hot combustion gases and soot at 3000 K that is 1.34 cm in diameter and 2.68 cm long. One end of the cylinder is adjacent to a solid propellant surface (producing the gases) that is at 1000 K with an emissivity of 1. The other surfaces of the cylinder are assumed to be bounded by a cold, nonemitting blackbody. The total pressure is 50 atm, the soot volume fraction is 10^{-5}, and the water vapor partial pressure is 20 atm. Assuming the soot properties can be characterized by $f(n, k) = 5$ [see Eq. (10-52)], calculate the net radiant heat flux to the surface of the propellant including the soot contribution and water vapor contribution (all bands). Repeat the calculation assuming no soot contribution.

3. A cubic combustion chamber, 5 m on a side, contains 10^{-5}% soot ($f_v = 10^{-7}$), 6% CO_2, 18% H_2O, and the rest N_2 (by volume). The gases are at 1400 K and 1 atm total pressure and the walls, which are black, are at 1000 K. Considering only the 2.7-μm bands and the soot, calculate the net radiant heat flux to the wall. Use the gray band approach for the gas bands. Assume L_m = constant. Take the soot spectral emittance factor as constant over the gas bands. Outside the bands account for the spectral dependence of soot in a manner consistent with the assumption of n, k = constant.

4. Show that the mean spectral emittance factor (or total emittance factor if gray) for an isothermal slab radiating to one or both boundaries is

$$\bar{\varepsilon} = 1 - 2E_3(t_L) \quad \text{where} \quad E_n(x) = \int_0^1 e^{-x/\mu}\mu^{n-2}\, d\mu$$

5. Write an integral expression that could be used to rigorously evaluate the mean total emissivity for a plane parallel chamber filled with an isothermal suspension of soot, analogous to the development carried out for a spherical chamber. Do not try to evaluate the integral. Assume n and k are weak functions of wavelength. Plot qualitatively total soot emissivity as a function of path length, showing the correct functional dependence for very small and very large path lengths. Explain any similarities and differences with the typical curve of growth for a gas band (A vs. path length).

6. Consider an isothermal slab of radiating gas (nonscattering) with thickness L.

(a) Use the exponential wide band model results to derive Eqs. (10-45) for the nondimensional geometric mean bandwidth of a band of strong overlapping lines ($\beta > 1$) as a function of $u_L = \rho L \alpha / \omega$.

(b) Show that this general result reduces to the correct limiting results for the two limiting cases represented by Eqs. (10-46) and (10-47) of (i) all slant paths optically thick ($u_L > 1$) and (ii) all slant paths optically thin ($u_L \to 0$).

7. A simple model of a radiant boiler is two black parallel walls separated by a distance L, with a gray, isothermal, nonscattering gas between the walls. Consider the question of whether it is better to put steam tubes on both walls or only one wall (with the other wall a refractory). Use total volumetric heat release (total heat transfer rate per unit volume of boiler) as the criterion for comparison. Does the answer depend on optical thickness t_L? Make the comparison for the limiting cases of optically thick $t_L \to \infty$ and optically thin $t_L \to 0$.

8. A radiant boiler is modeled as two black, parallel walls separated by a distance $L = 10$ cm with a nonscattering, isothermal gas at $T_g = 1500$ K between. One wall is a heat sink at $T_1 = 1000$ K and the other wall is adiabatic. Assume that the heat transfer is dominated by soot radiation with $f_v = 10^{-7}$ and $f(n, k) = 5$ [see Eq. (10-52)]. Calculate the net radiant heat flux to the sink.

9. Two black, parallel, infinite flat plates at temperatures $T_1 = 2000$ K and $T_2 = 1000$ K are separated by a distance $L = 1$ cm. An adiabatic, participating gas exists between the two plates. Assume that the gas has only one vibration-rotation band centered at 3 μm, the lines are overlapping ($\beta > 1$), and the band is a fundamental band (i.e., $\alpha =$ const $= \alpha_0$). Assume the following properties for the gas:

$$\alpha_0 = 22.4 \text{ cm}^{-1}/(\text{g m}^{-2})$$

$$\omega_0 = 60 \text{ cm}^{-1} \ (T_0 = 100 \text{ K})$$

$$\rho_a = 1076 \text{ g/m}^3$$

Calculate the total net radiant heat flux from the hot plate to the cold plate.

10. Derive the exact expression for the mean emittance factor (spectral or total gray) for an isothermal sphere of a nonscattering participating medium radiating to an isothermal wall as a function of optical depth based on diameter D.

11. Consider radiation from an optically thin, homogeneous, nonscattering medium with volume V, surface area A_w, and temperature T_g to cold, black surroundings at $T_w \to 0$ K.

(a) Use the divergence theorem

$$\int_V \nabla \cdot \mathbf{q}_{r_\lambda} \, dV = \int_{A_w} \mathbf{q}_{r_\lambda} \cdot \mathbf{e}_w \, dA_w$$

with the appropriate results from Chapter 7 [Eqs. (7-44a, b)] and Chapter 10 [Eq. (10-19)] to derive the geometric mean beam length for an isothermal gas surrounded by an isothermal wall, Eq. (10-31).

(b) Show that the appropriate total emission coefficient for emission by an optically thin isothermal volume of gas is the Planck mean emission coefficient [Eq. (7-44b)]. Hint: Show that the emitted flux to the cold wall is $L_{m_0} \sigma T_g^4 K_{\text{em}_p}(T_g)$.

(c) Show that if the surrounding wall is at finite temperature T_w (but still black) the total radiative source term is given by Eq. (7-44c), where $T = T_g$ and $T_e = T_w$.

12. Consider radiation from an optically thin, homogeneous, nonscattering medium with volume V, surface area A_w, and temperature T_g to a gray wall at T_w surrounding the medium. Use the divergence theorem (see Problem 11) with the appropriate results from Chapter 7 [Eqs. (7-44a, b)] and Chapter 10 [Eq. (10-19)] to derive an expression for the spectral intensity in the medium I_λ as a function of the spectral Planck function based on T_g and T_w and the dimensionless parameters ε_w and $4VK_{a\lambda}/A_w$. Discuss the behavior of I_λ for various limiting values of ε_w and $4VK_{a\lambda}/A_w$. Is an enclosed, isothermal surface necessarily a blackbody, independent of ε_w, if it contains a participating medium at a different temperature? Explain.

13. An isothermal mixture of water vapor and nitrogen at 500 K and 1 atm is contained between two infinite parallel walls at 300 K. The walls are diffuse and gray with a total hemispherical emissivity of 0.8. The distance between the walls is 0.6 m. The mole fraction of water is 30%. What is the net radiative heat flux to the walls in the spectral region between 4 and 10 μm (i.e., due to the 6.3-μm band only)?

CHAPTER 11

NONISOTHERMAL TRANSFER: RADIATIVE EQUILIBRIUM AND DIFFUSION WITH ISOTROPIC SCATTERING

A common heat transfer problem involving a participating medium is that in which heat is transferred radiatively from one surface (a source) to another surface (a sink) across an intervening participating medium. The medium is adiabatic and acts as a barrier to the heat transfer. If conduction and convection in the participating medium are negligible the condition is known as *radiative equilibrium*. If most of the slant paths through the medium are optically thin, the medium will remain nearly isothermal and the methods of Chapter 10 can be applied to determine the heat transfer rate and medium temperature. On the other hand, if most of the slant paths through the medium are optically thick, the medium will not remain isothermal. Heat will be transferred in a sequential fashion from the hotter surface to the adjacent medium and from there to the neighboring cooler medium and eventually to the colder surface. This process resembles a diffusion process, like conduction, and is known as *radiative diffusion*. This chapter considers both radiative equilibrium and radiative diffusion. In both of these conditions the process of scattering can be easily included without additional complexity provided the scattering is isotropic. As will be seen, isotropic scattering under these conditions is equivalent to absorption and re-emission. Therefore isotropic scattering will also be included in the present treatment.

RADIATIVE EQUILIBRIUM IN A SLAB

The simplest configuration for radiative equilibrium is the slab geometry (Fig. 11-1). Consider radiative heat transfer from a diffuse, gray source wall (1) across a nonscattering medium to a sink wall (2) (isotropic scattering is added later). Conduction and convection are negligible and the system is in steady state. The walls have prescribed temperatures T_1 and T_2, and emissivities of ε_1 and ε_2, and the medium is assumed to be adiabatic with an absorption coefficient of K_a. The spacing between the plates is L and the optical depth t is based on the coordinate normal to the plates, x.

$$t = \int_0^x K_a \, dx' \quad \text{(gray)} \tag{11-1}$$

$$x = s\mu \tag{11-2}$$

$$\mu = \cos\theta \tag{11-3}$$

The intensity in the medium is a function of both position t and direction μ.

$$I(t, \mu) = \begin{cases} I^+ & \mu > 0 \\ I^-, & \mu < 0 \end{cases} \tag{11-4}$$

A + notation is used to indicate intensity in the forward hemisphere ($\mu > 0$) and the − indicates intensity in the backward hemisphere ($\mu < 0$).

Since the change in intensity along an arbitrary slant path (s) is $\cos\theta$ times the change along the normal path (x) ($dI/ds = \mu \, dI/dx$), the transfer equation along a slant path for this problem becomes*

$$\mu \frac{dI}{dt} = -I + I_b \tag{11-5}$$

The boundary conditions at the diffuse walls are

$$I^+(t = 0) = \frac{q_1^+}{\pi} \tag{11-6}$$

$$I^-(t = t_L) = \frac{q_2^-}{\pi} \tag{11-7}$$

where q_1^+ and q_2^- are the radiosities at the two walls. Note that the +/− notation here for radiosity is consistent with the convention of Eq. (11-4).

*In using the Planck function for the emission source function, it is assumed that local thermodynamic equilibrium LTE prevails. It is often the case, however [Goody and Yung, 1989], that for a system in radiative equilibrium the energy state populations may not be determined solely by collisional transitions but also by radiative transitions in which case LTE would break down and the Planck function would no longer be the appropriate source function for emission. A detailed analysis of this situation involves nonequilibrium concepts, which are beyond the present scope. To illustrate the transfer problem, however, LTE is assumed.

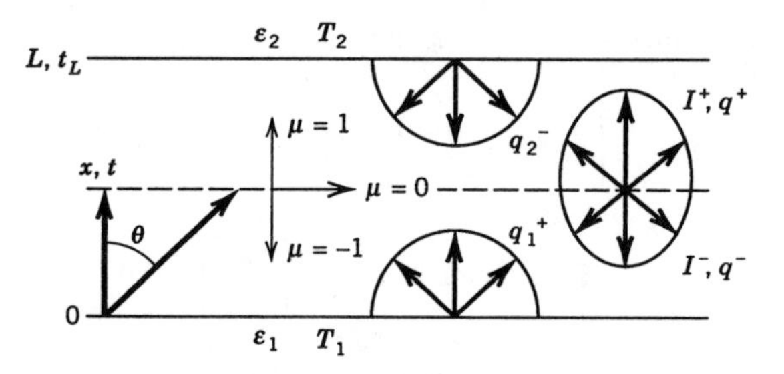

Figure 11-1 Slab geometry.

This convention agrees with the $+/-$ convention for radiosity and irradiation ($+ =$ radiosity; $- =$ irradiation) of Chapter 1, Eqs. (1-12) and (1-13), for the lower wall (1) but is opposite for the upper wall (2). The solution of (11-5), obtained using the integrating factor $\exp[t/\mu]$, is

$$I^+(t,\mu) = \frac{q_1^+}{\pi}\exp\left[-\frac{t}{\mu}\right] + \int_0^t I_b(t')\exp\left[\frac{t'-t}{\mu}\right]\frac{dt'}{\mu} \tag{11-8}$$

$$I^-(t,\mu) = \frac{q_2^-}{\pi}\exp\left[\frac{t_L-t}{\mu}\right] - \int_t^{t_L} I_b(t')\exp\left[\frac{t'-t}{\mu}\right]\frac{dt'}{\mu} \tag{11-9}$$

The first terms on the right-hand side of these two equations represent the contribution to the intensity at location t in direction μ from the wall. The q/π terms represent the (diffuse) intensity leaving the walls and the exponential terms are the transmissivities along the slant path. The second terms on the right-hand side represent the contribution to the intensity at location t in direction μ from emission along the path between location t and the wall. The hemispherical fluxes at location t are obtained by integrating the intensity field, using azimuthal symmetry ($d\Omega = -2\pi\, d\mu$).

$$q^+ = 2\pi\int_0^1 I^+\mu\, d\mu \tag{11-10}$$

$$q^- = 2\pi\int_0^{-1} I^-\mu\, d\mu \tag{11-11}$$

Substituting (11-8) and (11-9) into (11-10) and (11-11) gives

$$q^+ = 2\int_0^1\left[q_1^+\exp\left(\frac{-t}{\mu}\right) + \int_0^t e_b(t')\exp\left(\frac{t'-t}{\mu}\right)\frac{dt'}{\mu}\right]\mu\, d\mu \tag{11-12}$$

$$q^- = 2\int_0^{-1}\left[q_2^-\exp\left(\frac{t_L-t}{\mu}\right) - \int_t^{t_L} e_b(t')\exp\left(\frac{t'-t}{\mu}\right)\frac{dt'}{\mu}\right]\mu\, d\mu \tag{11-13}$$

The net radiative flux at location t then becomes

$$q = q^+ - q^- = 2\left\{ q_1^+ E_3(t) + \int_0^t e_b(t') E_2(t - t')\, dt' \right.$$
$$\left. - q_2^- E_3(t_L - t) - \int_t^{t_L} e_b(t') E_2(t' - t)\, dt' \right\} \quad (11\text{-}14)$$

In Eq. (11-14) the exponential integral functions E_n have been used. These functions are defined as

$$E_n(t) = \int_0^1 \mu^{n-2} \exp\left(\frac{-t}{\mu}\right) d\mu \quad (11\text{-}15)$$

They have the property that the derivative of the nth-order exponential integral for $n \geq 2$ gives the next lower order exponential integral with a sign change.

$$\frac{dE_n(t)}{dt} = -E_{n-1}(t) \quad (11\text{-}16)$$

A brief table of exponential integrals is provided in Appendix G.

At this point, if the temperature profile were known, it could be substituted into (11-14), and the net radiative flux at any location $q(t)$ could be computed. Since the temperature profile is not known, it is necessary to solve the energy equation coupled (through temperature) with the net radiative flux equation, Eq. (11-14). The general form of the thermal energy equation for a continuum is given in Eq. (11-17).

$$\rho \mathbf{V} \cdot \nabla h = \mathbf{V} \cdot \nabla P - \nabla \cdot [-k \nabla T + \mathbf{q}] + \Phi + q''' \quad (11\text{-}17)$$

The term on the left-hand side represents advection (or convection) of thermal energy (plus flow work). Conduction is given by the $k\nabla T$ term, $\mathbf{q}$ is the net radiative heat flux vector, Φ is the dissipation function, and q''' is the source/sink term. The pressure gradient term $\mathbf{V} \cdot \nabla P$ is negligible for most low speed flows. For a system in radiative equilibrium, conduction, convection, dissipation, and other sources and sinks are all negligible, leaving only

$$\nabla \cdot \mathbf{q} = 0 \quad (11\text{-}18)$$

In the current one-dimensional slab problem Eq. (11-18) becomes

$$\frac{dq}{dx} = 0 \quad (11\text{-}19)$$

Equation (11-19) indicates that the net radiative flux is constant through the medium.

The radiative analysis up to this point is valid both on a spectral basis for nongray media as well as on a total basis for gray media. The important distinction is that Eq. (11-19) calls for the total heat flux and in the case of a nongray medium Eq. (11-14) would have to be written on a spectral basis and integrated over all wavelengths before substituting for q in (11-19).

$$\frac{d}{dx}\int_0^\infty q_\lambda \, d\lambda = 0 \quad \text{(nongray)} \tag{11-20}$$

Thus, nongray behavior significantly complicates the solution of the problem. Often, in the case of nongray media in radiative equilibrium, the assumption of spectral radiative equilibrium is made, and Eq. (11-19) is applied on a spectral basis. This assumption underestimates the divergence of the spectral radiative heat flux vector at certain wavelengths and overestimates it at others, but generally yields reasonable estimates for the total radiative heat flux.

Returning to the gray case, Eq. (11-1) can be used to write (11-19) as

$$\frac{dq}{dt} = 0 \quad \text{(gray)} \tag{11-21}$$

Differentiating (11-14) and setting it equal to zero, according to (11-21) gives

$$0 = \frac{dq}{dt} = -2\bigg\{q_1^+ E_2(t) + \int_0^t e_b(t')E_1(t-t')\,dt'$$
$$-2e_b(t) + q_2^- E_2(t_L - t) + \int_t^{t_L} e_b(t')E_1(t'-t)\,dt'\bigg\} \tag{11-22}$$

In obtaining (11-22) Leibnitz' rule has been used as well as the result that $E_2(0) = 1$, leading to the term $2e_b(t)$. Next a nondimensional blackbody hemispherical flux e_b^* is defined using the radiosities at the walls.

$$e_b^*(t) = \frac{e_b(t) - q_2^-}{q_1^+ - q_2^-} \tag{11-23}$$

Substituting (11-23) into (11-22) gives

$$2e_b^*(t) = E_2(t) + \int_0^{t_L} e_b^*(t')E_1(|t-t'|)\,dt' \tag{11-24}$$

which is a linear, nonhomogeneous, Fredholm integral equation of the

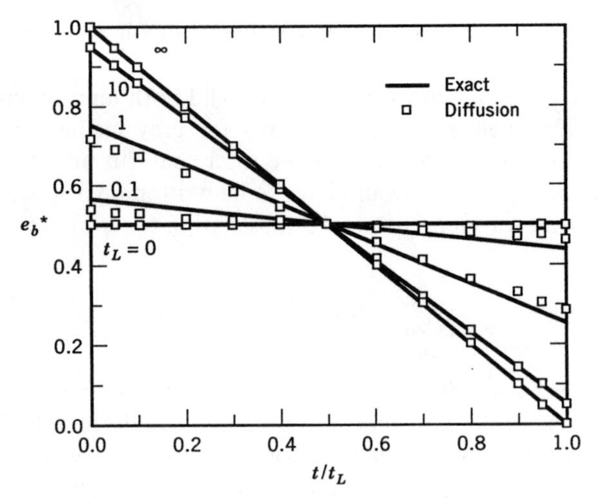

Figure 11-2 Nondimensional temperature profile for a gray slab in radiative equilibrium.

second kind. A nondimensional heat flux is also defined as

$$q^* = \frac{q}{q_1^+ - q_2^-} \tag{11-25}$$

which, combined with (11-14) gives

$$q^* = 1 - 2\int_0^{t_L} e_b^*(t') E_2(t')\, dt' \tag{11-26}$$

Equations (11-24) and (11-26) have been solved for e_b^* and q^* [Usiskin and Sparrow, 1960; Viskanta and Grosh, 1961; Heaslet and Warming, 1965], and these results are plotted as "exact" in Figs. 11-2 and 11-3.

In order to interpret the results of Fig. 11-2 assume for a moment that the walls are black $\varepsilon_1 = \varepsilon_2 = 1$. In this case the wall radiosities are identical to the blackbody hemispherical fluxes based on wall temperature $q_1^+ = e_{b_1}$ and $q_2^- = e_{b_2}$. Thus the nondimensional blackbody flux becomes

$$e_b^*(t) = \frac{e_b(t) - e_{b_2}}{e_{b_1} - e_{b_2}} \quad (\varepsilon_1 = \varepsilon_2 = 1)$$

By definition, the value of e_b^* at the temperature of wall 1 is $e_b^*(T_1) = 1$, and

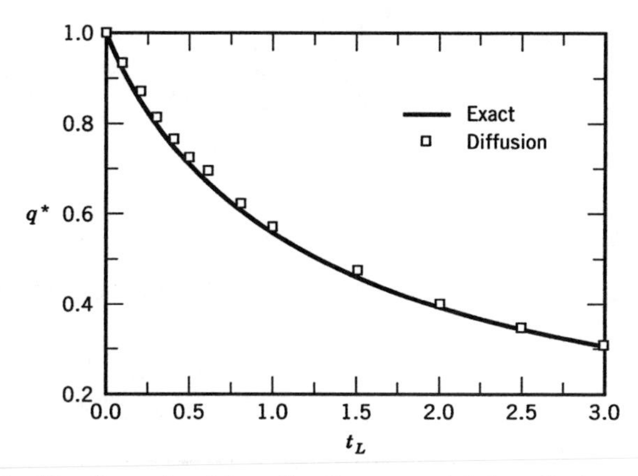

Figure 11-3 Nondimensional heat flux for a gray slab in radiative equilibrium.

the value of e_b^* at the temperature of wall 2 is $e_b^*(T_2) = 0$. Therefore the value of $e_b^* = 1$ represents the wall temperature condition for wall 1 and the value of $e_b^* = 0$ represents the wall temperature condition for wall 2 for black walls.

Information about the gas temperature is given in Fig. 11-2. From Fig. 11-2 it can be seen that the value of $e_b^*(t = 0)$ does not necessarily equal 1 and the value of $e_b^*(t = t_L)$ does not necessarily equal 0. This means that there is a temperature slip at the walls. That is, the gas temperature at the wall does not necessarily equal the wall temperature. Furthermore, the magnitude of the slip increases as t_L decreases. This temperature slip is compatible with the formulation of the problem in which conduction was neglected. In the limit as $t_L \to \infty$ the gas temperature at the wall approaches the wall temperature. This corresponds to the limiting behavior of a photon continuum where the no slip condition applies. The heat transfer process in this limit resembles a diffusion process. In the optically thin limit ($t_L \to 0$) the gas temperature approaches a constant that, for black walls, is equal to $[0.5(T_1^4 + T_2^4)]^{1/4}$. In this limit the methods of Chapter 10 would also be applicable to describe the heat transfer

The results for heat flux are given in Fig. 11-3. These results show that as the optical thickness t_L increases the heat transfer rate decreases. Thus the participating medium acts as a barrier to the heat transfer.

For nonblack walls the temperature profile and heat flux solutions are given by Eqs. (11-23) and (11-25), respectively. These relations involve the unknown radiosities q_1^+ and q_2^-. To eliminate the unknown radiosities in favor of the natural boundary conditions e_{b_1} and e_{b_2} the diffuse surface relations from Eq. (3-40) are introduced. In the present problem these relations are

$$q = \frac{\varepsilon_1}{1 - \varepsilon_1}(e_{b_1} - q_1^+) \tag{11-27}$$

$$-q = \frac{\varepsilon_2}{1 - \varepsilon_2}(e_{b_2} - q_2^-) \tag{11-28}$$

Introducing Eqs. (11-27) and (11-28) into (11-23) and (11-25) and eliminating the radiosities gives

$$\frac{e_b(t) - e_{b_2}}{e_{b_1} - e_{b_2}} = \frac{e_b^*(t) + \dfrac{1 - \varepsilon_2}{\varepsilon_2} q^*}{1 + q^*\left(\dfrac{1}{\varepsilon_1} + \dfrac{1}{\varepsilon_2} - 2\right)} \tag{11-29}$$

$$q = \frac{q^*(e_{b_1} - e_{b_2})}{1 + \left(\dfrac{1 - \varepsilon_1}{\varepsilon_1} + \dfrac{1 - \varepsilon_2}{\varepsilon_2}\right) q^*} \tag{11-30}$$

The temperature and heat flux for a case involving nonblack walls can be obtained from Eqs. (11-29) and (11-30) by substituting the values of e_b^* and q^*, which can be obtained from Figs. 11-2 and 11-3 or from some other approximate solution.

Slab in Radiative Equilibrium with Isotropic Scattering

Consider next the changes introduced into the previous analysis if isotropic scattering is allowed. The thermal emission source function $I_b(t)$ in Eq. (11-5) should be replaced by the source function $J(t)$, which includes both thermal emission and in-scattering from other directions

$$J(t) = (1 - \omega_0) I_b(t) + \frac{\omega_0}{2} \int_{-1}^{1} I(\mu')\, d\mu' \tag{11-31}$$

The resulting transfer equation is

$$\mu \frac{dI}{dt} = -I + J \tag{11-32}$$

Without solving the entire problem again it can be shown that this problem is equivalent to the previous one by integrating the transfer equation over all directions and using the energy equation. Integrating (11-32) with respect to μ over $-1 < \mu < 1$ and interchanging the order of integration and differentiation on the left-hand side gives

$$\frac{d}{dt}\int_{-1}^{1} I\mu \, d\mu = -\int_{-1}^{1} I \, d\mu + \int_{-1}^{1} J \, d\mu \qquad (11\text{-}33)$$

The integral on the left-hand side is simply the heat flux q divided by 2π. For radiative equilibrium $(dq/dt = 0)$ the left-hand side is zero. Substituting (11-31) into (11-33) gives

$$(1 - \omega_0)\left[I_b(t) - \frac{1}{2}\int_{-1}^{1} I(\mu')\, d\mu'\right] = 0 \qquad (11\text{-}34)$$

If the pure scattering case $(\omega_0 = 1)$ is excluded, (11-34) indicates that the average intensity at any location t is equal to the local blackbody intensity, independent of albedo.

$$\frac{1}{2}\int_{-1}^{1} I(\mu')\, d\mu' = I_b(t) \qquad (\omega_0 \neq 1) \qquad (11\text{-}35)$$

Substituting (11-35) back into (11-31) shows that the source function is also equivalent to the local blackbody intensity.

$$J(t) = I_b(t) \qquad (\omega_0 \neq 1) \qquad (11\text{-}36)$$

Since the source function can be replaced by the blackbody intensity, irrespective of the value of ω_0 (as long as $\omega_0 \neq 1$), the isotropic scattering case is identical to the nonscattering case and ω_0 is not a parameter. In other words *isotropic scattering by a gray medium in radiative equilibrium is equivalent to absorption and re-emission*. The medium can be given any value of ω_0 (except 1) including $\omega_0 = 0$ for simplicity.

This situation is analogous to the case of a diffuse, gray, radiatively adiabatic surface from Chapter 3. Diffuse reflection by a gray, radiatively adiabatic surface is equivalent to absorption and re-emission. Emissivity is not a parameter and the blackbody hemispherical flux, radiosity, and irradiation are all equivalent $(e_b = q^+ = q^-)$ as long as $\varepsilon \neq 1$. In the case that $\varepsilon = 1$, e_b is indeterminate.

Similarly, for the gray participating medium, the average intensity, source function, and blackbody intensity are all equivalent if $\omega_0 \neq 1$ [see Eqs. (11-35) and (11-36)]. If $\omega_0 = 1$, the medium temperature becomes indeterminant since radiative transfer is independent from the energy equation.

The focus of the discussion will now shift from problems restricted to radiative equilibrium (but unrestricted with respect to optical thickness) to problems that are unrestricted with respect to the energy equation (i.e., conduction, convection, and internal generation are allowed) but are limited to optically thick conditions.

ONE-DIMENSIONAL DIFFUSION APPROXIMATION WITH ISOTROPIC SCATTERING

One of the most important limiting cases of radiative transfer is that which occurs in an optically thick medium. Under optically thick conditions the net radiative flux at a particular point in space is no longer influenced by spatially remote conditions. The local flux is only a function of local conditions, and the transport rate is described by a diffusion model, similar to Fourier's law for conduction.

Consider again the one-dimensional transfer equation in an isotropic scattering medium:

$$\frac{\mu}{K_e}\frac{dI}{dx} = -I + J \tag{11-37}$$

$$J = (1 - \omega_0) I_b(x) + \frac{\omega_0}{2}\int_{-1}^{1} I(\mu')\, d\mu' \tag{11-38}$$

These relations hold on either a spectral basis if the medium is nongray or a total basis if the medium is gray.

Nondimensionalizing the intensity by some characteristic value I_0 and the spatial coordinate x by L

$$I^* = \frac{I}{I_0} \tag{11-39}$$

$$x^* = \frac{x}{L} \tag{11-40}$$

gives the following nondimensional form of the transfer equation:

$$\mu\left(\frac{1}{t_L}\right)\frac{dI^*}{dx^*} = -I^* + (1 - \omega_0) I_b^* + \frac{\omega_0}{2}\int_{-1}^{1} I^*(\mu')\, d\mu' \tag{11-41}$$

The optical depth based on L is

$$t_L = \int_0^L K_e\, dx \sim K_e L = \frac{L}{\text{MFP}} \tag{11-42}$$

In this context L is the length scale for intensity change such that the nondimensional derivative is of order one $dI^*/dx^* \sim O(1)$. In radiative equilibrium L corresponds to some physical dimension of the system such as wall or boundary spacing. When convective transport establishes a thermal boundary layer, L corresponds to the thickness of the thermal boundary layer.

When the length scale for intensity change is much greater than the mean free photon path ($L \gg$ MFP), the medium is optically thick.

$$t_L \gg 1 \quad \text{(optically thick medium)} \tag{11-43}$$

The appearance of a small parameter $(1/t_L)$ in the governing equation (11-41) suggests that a series solution in terms of the small parameter be tried.

$$I^* = I^{*(0)} + \left(\frac{1}{t_L}\right) I^{*(1)} + \left(\frac{1}{t_L^2}\right) I^{*(2)} + \cdots \tag{11-44}$$

Substituting (11-44) into (11-41) and collecting terms of like order gives

$$\left(\frac{1}{t_L}\right)^0 : I^{*(0)} = (1 - \omega_0) I_b^* + \frac{\omega_0}{2} \int_{-1}^{1} I^{*(0)} \, d\mu' \tag{11-45}$$

$$\left(\frac{1}{t_L}\right)^1 : \mu \frac{dI^{*(0)}}{dx^*} = -I^{*(1)} + \frac{\omega_0}{2} \int_{-1}^{1} I^{*(1)} \, d\mu' \tag{11-46}$$

Integrating (11-45) with respect to μ over $-1 < \mu < 1$ results in

$$(1 - \omega_0)\left(I_b^* - \frac{1}{2} \int_{-1}^{1} I^{*(0)} \, d\mu'\right) = 0 \tag{11-47}$$

which gives

$$I_b^* = \frac{1}{2} \int_{-1}^{1} I^{*(0)} \, d\mu', \qquad (\omega_0 \neq 1) \tag{11-48}$$

when the albedo is not equal to one. Substituting (11-48) into (11-45) gives the first-order solution

$$I^{*(0)} = I_b^*, \qquad (\omega_0 \neq 1) \tag{11-49}$$

The second-order solution $I^{*(1)}$ can be found by using (11-49) in (11-46) and integrating over $-1 < \mu < 1$. The result is

$$I^{*(1)} = -\mu \frac{dI_b^*}{dx^*} \tag{11-50}$$

The solution for I^* to order $(1/t_L)$ from (11-44), (11-49), and (11-50) is then

$$I^* = \underset{O(1)}{I_b^*} - \underset{O\left(\frac{1}{t_L}\right)}{\mu \frac{1}{t_L} \frac{dI_b^*}{dx^*}} + \cdots \tag{11-51}$$

In dimensional form Eq. (11-51) becomes

$$I = I_b - \frac{\mu}{K_e} \frac{dI_b}{dx} + \cdots \tag{11-52}$$

Since the first-order term is isotropic, (11-52) indicates that the intensity field under optically thick conditions is quasi-isotropic. If Eq. (11-52) is used in (11-38) to evaluate the source function, the result is

$$J = I_b \qquad (\omega_0 \neq 1) \tag{11-53}$$

and the albedo falls out of the problem. The implication of (11-53) is that *isotropic scattering under optically thick conditions is equivalent to absorption and re-emission*. The net flux can be evaluated using (11-10) and (11-11)

$$q = -\frac{4}{3K_e} \frac{de_b}{dx} \qquad (\omega_0 \neq 1, t_L \gg 1) \tag{11-54}$$

This is known as the *Rosseland diffusion approximation* [Rosseland, 1936]. Equation (11-54) indicates that for optically thick conditions the net radiative flux depends only on local conditions, similar to conduction.

In obtaining Eq. (11-54) no assumption of radiative equilibrium is necessary and no assumption of gray properties is made. Only isotropic scattering and $\omega_0 \neq 1$ have been assumed. Equation (11-54) is valid on a spectral basis (nongray) as well as a total basis (gray). A method for converting anisotropic scattering properties to equivalent isotropic scattering properties for optically thick conditions is discussed in Chapter 12 [see Eq. (12-95)].

Diffusion Slip Temperature Condition

The diffusion approximation is a simple yet effective representation of radiative transfer when the length scale for intensity change is much greater than the photon mean free path. However, near boundaries of the medium, the diffusion approximation can break down. This is due to the fact that the length scale for intensity change (which can vary through the medium) can become on the order of the photon mean free path near a boundary, particularly a cold, nonreflecting boundary. To compensate for the error incurred in applying the diffusion approximation near such a boundary, an artificial boundary condition, called a slip condition, is used that results in the correct heat flux [Deissler, 1964; Howell, 1967].

The slip condition can be obtained by considering the radiative balance at the lower wall of Fig. 11-1. The net radiative flux is given in terms of the blackbody hemispherical flux at the wall temperature e_{b_1} and the flux incident on the wall as

$$q_1 = \varepsilon_1 \left(e_{b_1} - q_1^- \right) \tag{11-55}$$

The irradiation q_1^- can be obtained from Eq. (11-13) by evaluating it at $t = 0$ and neglecting the first term on the right-hand side on the grounds that $t_L \gg 1$. [It should also be noted that (11-13) was originally derived for a nonscattering medium, but it was subsequently shown in Eq. (11-53) that $\pi J(t)$ could be replaced by $e_b(t)$ for isotropic scattering if $t_L \gg 1$ and $\omega_0 \neq 1$.] Equation (11-13) thus yields

$$q_1^- = 2 \int_0^\infty e_b(t') E_2(t')\, dt' \tag{11-56}$$

for the irradiation at the wall. Expanding $e_b(t)$ in (11-56) in a Taylor's series about zero gives

$$e_b(t) = e_b(0) + \left. \frac{de_b}{dt} \right|_0 (t - 0) + \frac{1}{2!} \left. \frac{d^2 e_b}{dt^2} \right|_0 (t - 0)^2 + \cdots \tag{11-57}$$

Using the following definite integrals

$$\int_0^\infty E_2(t')\, dt' = \frac{1}{2} \tag{11-58a}$$

$$\int_0^\infty t' E_2(t')\, dt' = \frac{1}{3} \tag{11-58b}$$

$$\int_0^\infty t'^2 E_2(t')\, dt' = \frac{1}{2} \tag{11-58c}$$

the resulting expression from (11-56) can be substituted into (11-55) to give

$$q_1 = \varepsilon_1 \left[e_{b_1} - e_b(0) - \frac{2}{3} \left. \frac{de_b}{dt} \right|_0 - \frac{1}{2} \left. \frac{d^2 e_b}{dt^2} \right|_0 + \cdots \right] \tag{11-59}$$

According to the diffusion approximation (11-54), the heat flux at the wall is also given by

$$q_1 = -\frac{4}{3}\frac{de_b}{dt}\bigg|_0 \tag{11-60}$$

Combining (11-59) and (11-60) gives the slip condition

$$e_{b_1} - e_b(0) = -\frac{4}{3}\left(\frac{1}{\varepsilon_1} - \frac{1}{2}\right)\frac{de_b}{dt}\bigg|_0 + \frac{1}{2}\frac{d^2e_b}{dt^2}\bigg|_0 + \cdots \tag{11-61}$$

An interesting observation about the magnitude of the slip can be seen by noting that on an order of magnitude basis (11-61) can be written as

$$\frac{e_{b_1} - e_b(0)}{e_b(0) - e_b(L)} \sim \left(\frac{1}{t_L}\right) + \left(\frac{1}{t_L}\right)^2 + \cdots \tag{11-62}$$

This result shows that to highest order the slip in blackbody hemispherical flux is of order $1/t_L$. Thus in the limit $t_L \to \infty$ the no slip (photon continuum) condition prevails.

The slip condition is often written in terms of a nondimensional slip coefficient ψ

$$\psi_1 = \frac{e_{b_1} - e_b(0)}{q_1} \tag{11-63a}$$

$$\psi_2 = \frac{e_b(t_L) - e_{b_2}}{q_2} \tag{11-63b}$$

where

$$q_{1,2} = -\frac{4}{3}\frac{de_b}{dt}\bigg|_{t=0,\,t_L} \tag{11-64}$$

In (11-63) and (11-64) the distinction has been made between e_b evaluated at the wall temperature $T_{1,2}$ and e_b evaluated at the temperature of the gas at the wall $T(0, t_L)$. This distinction is made necessary by the discontinuity in temperature introduced by the slip condition.

To highest order, the slip coefficient corresponding to Eq. (11-61) becomes

$$\psi_{1,2} = \frac{1}{\varepsilon_{1,2}} - \frac{1}{2} \tag{11-65}$$

The slip condition is thus introduced to compensate for the error in using the diffusion approximation near a boundary. It works by imposing an artificial temperature profile near the boundary such that the correct radiative heat flux is predicted. Another important feature of the slip condition is that it introduces the wall emissivity into the problem as a parameter (which otherwise would not appear). To conclude this section it should be noted that in the diffusion approximation, there has been no other approximation introduced other than isotropic scattering and optically thick conditions, namely, $t_L \gg 1$ where L is the characteristic length scale for intensity change. Radiative equilibrium, for example, has not been assumed.

Slab in Radiative Equilibrium using Diffusion Approximation

It is instructive now to compare the diffusion approximation with the previously considered exact solution for a gray, isotropically scattering slab in radiative equilibrium between diffuse, gray walls. The appropriate equations are the diffusion radiation approximation

$$q = -\frac{4}{3}\frac{de_b}{dt} \tag{11-66}$$

the energy equation for radiative equilibrium

$$\frac{dq}{dt} = 0 \qquad (q = \text{constant}) \tag{11-67}$$

and the slip conditions

$$e_{b_1} - e_b(0) = \left(\frac{1}{\varepsilon_1} - \frac{1}{2}\right)q \tag{11-68a}$$

$$e_b(t_L) - e_{b_2} = \left(\frac{1}{\varepsilon_2} - \frac{1}{2}\right)q \tag{11-68b}$$

Combining (11-66) through (11-68) yields a nondimensional temperature profile [see Eq. (11-23)] of

$$e_b^*(t, t_L) = \frac{1}{1 + 4/(3t_L)}\left[1 + \frac{2}{3t_L} - \frac{t}{t_L}\right] \tag{11-69}$$

and a nondimensional heat flux q^* [see Eq. (11-25)] of

$$q^* = \frac{1}{1 + 3t_L/4} \tag{11-70}$$

These solutions are plotted in Figs. 11-2 and 11-3 with the exact results and the comparison is remarkably good, even for optically thin conditions ($t_L \to 0$).

The fact that the diffusion approximation does so well for the optically thick case is expected, but the agreement at all values of t_L (even $t_L \to 0$) is surprising and in fact is fortuitous. The reason for the agreement is that in radiative equilibrium the second derivative term in (11-61) and all higher order derivatives, terms that would blow up as $t_L \to 0$, instead vanish identically due to the energy condition. Thus the agreement at small optical thicknesses is fortuitous. It should not be expected that the diffusion approximation with the slip boundary condition will work well under optically thin conditions in general. Use of the diffusion approximation should be restricted to the conditions for which it was derived, namely length scale for intensity change much larger than photon mean free path.

Example 11-1. Model of Earth's Atmosphere

A rudimentary model of the earth's atmosphere is a semigray, plane, parallel, nonscattering, medium in radiative equilibrium that is transparent to solar radiation ($\lambda < 5\ \mu\text{m}$) and gray for infrared terrestrial radiation ($\lambda > 5\ \mu\text{m}$). Based on such a model, (a) evaluate the temperature at the outer edge of the atmosphere $x = L \to \infty$ (skin temperature) and (b) obtain an expression for the temperature distribution through the atmosphere as a function of altitude x. Assume that the optical depth as a function of altitude can be expressed as

$$t = t_L[1 - \exp(-x/L_a)] \tag{11-71}$$

where L_a is the characteristic length describing the distribution of absorbing gases. For H_2O (the principal absorbing gas in earth's atmosphere) $L_a =$ 2 km [Goody and Yung, 1989]. Assume the earth's surface is black at terrestrial wavelengths and that the effective gray optical depth of the atmosphere is $t_L = 4$.

An analytic solution of this problem has already been obtained, using the diffusion approximation and is given by Eqs. (11-29), (11-30), (11-69), and (11-70). Combining Eqs. (11-29) and (11-69) and setting $\varepsilon_1 = \varepsilon_2 = 1$ and $e_{b_2} = 0$, the distribution of $e_b(t) = \sigma T(t)^4$ as a function of t can be written as

$$e_b(t) = e_{b_1}\frac{1}{1 + 4/(3t_L)}\left[1 + \frac{2}{3t_L} - \frac{t}{t_L}\right] \tag{11-72}$$

Similarly, combining Eqs. (11-30) and (11-70) and setting $\varepsilon_1 = \varepsilon_2 = 1$ and

$e_{b_2} = 0$, gives

$$e_{b_1} = q\left[1 + \frac{3t_L}{4}\right] \tag{11-73}$$

Combining (11-72) and (11-73) gives $e_b(t)$ in terms of the net infrared flux q and the optical depth of the atmosphere t_L.

$$e_b(t) = q\frac{1 + 3t_L/4}{1 + 4/(3t_L)}\left[1 + \frac{2}{3t_L} - \frac{t}{t_L}\right] \tag{11-74}$$

The discontinuity in e_b at the boundaries can be obtained by evaluating (11-74) at $t = 0$ and $t = t_L$. After some algebraic manipulation and making use of (11-73) the discontinuities in e_b at the two boundaries are found to be equal and given by

$$e_{b_1} - e_b(0) = e_b(t_L) - 0 = \frac{q}{2} \tag{11-75}$$

An alternative expression for the temperature of the medium at the lower boundary can be obtained by combining (11-75) with (11-73) to give

$$\sigma T(0)^4 = e_b(0) = \frac{q}{2}\left[1 + \frac{3t_L}{2}\right] \tag{11-76}$$

The infrared flux q can be estimated by considering an energy balance on the earth. The collimated solar flux is 1376 W/m^2. The average flux incident on the atmosphere, averaging out variations due to latitude and rotation of the earth, is 1376/4 = 344 W/m^2, where the factor of $\frac{1}{4}$ is the ratio of the projected area to the surface area of the earth. The effective reflectivity of the earth to solar radiation (the solar albedo*) is 0.3. Therefore the average solar flux absorbed by the earth is (0.7)(344) = 240 W/m^2. Since absorption of solar radiation by the atmosphere is neglected, the solar flux absorbed by the ground and re-emitted as infrared radiation is also 240 W/m^2. Thus the infrared flux q, which is transferred from the surface of the earth through the atmosphere and emitted into space, is constant with altitude and equal to 240 W/m^2. Substituting this value into (11-75) and solving for the skin temperature gives $T(t_L) = 214$ K. This value happens to be in close agreement with the average observed temperature in the lower stratosphere.

*This albedo includes effects of multiple scattering by clouds (67%), gas molecules (20%) as well as reflection by the ground (13%) [Gill, 1982]. Clouds have a pronounced influence on the earth's energy balance yet the formation of clouds is still very difficult to predict.

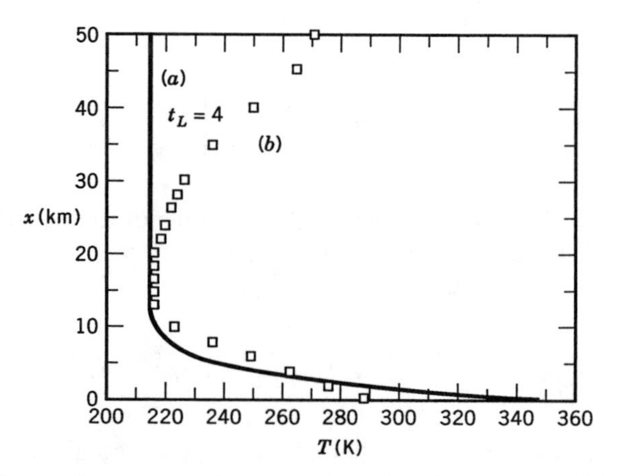

Figure 11-4 Atmospheric temperature as a function of altitude for (*a*) gray radiative equilibrium model and (*b*) U.S. Standard Atmosphere.

The effective gray optical depth of the atmosphere is assumed to be $t_L = 4$. (Due to nongray gas band behavior, the optical thickness of the atmosphere is actually very large at some wavelengths and very small at others.) Using this value in Eqs. (11-73) and (11-76) gives the temperature of the lower boundary (earth's surface) as $T_1 = 360$ K and the temperature of the medium at the lower boundary as $T(0) = 349\ K$. This temperature is somewhat higher than the average observed temperature on the earth's surface 290 K. The temperature profile as a function of altitude can be evaluated using Eq. (11-71) and either (11-72) or (11-74). The results are plotted in Fig. 11-4 along with a more realistic representation from the U.S. Standard Atmosphere [1976].

The most significant difference between the gray, radiative equilibrium model and the U.S. Standard Atmosphere at the lower altitudes of the troposphere ($x < 10$ km) is in the slope of the temperature profile, also called the lapse rate. The lapse rate determines whether natural convection will occur. If the radiative equilibrium lapse rate exceeds a critical value, called the adiabatic lapse rate* by a significant amount, a hydrodynamically

*The adiabatic lapse rate is determined by considering the temperature change experienced by a fluid particle of air as it moves up or down isentropically (reversibly and adiabatically) in the atmospheric hydrostatic pressure field. As the particle moves up, pressure decreases, the particle expands and its temperature decreases. The rate at which temperature decreases with increasing altitude in a dry atmosphere is the dry adiabatic lapse rate and has a value of 10 K/km. The corresponding rate for a moist atmosphere (moist adiabatic lapse rate) depends on relative humidity with typical values between 3 and 5 K/km. The observed average lapse rate in the troposphere is 6.5 K/km, slightly greater than the moist adiabatic lapse rate.

unstable situation exists and convection will occur. The radiative equilibrium model of this example results in a lapse rate at the ground of 37 K/km. The fact that this rate exceeds the observed lapse rate of 6.5 K/km in the troposphere indicates that convection effects, which are ignored in the radiative equilibrium model, play an important role in determining the temperature distribution.

There is also a significant difference between the gray, radiative equilibrium model and the U.S. Standard Atmosphere at the higher altitudes of the middle and upper atmosphere ($x > 10$ km). The gray radiative equilibrium model predicts a monotonically decreasing temperature with increasing altitude whereas the temperature of the atmosphere actually begins increasing again above 20 km. A mechanism for causing higher temperatures in the upper atmosphere, which has been neglected in the simple model, is the absorption of solar radiation. Ozone, in particular, plays an important role in absorbing solar radiation. The warm layer between 30 and 60 km is attributable to absorption of solar radiation by ozone. At altitudes above 75 km thermodynamic nonequilibrium effects become important and comparison with simple models is not very meaningful.

The assumptions that have been made in Example 11-1 are clearly rather severe ones that can easily be relaxed. To refine the analysis, the nongray gas band behavior should be represented. A starting point would be to include the most important single infrared band, the wide water vapor rotational band. The semigray analysis of Example 11-1 could be adapted for this purpose by shifting the cutoff wavelength from around 5 μm to somewhere between 15 and 20 μm, representative of the edge of the rotational water band. This analysis is left as an exercise (see Problem 11-2).

Rosseland Extinction Coefficient and Radiative Conductivity

Because of its simple form, the diffusion approximation is amenable to defining an effective radiative conductivity, analogous to the thermal conductivity defined by Fourier's law of conduction. To do so, the diffusion approximation

$$q_\lambda = -\frac{4}{3K_{e_\lambda}}\frac{de_{b_\lambda}}{dx} \tag{11-77}$$

must first be integrated over all wavelengths giving

$$q = -\frac{4}{3K_{e_R}}\frac{de_b}{dx} \tag{11-78}$$

where the *Rosseland mean extinction coefficient* K_{e_R} is defined* as

$$\frac{1}{K_{e_R}} = \frac{\int_0^\infty \frac{1}{K_{e_\lambda}} \frac{\partial e_{b_\lambda}}{\partial T} d\lambda}{\int_0^\infty \frac{\partial e_{b_\lambda}}{\partial T} d\lambda} = \int_0^1 \frac{1}{K_{e_\lambda}} df_i(\lambda T) \tag{11-79a}$$

$$f_i(\lambda T) = \frac{1}{4\sigma T^3} \int_0^\lambda \frac{\partial e_{b_\lambda}}{\partial T} d\lambda \tag{11-79b}$$

The Rosseland mean extinction coefficient is a spectrally averaged extinction coefficient that is valid for optically thick radiative transfer with nongray properties [Rosseland, 1936]. It is the counterpart in the optically thick limit of the Planck mean emission coefficient for optically thin radiative transfer [see Eq. (7-44b) and Problem 10-11]. The function f_i is the *internal fractional function* and is analogous to the fractional function. A table of the internal fractional function is provided in Appendix H.

By analogy with Fourier's law of heat conduction

$$q = -k_r(T) \frac{dT}{dx} \tag{11-80}$$

a radiative conductivity can be defined for optically thick media using the Rosseland mean extinction coefficient.

$$k_r(T) = \frac{16\sigma T^3}{3K_{e_R}} \tag{11-81}$$

Thus the optically thick diffusion of radiation is equivalent to conduction with a temperature-dependent conductivity.

Radiation Diffusion in Molecular Gases

When the participating medium is a molecular gas, the radiative properties are dominated by nongray banded behavior. Equation (11-77) holds spectrally with $K_{e_\lambda} = K_{a_\lambda}$ in the band regions if

$$\frac{\alpha \rho_a L}{\omega} \gg 1 \quad \text{(optically thick gas band)} \tag{11-82}$$

*Since isotropic scattering has been assumed here and throughout this chapter, the extinction coefficient in (11-79a) is an effective value for isotropic scattering, K_{eI_λ}. For anisotropic scattering, the effective isotropic extinction coefficient is defined in terms of the actual extinction coefficient K_{e_λ} and an asymmetry parameter p^* as $K_{eI_\lambda} = K_{e_\lambda}(1 - \omega_0 p^*)$; [see Eqs. (12-95) and (12-96)].

where L is the characteristic length for temperature change. Outside the bands, however, $K_{a_\lambda} \to 0$ and (11-77) is not valid. In such cases it is recommended that the radiative analysis be done spectrally on a band-by-band basis with a nonparticipating analysis between the bands. To do so requires some knowledge of the spectral absorption coefficient K_{a_λ}. This can be obtained from a narrow band model or estimated from wide band model results. For example, a spectral integration could be performed using the exponential wide band model for the spectrally smoothed absorption coefficient, Eq. (8-62). Such an analysis would require one to estimate the spectral region over which a given band was optically thick since the diffusion approximation cannot be used outside this region. A reasonable estimate of this region is given by the nonhomogeneous, effective bandwidth $A(L)$ based on the characteristic length scale for temperature change.

Another approach for estimating K_{a_λ} for optically thick radiation diffusion in molecular gases is to assume K_{a_λ} is constant at a value of $\rho_a \alpha / A(L)$ over an equivalent bandwidth of $A(L)$ centered at η_0 and zero outside this band. This approximation is shown in Eq. (11-83) and Fig. 11-5.

$$K_{a_\lambda} = \begin{cases} \rho_a \dfrac{\alpha}{A(L)}, & \eta_0 - \dfrac{A(L)}{2} < \eta < \eta_0 + \dfrac{A(L)}{2} \\ 0, & \eta < \eta_0 - \dfrac{A(L)}{2}, \eta > \eta_0 + \dfrac{A(L)}{2} \end{cases} \qquad (11\text{-}83)$$

The effective bandwidth $A(L)$ is based on L, which is the length scale for temperature (i.e., intensity) change in accordance with the assumptions of the diffusion approximation. For heat transfer across an adiabatic gas layer, L would be the thickness of the layer. While this approach is only a first approximation, it avoids complicated spectral integration and yields reasonable engineering estimates of heat transfer.

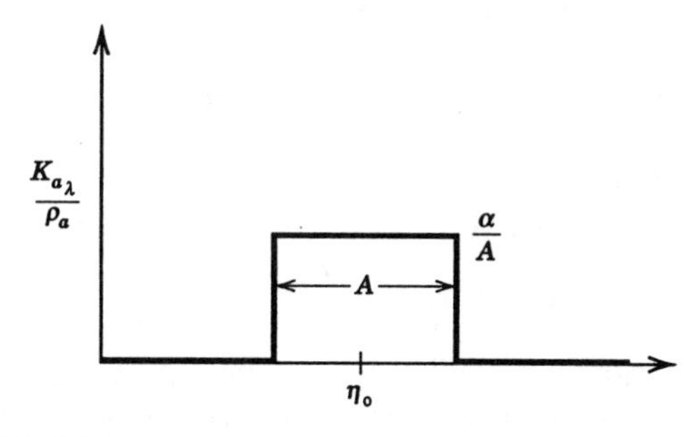

Figure 11-5 Model of gas properties for radiative diffusion in molecular gases.

Radiative Equilibrium in a Molecular Gas Slab

Consider a slab of molecular gas between diffuse, gray walls (see Fig. 11-1) that is in radiative equilibrium. Assume that the gas properties are dominated by a single band. In this case the radiative equilibrium condition also applies spectrally [recall discussion associated with Eqs. (11-19) and (11-20)]. Using the diffusion approximation [Eq. (11-70)] the spectral heat flux inside the band can be given by combining the spectral form of Eqs. (11-30) and (11-70).

$$q_\eta = \frac{e_{b_\eta}(T_1) - e_{b_\eta}(T_2)}{3t_L/4 + 1/E}, \qquad \eta_0 - \frac{A}{2} < \eta < \eta_0 + \frac{A}{2} \tag{11-84}$$

$$E = \frac{1}{1/\varepsilon_1 + 1/\varepsilon_2 - 1} \tag{11-85}$$

$$t_L = \int_0^L K_{a_\lambda}\, dx = \frac{1}{A}\int_0^L \rho_a \alpha\, dx = \frac{\alpha_e(0, L)X(0, L)}{A}, \qquad A = A(0, L) \tag{11-86}$$

Here the wide band scaling Eq. (8-102) has been utilized to express the optical thickness of the nonhomogeneous slab t_L. The spectral heat flux outside the band comes from a nonparticipating analysis (see Chapter 3).

$$q_\eta = E\left[e_{b_\eta}(T_1) - e_{b_\eta}(T_2)\right]; \qquad \eta < \eta_0 - \frac{A}{2}, \quad \eta > \eta_0 + \frac{A}{2} \tag{11-87}$$

Using the band approximation to integrate over all wavenumbers gives the total heat flux as

$$q = E\left\{\sigma\left(T_1^4 - T_2^4\right) - A\left[e_{b_\eta}(T_1) - e_{b_\eta}(T_2)\right]\left[1 - \frac{1}{1 + 3t_L E/4}\right]\right\} \tag{11-88}$$

It is possible to account for multiple bands by summing over the band terms giving

$$q = E\left\{\sigma\left(T_1^4 - T_2^4\right) - \sum_{k \text{ bands}} A_k\left[e_{b_{\eta k}}(T_1) - e_{b_{\eta k}}(T_2)\right]\left[1 - \frac{1}{1 + 3t_{L_k} E/4}\right]\right\} \tag{11-89}$$

This assumes that radiative equilibrium applies spectrally to each band. As noted earlier this assumption is necessary as a matter of expediency in obtaining a solution.

Example 11-2

Consider a molecular gas in radiative equilibrium between two black parallel plates that are at $T_1 = 2000$ K and $T_2 = 1000$ K. The slab thickness is $L = 10$ cm. The gas has only one vibration-rotation band centered at 3 μm, the lines are overlapping ($\beta > 1$), and the band is a fundamental band (i.e., $\alpha = \text{const} = \alpha_0$) with the following properties:

$$\alpha_0 = 22.4 \text{ cm}^{-1}/(\text{g m}^{-2})$$

$$\omega_0 = 60 \text{ cm}^{-1} \ (T_0 = 100 \text{ K})$$

$$P_a = 7.87 \text{ atm}$$

$$\text{Molecular weight} = 18$$

Calculate the total net radiant heat flux between the two plates. For estimating the gas properties use a two-step approximation to represent the non-isothermal temperature profile.

The appropriate equation to determine the total heat flux is Eq. (11-88) with $E = 1$. To evaluate Eq. (11-88) it is necessary to evaluate the effective bandwidth A and the average optical thickness t_L. Since the slab is in radiative equilibrium, it will be nonisothermal (unless it is optically thin). This means that the band properties should be evaluated using the wide band scaling rules Eqs. (8-102) to (8-106). Since the band is a fundamental, the band intensity is easily evaluated since it is constant throughout the slab $\alpha_e = \alpha_0$. However, further evaluation of gas properties requires an estimate of the gas temperature profile. Since the temperature profile is unknown it must be guessed and verified later. Anticipating that the slab is optically thick in the gas band, the following linear profile in $e^*_{b_\eta}(t)$ is assumed from Eq. (11-69) (see also Fig. 11-2).

$$e^*_{b_\eta}(t) = 1 - t/t_L$$

Imposing a two-step approximation on this linear function gives

$$e^*_{b_\eta}(t) = \begin{cases} 0.75, & 0 < t/t_L < 0.5 \\ 0.25, & 0.5 < t/t_L < 1 \end{cases}$$

as shown in Fig. 11-6.

Solving the spectral form of Eq. (11-29) with $\varepsilon_1 = \varepsilon_2 = 1$, $e_{b_\eta}(T_1) = 139$ W/(m^2 cm^{-1}) and $e_{b_\eta}(T_2) = 12$ W/(m^2 cm^{-1}) gives

$$e_{b_\eta}(t) = \begin{cases} 107 \text{ W}/(\text{m}^2 \text{ cm}^{-1}) & 0 < t/t_L < 0.5 \\ 44 \text{ W}/(\text{m}^2 \text{ cm}^{-1}) & 0.5 < t/t_L < 1 \end{cases}$$

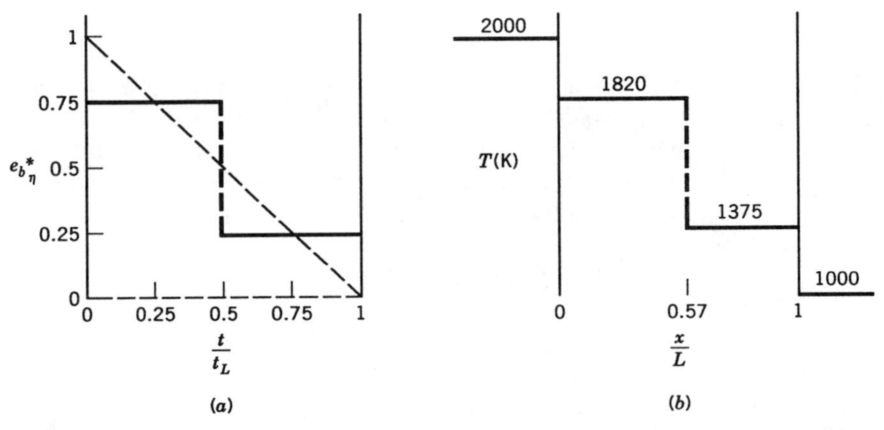

Figure 11-6 Two-step approximation to profile of (*a*) $e^*_{b_\eta}$ versus t/t_L and (*b*) T versus x/L for molecular gas slab in radiative equilibrium.

which can be inverted using the Planck function to give

$$T(x) = \begin{cases} 1820 \text{ K} & 0 < x/L < 0.57 \\ 1375 \text{ K} & 0.57 < x/L < 1 \end{cases}$$

The value of $x/L = 0.57$ (which corresponds to $t/t_L = 0.5$) was determined from

$$\frac{t}{t_L} = \frac{\int_0^x \rho_a \, dx'}{\int_0^L \rho_a \, dx'} = \frac{\int_0^x (1/T) \, dx'}{\int_0^L (1/T) \, dx'} = \frac{x/1820}{x/1820 + (L-x)/1375} = 0.5$$

From Eq. (8-71) the homogeneous bandwidth parameters are $\omega(0, x) = 256 \text{ cm}^{-1}$ and $\omega(x, L) = 222 \text{ cm}^{-1}$. Then the scaled bandwidth parameter can be determined from Eq. (8-103) as

$$\omega_e(0, L) = \frac{\int_0^L \rho_a \omega \, dx}{\int_0^L \rho_a \, dx} = \frac{\int_0^x \frac{\omega}{T} \, dx}{\int_0^L \frac{1}{T} \, dx}$$

$$= \frac{\frac{(256)(0.57)}{1820} + \frac{(222)(0.43)}{1357}}{0.57/1820 + 0.43/1375} = 239 \text{ cm}^{-1}$$

The mass path length is

$$X(0, L) = \int_0^x \rho_a \, dx' + \int_x^L \rho_a \, dx' = \frac{P_a M}{R}\left[\frac{x}{T(0, x)} + \frac{L - x}{T(x, L)}\right]$$

$$= \frac{(7.87 \text{ atm})(18 \text{ g/mole})(1.01 \times 10^5 \text{ J/m}^3 \text{ atm})}{8.314 \text{ J/mole K}}$$

$$\times \left(\frac{0.057 \text{ m}}{1820 \text{ K}} + \frac{0.043 \text{ m}}{1375 \text{ K}}\right)$$

$$= 53.9 + 53.9 = 108 \text{ g/m}^2$$

and the band center optical depth parameter is

$$u_L = \frac{\alpha_e X}{\omega_e} = \frac{(22.4 \text{ cm}^{-1}/\text{g m}^{-2})(108 \text{ g/m}^2)}{239 \text{ cm}^{-1}} = 10$$

The effective bandwidth is thus

$$A = \omega_e(\ln u_L + 1) = (239)(3.31) = 792 \text{ cm}^{-1}$$

and the average optical depth over the band is

$$t_L = \frac{\alpha_e X}{A} = \frac{u_L \omega_e}{A} = \frac{u_L}{\ln u_L + 1} = 3.1$$

At this point the initial assumption of $t_L \gg 1$, which has made to estimate the gas properties, can be checked. From Eq. (11-69) and Fig. 11-2 it can be seen that the temperature of $T(0, x) = 1820$ K (based on $t_L \gg 1$) somewhat overestimates the value corresponding to $t_L = 3.1$ which is 1765 K. On the other hand $T(x, L) = 1375$ K (for $t_L \gg 1$) somewhat underestimates the value corresponding to $t_L = 3.1$, which is 1515 K. Since these errors tend to offset each other, another iteration is not warranted. Thus the total heat flux can be calculated from Eq. (11-88) as

$$q = 5.67(2^4 - 1^4) - (792)(0.0139 - 0.00115)\left[1 - \frac{1}{1 + 0.75(3.1)}\right]$$

$$= 85.0 - 10(1.0 - 0.3)$$

$$= 85 - 10 + 3 = 78.0 \frac{\text{W}}{\text{cm}^2}$$

The flux of 85 W/cm^2 is the heat flux over the entire spectrum assuming no participating medium. The flux of 10 W/cm^2 is the flux over the spectral interval covered by the band. If the band were totally opaque, then 10 W/cm^2 would be the flux blocked by the gas; 3 W/cm^2 is the flux transferred by the gas over the band region, which is optically thick but not opaque.

To complete this example, an estimate of the convective heat flux should be made to verify the assumption of radiative equilibrium. It will be assumed that the layer is oriented vertically in a $1g$ gravity field. Since the "hypothetical" gas band in this example is a thinly veiled reference to the strong, asymmetric stretching fundamental band of water at 2.7 μm, the properties of steam at 1600 K and 7.87 atm will be used to estimate the convective heat flux.

$$\alpha = \nu = 50 \times 10^{-6} \text{ m}^2/\text{s} \qquad k = 0.13 \text{ W/m K}$$

The Rayleigh number is

$$\text{Ra}_L = \frac{g\beta\,\Delta T L^3}{\alpha\nu} = \frac{(9.8 \text{ m/s}^2)(1/1600 \text{ K})(1000 \text{ K})(0.1 \text{ m})^3}{(50 \times 10^{-6} \text{ m}^2/\text{s})(50 \times 10^{-6} \text{ m}^2/\text{s})} = 2.4 \times 10^6$$

The Nusselt number is obtained from the correlation [Incropera and DeWitt, 1985]

$$\overline{\text{Nu}_L} = \frac{\bar{h}L}{k} = 0.046 \text{ Ra}_L^{1/3} = 0.046(2.4 \times 10^6)^{1/3} = 6.2$$

The heat transfer coefficient is thus

$$\bar{h} = \frac{(6.2)(0.13)}{0.1} = 8 \frac{\text{W}}{\text{m}^2 \text{ K}}$$

and the convective heat flux is

$$q_c = \bar{h}\Delta T = \left(8 \frac{\text{W}}{\text{m}^2 \text{ K}}\right)(1000 \text{ K}) = 8000 \frac{\text{W}}{\text{m}^2} = 0.8 \frac{\text{W}}{\text{cm}^2} < 3 \frac{\text{W}}{\text{cm}^2}$$

Since the convective flux is a factor for 4 smaller than the radiative flux transferred by the gas at the band (3 W/cm^2), the assumption that radiative transfer establishes the temperature profile (radiative equilibrium) is reasonable for this problem.

SUMMARY

Two classic problems illustrate important features of radiative transfer in nonisothermal participating media. One problem is radiative transfer in an isotropically scattering slab in radiative equilibrium with arbitrary optical thickness. The other problem is radiative transfer in an optically thick, isotropically scattering slab not necessarily in radiative equilibrium. In both of these cases, the transfer equation can be solved without including the in-scattering integral term. The albedo falls out of the problem and does not influence the heat transfer rate. Hence, isotropic scattering is equivalent to absorption and re-emission under both optically thick conditions and in radiative equilibrium. The main radiative property influencing the transfer rate is the extinction coefficient. Although isotropic scattering is assumed, these two problems can be applied to a wider range of problems involving anisotropic scattering through the use of equivalent isotropic scattering properties (see Chapter 12). These two problems also illustrate the simplest of a class of problems involving coupled radiative transfer and energy equations. More complex problems involving combined modes of conduction, convection, and radiation are addressed in Chapter 13.

REFERENCES

Deissler, R. G. (1964), "Diffusion Approximation for Thermal Radiation in Gases with Jump Boundary Condition," *J. Heat Transfer*, Vol. 86, No. 2, pp. 240–246.

Gill, A. E. (1982), *Atmosphere-Ocean Dynamics*, International Geophysics Series, Vol. 30, Academic Press, London.

Goody, R. M. and Yung, Y. L. (1989), *Atmospheric Radiation Theoretical Basis*, 2nd ed., Oxford University Press. New York.

Heaslet, Max A. and Warming, R. F. (1965), "Radiative Transport and Wall Temperature Slip in an Absorbing Planar Medium," *Int. J. Heat Mass Transfer*, Vol. 8, pp. 979–994.

Howell, J. R. (1967), "On the Radiation Slip between Absorbing-Emitting Regions with Heat Sources," *Int. J. Heat Mass Transfer*, Vol. 10, No. 3, pp. 401–402.

Incropera, F. P. and DeWitt, D. P. (1985), *Fundamentals of Heat and Mass Transfer*, 2nd ed., Wiley, New York.

Rosseland, S. (1936), *Theoretical Astrophysics: Atomic Theory and the Analysis of Stellar Atmospheres and Envelopes*, Clarendon Press, Oxford.

U.S. Standard Atmosphere, (1976), Publication NOAA-S/T76-1562, U.S. Government Printing Office, Washington D.C.

Usiskin, C. M. and Sparrow, E. M. (1960), "Thermal Radiation between Parallel Plates Separated by an Absorbing-Emitting Nonisothermal Gas," *Int. J. Heat Mass Transfer*, Vol. 1, pp. 28–36.

Viskanta, R. and Grosh, R. J. (1961), "Heat Transfer in a Thermal Radiation Absorbing and Scattering Medium," *Proceedings of International Heat Transfer Conference*, Boulder, Colorado, p. 820.

REFERENCES FOR FURTHER READING

Cess, R. D. and A. E. Sotak, "Radiation Heat Transfer in an Absorbing Medium Bounded by a Specular Reflector," *ZAMP*, Vol. 15, 1964, pp. 642–647.

Probstein, R. F., "Radiation Slip," *AIAA J.*, Vol. 1, 1963, pp. 1202–1204.

Yuen, W. W. and L. W. Wong, "Analysis of Radiative Equilibrium in a Rectangular Enclosure with Gray Medium," *J. Heat Transfer*, Vol. 106, 1984, pp. 433–440.

PROBLEMS

1. Derive the expressions for nondimensional heat transfer q^*, Eq. (11-70), and temperature profile e_b^*, Eq. (11-69), for a gray slab in radiative equilibrium according to the diffusion approximation.

2. Develop a simple model of radiative transfer in the earth's atmosphere. Include the following features and assumptions. The average incident solar flux, including rotation and curvature effects, is 344 W/m^2 and is concentrated at short wavelengths ($< 5\ \mu m$). The earth albedo for solar radiation is 0.3. Neglect absorption of solar radiation by the atmosphere. The atmosphere is a planar slab in monochromatic radiative equilibrium and the optical properties can be modeled by considering only the H_2O rotation band. The partial density distribution of water vapor in the earth's atmosphere can be approximated as an exponential function of altitude (x)

$$\rho_a(x) = \rho_{a0}\exp(-x/L_a)$$

where $L_a = 2$ km and $\rho_{a0} = 2\ g/m^3$. Assume that the temperature variation $T(x)$ is small enough that the integrated band intensity α, bandwidth parameter ω, and line width parameter γ can be taken as constant at $T_c = 250$ K. Also assume that the average pressure is 0.9 atm for calculating the equivalent broadening pressure P_e. Assume the surface of the earth is black and adiabatic in the infrared region. Other than the solar flux, outer space can be taken as a blackbody at 0 K.

(a) Calculate the average temperature of the earth's surface.

(b) Compare the average temperature obtained for this model with that obtained for no atmosphere. Comment on the effect of increasing A.

(c) Calculate the temperature of the outer edge of the atmosphere (skin temperature) according to the given assumptions.

3. Repeat Problem 2 assuming the atmosphere is isothermal instead of in monochromatic radiative equilibrium.

(a) Calculate the average temperature of the earth's surface.

(b) Calculate the average temperature of the atmosphere.

4. Utilizing the assumptions of Example 11-1 (gray atmosphere in radiative equilibrium with no solar absorption), obtain an expression for the lapse rate at the earth's surface that shows that the maximum tendency for hydrodynamic instability occurs in the lowest layers of the atmosphere. Evaluate the lapse rate for $t_L = 4$, $q = 240\ \text{W/m}^2$, $L_a = 2$ km at $x = 0$.

5. Repeat Example 11-2 for $L = 1$ cm. Compare with the result of Problem 10-9. Note: a single-step approximation for the temperature profile is sufficient to estimate the gas properties.

6. A slab of a molecular gas (thickness L) is bounded on both sides by a cold, nonparticipating medium. The gas has one primary absorption band at wavenumber η. The blackbody intensity at wavenumbers near η varies linearly with optical depth through the slab

$$I_{b_\eta}(t) = I_{b_\eta}(T_0) + \left[I_{b_\eta}(T_L) - I_{b_\eta}(T_0)\right]\frac{t_\eta}{t_{L_\eta}}$$

(a) Calculate the nondimensional total intensity emitted normal to the slab at $x = L$, $I(L)/\omega_e I_{b_\eta}(T_L)$ as a function of $u_L = \alpha_e X/\omega_e$ and $I_r = I_{b_\eta}(T_0)/I_{b_\eta}(T_L)$ for u_L between 1 and 1000 and $I_r = 1, 5, 10$, and 20 by three methods and compare the results:

(i) Diffusion approximation Eq. (11-52) with Eq. (11-83)

(ii) Complete transfer equation solution (see Problem 7-2)

(iii) Diffusion approximation assuming the spectral absorption coefficient K_{a_λ} is constant at a value of $\rho_a \alpha/\omega_e$ over an equivalent bandwidth of ω_e centered at η and zero outside the band.

Comment on the agreement between the various methods. In (ii) assume the absorption coefficient is given by the rigid rotator, harmonic oscillator model and assume that ω is constant across the slab and equal to ω_e. Utilize the band approximation in integrating the spectral intensity. Assume overlapping lines.

(b) Discuss why the intensity formulation of the diffusion approximation (11-52) should be valid to predict the emitted intensity without a slip condition while the net flux formulation (11-54) would not be valid near the boundary in this situation.

7. Consider a plane parallel slab of thickness L consisting of a gray, absorbing, emitting, isotropic scattering medium that is optically thick ($t_L > 1$). The extinction coefficient K_e is uniform over the slab. Energy is being generated uniformly inside the medium at a local volumetric rate of q'''. Conduction and convection are negligible. The walls are diffuse and gray with emissivity ε_w and temperature T_w. Apply the diffusion approximation to obtain an equation for the temperature distribution in the medium as a function of t_L and ε_w. What is the effect of these parameters on the maximum temperature in the medium? What is the heat flux at each wall?

8. Verify that isotropic scattering is equivalent to absorption and re-emission in an optically thick medium by showing that the source function for a one-dimensional, optically thick, isotropically scattering medium is the Planck function, independent of albedo [i.e., show that Eq. (11-53) follows from substituting (11-52) into (11-38)].

9. Obtain an expression for the Rosseland and Planck mean absorption-emission coefficients for an elemental volume of soot particles at temperature T. Assume n and k are constant. Show that both coefficients are proportional to soot temperature.

CHAPTER 12

RADIATIVE TRANSFER WITH ANISOTROPIC, MULTIPLE SCATTERING

Radiative transfer in particle dispersions often involves anisotropic, multiple scattering. *Multiple* scattering means that photons that are emitted by the particle medium or are incident from the boundaries of the medium are scattered more than once before being reabsorbed elsewhere in the medium or at a boundary. A complete description of multiple scattering requires that the transfer equation be solved along all optical paths through the scattering medium simultaneously to account for in-scattering from other directions into a particular line of sight. This simultaneous solution of all lines of sight with multiple scattering represents a significant increase in the complexity of the analysis over that for single-scattering or nonscattering environments.

Anisotropic scattering means that scattered radiant energy is not distributed uniformly in all directions. Single scattering of radiation by particle dispersions is always anisotropic, to one degree or another. This anisotropy of scattered radiation also introduces a significant increase in complexity into the analysis of radiative transfer.

This chapter introduces two techniques for solving radiative transfer problems with anisotropic, multiple scattering, the *flux method* and the *discrete ordinate method*. The slab geometry is used to illustrate the anisotropic, multiple scattering formulation. Three specific example problems are considered: (1) a gray slab in radiative equilibrium between diffuse gray walls, (2) a semi-infinite gray slab with prescribed temperature (nonradiative equilibrium), and (3) and a cold slab with nonreflecting boundaries subject to collimated incident radiation.

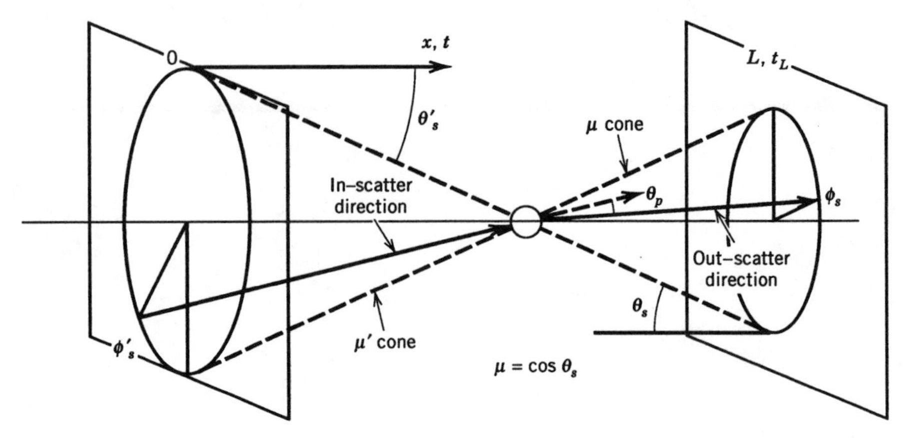

Figure 12-1 Slab and single-particle scattering angles.

MULTIPLE SCATTERING IN A PLANAR SLAB

Consider a planar slab composed of anisotropic scattering particles as shown in Fig. 12-1. The transfer equation along an arbitrary line of sight defined by the slab polar and azimuthal angles θ_s and ϕ_s is

$$\mu \frac{dI(t, \mu, \phi_s)}{dt} = -I(t, \mu, \phi_s) + (1 - \omega_0) I_b(t) + \frac{\omega_0}{4\pi} \int_0^{2\pi} \int_{-1}^{1} I(t, \mu', \phi'_s) p(\mu, \phi_s; \mu', \phi'_s) \, d\mu' \, d\phi'_s \quad (12\text{-}1)$$

where $\mu = \cos\theta_s$ is the cosine of the polar angle in the slab geometry. The in-scattering angles are indicated by a "prime" superscript (i.e., $\mu' = \cos\theta'_s, \phi'_s$) and the out-scattering angles have no prime (θ_s, ϕ_s). Thus the notation $p(\mu, \phi_s; \mu', \phi'_s)$ indicates the phase function for scattering from the μ', ϕ'_s direction into the μ, ϕ_s direction. The single-scattering phase function is typically known as a function of the particle single-scattering polar angle θ_p, which is the angle between the in-scattering direction and the out-scattering direction* (see Fig. 12-1). A relation that gives the particle single-scattering polar angle θ_p as a function of the slab angles θ'_s, ϕ'_s, θ_s, and ϕ_s can be obtained from analytic geometry [Ozisik, 1985].

$$\cos\theta_p = \mu\mu' + \sqrt{1 - \mu^2}\sqrt{1 - \mu'^2}\cos(\phi_s - \phi'_s) \quad (12\text{-}2)$$

Given the angles θ'_s, ϕ'_s, θ_s, and ϕ_s, the single-scattering particle polar angle θ_p can be determined using Eq. (12-2). Then the phase function can be

*This holds for spherical and randomly oriented nonspherical particles. For aligned nonspherical particles the single-scattering phase function will also be a function of the particle single-scattering azimuthal angle ϕ_p.

determined for use in Eq. (12-1), assuming $p(\theta_p)$ is known (e.g., from Mie theory).

Multiple Scattering in a Planar Slab with Azimuthal Symmetry

If the particles are randomly oriented and the boundary conditions (e.g., incident radiation) are azimuthally symmetric, then the intensity field in the slab will be azimuthally symmetric and intensity will be a function of only optical depth t and slab polar angle μ, $I = I(t, \mu)$. The phase function in the in-scattering term can be averaged over ϕ_s' and the transfer equation (12-1) becomes

$$\mu \frac{dI(t,\mu)}{dt} = -I(t,\mu) + (1-\omega_0)I_b(t) + \frac{\omega_0}{2}\int_{-1}^{1} I(t,\mu')p(\mu,\mu')\,d\mu' \tag{12-3}$$

where the phase function

$$p(\mu,\mu') = \frac{1}{2\pi}\int_0^{2\pi} p(\mu,\phi_s,\mu',\phi_s')\,d\phi_s' \tag{12-4}$$

is the phase function for scattering from the μ' direction (or more precisely, cone of directions) into the μ direction (or cone of directions). By symmetry Eq. (12-4) reduces to

$$p(\mu,\mu') = \frac{1}{\pi}\int_0^{\pi} p(\mu,\phi_s = 0,\mu',\phi_s')\,d\phi_s' \tag{12-5}$$

Equation (12-5) can be expressed as an integration with respect to θ_p by implicitly differentiating (12-2) holding μ and μ' constant,

$$d\phi_s' = \frac{\sin\theta_p\,d\theta_p}{\left[(1-\mu^2)(1-\mu'^2) - (\cos\theta_p - \mu\mu')^2\right]^{1/2}} \tag{12-6}$$

to give

$$p(\mu,\mu') = \frac{1}{\pi}\int_{\mu_{p_\pi}}^{\mu_{p_0}} \frac{p(\mu_p)\,d\mu_p}{\left[(1-\mu^2)(1-\mu'^2) - (\mu_p - \mu\mu')^2\right]^{1/2}} \tag{12-7}$$

where $\mu_p = \cos\theta_p$ and

$$\mu_{p_0} = \mu\mu' + \sqrt{1-\mu^2}\sqrt{1-\mu'^2} \tag{12-8a}$$

$$\mu_{p_\pi} = \mu\mu' - \sqrt{1-\mu^2}\sqrt{1-\mu'^2} \tag{12-8b}$$

Equation (12-7) allows the slab phase function $p(\mu, \mu')$ to be calculated from the single-scattering phase function $p(\mu_p)$. For spheres, the phase function $p(\mu_p)$ comes from Mie theory. Note the symmetry that exists in the slab phase function between the in-scattering and out-scattering directions, $p(\mu, \mu') = p(\mu', \mu)$. The phase function for scattering from the μ' direction into the μ direction is the same as the scattering from the μ direction into the μ' direction.

It is also possible to apply Eq. (12-3) to situations in which azimuthal symmetry is not obtained. An example of such a case is a slab (e.g., a plane parallel atmosphere) with collimated radiation incident at an off-normal direction. In such cases $I(t, \mu)$ is viewed as the azimuthally *averaged* intensity rather than the azimuthally independent intensity. While the solution of Eq. (12-3) does not yield azimuthally resolved intensity information, it is considerably less complicated than that of the azimuthally dependent Eq. (12-1), and is usually adequate for heat transfer purposes.

With the inclusion of anisotropic scattering it becomes necessary to represent the angular distribution of scattered energy. This is accomplished by the use of the phase function. What makes this representation seem complicated is the fact that in transfer problems with multiple scattering there are really two phase functions to speak of. There is the phase function for scattering of radiation in the single-particle geometry $p(\mu_p)$, which was discussed in Chapter 9. And there is the phase function for scattering of radiation in the slab geometry $p(\mu, \mu')$. The relation between these two phase functions is given in Eq. (12-7). Thus it is necessary in general to transform the single-particle phase function to the slab phase function using Eq. (12-7). Another method for transforming the single-particle phase function to the slab phase function that does not require explicit integration of Eq. (12-7) is the Legendre polynomial method.

Legendre Polynomial Representation of Phase Function

The Legendre polynomial method consists of representing the single-particle phase function as a series of Legendre polynomials

$$p(\mu_p) = \sum_{i=0}^{N_L} a_i P_i(\mu_p) \tag{12-9}$$

$$P_0(\mu_p) = 1 \tag{12-9a}$$

$$P_1(\mu_p) = \mu_p \tag{12-9b}$$

$$P_2(\mu_p) = \tfrac{1}{2}\left(3\mu_p^2 - 1\right) \tag{12-9c}$$

To satisfy the phase function normalization condition, the first coefficient a_0 is set equal to 1. The higher the value of N_L that is taken the higher the

TABLE 12-1 Legendre Polynomial Representation of a Few Special Single-Scattering Phase Functions

Type of Scattering	N_L	Phase Function $p(\mu_p)$	Conditions
Isotropic	0	1	
Linear anisotropic	1	$1 + a_1\mu_p$	$-1 \leq a_1 \leq 1$
Rayleigh	2	$1 + \frac{1}{4}(3\mu_p^2 - 1)$	$a_1 = 0, a_2 = \frac{1}{2}$

degree is of anisotropy that can be described. For example $N_L = 0$ corresponds to isotropic scattering (Table 12-1). The value of $N_L = 1$ corresponds to linear anisotropic scattering. And $N_L = 2$ gives quadratic anisotropic scattering. A special case of quadratic anisotropic scattering is when $a_1 = 0$ and $a_2 = 0.5$, which results in the Rayleigh scattering phase function.

A technique for curve-fitting Legendre polynomials to arbitrary Mie scattering phase functions was developed by Chu and Churchill [1955]. For moderate size parameters ($x \lesssim 5$) this technique works well. However, for larger size parameters an inordinately high number of terms is necessary to accurately represent the highly angular dependent forward scattering lobe, which corresponds to diffraction in the limit of $x \to \infty$.

The real utility of using Legendre polynomials to represent the phase function is not for the single-scattering phase function $p(\mu_p)$ but for the slab phase function $p(\mu, \mu')$. When Eq. (12-2) is substituted in (12-9), the addition theorem for Legendre polynomials

$$P_i(\cos\theta_p) = P_i(\mu)P_i(\mu') + 2\sum_{j=1}^{\infty}\frac{(i-j)!}{(i+j)!}P_i^j(\mu)P_i^j(\mu')\cos j(\phi_s - \phi_s') \tag{12-10}$$

can be used [Ozisik, 1985] to give

$$p(\mu, \phi_s; \mu', \phi_s') = \sum_{i=0}^{N_L} a_i P_i(\mu)P_i(\mu') + 2\sum_{i=0}^{N_L} a_i \sum_{j=1}^{\infty}\frac{(i-j)!}{(i+j)!}P_i^j(\mu)P_i^j(\mu')\cos j(\phi_s - \phi_s') \tag{12-11}$$

where P_i^j are the associated Legendre functions. Substituting (12-11) into (12-4) and noting that

$$\int_0^{2\pi}\cos j(\phi - \phi')\,d\phi' = 0 \qquad (j = 1, 2, \ldots) \tag{12-12}$$

TABLE 12-2 Legendre Polynomial Representation of a Few Special Slab Scattering Phase Functions

Type of Scattering	N_L	Phase Function $p(\mu, \mu')$	Conditions
Isotropic	0	1	
Linear anisotropic	1	$1 + a_1\mu\mu'$	$-1 \leq a_1 \leq 1$
Rayleigh	2	$1 + \frac{1}{8}(3\mu^2 - 1)(3\mu'^2 - 1)$	$a_1 = 0, a_2 = \frac{1}{2}$

for integer values of j gives

$$p(\mu, \mu') = \sum_{i=0}^{N_L} a_i P_i(\mu) P_i(\mu') \tag{12-13}$$

Thus for a single-scattering phase function that can be represented by a series of Legendre polynomials with argument $\mu_p(= \cos\theta_p)$, the transformation to the slab phase function can be readily made according to (12-13). A few special cases are listed in Table 12-2.

SOLUTION OF THE TRANSFER EQUATION

The primary difficulty in solving the transfer equation is the presence of the in-scattering integral term. The way in which the in-scattering term is handled represents the main distinction between the various methods for solving Eq. (12-1) or (12-3). In the following section two approaches for integrating the in-scattering term will be described. These two approaches are the flux model approach and the discrete ordinate method. The approach of the flux models is to assume that intensity is constant over small intervals of solid angle and integrate the in-scattering term accordingly. The approach of the discrete ordinate method is to replace the in-scattering integral with a finite-sum numerical quadrature formula. Although the basic assumptions involved are fundamentally different, the system of equations obtained, and hence the solutions, are quite similar. In addition to these two techniques there are many others that have been developed such as the spherical harmonics method (or P_N approximation) and the differential approximation (or moment method). However, these techniques are equivalent to the flux and discrete ordinate methods [Krook, 1955] and are not discussed here.

Two-Flux Model

The approach of the flux models is to assume that $I(t, \mu)$ is constant over various subintervals of the range $-1 < \mu < 1$ and integrate the in-scattering

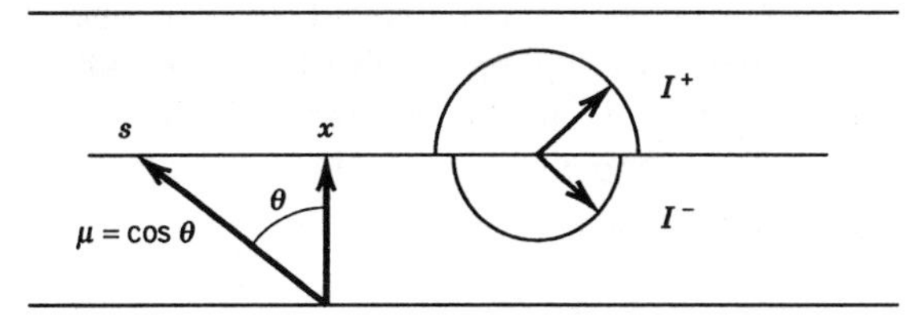

Figure 12-2 Two-flux model.

term accordingly. In the two-flux model (also known as the Schuster-Schwarzchild approximation) [Chandrasekhar, 1960] the intensity distribution is assumed to be semi-isotropic as pictured in Fig. 12-2.

$$I(t,\mu) = \begin{cases} I^+(t) & 0 < \mu < 1 \\ I^-(t) & -1 < \mu < 0 \end{cases} \tag{12-14}$$

Introducing (12-14) into the integral term in (12-3) and integrating (12-3) over $-1 \le \mu \le 1$ yields two coupled, first-order ordinary differential equations.

$$\frac{1}{2}\frac{dI^+}{dt} = -(1 - \omega_0 + \omega_0 B)I^+ + \omega_0 B I^- + (1 - \omega_0) I_b \tag{12-15a}$$

$$-\frac{1}{2}\frac{dI^-}{dt} = -(1 - \omega_0 + \omega_0 B)I^- + \omega_0 B I^+ + (1 - \omega_0) I_b \tag{12-15b}$$

where

$$B = \frac{1}{2}\int_0^1 \int_{-1}^0 p(\mu,\mu')\, d\mu'\, d\mu \tag{12-16}$$

is the back-scattering fraction for diffuse incident radiation. Physically, B represents the fraction of hemispherical flux ($q^+ = \pi I^+$ or $q^- = \pi I^-$) that is scattered into the opposite hemisphere. Upon solving (12-15) for $I^+(t)$ and $I^-(t)$ subject to some boundary conditions, the net flux is obtained by integrating intensity over all directions as

$$q = q^+ - q^- = \pi(I^+ - I^-) \tag{12-17}$$

The two flux equations (12-15) can be solved by differentiating (12-15a), substituting for dI^-/dt using (12-15b), and eliminating I^- by solving (12-15a) for I^- and substituting into the differentiated form of (12-15a). The result of this process will be a single second-order ordinary differential equation for I^+ that can readily be solved. An alternative method for solving the coupled first-order flux equations is to solve the coupled system of equations as an eigenvalue problem. This approach is more appropriate when the number of fluxes is greater than two, since algebraic substitution becomes unwieldy. The method for formulating the eigenvalue approach is discussed in the subsequent section on the discrete ordinate method, but applies to the solution of the flux model equations as well.

Method of Discrete Ordinates

Another popular technique for solving radiative transfer with anisotropic multiple scattering is the method of discrete ordinates of Chandrasekhar [1960]. This method has the appearance of a multiflux model but the basic assumption is different. In the method of discrete ordinates the in-scattering integral is approximated by the finite summation quadrature formula known as Gaussian quadrature. If $f(\mu)$ is a function to be integrated over the interval $-1 \leq \mu \leq 1$, the Gaussian quadrature formula is

$$\int_{-1}^{1} f(\mu)\, d\mu \cong \sum_{j=1}^{N} w_j f(\mu_j) \tag{12-18}$$

where the weight factors are

$$w_j = \frac{1}{P'_N(\mu_j)} \int_{-1}^{1} \frac{P_N(\mu)}{\mu - \mu_j}\, d\mu \tag{12-19}$$

and the discrete ordinates μ_j are the zeroes of the Legendre polynomial of order N

$$P_N(\mu_j) = 0 \tag{12-20}$$

which occur in equal and opposite pairs. A brief listing of the zeroes and weight factors is given in Table 12-3. A more complete tabulation is given by Abramowitz and Stegun [1972].

The advantage of the Gaussian formula is that it is exact if the function f is a polynomial of order less than $2N$. By incorporating Eq. (12-18) into (12-3) the transfer equation along a particular discrete ordinate direction μ_i becomes

$$\mu_i \frac{dI}{dt}(t, \mu_i) = -I(t, \mu_i) + \frac{\omega_0}{2} \sum_{j=1}^{N} I(t, \mu_j) p(\mu_i, \mu_j) w_j + (1 - \omega_0) I_b(t) \tag{12-21}$$

There are N of these equations, one for each of the ordinates pictured in Fig. 12-3, and they are all coupled through the in-scattering summation term.

TABLE 12-3 Zeroes and Weight Factors for Gaussian Quadrature

N	$\pm\mu_j$	w_j
2	0.57735	1.00000
4	0.33998	0.65214
	0.86113	0.34785
6	0.23861	0.46791
	0.66120	0.36076
	0.93246	0.17132
8	0.18343	0.36268
	0.52553	0.31370
	0.79666	0.22238
	0.96028	0.10122

It is important to note that the assumption made in the discrete ordinate method is fundamentally different than that of the flux models. In the discrete ordinate method the intensity is not assumed constant over some finite solid angle as in the flux models. Instead a quadrature formula is introduced that discretizes the intensity field into a finite number of streams or ordinates.

Using matrix notation to simplify (12-21), the transfer equations can also be written as

$$\frac{dI_i}{dt} = \sum_{j=1}^{N} M_{ij} I_j + \frac{(1-\omega_0)}{\mu_i} I_b(t) \tag{12-22}$$

where

$$M_{ij} = \frac{-\delta_{ij}}{\mu_i} + \frac{\omega_0 w_j p(\mu_i, \mu_j)}{2\mu_i} \tag{12-23}$$

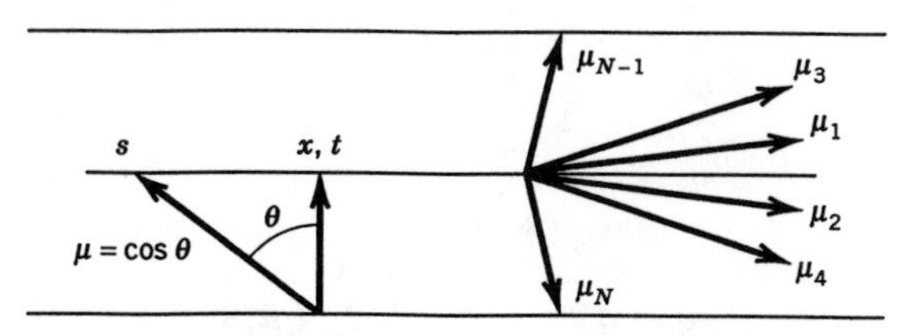

Figure 12-3 Discrete ordinate method.

The homogeneous problem

$$\frac{dI_i}{dt} = \sum_{j=1}^{N} M_{ij} I_j \tag{12-24}$$

is solved by assuming a solution of the form

$$I_{i_h} = V^{(i)} e^{\lambda t} \tag{12-25}$$

and substituting into (12-24). This leads to the eigenvalue problem

$$\sum_{j} (M_{ij} - \lambda \delta_{ij}) V^{(j)} = 0 \tag{12-26}$$

Solving (12-26) for the eigenvalues λ_j and corresponding eigenvectors $V_j^{(i)}$ gives the homogeneous solution as the sum of all the linearly independent solutions of (12-24)

$$I_{i_h} = \sum_{j=1}^{N} V_j^{(i)} e^{\lambda_j t} \tag{12-27}$$

A particular solution I_{i_p} can be found, for the case of an emitting medium, by using the method of variation of parameters and an assumed temperature profile [Roux, et al., 1975; Rish and Roux, 1986]. The general solution is then

$$I_i = \sum_{j=1}^{N} C_j V_j^{(i)} e^{\lambda_j t} + I_{i_p} \tag{12-28}$$

where the constants C_j can be determined using the boundary conditions to give the complete solution. The net flux is then given by

$$q = 2\pi \int_{-1}^{1} I(t, \mu) \mu \, d\mu \tag{12-29}$$

which is approximated by Gaussian quadrature as

$$q^{+} \cong 2\pi \sum_{i=1,\text{odd}}^{N} I_i \mu_i w_i \tag{12-30a}$$

$$q^{-} \cong 2\pi \sum_{i=1,\text{even}}^{N} I_i \mu_i w_i \tag{12-30b}$$

$$q = q^{+} - q^{-} \cong 2\pi \sum_{i=1}^{N} I_i \mu_i w_i \tag{12-30c}$$

Two Discrete Ordinates Method

The simplest discrete ordinate model is that of two discrete ordinates. Setting $N = 2$ in Eq. (12-22) gives the following equations for I_1 and I_2:

$$\frac{dI_1}{dt} = \frac{1}{\mu_1}\left[-1 + \frac{\omega_0}{2}p(\mu_1, \mu_1)\right]I_1 + \frac{1}{\mu_1}\left[\frac{\omega_0}{2}p(\mu_1, \mu_2)\right]I_2 + \frac{1 - \omega_0}{\mu_1}I_b \tag{12-31a}$$

$$\frac{dI_2}{dt} = \frac{1}{\mu_2}\left[-1 + \frac{\omega_0}{2}p(\mu_2, \mu_2)\right]I_2 + \frac{1}{\mu_2}\left[\frac{\omega_0}{2}p(\mu_2, \mu_1)\right]I_1 + \frac{1 - \omega_0}{\mu_2}I_b \tag{12-31b}$$

As written, Eqs. (12-31) call for the phase function to be evaluated at the discrete ordinate directions $\mu_{1,2}$. However, this procedure may not give the best representation of the scattering over other directions. Therefore it is customary to replace the phase function terms with terms that try to account for the scattering distribution over all directions and not just the discrete ordinate directions as follows:

$$p(\mu_1, \mu_1) = p(\mu_2, \mu_2) = 1 + g \tag{12-32a}$$

$$p(\mu_1, \mu_2) = p(\mu_2, \mu_1) = 1 - g \tag{12-32b}$$

The parameter g can be thought of as an asymmetry parameter for anisotropic *multiple* scattering in the slab geometry [as opposed to the anisotropic *single* scattering asymmetry parameter $\langle p \rangle$ defined in Eq. (9-5)]. Its value ranges over the interval $-1 < g < 1$. A value of $g = 0$ indicates balanced forward and backward scattering such as isotropic scattering or Rayleigh scattering. Negative values of g indicate net backward scattering while positive values of g indicate net forward scattering. Incorporating these changes into Eqs. (12-31) and setting $\mu = \mu_1 = -\mu_2\ (= 1/\sqrt{3})$ gives

$$\mu\frac{dI_1}{dt} = -\left[1 - \frac{\omega_0}{2}(1 + g)\right]I_1 + \left[\frac{\omega_0}{2}(1 - g)\right]I_2 + (1 - \omega_0)I_b \tag{12-33a}$$

$$-\mu\frac{dI_2}{dt} = -\left[1 - \frac{\omega_0}{2}(1 + g)\right]I_2 + \left[\frac{\omega_0}{2}(1 - g)\right]I_1 + (1 - \omega_0)I_b \tag{12-33b}$$

Comparing the two discrete ordinates equations (12-33) with the two-flux equations (12-15) shows that the mean direction of the rays in the two-flux model corresponds to $\mu = \frac{1}{2}$ (60° from the normal direction). The two formulations are equivalent if the effective value of $\mu = \frac{1}{2}$ in the two-flux model is changed to $\mu = 1/\sqrt{3}$ and the asymmetry parameter g is interpreted as $1 - 2B$. Hence the two-flux model offers one means of evaluating the multiple scattering asymmetry parameter g. Comparisons between the two-ordinate predictions of Eqs. (12-33) and exact solutions have shown that this method of evaluating g (i.e., $g = 1 - 2B$) is generally most appropriate when the slab is optically thin.

Another method for evaluating g which is appropriate when the slab is optically thick follows from the diffusion approximation. By substituting the diffusion approximation Eq. (11-52) into the transfer equation (12-3), integrating over the intervals $-1 < \mu < 1$ and $0 < \mu < 1$, and using the phase function normalization condition, it can be shown that the appropriate expression for the slab multiple scattering asymmetry parameter under optically thick conditions is

$$p^* = \int_0^1 \int_{-1}^1 p(\mu, \mu')\mu' \, d\mu' \, d\mu \tag{12-34a}$$

Combining Eq. (12-34a) with (12-7) gives

$$p^* = \frac{1}{\pi} \int_0^1 \int_{-1}^1 \int_{\mu_{p\pi}}^{\mu_{p0}} \frac{p(\mu_p)\, d\mu_p}{\left[(1 - \mu^2)(1 - \mu'^2) - (\mu_p - \mu\mu')^2\right]^{1/2}} \mu' \, d\mu' \, d\mu \tag{12-34b}$$

In spite of apparent differences in their definitions, it can be shown by numerical evaluation that p^* is equivalent to $\langle p \rangle$ [Eq. (9-5)].

$$\langle p \rangle = \frac{1}{2} \int_{-1}^1 p(\mu_p)\mu_p \, d\mu_p \tag{12-35}$$

Furthermore, it can be shown that $p^* = \langle p \rangle = a_1/3$ by assuming that $p(\mu_p)$ can be represented by a Legendre polynomial series [Eqs. (12-9) and (12-13)] and using the orthogonality properties of Legendre polynomials, where a_1 is the first coefficient in the series.

To summarize the issue of evaluating the multiple scattering asymmetry parameter, the choice of g depends on the optical thickness of the slab according to the following recommendation.

$$g = \begin{cases} 1 - 2B & \text{optically thin; Eq. (12-16)} \\ p^*(= \langle p \rangle) & \text{optically thick; Eqs. (12-34, 35)} \end{cases} \tag{12-36}$$

Now let us return to the problem of solving the two discrete ordinate equations. Equation (12-26) represents a homogeneous system of linear algebraic equations. The only nontrivial solution exists when the determinant of the coefficient matrix is zero. In this case all of the algebraic equations are not linearly independent and some of the $V^{(j)}$ must be arbitrarily specified. The eigenvalues are obtained by setting the determinant of the coefficient matrix in Eq. (12-26) to zero.

$$\begin{vmatrix} M_{11} - \lambda & M_{12} \\ M_{21} & M_{22} - \lambda \end{vmatrix} = 0 \tag{12-37}$$

With the elements of the M matrix given by

$$M_{11} = -M_{22} = \frac{-1}{\mu} + \frac{\omega_0(1+g)}{2\mu} \tag{12-38}$$

$$M_{12} = -M_{21} = \frac{\omega_0(1-g)}{2\mu} \tag{12-39}$$

the eigenvalues are obtained from Eq. (12-37) as

$$\lambda_{1,2} = \pm\frac{1}{\mu}\sqrt{(1-\omega_0)(1-\omega_0 g)} \tag{12-40}$$

The first elements of the eigenvectors $V_{1,2}^{(1)}$ are arbitrarily set to one, allowing the second elements to be obtained from Eq. (12-26). The eigenvectors are thus

$$V_{1,2}^{(1)} = 1 \tag{12-41}$$

$$V_{1,2}^{(2)} = \frac{1 \pm \zeta}{1 \mp \zeta} \qquad \zeta = \sqrt{\frac{1-\omega_0}{1-\omega_0 g}} \tag{12-42}$$

It remains to find particular solutions $I_{1,2p}$ and integration constants $C_{1,2}$. The particular solutions depend on the nature of the nonhomogeneous terms appearing in the transfer equation. Typical nonhomogeneous terms are source terms representing emission and collimated incident radiation. The treatment of these two types of source terms is demonstrated in subsequent examples. The integration constants depend on the nature of the boundary conditions. Typical boundary conditions include nonemitting, nonreflecting boundaries and diffusely emitting, diffusely reflecting boundaries.

The solution of anisotropic multiple scattering in a slab is next demonstrated for three different problems: (1) a gray slab in radiative equilibrium between diffusely emitting, diffusely reflecting, gray walls; (2) a semi-infinite gray slab with prescribed temperature (nonradiative equilibrium) adjacent to a diffuse, gray wall; and (3) and a cold slab with nonreflecting, nonemitting boundaries subject to collimated incident radiation.

GRAY SLAB IN RADIATIVE EQUILIBRIUM

Consider a plane parallel layer in radiative equilibrium between diffuse gray walls at T_1 and T_2 as shown in Fig. 11-1. This problem was solved assuming isotropic scattering in Chapter 11. Here anisotropic scattering effects are included. The solution is formulated using a general notation that applies to both the two-flux model and the two discrete ordinates model. The $\pm$ intensity notation of the two-flux model is used; however, it should be noted that this formulation is equivalent to the discrete ordinate formulation with $I_1 = I^+$ and $I_2 = I^-$.

In this particular case the net flux can be obtained from the transfer equations without obtaining particular solutions and without solving for I^+ and I^- individually. Taking the sum and difference of Eqs. (12-33a) and (12-33b) gives, respectively

$$\mu \frac{d}{dt}(I^+ - I^-) = (1 - \omega_0)[2I_b - (I^+ + I^-)] \tag{12-43a}$$

$$\mu \frac{d}{dt}(I^+ + I^-) = -(1 - \omega_0 g)(I^+ - I^-) \tag{12-43b}$$

Combining the radiative equilibrium energy equation

$$\frac{dq}{dt} = 2\pi\mu \frac{d}{dt}(I^+ - I^-) = 0 \tag{12-44}$$

with (12-43) gives $I_b = 0.5(I^+ + I^-)$ for $\omega_0 \neq 1$ and

$$q = -\frac{4\mu^2}{1 - \omega_0 g} \frac{de_b}{dt} = \text{constant} \tag{12-45}$$

This expression, which includes anisotropic scattering ($g \neq 0$), can be expressed in terms of an equivalent isotropic scattering optical depth by defining the isotropic optical depth

$$t_I = \int_0^x (1 - \omega_0 g) K_e \, dx; \qquad dt_I = (1 - \omega_0 g)\, dt \tag{12-46}$$

to give

$$q = -4\mu^2 \frac{de_b}{dt_I} = \text{constant} \tag{12-47}$$

In integrating (12-47) to obtain $e_b(t_I)$ it is necessary to use an appropriate slip condition since radiative equilibrium has been assumed. Analogous to the derivation of the slip condition for the diffusion approximation in Chapter 11, a slip condition for the two-flux model with arbitrary optical thickness t_{I_L} and radiative equilibrium can be derived as

$$e_{b_1} - e_b(0) = \frac{1}{1 - \exp(-t_{I_L})} \left\{ -(e_{b_1} - e_{b_2}) \exp(-t_{I_L}) + q\left[\frac{1}{\varepsilon_1} - \frac{1}{2} + \left(\frac{t_{I_L}}{2} - \frac{1}{2} + \frac{1}{\varepsilon_2} \right) \exp(-t_{I_L}) \right] \right\} \quad (12\text{-}48)$$

Using (12-48) for the wall at $t_I = 0$ and a similar expression for the other wall at $t_I = t_{I_L}$ gives [Wang and Tien, 1983]

$$q = \frac{e_{b_1} - e_{b_2}}{\dfrac{1}{\varepsilon_1} + \dfrac{1}{\varepsilon_2} - 1 + \dfrac{t_{I_L}}{4\mu^2}} \quad (12\text{-}49)$$

The nondimensional heat flux is

$$q^* = \left. \frac{q}{e_{b_1} - e_{b_2}} \right|_{\varepsilon_1 = \varepsilon_2 = 1} = \frac{1}{1 + \dfrac{t_{I_L}}{4\mu^2}}, \qquad \mu = \begin{cases} 1/\sqrt{3} & \text{2 ordinate} \\ 1/2 & \text{2 flux} \end{cases} \quad (12\text{-}50)$$

Comparison of (12-50) with the diffusion model solution Eq. (11-70) (which assumes isotropic scattering and is exact for $t_{I_L} \gg 1$) shows that the two-flux model ($\mu = \frac{1}{2}$) underpredicts q^* by a factor of $\frac{3}{4}$ as $t_{I_L} \to \infty$. This discrepancy can be related to the semi-isotropic intensity assumption of the two-flux model, which inherently ignores the fact that path lengths normal to the slab are shorter than paths slanted at an angle. The accuracy can be improved by increasing the number of assumed fluxes from two to three, four, six, or more.

While the two-flux model is not accurate in the optically thick limit, Eq. (12-50) shows that the two discrete ordinates model ($\mu = 1/\sqrt{3}$) matches the diffusion model Eq. (11-70) exactly in this limit. The reason is that in the optically thick diffusion limit, the intensity distribution is a first-order polynomial in μ; [recall Eq. (11-52) and the fact that Gaussian quadrature is exact for polynomials of order less than $2N$, which is 2 in this case].

The effect of anisotropic scattering on the heat flux is demonstrated by Eq. (12-50). Consider the common situation of highly forward anisotropic scattering ($0 < g < 1$). For this case the anisotropic optical depth is less than the equivalent isotropic optical depth.

$$t_{I_L} = \int_0^L (1 - \omega_0 g) K_e \, dx < t_L = \int_0^L K_e \, dx \quad \text{(forward scattering; } 0 < g < 1\text{)} \tag{12-51}$$

Therefore according to Eq. (12-50) the heat flux for anisotropic forward scattering is greater than that which would be predicted assuming isotropic scattering for the same extinction coefficient. The reason is that anisotropic forward scattering particles are not as efficient at blocking heat transfer from wall 1 to wall 2 as isotropic scattering particles with the same extinction coefficient. The role of albedo is also clear from these results. As ω_0 decreases, the effect of anisotropic scattering becomes less important as indicated by Eq. (12-51).

Example 12-1

Reconsider the semigray, radiative equilibrium model of the earth's atmosphere of Example 11-1. Include anisotropic scattering by particulates in the atmosphere and determine the temperature at the earth's surface. Assume that the product of albedo and asymmetry is constant at $\omega_0 g = 0.5$ over all altitudes. Use the same effective optical thickness $t_L = 4$ as Example 11-1.

The appropriate equation for determining the temperature of the atmosphere at the earth's surface is Eq. (11-76). In that equation the optical thickness t_L should be replaced by the equivalent isotropic optical thickness t_{I_L} since isotropic scattering was assumed in Chapter 11.

$$\sigma T(0)^4 = e_b(0) = \frac{q}{2}\left[1 + \frac{3t_{I_L}}{2}\right] \tag{12-52}$$

From Eq. (12-51) the effective isotropic optical thickness can be calculated as

$$t_{I_L} = (1 - \omega_0 g) t_L = (1 - 0.5)(4) = 2$$

and $T(0)$ can be determined from Eq. (12-52) as

$$T(0) = \left\{\frac{240 \text{ W/m}^2}{(2)(5.67 \times 10^{-8} \text{ W/m}^2 \text{ K}^4)}\left[1 + \frac{(3)(2)}{2}\right]\right\}^{1/4} = 303 \text{ K}$$

The earth temperature can be calculated from the slip condition Eq. (11-75) as $T_1 = 320$ K. The corresponding values for isotropic scattering predicted in Example 11-1 were $T(0) = 349$ K and $T_1 = 360$ K. Thus the effect of anisotropic forward scattering in this example is to reduce by 40 K the temperature of the earth necessary to reject a fixed heat flux of 240 W/m^2 to space. Again it should be noted that this is not a very good model of radiative transfer in the atmosphere since the atmosphere is actually optically thick at certain wavelengths and optically thin at others. Thus the effect of anisotropic scattering on the earth's energy balance is not as great as this example would indicate. However, the principle is the same. Furthermore the equivalent isotropic concept is still valid and can be used with confidence if in any given problem accurate optical properties are used and correct spectral variations are accounted for.

SEMI-INFINITE SLAB WITH PRESCRIBED TEMPERATURE

As another application of the methods presented in this chapter consider the problem of radiative transfer from an emitting, absorbing, anisotropic scattering, semi-infinite medium to an adjacent opaque surface (Fig. 12-4). This problem is a reasonable model of heat transfer to a submerged surface in a particulate medium such as a tube in a fluidized bed.

In Fig. 12-4, I_w^+ is the intensity leaving the wall, T_0 is the wall temperature and T_∞ is the constant temperature of the medium at $t \to \infty$. It should be noted that radiative equilibrium is not a valid energy equation for a semi-infinite medium in steady state. Otherwise, if the net radiative flux were constant, as radiative equilibrium prescribes, then it must be zero, which is the value at $t \to \infty$, throughout the medium, which it cannot be. Instead of applying the energy equation, the temperature profile is assumed to be known; the characteristic length for temperature change near the wall is δ. The appropriate governing equation for this problem is Eq. (12-3) with the boundary conditions

$$I(t = 0, \mu) = I_w^+ \qquad \mu > 0 \tag{12-53}$$

$$\left.\frac{dI}{dt}\right|_{t \to \infty} = 0 \qquad \mu < 0 \tag{12-54}$$

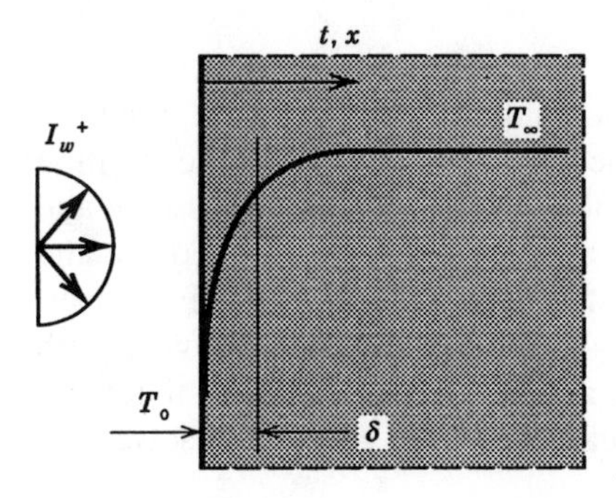

Figure 12-4 Semi-infinite medium with prescribed temperature profile.

This problem has two nonhomogeneous terms in it, the emission term $(1 - \omega_0)I_b$ in the transfer equation and the incident intensity I_w^+ in the boundary condition. An expedient way to solve this problem is to use superposition to separate the nonhomogeneous terms and consider two simpler problems.

Effective Emissivity Problem

The first problem is the effective emissivity problem, which retains the nonhomogeneous emission term in the transfer equation but assumes the wall at $t = 0$ is nonemitting and nonreflecting. The solution is obtained from (12-3), (12-53), and (12-54) by setting $I_w^+ = 0$. The result is expressed as an effective emissivity of the medium based on the temperature T_∞

$$\varepsilon_{\text{eff}} = \frac{q^-(0)}{2\pi\int_0^{-1} I_b(T_\infty)\mu\, d\mu} = \frac{2\pi\int_0^{-1} I(0,\mu)\mu\, d\mu}{2\pi\int_0^{-1} I_b(T_\infty)\mu\, d\mu} \tag{12-55}$$

Effective Absorptivity Problem

The second problem is the effective absorptivity problem, which retains the nonhomogeneous term in the boundary condition I_w^+ but assumes a cold, nonemitting medium. The solution is obtained from (12-3), (12-53), and (12-54) by setting $I_b = 0$ in (12-3). The result is expressed as an effective absorptivity of the medium

$$\alpha_{\text{eff}} = 1 - \frac{q^-(0)}{q^+(0)} = 1 - \frac{2\pi\int_0^{-1} I(0,\mu)\mu\, d\mu}{2\pi\int_0^{1} I_w^+\mu\, d\mu} \tag{12-56}$$

If the properties are assumed to be independent of temperature, then the equations and boundary conditions are linear and the powerful superposition principle holds. This means that the solutions of the effective emissivity and absorptivity problems can be added. The transfer between the medium and the wall can thus be represented as transfer between two parallel surfaces, the wall with properties T_0 and ε_w $(= \alpha_w)$ and an effective wall with properties T_∞, ε_{eff}, and α_{eff}. The radiosity and irradiation relations that were used in Chapter 3 can be written (in this case with $\alpha_{\text{eff}} \neq \varepsilon_{\text{eff}}$) and solved for the net radiative flux to the wall.

$$q = \frac{\dfrac{\varepsilon_{\text{eff}}}{\alpha_{\text{eff}}} e_b(T_\infty) - e_b(T_0)}{\dfrac{1}{\alpha_{\text{eff}}} + \dfrac{1}{\alpha_w} - 1} \tag{12-57}$$

For the isothermal case it can be shown that Kirchhoff's law holds for the medium ($\varepsilon_{\text{eff}} = \alpha_{\text{eff}}$) and (12-57) reduces to the same expression as that obtained in Chapter 3 for isothermal, parallel surfaces [Eq. (3-46)]. Thus α_{eff} can be obtained by solving either the effective absorptivity problem or the effective emissivity problem with $T(x)$ = constant.

For the nonisothermal case the effective emissivity and absorptivity are not equal ($\varepsilon_{\text{eff}} \neq \alpha_{\text{eff}}$), and the temperature profile must be specified to solve for ε_{eff}. This requires an estimate of the temperature profile to be made from the energy equation based on the flow field of the particular system involved.

Example 12-2

Determine the effective emissivity and absorptivity of a semi-infinite medium that has an exponential temperature profile near the boundary as given by Eq. (12-58).

$$\frac{T(x) - T_\infty}{T_0 - T_\infty} = \exp\left[-\frac{x}{\delta}\right] \tag{12-58}$$

The medium is composed of gray, anisotropic scattering particles that are uniformly distributed. Use the two flux–two discrete ordinates model.

The governing equations are (12-33a) and (12-33b), which are written here using the I^+ and I^- notation instead of the I_1 and I_2 notation.

$$\mu \frac{dI^+}{dt} = -\left[1 - \frac{\omega_0}{2}(1+g)\right] I^+ + \left[\frac{\omega_0}{2}(1-g)\right] I^- + (1-\omega_0) I_b \tag{12-59a}$$

$$-\mu \frac{dI^-}{dt} = -\left[1 - \frac{\omega_0}{2}(1+g)\right] I^- + \left[\frac{\omega_0}{2}(1-g)\right] I^+ + (1-\omega_0) I_b \tag{12-59b}$$

The boundary conditions for the effective emissivity problem are

$$I^+(0) = 0 \tag{12-60a}$$

$$\left.\frac{dI^-}{dt}\right|_{t\to\infty} = 0 \tag{12-60b}$$

The emission term is given in terms of the exponential temperature profile as

$$I_b = \frac{\sigma T^4}{\pi} = \frac{\sigma}{\pi}\left\{T_\infty + (T_0 - T_\infty)\exp\left[\frac{-t}{t_\delta}\right]\right\}^4 \qquad (12\text{-}61)$$

Here the assumption that the radiative properties are uniform has been used to say that optical depth t is proportional to distance x.

$$\frac{t}{t_\delta} = \frac{\int_0^x K_e\,dx}{\int_0^\delta K_e\,dx} = \frac{K_e x}{K_e \delta} = \frac{x}{\delta} \quad (K_e = \text{constant}) \qquad (12\text{-}62)$$

The general solution of Eqs. (12-59) from Eqs. (12-28) and (12-38) through (12-42) is

$$I^+(t) = C_1\exp(\lambda_1 t) + C_2\exp(\lambda_2 t) + I_p^+ \qquad (12\text{-}63a)$$

$$I^-(t) = C_1 V_1\exp(\lambda_1 t) + C_2 V_2\exp(\lambda_2 t) + I_p^- \qquad (12\text{-}63b)$$

where the eigenvalues are

$$\lambda_{1,2} = \pm\frac{1}{\mu}\sqrt{(1-\omega_0)(1-\omega_0 g)} \qquad (12\text{-}64)$$

and the eigenvector elements are

$$V_{1,2} = \frac{1 \pm \zeta}{1 \mp \zeta} \qquad \zeta = \sqrt{\frac{1-\omega_0}{1-\omega_0 g}} \qquad (12\text{-}65)$$

The form of the particular solutions in this problem can be guessed by inspection from the nonhomogeneous emission term as

$$I_p^+ = a_0 + a_1\exp\left[\frac{-t}{t_\delta}\right] + a_2\exp\left[\frac{-2t}{t_\delta}\right] + a_3\exp\left[\frac{-3t}{t_\delta}\right] + a_4\exp\left[\frac{-4t}{t_\delta}\right] \qquad (12\text{-}66a)$$

$$I_p^- = b_0 + b_1\exp\left[\frac{-t}{t_\delta}\right] + b_2\exp\left[\frac{-2t}{t_\delta}\right] + b_3\exp\left[\frac{-3t}{t_\delta}\right] + b_4\exp\left[\frac{-4t}{t_\delta}\right] \qquad (12\text{-}66b)$$

The constants a_{0-4} and b_{0-4} are determined by substituting Eqs. (12-66) into (12-59), collecting like terms, and solving for the a's and b's. The first integration constant is determined by the second boundary condition (12-60b) as $C_1 = 0$. The second integration constant is determined by the first boundary condition (12-60a). The algebra can be simplified by nondimensionalizing intensity by $\sigma T_\infty^4/\pi$ as suggested by the form of the effective emissivity.

$$\varepsilon_{\text{eff}} = \frac{2\pi \int_0^{-1} I(0, \mu)\mu \, d\mu}{2\pi \int_0^{-1} I_b(T_\infty)\mu \, d\mu} = \frac{\pi I^-(0)}{\sigma T_\infty^4} \tag{12-67}$$

After some algebra, the solution is obtained as

$$\varepsilon_{\text{eff}} = \frac{2\zeta^2}{1+\zeta} \sum_{n=0}^{4} \frac{4!}{n!(4-n)!} \frac{(\eta - 1)^n}{\left(\dfrac{\mu n}{t_{\delta_I}} + \zeta\right)}, \qquad \mu = \begin{cases} 1/\sqrt{3} & \text{2 ordinate} \\ 1/2 & \text{2 flux} \end{cases} \tag{12-68}$$

where*

$$\eta = \frac{T_0}{T_\infty} \tag{12-69}$$

$$\zeta = \sqrt{\frac{1 - \omega_0}{1 - \omega_0 g}} \tag{12-70}$$

$$t_{\delta_I} = t_\delta(1 - \omega_0 g) \tag{12-71}$$

By setting $\eta = 1$ the effective absorptivity is obtained from (12-68) as

$$\alpha_{\text{eff}} = \frac{2\zeta}{1 + \zeta} \tag{12-72}$$

Equation (12-68) indicates that ε_{eff} is a function of three mathematically independent variables ζ, η, and t_{δ_I} and four physically independent variables

*The albedo can also be scaled to an effective isotropic value as shown by Lee and Buckius [1982, 1983]. See Problem 12-11.

$$\omega_{0_I} = \frac{\omega_0(1-g)}{1 - \omega_0 g} \qquad \zeta = \sqrt{1 - \omega_{0_I}}$$

as shown in Eq. (12-73).

$$\varepsilon_{\text{eff}} = \varepsilon_{\text{eff}}\left(t_\delta, \omega_0, g, \frac{T_0}{T_\infty}\right) \tag{12-73}$$

Equation (12-72) shows that α_{eff} is a function of one mathematically independent variable ζ and two physical variables as shown in Eq. (12-74).

$$\alpha_{\text{eff}} = \alpha_{\text{eff}}(\omega_0, g) \tag{12-74}$$

Figure 12-5 shows a plot of the effective emissivity versus temperature ratio (T_0/T_∞) for isotropic scattering ($g = 0$; $B = 0.5$). Two optical thicknesses are shown, an optically thin case ($t_\delta = 0.1$) and an optically intermediate case ($t_\delta = 1$). For a relatively cold wall and hot medium ($T_0/T_\infty < 1$), the insulating effect of the cold boundary layer is clearly evident in Fig. 12-5 as ε_{eff} decreases rapidly with decreasing T_0/T_∞ for moderate and large values of t_δ. For moderate optical thicknesses (t_δ of order one) and small values of T_0/T_∞ the most important parameter is t_δ and ω_0 is secondary.

Figure 12-6 shows the effective absorptivity of a semi-infinite scattering medium versus $1 - \omega_0$, which can be thought of as the particle emissivity. The plot of α_{eff} in Fig. 12-6 is also a plot of ε_{eff} for an isothermal medium. Figure 12-6 indicates that the effective emissivity of an isothermal, scattering medium is always greater than the intrinsic emissivity of the individual

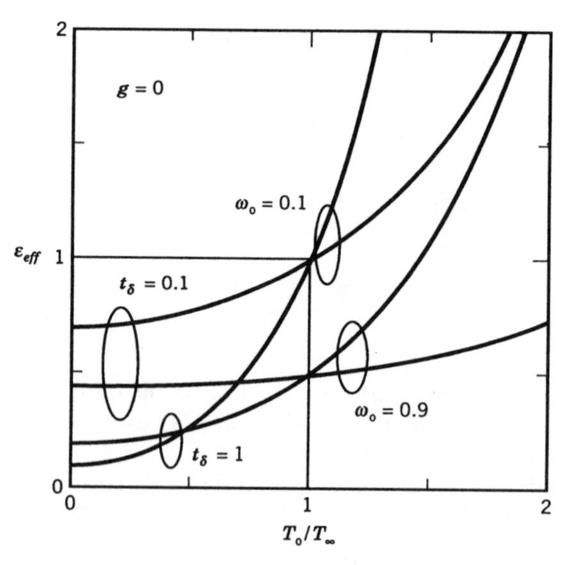

Figure 12-5 Effective emissivity of a semi-infinite slab with exponential temperature profile ($\mu = \frac{1}{2}$; two flux).

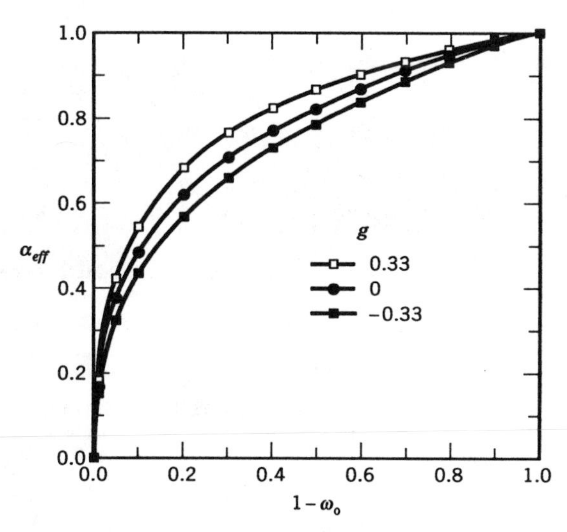

Figure 12-6 Effective absorptivity and isothermal emissivity of a semi-infinite slab.

particles $(1 - \omega_0)$. The effect of multiple scattering is to increase the effective absorptivity. This effect is the same one responsible for the behavior of an isothermal cavity blackbody.

MULTIPLE SCATTERING IN A PLANAR SLAB WITH COLLIMATED INCIDENT RADIATION

The preceding examples all consider radiative transfer in a slab with diffuse boundary conditions and azimuthally symmetric intensity fields. A notable case of radiative transfer in a medium that is not azimuthally symmetric occurs when collimated radiation is incident at the boundaries of a slab at an off-normal angle. Such a case occurs for solar radiation incident on a planetary atmosphere. Consider the problem of a collimated flux of magnitude q_0 incident on a slab with nonreflecting boundaries (Fig. 12-7).

The direction of incidence is $\mu_0 = \cos\theta_{s_0}$ (< 0). The collimated incident radiation can be treated either as a boundary condition or as a source term in the transfer equation. Following the latter approach, the intensity field is written as the sum of a scattered intensity term plus an unscattered intensity term.

$$\text{Total intensity} = I(t, \mu, \phi_s) + q_0 \exp\left[\frac{t_L - t}{\mu_0}\right] \delta(\mu - \mu_0)\delta(\phi_s - \phi_{s_0}) \tag{12-75}$$

Substituting (12-75) into (12-1) for the total intensity gives the transfer

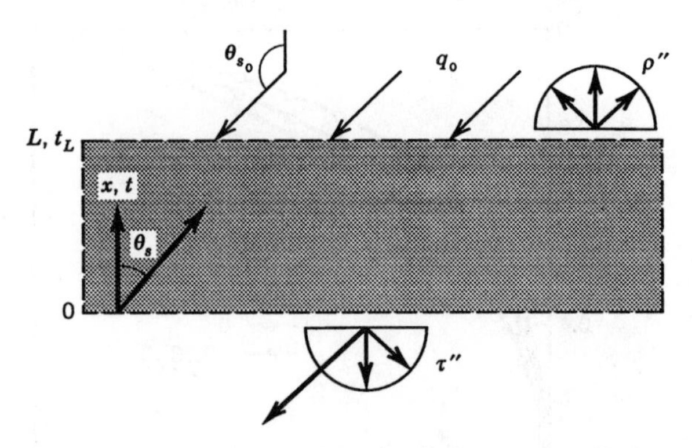

Figure 12-7 Plane layer subject to collimated irradiation.

equation for collimated incident radiation.

$$\mu \frac{dI(t,\mu,\phi_s)}{dt} = -I(t,\mu,\phi_s) + (1-\omega_0)I_b(t) + \frac{\omega_0}{4\pi}\int_0^{2\pi}\int_{-1}^{1} I(t,\mu',\phi_s')p(\mu,\phi_s;\mu',\phi_s')\,d\mu'\,d\phi_s' + \frac{\omega_0}{4\pi}q_0 p(\mu,\phi_s;\mu_0,\phi_{s_0})\exp\left[\frac{t_L - t}{\mu_0}\right] \qquad (12\text{-}76)$$

The last term in the equation represents the energy scattered from the collimated direction (μ_0, ϕ_{s_0}) into an arbitrary line of sight (μ, ϕ_s) at location t. The boundary conditions for the noncollimated intensity component are

$$I(0, \mu > 0) = 0 \qquad (12\text{-}77a)$$

$$I(t_L, \mu < 0) = 0 \qquad (12\text{-}77b)$$

(Recall that the incident intensity is accounted for by a source term in the transfer equation.) Although the intensity field is clearly azimuthally dependent if the collimated radiation is not normal to the slab ($\mu_0 \neq -1$), it is often sufficient (and much simpler) for heat transfer purposes to consider the azimuthally averaged intensity. Integrating Eq. (12-76) over azimuthal angle gives the azimuthally independent transfer equation.

$$\mu \frac{dI(t,\mu)}{dt} = -I(t,\mu) + (1-\omega_0)I_b(t) + \frac{\omega_0}{4\pi}\int_0^{2\pi}\int_{-1}^{1} I(t,\mu')p(\mu,\mu')\,d\mu' + \frac{\omega_0}{4\pi}q_0 p(\mu,\mu_0)\exp\left[\frac{t_L - t}{\mu_0}\right] \qquad (12\text{-}78)$$

In many cases the emission term in (12-78) is negligible. Such cases include scattering of solar radiation by planetary atmospheres and scattering of laser light by fogs, aerosols, and other particle dispersions. In these situations the only source of radiation is the external collimated incident flux and the unknowns in this problem are the scattered intensities leaving the top and bottom of the layer. These results can be expressed as an effective bi-directional transmissivity and reflectivity for the layer.

$$\rho'' = \frac{\pi I(t_L, \mu > 0)}{-\mu_0 q_0} \tag{12-79}$$

$$\tau'' = \frac{\pi I(0, \mu < 0)}{-\mu_0 q_0} \tag{12-80}$$

Example 12-3

Use the two discrete ordinates model to evaluate the transmitted and reflected flux from an anisotropic scattering layer of clouds with optical thickness t_L subject to collimated incident flux q_0 as shown in Fig. 12-7. Investigate the effects of optical thickness t_L and multiple scattering asymmetry g for conservative scattering ($\omega_0 = 1$) and normal incidence ($\mu_0 = -1$).

The governing transfer equations are obtained by applying the two discrete ordinates approximation to Eq. (12-78). The resulting equations are

$$\mu_1 \frac{dI_1}{dt} = -\left[1 - \frac{\omega_0}{2}(1+g)\right] I_1 + \left[\frac{\omega_0}{2}(1-g)\right] I_2 + S_1 \exp\left[\frac{t_L - t}{\mu_0}\right] \tag{12-81a}$$

$$-\mu_1 \frac{dI_2}{dt} = -\left[1 - \frac{\omega_0}{2}(1+g)\right] I_2 + \left[\frac{\omega_0}{2}(1-g)\right] I_1 + S_2 \exp\left[\frac{t_L - t}{\mu_0}\right] \tag{12-81b}$$

with

$$S_{1,2} = \frac{\omega_0}{4\pi} q_0 (1 \pm 3g\mu_1\mu_0) \tag{12-82}$$

Comparing these equations with Eqs. (12-33) shows that the emission source term has been replaced by the collimated incident flux source terms involving $S_{1,2}$. Again the phase function has not been evaluated precisely at the discrete ordinate directions. Instead an attempt is made to account for scattering from the collimated direction into all directions by using the multiple scattering asymmetry parameter g weighted by the direction cosines $\mu_1 = 1/\sqrt{3}$ and μ_0. Since the homogeneous problem is the same as before, the eigenvalues and eigenvectors are given by Eqs. (12-40), (12-41), and

(12-42). The particular solutions are obtained by assuming the form

$$I_{1p} = a \exp\left(\frac{t_L - t}{\mu_0}\right) \tag{12-83a}$$

$$I_{2p} = b \exp\left(\frac{t_L - t}{\mu_0}\right) \tag{12-83b}$$

Substituting (12-83) into (12-81) and solving for a and b gives

$$a = \frac{-S_1(\mu_1/\mu_0 + r) - sS_2}{s^2 - (-\mu_1/\mu_0 + r)(\mu_1/\mu_0 + r)} \tag{12-84a}$$

$$b = \frac{-S_2(-\mu_1/\mu_0 + r) - sS_1}{s^2 - (-\mu_1/\mu_0 + r)(\mu_1/\mu_0 + r)} \tag{12-84b}$$

$$r = 1 - \frac{\omega_0}{2}(1 + g) \tag{12-85a}$$

$$s = \frac{\omega_0}{2}(1 - g) \tag{12-85b}$$

The integration constants are obtained by using the boundary conditions

$$I_1(0) = 0 \tag{12-86a}$$

$$I_2(t_L) = 0 \tag{12-86b}$$

giving

$$C_1 = \frac{-a \exp(t_L/\mu_0) V_2 \exp(\lambda_2 t_L) + b}{V_2 \exp(\lambda_2 t_L) - V_1 \exp(\lambda_1 t_L)} \tag{12-87}$$

$$C_2 = \frac{a \exp(t_L/\mu_0) V_1 \exp(\lambda_1 t_L) - b}{V_2 \exp(\lambda_2 t_L) - V_1 \exp(\lambda_1 t_L)} \tag{12-88}$$

The complete solution is thus

$$I_1(t) = C_1 \exp(\lambda_1 t) + C_2 \exp(\lambda_2 t) + I_{1p} \tag{12-89}$$

$$I_2(t) = C_1 V_1 \exp(\lambda_1 t) + C_2 V_2 \exp(\lambda_2 t) + I_{2p} \tag{12-90}$$

where the eigenvalues $\lambda_{1,2}$ are given by Eq. (12-64) and the second elements of the eigenvectors $V_{1,2}$ are given by Eq. (12-65). The reflected and transmitted fluxes are

$$q_{\text{refl}} = q^+(t_L) = 2\pi\mu_1 I_1(t_L) \tag{12-91}$$

$$q_{\text{trans}} = q^-(0) = 2\pi\mu_1 I_2(0) - \mu_0 q_0 \exp\left[\frac{t_L}{\mu_0}\right] \tag{12-92}$$

The transmitted flux includes the diffuse component as well as the unscattered, collimated component. The results are most conveniently expressed in terms of the diffuse reflectivity ρ', diffuse transmissivity τ_d', and collimated transmissivity τ_c', defined as follows:

$$\rho' = \frac{2\pi\mu_1 I_1(t_L)}{-\mu_0 q_0} \qquad \tau_d' = \frac{2\pi\mu_1 I_2(0)}{-\mu_0 q_0} \qquad \tau_c' = \exp\left(\frac{t_L}{\mu_0}\right) \tag{12-93}$$

Figure 12-8 shows a plot of the results for normal incidence ($\mu_0 = -1$) and conservative ($\omega_0 = 1$), isotropic ($g = 0$), and anisotropic ($g = 0.5$) scattering. As optical thickness of the cloud layer increases the collimated transmissivity decays exponentially to zero, in accordance with Beer's law. At small optical thicknesses both the diffuse transmissivity and reflectivity are small. As optical thickness increases, the reflectivity increases and approaches a value of one while the diffuse transmissivity passes through a maximum value and then decreases to zero. Thus even nonabsorbing water droplets can prevent sunlight from reaching the earth's surface if the cloud layer is sufficiently thick, although this does not often happen.

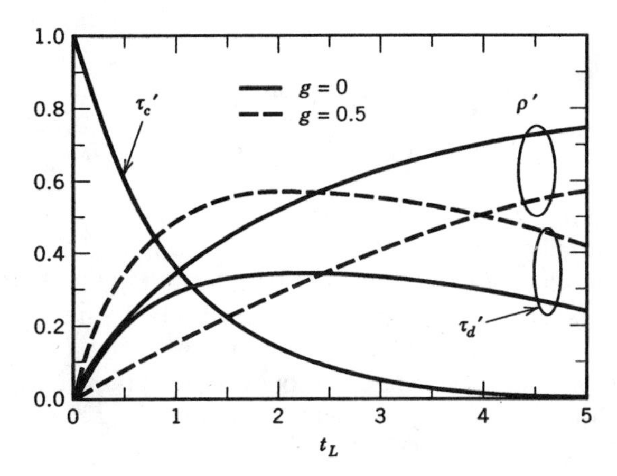

Figure 12-8 Reflectivity and transmissivity for a slab subject to collimated incident radiation with conservative isotropic ($g = 0$) and anisotropic ($g = 0.5$) scattering ($\omega_0 = 1, \mu_0 = -1$).

Figure 12-8 also shows the effect of anisotropic scattering. As g increases from 0 (isotropic scattering) to 0.5 (forward anisotropic scattering), the diffuse transmissivity increases at all values of t_L. This behavior is accompanied by a corresponding decrease in reflectivity to satisfy conservation of energy (since τ_c' is independent of g).

$$1 = \rho' + \tau_d' + \tau_c' \qquad (\omega_0 = 1) \tag{12-94}$$

The preceding example was concerned with an azimuthally symmetric intensity field since the incident radiation was normal to the slab. When the incident radiation is not normal to the slab, the intensity field will be a function of azimuthal angle ϕ_s as well as polar angle μ. As noted earlier, azimuthally averaged information is often sufficient for heat transfer purposes. However, if azimuthally resolved intensity information is desired, it is necessary to solve the azimuthally dependent transfer equation, Eq. (12-76). The solution of this problem is beyond the scope of this text but is discussed further by Chandrasekhar [1960] and Evans, et al. [1965].

ANISOTROPIC SCATTERING IN THE DIFFUSION LIMIT

The assumption of isotropic scattering is a useful simplification in many heat transfer analyses. In particular, isotropic scattering is an inherent assumption in the diffusion approximation (see Chapter 11). However, as has been shown in Chapter 9, real particles all scatter anisotropically to a greater or lesser degree. Therefore a method is needed for converting the anisotropic Mie scattering properties into effective isotropic scattering values, which, for the purpose of calculating radiant heat flux, are equivalent.

In Eq. (12-46) a method was demonstrated for converting anisotropic scattering properties to equivalent isotropic scattering properties for a slab in *radiative equilibrium*. The method consisted of modifying the extinction coefficient by multiplying by the anisotropic scattering factor $(1 - \overline{\omega}_0 \overline{g})$.

$$K_{e_I} = K_e(1 - \overline{\omega}_0 \overline{g}) \tag{12-95}$$

It can be shown that this same expression is the appropriate scaling to convert anisotropic to equivalent isotropic scattering in the *diffusion approximation*. The development is carried out by substituting the diffusion approximation Eq. (11-52) into the transfer equation (12-3), integrating over the intervals $-1 < \mu < 1$ and $0 < \mu < 1$, and using the phase function normalization condition. This development also gives the appropriate expression for the slab multiple scattering asymmetry parameter g for optically thick radiation diffusion [see Eqs. (12-34, 35)] as

$$\overline{g} = \overline{p}^* = \int_0^1 \int_{-1}^1 \overline{p}(\mu, \mu')\mu' \, d\mu' \, d\mu = \langle \overline{p} \rangle = \frac{a_1}{3} \tag{12-96}$$

Here the overbar has been included to indicate that averaging over particle size distribution has been done if necessary (as discussed in Chapter 9). Thus it is possible to include anisotropic scattering in optically thick diffusion radiative transfer in a rigorous way. The corresponding equivalent isotropic extinction efficiency follows from Eq. (12-95) as

$$\overline{Q}_{e_I} = \overline{Q}_e\left[1 - \overline{\omega}_0 \overline{p}^*\right] \quad \text{(optically thick)} \tag{12-97}$$

where

$$K_{e_I} = \frac{1.5\overline{Q}_{e_I} f_v}{d_{32}} \tag{12-98}$$

$$\overline{\omega}_0 = \frac{K_s}{K_e} = \frac{\overline{Q}_s}{\overline{Q}_e} \tag{12-99}$$

The effective spectral, isotropic scattering efficiency $\overline{Q}_{e_I}$ can be calculated for a monomodal polydispersion of spherical particles from the basic parameters of refractive index $n - ik$, particle size parameter based on mean optical size $x_{32} = \pi d_{32}/\lambda$, and ratio of most probable to mean optical size x_{mp}/x_{32}. The effective isotropic results are qualitatively similar to the anisotropic results discussed in Chapter 9. At small values of x_{32}, $\overline{Q}_e$ and $\overline{Q}_{e_I}$ are the same since scattering is small, and what scattering occurs is balanced in the forward and backward hemispheres. However, at large values of x_{32} scattering becomes anisotropic and $\overline{Q}_e$ approaches 2 (including a contribution of $\overline{Q}_s = 1$ for diffraction) while $\overline{Q}_{e_I}$ approaches 1 since forward diffraction scattering is effectively removed from consideration.

Rosseland Extinction Efficiency

Using the previous results for effective isotropic spectral scattering quantities, an effective Rosseland extinction efficiency can be calculated from the definition of Rosseland extinction coefficient.

$$\overline{Q}_{e_R}^{-1} = \frac{\int_0^\infty \overline{Q}_{e_I}^{-1} \frac{\partial e_{b_\lambda}}{\partial T} d\lambda}{\int_0^\infty \frac{\partial e_{b_\lambda}}{\partial T} d\lambda} = \int_0^1 \overline{Q}_{e_I}^{-1} df_i(\lambda T) \tag{12-100}$$

The Rosseland mean extinction coefficient can be calculated from the efficiency using Eq. (12-101)

$$K_{e_R} = \frac{1.5 f_v \overline{Q}_{e_R}}{d_{32}} \tag{12-101}$$

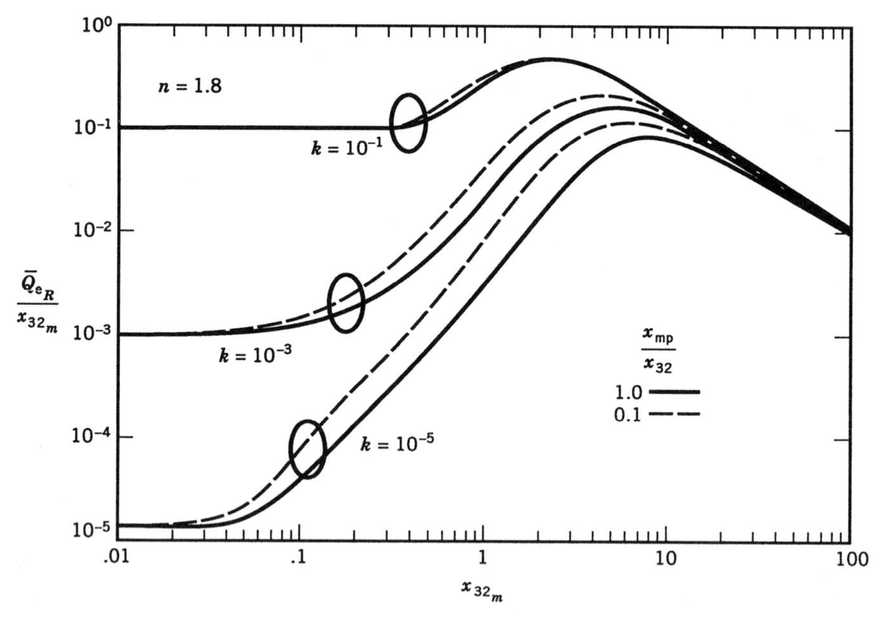

Figure 12-9 Rosseland extinction efficiency.

Equation (12-100) has been integrated [Brewster, 1989] assuming n and k are constant (independent of wavelength) and the results are plotted in Fig. 12-9 as a function of x_{32_m} where

$$x_{32_m} = \pi d_{32}/\lambda_m \tag{12-102}$$

$$\lambda_m T = C_3 = 2898 \ \mu\text{m K} \tag{12-103}$$

The general behavior of $\overline{Q}_{e_R}$ with x_{32_m} is similar to the behavior of Q_{e_l} with x_{32}. At small values of x_{32_m} (small values of d_{32} and/or low temperatures) $\overline{Q}_{e_R}$ is proportional to x_{32_m} such that K_{e_R} approaches a constant whose value depends on the magnitude of k. At large values of x_{32_m} (large values of d_{32} and/or high temperatures) $\overline{Q}_{e_R}$ approaches 1 such that K_{e_R} decreases with increasing particle size as $1/d_{32}$.

DIFFERENTIAL (P_N) APPROXIMATION

Noting that there are several equivalent ways the transfer equation can be converted to an approximate second-order differential equation for net radiative flux q, a brief discussion of transfer formulations follows. One such way is the discrete ordinates method introduced in this chapter [Chandrasekhar, 1960]. Another way is the lowest order spherical harmonics method (the P_1 approximation [Ozisik, 1985]). Still another way is the lowest order moment method (differential approximation or Eddington approximation [Eddington, 1959]). It has been shown by Krook [1955] that these methods are all equivalent. The governing differential equation obtained by each of these methods (see Problem 12-11) is

$$\frac{d^2q}{dt_I^2} - 3\zeta^2 q = 4\zeta^2 \frac{de_b}{dt_I} \tag{12-104}$$

where the effective isotropic optical depth is the same as that defined in Eq. (12-46)

$$dt_I = (1 - \omega_0 g)\, dt \tag{12-105}$$

and the parameter ζ is the same as that defined in Eq. (12-42)

$$\zeta = \sqrt{\frac{1 - \omega_0}{1 - \omega_0 g}} = \sqrt{1 - \omega_{0_I}} \tag{12-106}$$

An effective isotropic scattering albedo can also be defined as shown by Lee and Buckius [1982, 1983] as

$$\omega_{0_I} = \frac{\omega_0(1 - g)}{1 - \omega_0 g} \tag{12-107}$$

The subject of the appropriate boundary conditions to apply to Eq. (12-104) has been debated somewhat [Shokair and Pomraning, 1981; Ozisik, 1985]. The difficulty is that the natural variable for specifying boundary conditions (intensity) has been replaced by net flux, which is not usually known. For diffuse boundaries with emissivities $\varepsilon_{1,2}$ the following conditions, which are equivalent to those given by Sparrow and Cess [1978], are recommended:

$$\left(\frac{1}{\varepsilon_1} - \frac{1}{2}\right) q(0) - \frac{1}{4}\left(\frac{dq}{dt_I}\right)_0 = e_{b_1} - e_b(0) \tag{12-108}$$

$$\left(\frac{1}{\varepsilon_2} - \frac{1}{2}\right) q(t_{I_L}) - \frac{1}{4}\left(\frac{dq}{dt_I}\right)_{t_{I_L}} = e_b(t_{I_L}) - e_{b_2} \tag{12-109}$$

As discussed by Sparrow and Cess, these boundary conditions are compatible with the slip conditions in the optically thick regime and under radiative equilibrium [see Eq. (11-63)].

SUMMARY

Radiative transfer in particle dispersions often involves multiple scattering whereby photons that are emitted by the medium or are incident from the boundaries of the medium are scattered more than once before being reabsorbed elsewhere in the medium or at a boundary. Accounting for multiple scattering in general requires that the transfer equation be solved along all slant paths through the scattering medium simultaneously to account for in-scattering from other directions into the line of sight. Two techniques for solving the transfer equation with multiple scattering are the flux method and the discrete ordinate method. Both of these techniques result in systems of coupled first-order, ordinary differential equations. In the limit as the number of fluxes or discrete ordinates increases to infinity, either of these methods can be considered to be exact.

Radiative transfer in most realistic particle dispersions also involves anisotropic single scattering. Anisotropic scattering may or may not be important depending on the particular problem, but it increases the complexity of the analysis considerably. For heat transfer calculations anisotropic scattering properties (e.g., Mie scattering properties) can usually be converted to equivalent effective isotropic scattering properties.

REFERENCES

Abramowitz, M. and Stegun, I. A. (1972), *Handbook of Mathematical Functions*, Dover Publications, New York.

Brewster, M. Q. (1982), "Examination of the Two-Flux Model for Radiative Transfer in Particulate Systems," *Int. J. Heat Mass Transfer*, Vol. 25, pp. 1905–1907.

Brewster, M. Q. (1989), "Radiation-Stagnation Flow Model of Aluminized Solid Rocket Motor Internal Insulator Heat Transfer," *J. Thermophysics and Heat Transfer*, Vol. 3, No. 2, pp. 132–139.

Chandrasekhar, S. (1960), *Radiative Transfer*, Dover Publications, New York.

Chu, C. M. and Churchill, S. W. (1955), "Representation of the Angular Distribution of Radiation Scattered by a Spherical Particle," *J. Opt. Soc. America*, Vol. 45, pp. 958–962.

Eddington, A. S. (1959), *The Internal Constitution of Stars*, Dover, New York.

Evans, L. B., Chu, C. M., and Churchill, S. W. (1965), "The Effect of Anisotropic Scattering on Radiant Transport," *J. Heat Transfer*, August, pp. 381–387.

Krook, M. (1955), "On the Solution of the Equation of Transfer, I," *Astrophy. J.*, Vol. 122, pp. 488–497.

Lee, H. and Buckius, R. O. (1982), "Scaling Anisotropic Scattering in Radiation Heat Transfer for a Planar Medium," *J. Heat Transfer*, Vol. 104, pp. 68–75.

Lee, H. and Buckius, R. O. (1983), "Reducing Scattering to Non-Scattering Problems in Radiation Heat Transfer," *Int. J. Heat Mass Transfer*, Vol. 26, pp. 1055–1062.

Ozisik, M. N. (1985), *Radiative Transfer and Interactions with Conduction and Convection*, Werbel and Peck, New York.

Rish III, J. W. and Roux, J. A. (1986), "Evaluation of Emission Integrals for the Radiative Transport Equation," *AIAA Journal*, Vol. 24, No. 12, Dec., pp. 2049–2052.

Roux, J. A., Smith, A. M., and Todd, D. C. (1975), "Radiative Transfer with Anisotropic Scattering and Arbitrary Temperature for Plane Geometry," *AIAA Journal*, Vol. 13, No. 9, Sept., pp. 1203–1211.

Shokair, I. R. and Pomraning, G. C. (1981), "Boundary Conditions for Differential Approximations," *J. Quant. Spectr. and Rad. Transfer*, Vol. 25, No. 4, pp. 325–337.

Sparrow, E. M. and Cess, R. D. (1978), *Radiation Heat Transfer*, Augmented Edition, McGraw-Hill and Hemisphere, New York.

Wang, K. Y. and Tien, C-L. (1983), "Radiative Transfer Through Opacified Fibers and Powders," *J. Quant. Spect. Rad. Transfer*, Vol. 30, No. 3, pp. 213–223.

REFERENCES FOR FURTHER READING

Brewster, M. Q., "Effective Absorptivity and Emissivity of Particulate Media with Application to a Fluidized Bed," *J. Heat Transfer*, Vol. 108, 1986, pp. 710–713.

Cheng, P., "Two-Dimensional Radiating Gas Flow by a Moment Method," *AIAA J.*, Vol. 2, No. 9, 1964, pp. 1662–1664.

Condiff, D. W., "Anisotropic Scattering in Three-Dimensional Differential Approximation for Radiation Heat Transfer," *Int. J. Heat Transfer*, Vol. 30, No. 7, 1987, pp. 1371–1380.

Crosbie, A. L. and R. G. Shrenker, "Multiple Scattering in a Two Dimensional Rectangular Medium," *J. Quant. Spect. Rad. Transfer*, Vol. 33, No. 2, 1985, pp. 101–125.

Dayan, A. and C-L. Tien, "Heat Transfer in a Gray Planar Medium with Linear Anisotropic Scattering," *J. Heat Transfer*, Vol. 97, 1975, pp. 391–396.

Domoto, G. A. and W. C. Wang, "Radiative Transfer in Homogeneous Nongray Gases With Nonisotropic Particle Scattering," *J. Heat Transfer*, August 1974, pp. 385–390.

Fiveland, W. A., "Discrete-Ordinate Solutions of the Radiative Transport Equation for Rectangular Enclosures," *J. Heat Transfer*, Vol. 106, 1984, pp. 613–619.

Fiveland, W. A., "Discrete Ordinate Methods for Radiative Heat Transfer in Isotropically and Anisotropically Scattering Media," *J. Heat Transfer*, Vol. 109, No. 3, 1987, pp. 809–812.

Howell, J. R., "Thermal Radiation in Participating Media: The Past, the Present, and Some Possible Futures," *J. Heat Transfer*, Vol. 110, 1988, pp. 1220–1229.

Hsia, H. M. and T. J. Love, "Radiative Heat Transfer Between Parallel Plates Separated by a Nonisothermal Medium with Anisotropic Scattering," *J. Heat Transfer*, Vol. 89c, 1967, pp. 197–204.

Kourganoff, V. *Basic Methods in Transfer Problems*, Dover Publications, New York, 1963.

Kumar, S. and J. D. Felske, "Radiative Transport in a Planar Medium Exposed to Azimuthally Unsymmetric Incident Radiation," *J. Quant. Spectr. Rad. Transfer*, Vol. 35, 1986, pp. 187–212.

Kumar, S., A. Majumdar, and C-L. Tien, "The Differential-Discrete-Ordinate Method for Solutions of the Equation of Radiative Transfer," *J. Heat Transfer*, Vol. 112, 1990, pp. 424–429.

Ma, Y. and H. S. Lee, "Surface Exchange Model of Radiative Heat Transfer From Anisotropic Scattering Layers," *J. Heat Transfer*, Vol. 111, 1989, pp. 1015–1020.

Menguc, M. P. and R. Viskanta, "Comparison of Radiative Transfer Approximations for Highly Forward Scattering Planar Medium," *J. Quant. Spectr. Rad. Transfer*, Vol. 29, 1983, pp. 381–394.

Menguc, M. P. and R. Viskanta, "Radiative Transfer in Three Dimensional Enclosures Containing Inhomogeneous, Anisotropically Scattering Media," *J. Quant. Spect. Rad. Transfer*, Vol. 33, No. 6, 1985, pp. 533–549.

Modest, M. R. and F. H. Azad, "The Influence and Treatment of Mie-Anisotropic Scattering in Radiative Heat Transfer," *J. Heat Transfer*, Vol. 102, 1980, pp. 92–98.

Raithby, G. D. and E. H. Chui, "A Finite-Volume Method for Predicting Radiant Heat Transfer in Enclosures with Participating Media," *J. Heat Transfer*, Vol. 112, 1990, pp. 415–423.

Ratzel, A. C., III and J. R. Howell, "Two Dimensional Radiation in Absorbing-Emitting Media Using the P-N Approximation," *J. Heat Transfer*, Vol. 105, 1983, pp. 333–340.

Skocypec, R. D., D. V. Walters, and R. O. Buckius, "Spectral Emission Measurements from Planar Mixtures of Gas and Particulates," *J. Heat Transfer*, Vol. 109, 1987, pp. 151–158.

Tan, Z. "Radiative Heat Transfer in Multidimensional Emitting, Absorbing, and Anisotropic Scattering Media—Mathematical Formulation and Numerical Method," *J. Heat Transfer*, Vol. 111, 1989, pp. 141–147.

Tong, T. and C-L. Tien, "Radiative Heat Transfer in Fibrous Insulations—Part I: Analytical Study," *J. Heat Transfer*, Vol. 105, No. 1, 1983, pp. 70–75.

Truelove, J. S., "The Two-Flux Model for Radiative Transfer with Strongly Anisotropic Scattering," *Int. J. Heat and Mass Transfer*, Vol. 27, 1984, pp. 464–466.

Truelove, J. S., "Discrete Ordinate Solutions of the Radiative Transport Equation," *J. Heat Transfer*, Vol. 109, 1987, pp. 1048–1051.

PROBLEMS

1. Derive the two-flux equations (12-15) from (12-3) and (12-14) by integrating over forward and backward hemispheres.

2. A sheet of sintered, porous alumina (Al_2O_3) at room temperature contains pores with an average size (diameter) of 1 μm and a volume fraction of 3%. The pore scattering efficiency is 2, and the multiple scattering asymmetry parameter is $g = 0.8$. Neglect the reflectance at the surface of the sheet. Take the optical constants of the alumina to be $n = 1.75$ and $k = 10^{-6}$ at 1-μm wavelength. The pores are nonabsorbing pockets of air ($n = 1$, $k = 0$). According to the two discrete ordinates model:

(a) What are the absorptivity and reflectivity of an opaque (semi-infinite) sheet of this material for collimated, normal incident radiation at 1 μm?

(b) How thick must the sheet be to be opaque? (Take the criterion of opaqueness to be diffuse transmissivity of 0.01.)

3. Derive the expressions for the hemispherical, spectral transmissivity and reflectivity of an absorbing, scattering plane layer of thickness L subject to diffuse irradiation using the two-flux model. Check agreement with the semi-infinite result for hemispherical absorptivity by looking at the case of $L \to \infty$.

4. Use the discrete ordinate method with two ordinates to find the hemispherical absorptivity of a semi-infinite, planar, anisotropic scattering medium subject to diffuse irradiation as a function of albedo and multiple scattering asymmetry parameter g. Compare the result with that of the two-flux model.

5. **(a)** Derive Eq. (12-68) of Example 12-2 for the effective emissivity of a semi-infinite slab with an exponential temperature distribution using the two discrete ordinates/two-flux model.

(b) Compare the effective emissivity of the two discrete ordinates model with that of the two-flux model for $t_\delta = 1$, $\omega_0 = 0.1$ and 0.9, $g = 0$ and 0.5, and $\eta = 0.5$, 0.7, 0.9, and 1. Comment on the results of the two discrete ordinates model versus the two-flux model. (Does one consistently predict higher than the other? Is this trend consistent with the results of the slab in radiative equilibrium?) Comment on the results for anisotropic scattering versus isotropic scattering.

6. Derive Eq. (12-57) for the net radiant heat flux from an absorbing, emitting, scattering medium to a wall. Start from the diffuse radiosity and irradiation equations applied to the boundary and to the medium.

7. Show that the two discrete ordinates model for conservative scattering by a plane layer subject to collimated irradiation satisfies the overall energy balance exactly by predicting that the layer absorptivity is zero [i.e., prove Eq. (12-94) from the results of Example 12-3].

8. Some studies indicate that the Sauter mean diameter of aluminum oxide smoke particles produced by combustion of aluminum droplets may be as small as $d_{32} = 0.1$ μm.

(a) Based on this estimate, predict the Rosseland mean extinction coefficient in an aluminized solid rocket motor (including only aluminum oxide radiation). Assume the average temperature is 2900 K, the optical constants of molten aluminum oxide at $\lambda_m = 1$ μm are $n = 1.6$ and $k = 10^{-2}$, the aluminum oxide volume fraction is 4×10^{-4} and n and k are independent of wavelength. Also assume that the nonscattering Rayleigh limiting results hold for the particle scattering properties.

(b) Discuss the accuracy of the latter assumption in terms of appropriate parameters and in terms of Fig. 12-9 (for this comparison assume that $n = 1.8$ and 1.6 are close enough that a curve for $k = 10^{-2}$ can be interpolated approximately halfway between the curves for $k = 10^{-1}$ and 10^{-3}).

9. Apply the diffusion approximation to the problem of transfer between a semi-infinite, anisotropic scattering, emitting medium and a black wall (Example 12-2) to obtain an expression for the net dimensionless radiative heat flux to the wall, $q/\sigma T_\infty^4$ as a function of $\eta = (T_0/T_\infty)^4$, t_δ, ω_0, and $g = 1 - 2B$. Compare the results of this model with those predicted by the two-flux model (Example 12-2) or discrete ordinate model (Problems 12-4 and 12-5) for isotropic scattering ($g = 0$), $\omega_0 = 0.1$ and 0.9, $t_\delta = 1$ and 5, and $\eta = 0.5$, 0.7, and 0.9.

10. Investigate the effect of albedo on scattering by a plane layer subject to collimated incident radiation. Assume isotropic scattering and normal incidence. Compare the layer diffuse transmissivity and reflectivity over $0 < t_L < 5$ for $\omega_0 = 1$ and 0.9. Also calculate the layer absorptivity for $\omega_0 = 0.9$.

11. (a) Show that under the two discrete ordinates formulation of radiative transfer in an absorbing, emitting, anisostropic scattering, one-dimensional plane layer, the net radiative flux is given by Eq. (12-104).

(b) Show that the effective isotropic scattering albedo is given by Eq. (12-107).

CHAPTER 13

RADIATIVE TRANSFER COUPLED WITH CONDUCTION AND CONVECTION

Radiative transport of energy in a participating medium is usually accompanied by conductive energy transfer or convective transfer or both. With multiple modes of energy transport possible, the question that confronts the heat transfer analyst is what are the relative magnitudes of the various modes of energy transport. Does radiative transfer alone dominate? If so, radiative equilibrium establishes the temperature field. If radiative and conductive transport are of similar magnitudes a radiative-conductive (i.e. radiative-diffusive) balance establishes the temperature field. If radiative, conductive, and convective transport are all of similar magnitudes, then a radiative, diffusive, convective balance establishes the temperature field.

The occurrence of combined modes significantly increases the mathematical complexity of the governing equations. In particular, the energy and radiative transfer equations are coupled and must be solved simultaneously. This coupling introduces nonlinearity into the mathematical formulation in addition to the already complex radiative transfer formulation. As a result of the increased mathematical complexity of these types of problems, it is prudent to introduce various simplifications into the radiative transport description so that all aspects of the problem are described with equal and appropriate levels of complexity. Such simplifications include linearization of the mathematical equations, use of the diffusion approximation under optically thick conditions, and use of the decoupled approximation under optically thin conditions.

This chapter offers an introduction to combined mode heat transfer. First, coupled conduction and radiation are discussed when energy transport by fluid flow is negligible. Some fundamental principles and parameters are introduced and then illustrated by considering the problem of combined conduction and radiation in a gray, planar medium. Next, flow situations are considered where all three modes, conduction, convection, and radiation, are important. Both external and internal flows are discussed. As an example of external, boundary layer-like flows, the problem of stagnation flow of a gray, optically thick fluid is considered. As an example of internal flows, the problem of thermally developing, turbulent flow of a nongray, molecular gas is considered.

While these few examples serve to illustrate some of the principles involved and approaches that can be adopted in combined mode heat transfer analysis, it should be recognized that this treatment is far from complete. Much progress is still being made in understanding combined mode heat transfer. This chapter provides only an introduction to a rather complex area of study. The reader interested in further details about a specific problem is invited to consult the list of references at the end of the chapter.

COUPLED RADIATION AND CONDUCTION

Consider a participating medium that can be described as a continuum for both radiative and conductive energy transport. If there are no sources or sinks of energy, no convective energy transport and no unsteadiness, the energy balance on a differential volume element of the medium is

$$\nabla \cdot (\mathbf{q}_c + \mathbf{q}_r) = 0 \tag{13-1}$$

where the conductive heat flux vector is given by Fourier's law.

$$\mathbf{q}_c = -k\nabla T \tag{13-2}$$

The radiative heat flux vector is given by

$$\mathbf{q}_r = \sum_{i=1}^{3} q_{r_i}\mathbf{e}_i \tag{13-3}$$

where $\mathbf{e}_i$ are the unit vectors in the coordinate system of interest. The components of the flux q_{r_i} are obtained from

$$q_{r_i} = \int_0^\infty \int_{4\pi} I_\lambda(\Omega)\mathbf{e}_\Omega \cdot \mathbf{e}_i \, d\Omega \, d\lambda \tag{13-4}$$

where $\mathbf{e}_\Omega$ is the unit vector in the direction of intensity I_λ and the components of $\mathbf{e}_\Omega$ are the direction cosines of I_λ. The intensity field is obtained from the solution of the transfer equation.

$$\frac{dI_\lambda(\Omega)}{dt} = -I_\lambda(\Omega) + (1 - \omega_0)I_{b_\lambda} + \frac{\omega_0}{4\pi}\int_{4\pi} I_\lambda(\Omega')p_\lambda(\Omega' \to \Omega)\, d\Omega' \quad (13\text{-}5)$$

Approximate Formulations of Coupled Radiation and Conduction

The solution of the preceding equations represents a formidable mathematical challenge, particularly for complex geometries. However, the level of detail included, especially in the radiative transfer formulation, is often not justified for making heat transfer calculations. For heat transfer calculations it is appropriate to introduce simplifications into the radiative transfer description, which make the precision of the analysis more consistent with the uncertainty in the known properties and the resolution of the desired result.

Most heat transfer problems involving coupled conduction and radiation can be adequately solved by characterizing the problem as being either optically thick or optically thin based on a characteristic dimension of the system (such as slab thickness for plane, parallel geometry). For optically thick systems, the diffusion approximation can be used to represent the radiative flux

$$\mathbf{q}_r = -k_r(T)\nabla T \quad \text{(optically thick)} \qquad (13\text{-}6)$$

with the slip boundary condition applied to the radiative temperature terms in the energy equation (13-1) and the no-slip condition applied to the conductive terms. The radiative conductivity is defined by Eq. (11-81).

For optically thin systems, the radiative and conductive terms in the energy equation can be decoupled giving

$$\nabla \cdot \mathbf{q}_r = 0 \qquad (13\text{-}7)$$

$$\nabla \cdot \mathbf{q}_c = 0 \qquad (13\text{-}8)$$

$$\mathbf{q} = \mathbf{q}_r + \mathbf{q}_c \quad \text{(optically thin)} \qquad (13\text{-}9)$$

Furthermore, rather than rigorously solve (13-7) with (13-3), (13-4), and (13-5) the radiative flux $\mathbf{q}_r$ can be obtained by treating the medium as isothermal and nonscattering, using the methods of Chapter 10. These approximate techniques will be demonstrated and compared with the exact formulation using the slab geometry.

Coupled Radiation and Conduction in a Slab — Exact Formulation

Consider a stagnant, gray, nonscattering medium between black, parallel walls as shown in Fig. 13-1. The medium has a constant thermal conductivity k and a constant absorption (or extinction) coefficient $K_a(= K_e)$. The medium is adiabatic (has no energy sources or sinks), and the walls are isothermal at temperatures T_1 and T_2. The exact formulation of the coupled radiative and conductive transfer in this problem is carried out as follows.

The conductive heat flux normal to the slab at any location y is

$$q_c = -k\frac{dT}{dy} \tag{13-10}$$

The radiative flux normal to the slab at any location t was formulated in Chapter 11 as

$$\begin{aligned} q_r &= 2\pi\int_{-1}^{1} I\mu\, d\mu \\ &= 2\left\{e_{b_1}E_3(t) - e_{b_2}E_3(t_L - t) + \int_0^{t_L} e_b(t')E_2(|t - t'|)\, dt'\right\} \end{aligned} \tag{13-11}$$

where $t = K_e y$ for K_e independent of wavelength and position. Substituting (13-10) and (13-11) into the energy equation

$$\frac{dq_r}{dy} + \frac{dq_c}{dy} = 0 \tag{13-12}$$

and introducing the nondimensional temperature

$$\tilde{T} = \frac{T}{T_1} \tag{13-13}$$

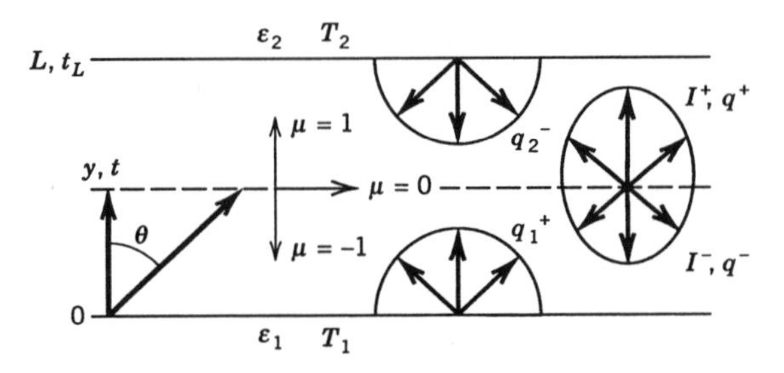

Figure 13-1 Slab geometry.

yields the nonlinear, integro-differential equation

$$N_1 \frac{d^2\tilde{T}}{dt^2} = \tilde{T}^4 - \tfrac{1}{2}\left\{E_2(t) + \tilde{T}_2^4 E_2(t_L - t) + \int_0^{t_L} \tilde{T}^4(t') E_1(|t - t'|)\, dt'\right\} \tag{13-14}$$

with boundary conditions $\tilde{T}(0) = 1$ and $\tilde{T}(t_L) = \tilde{T}_2$. The parameter N_1 appearing in (13-14) is the *conduction-radiation parameter* based on T_1.

$$N_1 = \frac{kK_e}{4\sigma T_1^3} \tag{13-15}$$

The conduction-radiation parameter is a measure of the relative importance of energy transport by radiation and conduction in a gray medium. For $N \to 0$ radiative transfer dominates (radiative equilibrium) and for $N \to \infty$ conduction dominates (nonparticipating medium). By recalling the definition of radiative conductivity under gray, optically thick conditions from Chapter 11, the conduction-radiation parameter can also be interpreted as a ratio of the molecular thermal conductivity to the radiative conductivity.

$$N = \frac{4}{3}\frac{k}{k_r} \tag{13-16}$$

Returning to the slab problem, once Eq. (13-14) has been solved numerically, the heat flux can be obtained by evaluating (13-10) and (13-11) at any location t (since q is constant). Evaluating at $t = 0$ gives

$$\frac{q}{\sigma T_1^4} = -4N_1 \left.\frac{d\tilde{T}}{dt}\right|_0 + 1 - 2\tilde{T}_2^4 E_3(t_L) + 2\int_0^{t_L} \tilde{T}(t) E_2(t)\, dt \tag{13-17}$$

This expression has been evaluated numerically by Viskanta and Grosh [1962] and the results for $\tilde{T}_2 = 0.5$ and $t_L = 0.1$, 1, and 10 are presented in Fig. 13-2 as discrete data points. From Fig. 13-2 it can be seen that for a given conduction-radiation parameter, as the optical thickness of the medium increases the dimensionless heat flux decreases.

Optically Thin Radiation and Conduction in a Slab — Decoupled Formulation

Under optically thin conditions ($t_L < 1$) the coupling between the radiative and conductive heat flux terms in the energy equation is weak, and the two contributions can be decoupled, calculated separately, and added together to give the total flux, as indicated in (13-7) through (13-9). The solution of (13-8)

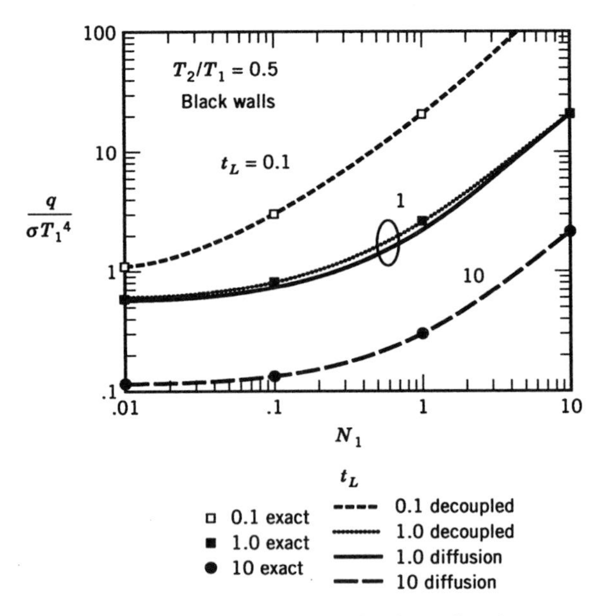

Figure 13-2 Combined conductive and radiative heat flux in a gray, nonscattering slab.

in the slab geometry gives the conductive flux as

$$q_c = \frac{k(T_1 - T_2)}{L} \tag{13-18}$$

Equation (13-7) was solved for the radiative heat flux for these conditions in Chapter 11 and the solution was given as

$$q_r = \frac{q_r^* \sigma (T_1^4 - T_2^4)}{1 + \left(\frac{1}{\varepsilon_1} + \frac{1}{\varepsilon_2} - 2\right) q_r^*} \tag{13-19}$$

The solution for q_r^* was obtained numerically and is given in Fig. 11-3. This numerical solution, however, adds undesirable complexity to the analysis and is not consistent with the level of approximation in the rest of the analysis. A simpler approach that is consistent with the optically thin approximation is to treat the medium as isothermal and use the techniques of Chapter 10 to obtain the radiative flux q_r.

Approximating the medium as isothermal, the two walls and medium can be represented as a three-node, source, sink, refractory system. The simultaneous algebraic equations can be solved using the network representation or by algebraic substitution, to give the net radiative flux from the source at T_1 to the sink at T_2 as

$$q_r = \frac{e_{b_1} - e_{b_2}}{\dfrac{1}{\varepsilon_1} + \dfrac{1}{\varepsilon_2} - 2 + \dfrac{1}{1 - \frac{1}{2}\bar{\varepsilon}_{12}}} \tag{13-20}$$

The mean emittance factor can be obtained from the concepts of Chapter 10 (see Problem 10-4) as

$$\bar{\varepsilon}_{12} = 1 - 2E_3(t_L) \tag{13-21}$$

Substituting (13-21) into (13-20) gives the net radiative flux as

$$q_r = \frac{\left[\frac{1}{2} + E_3(t_L)\right]\sigma\left(T_1^4 - T_2^4\right)}{1 + \left(\dfrac{1}{\varepsilon_1} + \dfrac{1}{\varepsilon_2} - 2\right)\left[\dfrac{1}{2} + E_3(t_L)\right]} \tag{13-22}$$

It is of interest at this point to compare the isothermal approximation for nondimensional radiative flux $[q_r^* = \frac{1}{2} + E_3(t_L)]$ with the exact numerical result. Table 13-1 shows that the isothermal result gives a good approximation to q_r^* for small values of t_L even up to $t_L = 1$ where the error in q_r^* is about 10%.

Combining (13-18) and (13-22) gives the total heat flux, nondimensionalized by σT_1^4 as

$$\frac{q}{\sigma T_1^4} = \frac{\left[\frac{1}{2} + E_3(t_L)\right]\left(1 - \tilde{T}_2^4\right)}{1 + \left(\dfrac{1}{\varepsilon_1} + \dfrac{1}{\varepsilon_2} - 2\right)\left[\dfrac{1}{2} + E_3(t_L)\right]} + \frac{4N_1\left(1 - \tilde{T}_2\right)}{t_L} \tag{13-23}$$

TABLE 13-1 Comparison between Exact Heat Flux and Isothermal Approximation for Optically Thin Slab in Radiative Equilibrium

t_L	Exact	Isothermal $[\frac{1}{2} + E_3(t_L)]$
0.1	0.9157	0.9163
0.2	0.8491	0.8519
0.3	0.7934	0.8000
0.4	0.7458	0.7573
0.5	0.7040	0.7216
0.6	0.6672	0.6916
0.8	0.6046	0.6443
1.0	0.5532	0.6097

Figure 13-2 shows the predictions for heat flux from the decoupled, optically thin analysis [Eq. (13-23)] for $t_L = 0.1$ and 1.0 and the comparison with the exact results (at least for $t_L \leq 1$) is very good.

One of the drawbacks of the decoupled analysis is that the temperature field cannot be readily determined except in the extreme cases where either radiative transfer or conductive transfer dominates. If conduction dominates ($N_1 \gg 1$), then obviously a linear, or nearly linear profile would be established. If radiation dominates ($N_1 \ll 1$), a nearly isothermal profile would be established.

Optically Thick Radiation and Conduction in a Slab — Nonlinear Diffusion Formulation

Next consider the planar participating medium under optically thick conditions ($t_L > 1$). In this case the diffusion approximation holds and the net radiative flux is given by

$$q_r = -k_r(T)\frac{dT}{dy} \tag{13-24}$$

where

$$k_r = \frac{16\sigma T^3}{3K_{e_R}} \quad \left(\text{gray: } K_{e_R} = K_e = \text{constant}\right) \tag{13-25}$$

The energy equation

$$\frac{d}{dy}\left[-(k_r + k)\frac{dT}{dy}\right] = 0 \tag{13-26}$$

can be integrated once to give the (constant) total net flux q,

$$-(k_r + k)\frac{dT}{dy} = C_1 = q \tag{13-27}$$

and again to give the temperature field.

$$\frac{4}{3}\frac{\sigma T^4}{K_e} + kT = -qy + C_2 \tag{13-28}$$

The constants of integration (q and C_2) can be determined by applying the boundary conditions. The no-slip conditions

$$T(y = 0) = T_1 \tag{13-29}$$

$$T(y = L) = T_2 \tag{13-30}$$

are applied to the conductive term (kT) and the slip conditions [Eqs. (11-63), (11-64), and (11-65)]

$$\psi_1 = \frac{1}{\varepsilon_1} - \frac{1}{2} = \frac{\sigma\left[T_1^4 - T(0)^4\right]}{q_{r_0}} \tag{13-31}$$

$$\psi_2 = \frac{1}{\varepsilon_2} - \frac{1}{2} = \frac{\sigma\left[T(L)^4 - T_2^4\right]}{q_{r_L}} \tag{13-32}$$

are applied to the radiative term (σT^4).

The radiative flux terms at the walls are obtained by evaluating (13-27) at 0 and L and introducing (13-16) and (13-24) to obtain

$$q_{r_0} = q\left(\frac{1}{1 + 3N_1/4}\right) \tag{13-33}$$

$$q_{r_L} = q\left(\frac{1}{1 + 3N_2/4}\right) \tag{13-34}$$

Substituting (13-33) and (13-34) into (13-31) and (13-32) gives the temperature slip boundary conditions as

$$T^4(0) = T_1^4 - \frac{q}{\sigma}\left(\frac{1}{\varepsilon_1} - \frac{1}{2}\right)\left(\frac{1}{1 + 3N_1/4}\right) \tag{13-35}$$

$$T^4(L) = T_2^4 + \frac{q}{\sigma}\left(\frac{1}{\varepsilon_2} - \frac{1}{2}\right)\left(\frac{1}{1 + 3N_2/4}\right) \tag{13-36}$$

Evaluating (13-28) using (13-29), (13-30), (13-35), and (13-36) gives two equations that can be solved for q and C_2. The resulting equation for the heat flux is

$$q = \frac{\sigma\left(T_1^4 - T_2^4\right) + \dfrac{k}{L}(T_1 - T_2)\dfrac{3t_L}{4}}{\dfrac{3t_L}{4} + \left(\dfrac{1}{\varepsilon_1} - \dfrac{1}{2}\right)\left(\dfrac{1}{1 + 3N_1/4}\right) + \left(\dfrac{1}{\varepsilon_2} - \dfrac{1}{2}\right)\left(\dfrac{1}{1 + 3N_2/4}\right)} \tag{13-37}$$

or in nondimensional form

$$\frac{q}{\sigma T_1^4} = \frac{\left(1 - \tilde{T}_2^4\right) + 3N_1\left(1 - \tilde{T}_2\right)}{\dfrac{3t_L}{4} + \left(\dfrac{1}{\varepsilon_1} - \dfrac{1}{2}\right)\left(\dfrac{1}{1 + 3N_1/4}\right) + \left(\dfrac{1}{\varepsilon_2} - \dfrac{1}{2}\right)\left(\dfrac{1}{1 + 3N_1/4\tilde{T}_2^3}\right)} \tag{13-38}$$

The limiting relations for radiative equilibrium and conductive equilibrium can be obtained by taking the limits $N_1 \to 0$ and $N_1 \to \infty$, respectively.

$$q = \frac{\sigma\left(T_1^4 - T_2^4\right)}{\frac{3}{4}t_L + 1/\varepsilon_1 + 1/\varepsilon_2 - 1} \qquad (N_{1,2} \ll 1) \tag{13-39}$$

$$q = \frac{k(T_1 - T_2)}{L} \qquad (N_{1,2} \gg 1) \tag{13-40}$$

A comparison between the results of (13-38) and the exact numerical results for black walls shows that (13-38) is accurate to within a few percent even at optical thicknesses as low as $t_L = 1$ (see Fig. 13-2).

Optically Thick Radiation and Conduction in a Slab — Linearized Diffusion Formulation

Although the nonlinear problem of optically thick radiation and conduction can be solved in closed form, as just demonstrated, there are many problems, especially those including convective transport, that cannot. These problems can be solved approximately, however, by linearization, and that approach is discussed next.

The approach taken is to linearize the T^4 radiation term of the diffusion approximation to match the conduction term that is linear in T. (The opposite approach of linearizing the conduction term to match the radiation term is not as often employed, but equally valid in principle.)

The diffusion radiation term T^4 is expanded in a truncated Taylor's series about a nearby reference temperature T_r.

$$T^4 \sim 4T_r^3 T - 3T_r^4 \tag{13-41}$$

Based on (13-41), the temperature profile for the slab in conductive-radiative equilibrium (13-28) can be written as

$$\frac{4}{3}\frac{\sigma}{K_e}\left[4T_r^3 T - 3T_r^4\right] + kT = -qy + C_2 \tag{13-42}$$

where $T_r = T_1$ near $y = 0$ and $T_r = T_2$ near $y = L$. The radiative slip conditions (13-31) and (13-32) become

$$\psi_1 = \frac{1}{\varepsilon_1} - \frac{1}{2} = \frac{4\sigma T_1^3[T_1 - T(0)]}{q_{r_0}} \tag{13-43}$$

$$\psi_2 = \frac{1}{\varepsilon_2} - \frac{1}{2} = \frac{4\sigma T_2^3[T(L) - T_2]}{q_{r_L}} \tag{13-44}$$

where the reference temperature T_r has been taken to be T_1 in (13-43) and T_2 in (13-44). Equations (13-33) and (13-34) are used to evaluate q_{r_0} and q_{r_L} in the slip conditions. The temperature equation (13-42) is evaluated at $y = 0$ ($T_r = T_1$) and at $y = L$ ($T_r = T_2$), using (13-29) and (13-30) for the conduction term and (13-43) and (13-44) for the radiation term. The result is that the same expression for heat flux Eq. (13-37) or (13-38) is obtained. Thus the same result for heat flux is obtained in this case from the linear and nonlinear formulations.

Uniform Boundary Conditions

In the preceding example the selective application of boundary conditions (i.e., no slip to the conduction and slip to the radiation terms) was straightforward, either on a nonlinear or linear basis. However in more complicated combined mode problems, particularly those involving convective transport, it is not convenient to selectively apply the boundary conditions to the various terms in the energy equation. It would be useful to have a method for applying a uniform boundary condition (i.e., same boundary condition applied to all terms in the energy equation), which gives the same total heat flux as in the selective boundary condition case. Such a method is next presented that will prove to be useful for solving combined convection, conduction, and radiation problems.

Consider the boundary conditions described by the following relations:

$$\psi_1 = \left(\frac{1}{\varepsilon_1} - \frac{1}{2}\right)\left(\frac{1}{1 + 3N_1/4}\right) = \frac{4\sigma T_1^3[T_1 - T(0)]}{q_{r_0}} \tag{13-45}$$

$$\psi_2 = \left(\frac{1}{\varepsilon_2} - \frac{1}{2}\right)\left(\frac{1}{1 + 3N_2/4}\right) = \frac{4\sigma T_2^3[T(L) - T_2]}{q_{r_L}} \tag{13-46}$$

Equations (13-45) and (13-46) are not the correct slip conditions as derived for optically thick radiative transport. An extra term involving the conduction-radiation parameter has been introduced into Eqs. (13-45) and (13-46). The correct radiative slip conditions are given by Eqs. (13-43) and (13-44).

However, if these slip boundary conditions (13-45) and (13-46) are applied uniformly (i.e., to every term) in (13-42), including the conduction term, the resulting expression for heat flux is again the same as that given by (13-37) or (13-38). Thus (13-45) and (13-46) represent effective, linearized, combined diffusion-radiation and conduction boundary conditions. These relations are also useful for flow problems involving convective transport with impermeable boundaries, since, in the region of the wall, the no-slip velocity condition holds that energy transport occurs only by conduction and radiation anyway.

COUPLED RADIATION, CONDUCTION, AND CONVECTION

Convective effects will next be included by considering a participating fluid medium, which can be described as a continuum. This discussion assumes the reader has a general background in convective heat transfer similar to the treatment presented by Kays and Crawford [1980]. The general equations for conservation of mass, momentum, and energy for a differential volumetric fluid element at steady state are

$$\nabla \cdot (\rho \mathbf{V}) = 0 \tag{13-47}$$

$$\rho \mathbf{V} \cdot \nabla \mathbf{V} = \mathbf{f} + \nabla \cdot \underline{\underline{\tau}} \tag{13-48}$$

$$\rho \mathbf{V} \cdot \nabla h = \mathbf{V} \cdot \nabla P - \nabla \cdot (\mathbf{q}_c + \mathbf{q}_r) + \Phi \tag{13-49}$$

where $\mathbf{V}$ is the fluid velocity vector, $\mathbf{f}$ is the body force per unit volume, $\underline{\underline{\tau}}$ is the stress tensor, h is the enthalpy, P is the pressure, and Φ is the viscous dissipation. The radiative flux vector $\mathbf{q}_r$ is obtained from the solution of the transfer equation, as discussed in the previous section.

Approximate Formulations of Coupled Radiation, Conduction, and Convection

For heat transfer calculations it is again appropriate to introduce simplifications that make the radiative transfer description consistent with the goal of obtaining a balanced formulation overall. It is again expedient to characterize the radiative transfer as being either optically thin or optically thick and use either an uncoupled, isothermal formulation or the diffusion approximation, as was discussed in the previous section for coupled radiation and conduction. However, the distinction to be made in this case is that the initial designation of optical thickness should be made on the basis of the thermal boundary layer thickness (δ_T; see Fig. 13-3), which is established by the convective, conductive, and radiative balance. If it is determined that the thermal boundary layer is optically thick ($t_{\delta_T} > 1$), then a coupled analysis using the diffusion approximation is appropriate. If it is determined that the thermal boundary layer is optically thin ($t_{\delta_T} < 1$), then a decoupled, isothermal analysis is appropriate. The important point is that *for predicting heat transfer at a boundary, the issue of whether the medium is optically thick or*

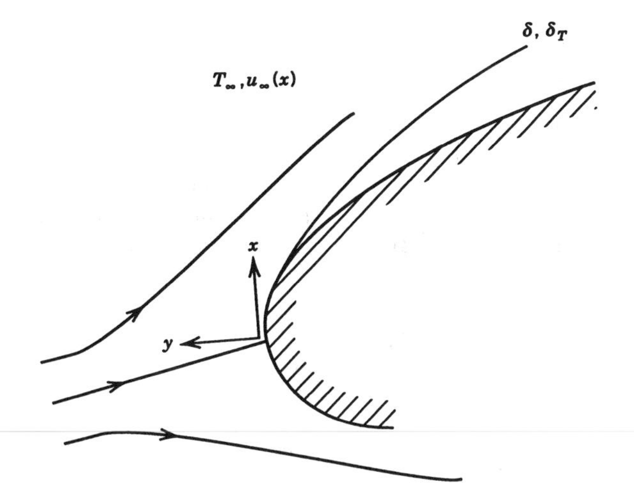

Figure 13-3 Two-dimensional (planar) boundary layer flow.

thin based on enclosure dimensions is secondary to the issue of the optical thickness based on thermal boundary layer.

RADIATIVE TRANSFER IN BOUNDARY LAYER FLOWS

Radiative transfer plays an important role in many boundary layer flows including atmospheric re-entry and metalized solid rocket internal and external flow fields. This class of problems will be illustrated here using steady, incompressible, two-dimensional, laminar equations. The governing equations for a two-dimensional (planar), steady, laminar boundary layer flow are [Ozisik, 1985]

$$\frac{\partial}{\partial x}(\rho u) + \frac{\partial}{\partial y}(\rho v) = 0 \tag{13-50}$$

$$\rho\left(u\frac{\partial u}{\partial x} + v\frac{\partial u}{\partial y}\right) = -\frac{dP}{dx} + \frac{\partial}{\partial y}\left(\mu\frac{\partial u}{\partial y}\right) \tag{13-51}$$

$$\rho C_p\left(u\frac{\partial T}{\partial x} + v\frac{\partial T}{\partial y}\right) = \frac{\partial}{\partial y}\left(k\frac{\partial T}{\partial y}\right) - \frac{\partial q_{r_y}}{\partial y} + u\frac{dP}{dx} + \mu\left(\frac{\partial u}{\partial y}\right)^2 \tag{13-52}$$

$$-\frac{dP}{dx} = \rho_\infty u_\infty \frac{du_\infty}{dx} \tag{13-53}$$

This flow configuration is illustrated in Fig. 13-3. It is used as an example here with the understanding that time dependence, three-dimensional effects, compressibility, and turbulence are fluid mechanical concerns outside the scope of this treatment, but which can be included when necessary.

Optically Thick, Incompressible Flow of a Gray Fluid over an Isothermal Wedge

To illustrate this class of problems consider optically thick, incompressible flow of a gray fluid at ambient temperature T_∞ over an isothermal wedge (wedge angle $= \pi\beta$), which is at T_0 (Fig. 13-4). Neglecting property temperature dependence, pressure gradient, and dissipation effects in the energy equation, and introducing the diffusion approximation (13-24), the governing equations become

$$\frac{\partial u}{\partial x} + \frac{\partial v}{\partial y} = 0 \tag{13-54}$$

$$u\frac{\partial u}{\partial x} + v\frac{\partial u}{\partial y} = u_\infty \frac{du_\infty}{dx} + \nu \frac{\partial^2 u}{\partial y^2} \tag{13-55}$$

$$u\frac{dT}{dx} + v\frac{\partial T}{\partial y} = \alpha\frac{\partial^2 T}{\partial y^2} + \frac{\alpha}{3NT^3}\frac{\partial^2 T^4}{\partial y^2} \tag{13-56}$$

where α is thermal diffusivity, ν is kinematic viscosity, and N is the conduction-radiation parameter. The boundary conditions are

$$x = 0: \quad u = u_\infty, v = v_\infty, T = T_\infty \tag{13-57}$$

$$y \to \infty: \quad u = u_\infty, v = v_\infty, T = T_\infty \tag{13-58}$$

$$y = 0: \quad u = 0, \ v = 0, \ T = T_0 \qquad \text{(conduction term)}$$

$$\psi_0 = \frac{1}{\varepsilon_w} - \frac{1}{2} = \frac{\sigma\left(T_0^4 - T^4\right)}{-\dfrac{4}{3}\dfrac{\sigma}{K_e}\dfrac{\partial T^4}{\partial y}} \qquad \text{(radiation term)} \tag{13-59}$$

where $u_\infty = Cx^m$ is the free stream velocity and $m = \beta/(2 - \beta)$ is the wedge angle parameter.

Upon introducing the stream function ψ, the nondimensional stream function variable F, the boundary layer similarity variable η, and the nondimensional temperature $\tilde{T}$,

$$\eta = y\sqrt{\frac{u_\infty}{\nu x}} \tag{13-60}$$

$$\psi = F(\eta)\sqrt{\nu x u_\infty} \tag{13-61}$$

$$\tilde{T} = \frac{T}{T_0} \tag{13-62}$$

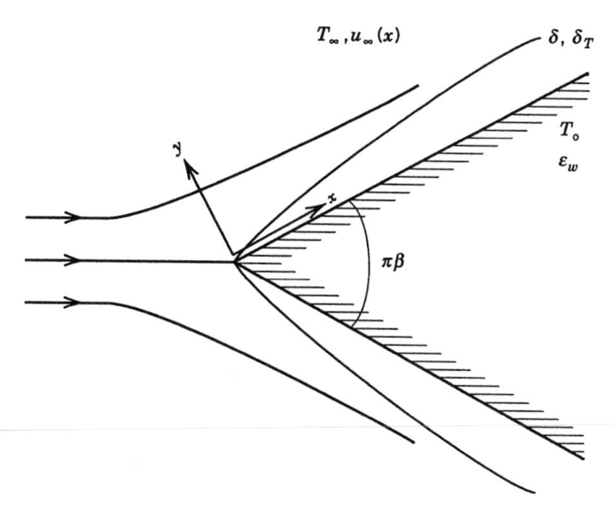

Figure 13-4 Wedge flow.

the governing equations (13-54) to (13-56) become

$$F''' + \frac{1+m}{2} FF'' + m(1 - F'^2) = 0 \tag{13-63}$$

$$\frac{1}{\Pr}\frac{d}{d\eta}\left[\frac{d\tilde{T}}{d\eta} + \frac{1}{3N_0}\frac{d\tilde{T}^4}{d\eta}\right] + \frac{1+m}{2} F\frac{d\tilde{T}}{d\eta} = 0 \tag{13-64}$$

where

$$\Pr = \frac{\nu}{\alpha} \quad \text{(Prandtl number)} \tag{13-65}$$

$$N_0 = \frac{kK_e}{4\sigma T_0^3} \quad \text{(conduction-radiation parameter)} \tag{13-66}$$

$$m = \frac{\beta}{2-\beta} \quad \text{(wedge angle parameter)} \tag{13-67}$$

The boundary conditions become

$$\eta \to \infty: \quad F = 1, \tilde{T} = \tilde{T}_\infty = \frac{T_\infty}{T_0} \tag{13-68}$$

$$\eta = 0: \quad F = F' = 0, \tilde{T} = 1 \tag{13-69}$$

where the no-slip condition has been imposed on even the radiation term in order to achieve similarity. (This restriction will be removed in the following example of stagnation flow.) For a given wedge angle parameter m, the momentum equation (13-63) can be numerically integrated to give $F(\eta)$. Then the energy equation (13-64) can be numerically integrated to give $\tilde{T}$. A closed-form solution can be obtained by linearizing the radiation term $\tilde{T}^4$ about 1

$$\tilde{T}^4 \sim 1 + 4(\tilde{T} - 1) + \cdots \cong 4\tilde{T} - 3 \tag{13-70}$$

giving, for the energy equation,

$$\frac{d^2\tilde{T}}{d\eta^2} + \left(\frac{1+m}{2}\right)\mathrm{Pr}_m F \frac{d\tilde{T}}{d\eta} = 0 \tag{13-71}$$

where

$$\mathrm{Pr}_m = \frac{\mathrm{Pr}}{1 + \dfrac{4}{3N_0}} \tag{13-72}$$

is the *radiation modified Prandtl number*. Equation (13-71) is the same equation that applies to a radiatively nonparticipating fluid with Pr replaced by Pr_m. Since N_0 is always greater than or equal to zero, the influence of radiative transport is to increase the effective fluid thermal conductivity or decrease the effective Prandtl number. Thus radiative transport thickens the thermal boundary layer relative to the case without radiation. If radiative transport is very strong ($N_0 \ll 1$), it is possible to obtain low Prandtl number flow behavior ($\delta_T \gg \delta$) even from a moderate Prandtl number fluid ($\mathrm{Pr} \sim 1$).

Optically Thick, Incompressible Stagnation Flow of a Gray Fluid on an Isothermal Wall

As a specific example of the wedge flow solutions, consider stagnation flow ($m = 1$) as pictured in Fig. 13-5. The outer potential flow field is given by

$$\psi_\infty(x, y) = \frac{C}{n-1} x^{n-1} y \tag{13-73}$$

$$u_\infty = \frac{1}{x^{n-2}} \frac{\partial \psi_\infty}{\partial y} = \frac{C}{n-1} x \tag{13-74}$$

$$v_\infty = -\frac{1}{x^{n-2}} \frac{\partial \psi_\infty}{\partial x} = -Cy \tag{13-75}$$

where n is a parameter designating either two-dimensional planar ($n = 2$) or

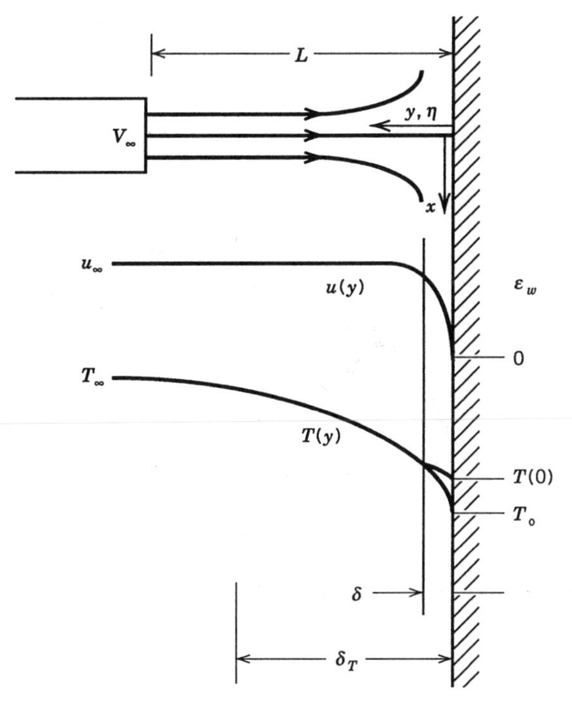

Figure 13-5 Radiating stagnation flow.

axisymmetric ($n = 3$) flow. The inner viscous flow field is given by

$$\psi(x, y) = \frac{x^{n-1}}{n-1}\sqrt{C\nu}\,F(\eta, n) \tag{13-76}$$

$$u = \frac{Cx}{n-1}F'(\eta, n) \tag{13-77}$$

$$v = -\sqrt{C\nu}\,F(\eta, n) \tag{13-78}$$

where

$$\eta = y\sqrt{\frac{C}{\nu}} = \frac{y}{\delta} \tag{13-79}$$

Equation (13-79) defines δ as the characteristic momentum boundary layer thickness

$$\delta = y|_{\eta=1} = \sqrt{\frac{\nu}{C}} \sim \sqrt{\frac{\nu L}{V_\infty}} \tag{13-80}$$

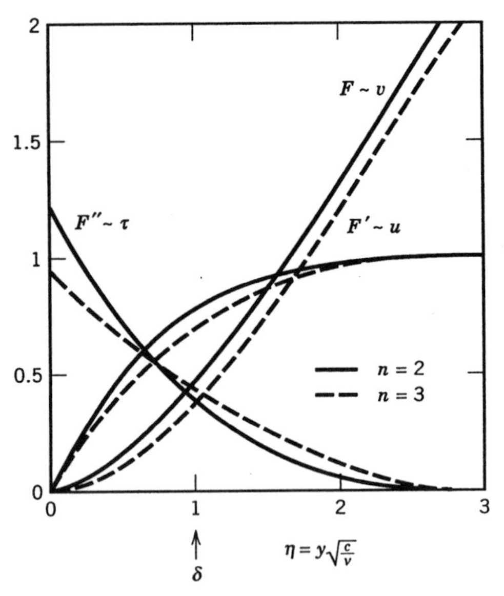

Figure 13-6 Incompressible stagnation flow field.

which can be related to a hypothetical nozzle separation distance L and exit velocity V_∞ as shown in (13-80). The solution of the momentum equation

$$F''' + FF'' + \frac{1}{n-1}(1 - F'^2) = 0 \tag{13-81}$$

$$F(0) = F'(0) = 0, F(\infty) = 1$$

is shown in Fig. 13-6 and can be found in standard viscous fluid flow textbooks [White, 1974].

The energy equation (13-71) can be reformulated in terms of the usual temperature nondimensionalization, since the radiation term has been linearized.

$$T^{*''} + F \, \mathrm{Pr}_m T^{*'} = 0 \tag{13-82}$$

$$T^* = \frac{T - T_0}{T_\infty - T_0} \tag{13-83}$$

$$\eta \to \infty: \quad T^* = 1 \tag{13-84}$$

$$\eta = 0: \quad \psi_0 = \left(\frac{1}{1 + \dfrac{3N_0}{4}} \right) \left(\frac{1}{\varepsilon_w} - \frac{1}{2} \right) = \frac{3}{4} \frac{T^*(0) t_\delta}{T^{*'}(0)} \tag{13-85}$$

Since the radiation and conduction terms cannot be conveniently separated to apply separate boundary conditions, the combined, linearized, effective slip condition from (13-45) is applied uniformly to both radiation and conduction terms in (13-82). This procedure was shown to be correct in the previous section for coupled conduction-radiation problems. It also holds in this case since convective energy transport at the wall goes to zero.

Equation (13-82) can be solved by applying the integrating factor $\exp[\mathrm{Pr}_m \int_0^\eta F(\eta')\,d\eta']$ and the boundary conditions (13-84) and (13-85). The result for the temperature field [Brewster, 1989] is

$$T^*(\eta)\left\{\left(\frac{1}{1+\dfrac{3N_0}{4}}\right)\left(\frac{1}{\varepsilon_w}-\frac{1}{2}\right)\frac{4}{3t_\delta}+\int_0^\eta \exp\left[-\mathrm{Pr}_m\int_0^{\eta'}F(\eta'')\,d\eta''\right]d\eta'\right\}\mathrm{Nu}_\delta \tag{13-86}$$

where

$$\mathrm{Nu}_\delta = T^{*\prime}(0) = \left\{\left(\frac{1}{1+\dfrac{3N_0}{4}}\right)\left(\frac{1}{\varepsilon_w}-\frac{1}{2}\right)\frac{4}{3t_\delta}+\int_0^\infty \exp\left[-\mathrm{Pr}_m\int_0^{\eta'}F(\eta'')\,d\eta''\right]d\eta'\right\}^{-1} \tag{13-87}$$

is the usual Nusselt number based on δ, defined as

$$\mathrm{Nu}_\delta = \frac{h\delta}{k} \tag{13-88}$$

and h is the convective heat transfer coefficient.

$$h = \frac{q_c}{T_\infty - T_0} = \frac{k\left.\dfrac{dT}{dy}\right|_{y=0}}{T_\infty - T_0} \tag{13-89}$$

The components of heat flux to the wall come from (13-10) and (13-24) as

$$q_c = k\left.\frac{dT}{dy}\right|_{y=0} \tag{13-90}$$

$$q_r = \frac{16}{3}\frac{\sigma T_0^3}{K_e}\left.\frac{dT}{dy}\right|_{y=0} \tag{13-91}$$

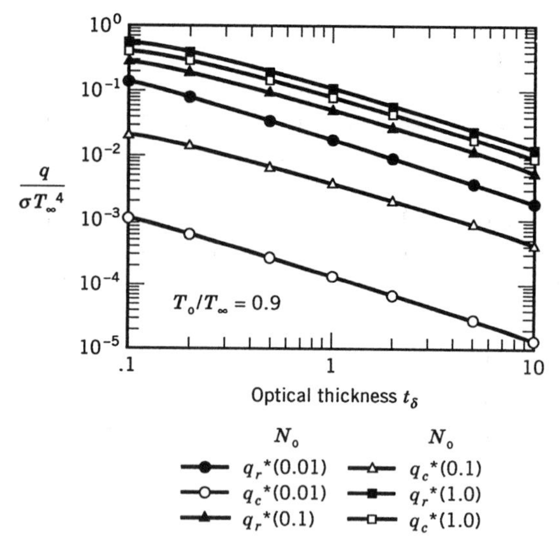

Figure 13-7 Radiative and conductive heat flux components in optically thick stagnation flow.

where the minus sign has been dropped in Eqs. (13-90) and (13-91) indicating that q_r and q_c represent heat flux in the negative y direction. Combining (13-90) and (13-91) and nondimensionalizing gives the total heat flux as

$$\frac{q_r + q_c}{\sigma T_\infty^4} = 4\left(\frac{4}{3} + N_0\right)\left(\frac{T_0}{T_\infty}\right)^3\left(1 - \frac{T_0}{T_\infty}\right)\frac{\mathrm{Nu}_\delta}{t_\delta} \tag{13-92}$$

The solutions for the heat flux components (13-92) are plotted in Fig. 13-7 for $\varepsilon_w = 1$, Pr $= 0.5$, $n = 2$, $N_0 = 0.01$, 0.1 and 1, $T_0/T_\infty = 0.9$, and t_δ from 0.1 to 10.

From Fig. 13-7 the nondimensional heat flux is seen to be a strong function of t_δ. As the boundary layer optical thickness t_δ increases, $q/\sigma T_\infty^4$ decreases approximately as $1/t_\delta$. As N_0 increases, $q/\sigma T_\infty^4$ increases approximately as $\sqrt{N_0}$. For small N_0 the radiative flux is larger than the conductive flux by a factor of $1/N_0$. A correlation for the total heat flux can be given as follows:

$$\frac{q}{\sigma T_\infty^4} = a\left(\frac{\sqrt{N_0}}{t_\delta}\right)^b \tag{13-93}$$

For a black wall and planar flow with Pr $= 0.5$ the following constants apply to Eq. (13-93).

T_0/T_∞	a	b
0.7	0.22	0.90
0.9	0.14	0.87

When the modified Prandtl number is very small (such as when Pr $\sim$ 1 and $N_0 \ll 1$), viscous effects are confined to a thin region next to the wall ($\delta \ll \delta_T$). In this situation the heat transfer can be adequately calculated using the potential flow field. Substituting $F(\eta) = \eta$ into (13-87) gives a closed-form expression for $q/\sigma T_\infty^4$ in terms of the error function. This derivation is left as an exercise (Problem 13-5).

Before concluding this problem, it is important to mention the criterion for optical thickness, since the solution has been based on the radiation diffusion approximation ($t_{\delta_T} > 1$). By comparing the momentum and energy transport length scales it can be shown that

$$\frac{\delta}{\delta_T} \sim \sqrt{\mathrm{Pr}_m} \tag{13-94}$$

or

$$t_{\delta_T} \sim \frac{t_\delta}{\sqrt{\mathrm{Pr}_m}} \tag{13-95}$$

where

$$t_\delta = K_e \delta \tag{13-96}$$

Thus the condition of optical thickness for applying the optically thick diffusion analysis is that t_{δ_T} evaluated from (13-95), (13-96), (13-80), and (13-72) must be approximately one or greater.

Optically Thin, Incompressible Stagnation Flow

When the thermal boundary layer in the radiating stagnation flow problem is optically thin ($t_{\delta_T} < 1$), the radiative transport can be decoupled from the conductive and convective transport and calculated independently.

$$q = q_c + q_r \tag{13-97}$$

The conductive contribution is evaluated from (13-87), setting $N_0 \to \infty$ and $\mathrm{Pr}_m \to \mathrm{Pr}$. The radiative contribution can be calculated as shown by

Eq. (12-57) from

$$q_r = \frac{\sigma\left(T_\infty^4 - T_0^4\right)}{\dfrac{1}{\varepsilon_{\text{eff}}} + \dfrac{1}{\varepsilon_w} - 1} \tag{13-98}$$

assuming that the medium is semi-infinite, that is, optically thick based on the enclosure or system dimensions. The effective emissivity of the fluid can be estimated from (12-72), noting that $\varepsilon_{\text{eff}} = \alpha_{\text{eff}}$ for an isothermal medium.

Example 13-1

Estimate the heat flux to a solid rocket motor internal insulator subject to stagnation flow under the following conditions. The gas flow has a molecular Prandtl number of Pr = 0.5 and is laden with radiating molten aluminum oxide particles with an optical mean size of $d_{32} = 1$ μm and a volume fraction of $f_v = 4 \times 10^{-4}$ (corresponding to complete combustion of a 16% aluminized propellant at 3.4 MPa). The gas has a bulk temperature of 3600 K, thermal conductivity of $k = 0.004$ W/(cm K), kinematic viscosity of $\nu = 1.0$ cm^2/s and the velocity field is such that the flow constant in (13-80) is $C = 6$ s^{-1}. The insulator is assumed to be black at 2500 K.

The characteristic wavelength range at these temperatures is $\lambda_m = 0.8$ to 1 μm from Eq. (12-103). The mean particle size parameter from (12-102) is thus

$$x_{32_m} = \frac{\pi(1\ \mu\text{m})}{1\ \mu\text{m}} = 3.1$$

At these wavelengths and temperatures the absorption index of molten aluminum oxide is estimated to be $k = 10^{-3}$ and the refractive index is 1.6 [Calia, et al., 1989; Parker, et al., 1989; Parry and Brewster, 1990]. Anticipating that the flow is optically thick, the effective gray extinction coefficient will be estimated using the Rosseland mean value. Ignoring the minor discrepancy between $n = 1.6$ and 1.8, the Rosseland mean extinction efficiency can be estimated from Fig. 12-9 assuming a wide particle size distribution ($x_{\text{mp}}/x_{32} = 0.1$) as

$$\frac{\overline{Q}_{eR}}{x_{32_m}} = 0.2, \qquad \overline{Q}_{e_R} = (0.2)(3.1) = 0.62$$

The Rosseland extinction coefficient comes from Eq. (12-101) as

$$K_e = K_{e_R} = \frac{1.5 f_v \overline{Q}_{e_R}}{d_{32}} = \frac{(1.5)(4 \times 10^{-4})(0.62)}{1 \times 10^{-4}\ \text{cm}} = 4\ \text{cm}^{-1}$$

and the conduction-radiation parameter from (13-66) is

$$N_0 = \frac{kK_e}{4\sigma T_0^3} = \frac{(0.004)(4)}{(4)(5.67 \times 10^{-12})(2500)^3} = 0.05$$

Thus it can be seen that since $N_0 \ll 1$, radiative transport plays a dominant role in the heat transfer to the insulator.

It remains to verify that the flow is optically thick and to calculate the heat flux. Both of these steps require that the boundary layer optical thickness be evaluated. The hydrodynamic layer thickness can be evaluated using Eq. (13-80)

$$\delta = \sqrt{\frac{\nu}{C}} = \sqrt{\frac{1.0\ \text{cm}^2/\text{s}}{6\ \text{s}^{-1}}} = 0.4\ \text{cm}$$

and the optical thickness of the hydrodynamic layer from (13-96) is

$$t_\delta = K_e \delta = (4\ \text{cm}^{-1})(0.4\ \text{cm}) = 1.6$$

Based on this value, it would not appear that the flow is extremely optically thick. However, the appropriate length scale to make this judgement is not the hydrodynamic layer thickness (δ) but the thermal layer thickness (δ_T). These two may be very different from each other for strongly radiating flows depending on the value of the radiation-modified Prandtl number. From (13-72) the radiation-modified Prandtl number is

$$\text{Pr}_m = \frac{\text{Pr}}{1 + \dfrac{4}{3N_0}} = \frac{0.5}{1 + \dfrac{4}{3(0.05)}} = 0.02$$

This indicates that due to the strong influence of radiation the flow behaves like that of a low Prandtl number fluid and the thermal boundary layer is much thicker than the hydrodynamic boundary layer. The optical thickness of the thermal layer can be estimated from (13-95) as

$$t_{\delta_T} \sim \frac{t_\delta}{\sqrt{\text{Pr}_m}} = \frac{1.6}{\sqrt{0.02}} = 11$$

Since $t_{\delta_T} \gg 1$, the flow is indeed optically thick over the characteristic length scale for temperature change.

Finally the heat flux can be estimated using Eq. (13-93) with $a = 0.22$ and $b = 0.90$ since temperature ratio is $T_0/T_\infty = 2500/3600 = 0.7$.

$$\frac{q}{\sigma T_\infty^4} = 0.22\left(\frac{\sqrt{0.05}}{1.6}\right)^{0.90} = 0.037$$

$$q = (0.037)(5.67 \times 10^{-12})(3600)^4 = 35 \text{ W/cm}^2$$

FULLY DEVELOPED INTERNAL FLOWS WITH RADIATION

As a final example of coupled convective, conductive, and radiative transport, consider flow of a participating fluid in a black, two-dimensional, rectangular duct as shown in Fig. 13-8. The flow is hydrodynamically fully developed. The initial fluid temperature is T_0 and the duct temperature is uniform at T_w. The bulk velocity is U as defined by mass conservation as

$$U = \frac{\dot{m}}{\rho A} \tag{13-99}$$

where ρ is the fluid density (assumed constant) and A is the duct cross-section area. The momentum equation for fully developed turbulent flow is

$$-\frac{dP}{dx} + \frac{d}{dy}\left[\rho(\nu + \varepsilon_m)\frac{d\bar{u}}{dy}\right] = 0 \tag{13-100}$$

where $\bar{u}$ is the mean component of the instantaneous x velocity u such that

$$u = \bar{u} + u' \tag{13-101}$$

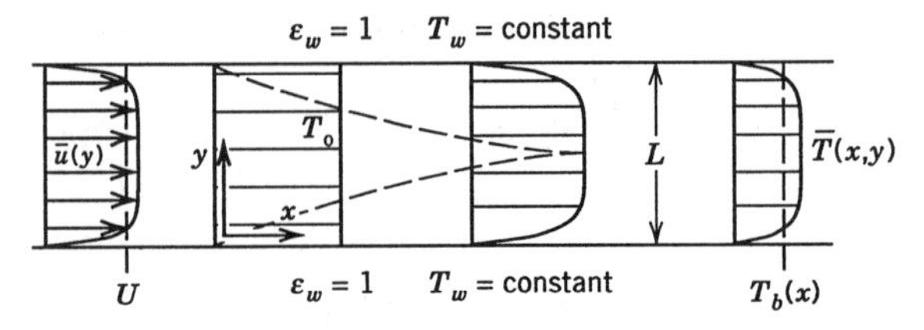

Figure 13-8 Duct flow of a radiating fluid.

and u' is the fluctuating component. The quantity ε_m, defined as

$$\varepsilon_m = \frac{-\overline{u'v'}}{\dfrac{d\bar{u}}{dy}} \tag{13-102}$$

is the turbulent momentum diffusivity. In Eqs. (13-101) and (13-102) the overbar represents time-averaging of fluctuating quantities. The energy equation is

$$\rho C_p \bar{u} \frac{\partial \overline{T}}{\partial x} = -\frac{\partial}{\partial y}\left[-\rho C_p(\alpha + \varepsilon_H)\frac{\partial \overline{T}}{\partial y} + q_r\right] \tag{13-103}$$

where $\overline{T}$ is the mean component of the fluctuating temperature

$$T = \overline{T} + T' \tag{13-104}$$

and

$$\varepsilon_H = \frac{\overline{v'T'}}{\dfrac{\partial \overline{T}}{\partial y}} \tag{13-105}$$

is the turbulent energy (or heat) diffusivity; v' is the fluctuating component of the y velocity. Further details on the formulation of convective heat transfer in turbulent flows are given by Kays and Crawford [1980]. The radiative heat flux is

$$q_r = \int_0^\infty q_{r_\lambda}\, d\lambda \tag{13-106}$$

where the spectral flux can be obtained from (13-11) for a nonscattering fluid. Equation (13-11) is based on a one-dimensional analysis (x variations only) and therefore neglects the effect of radiative transport in the axial direction. The neglect of axial radiation is reasonable in most cases, just as the neglect of axial conduction is usually justified based on large Peclet number (Pe = Re Pr).

The coupled solution of (13-11), (13-103), and (13-106) represents a significant mathematical undertaking, especially if nongray gas properties are included. Balakrishnan and Edwards [1979] have demonstrated how this solution can be carried out numerically for a molecular radiating gas using the exponential wide band model. Their analysis represents a significant contribution in analyzing combined convective and radiative heat transfer since it includes real, nongray gas properties. They showed that the boundary layer can have a significant shielding effect. For a hot gas and cold wall this shielding effect means that self-absorption significantly reduces the heat flux to the wall that would be expected based on radiation from an isothermal gas at the bulk temperature of the flow. This effect increases as the thermal development of the flow progresses and the boundary layers thicken.

In addition to their detailed numerical calculations Balakrishnan and Edwards also presented an engineering correlation for average Nusselt number. In the following correlations the overbar means averaging along the x direction of the duct. The total Nusselt number is represented as a sum of a convective Nusselt number and a radiative Nusselt number.

$$\overline{\mathrm{Nu}} = \overline{\mathrm{Nu}_c} + \overline{\mathrm{Nu}_r} \tag{13-107}$$

$$\overline{\mathrm{Nu}_c} = 0.020\ \mathrm{Re}^{0.8}\ \mathrm{Pr}^{0.33}\ \bar{F} \tag{13-108}$$

$$\mathrm{Re} = \frac{UD_h}{\nu} \tag{13-109}$$

$$\bar{F} = 1 + \frac{0.88}{1 + 0.4\dfrac{x}{D_h}} \tag{13-110}$$

$$D_h = \frac{4A}{P} = \text{hydraulic diameter } (A = \text{area},\ P = \text{perimeter}) \tag{13-111}$$

$$\overline{\mathrm{Nu}_r} = \frac{\bar{h}_r D_h}{k} \tag{13-112a}$$

$$\bar{h}_r = \sum_i A_i(D_h)\bar{\tau}_{\mathrm{WL},i}\, e'_{b_\eta} \tag{13-112b}$$

$$\bar{\tau}_{\mathrm{WL},i} = 1 - \frac{\ln\left(1 + u_{D_h}\bar{r}_{\mathrm{WL}}\right)}{\ln\left(1 + u_{D_h}\right)} \tag{13-113}$$

$$\bar{r}_{\mathrm{WL}} = \frac{1.37}{\mathrm{Re}^{0.37}}\left\{1 - \exp\left[-\left(\frac{x}{10D_h}\right)^{0.6}\right]\right\} \tag{13-114}$$

$$e'_{b_\eta} = \frac{\partial e_{b_\eta}}{\partial T} \tag{13-115}$$

This correlation is based on computing the radiative properties for an isothermal gas at the bulk temperature and correcting for nonisothermal wall layer effects. The effective bandwidths A_i can be determined by using the exponential wide band correlations, Eq. (8-65), based on the geometric mean beam length, which is equivalent to hydraulic diameter ($L_{mo} = D_h$) for a long duct. The term $\bar{\tau}_{\mathrm{WL},i}$ is an average wall layer transmissivity that accounts for self-absorption effects by the fluid near the wall. It is defined as the ratio of the radiative flux at the wall to that expected based upon an isothermal gas at the bulk temperature. Its value decreases as the length of the duct (x) increases due to the thickening of the thermal boundary layers.

The use of the derivative of the Planck function reflects the fact that the radiative transfer analysis has been linearized, under the assumption that the temperature difference $|T_0 - T_w|$ is small compared with the absolute temperature level $T_{0,w}$.

It has been suggested [Balakrishnan and Edwards, 1979] that these correlations can be applied to pipe flows as well as duct flows. In pipe flows the hydraulic diameter is simply the pipe inside diameter (D) whereas in a two-dimensional duct the hydraulic diameter is twice the duct thickness ($2L$). The following example illustrates the application of these correlations to a pipe flow calculation.

Example 13-2

Steam at 7.5 atm enters a black, 0.25-m-diameter pipe at a flow rate of 0.63 kg/s. Assuming an average bulk temperature of 600 K, determine the asymptotic Nusselt number for a long pipe.

The properties of steam at 7.5 atm and 600 K are

$$\rho = 2.74 \text{ kg/m}^3 \quad \nu = 7.5 \times 10^{-6} \text{ m}^2\text{/s} \quad k = 0.045 \text{ W/(mK)} \quad \text{Pr} = 1$$

Using these properties, the bulk velocity from (13-99) is

$$U = \frac{0.63 \text{ kg/s}}{(2.74 \text{ kg/m}^3)(\pi/4)(0.25 \text{ m}^2)} = 4.68 \text{ m/s}$$

and Reynold's number from (13-109) is

$$\text{Re} = \frac{(4.68)(0.25)}{7.5 \times 10^{-6}} = 1.56 \times 10^5$$

Since the duct is very long, the axially averaged properties (overbar) are equivalent to the asymptotic local values. Thus the entry region factors from (13-110) and (13-114) are

$$\bar{r}_{\text{WL}} = \frac{1.37}{\text{Re}^{0.37}} \qquad \bar{F} = 1$$

The band properties are calculated using the exponential wide band correlations with a pressure of 7.5 atm, a temperature of 600 K, and a path length of 0.25 m. The results are shown in Table 13-2. For the conditions given, all of the bands are optically thick with overlapped lines. Thus the effective

TABLE 13-2 Band Properties for Turbulent, Radiating Pipe Flow

λ (μm)	ω (cm^{-1})	u_{D_h}	β	A (cm^{-1})	$\bar{\tau}_{WL}$	$e'_{b_\eta}\left(\frac{W}{m^2\ cm^{-1}\ K}\right)$	$A\bar{\tau}_{WL}e'_{b_\eta}\left(\frac{W}{m^2\ K}\right)$
71.4	140	6600	1.8	1370	0.466	0.00050	1.890*
6.25	138	204	1.8	872	0.724	0.02207	13.933
2.66	147	116	4.2	845	0.776	0.00363	2.380
1.87	106	19.8	1.6	421	0.907	0.00033	0.126
1.38	78	21.8	2.0	320	0.902	0.00001	0.003
				Radiative heat transfer coefficient $\bar{h}_r$ = 18.32			

*Ae'_{b_η} replaced by $4\sigma T^3 \Delta f_i$ [see Eq. (11-79b)].

bandwidths are all calculated using the correlation from (8-65)

$$A = \ln u_{D_h} + 1$$

The wall layer transmissivities are smallest for the most optically thick bands. Based on the last column in the table, the most important bands in decreasing order of importance are the 6.25-μm, 2.66-μm, and 71.4-μm (rotational) bands. Adding up the last column in Table 13-2 gives the radiative heat transfer coefficient. Converting this value to a radiative Nusselt number gives

$$\overline{\mathrm{Nu}}_r = \frac{(0.25\ \mathrm{m})(18.32\ \mathrm{W/m^2\ K^4})}{0.045\ \mathrm{W/m\ K}} = 102$$

The convective Nusselt number is

$$\overline{\mathrm{Nu}}_c = (0.020)(156{,}000)^{0.8}(1)^{0.33} = 285$$

Thus radiation contributes 26% of the total heat flux.

It should be noted that in Table 13-2 the band approximation has not been used to calculate the contribution of the rotational band. This is because the rotational band is so wide that the band approximation is not accurate. Instead the spectral integration of the Planck function derivative is carried out using the internal fractional function, which is defined in Eq. (11-79b).

These approximate correlations represent a decoupled approach that assumes that the convective and radiative transfer do not influence each other. Detailed numerical solutions verify that this is the case except when the radiative transfer is so strong that it dominates anyway. Thus a decoupled analysis is seen to yield accurate engineering predictions in this situation even when the bands are optically thick.

Radiative Heat Transfer Coefficient

As demonstrated in the preceding example, the concept of a radiative heat transfer coefficient is often useful, particularly in combined radiation and convection problems. Here the concept is given formal definition.

In many radiative heat transfer problems there is only a single heat transfer rate of interest, that from a source to a sink. Such problems include two-node, source(1)–sink(2) problems [see Eqs. (3-44) to (3-47) and (10-19)] and three-node source(1)–sink(2)–refractory(3) problems [see Eqs. (3-55) to (3-58) and (10-48)]. In all such cases the net, spectral radiative heat flux is given by the following equation:

$$q_{r_{\lambda,\eta}} = c_{\lambda,\eta}\left[e_{b_{\lambda,\eta}}(T_1) - e_{b_{\lambda,\eta}}(T_2)\right] \tag{13-116}$$

where $c_{\lambda,\eta}$ is a function of only the enclosure geometry and the system radiative properties (λ, η indicates that the spectral variable can be either wavelength or wavenumber), and T_1 and T_2 are the source and sink temperatures, respectively. If the absolute value of the temperature difference between the source and the sink is small compared to the absolute temperatures in the system $|T_1 - T_2| \ll T_{1,2}$, then it is expedient to linearize the radiation problem in temperature by expanding $e_{b_{\lambda,\eta}}(T)$ in a Taylor series about some temperature in the system, say T_1

$$e_{b_{\lambda,\eta}}(T) \approx e_{b_{\lambda,\eta}}(T_1) + \left.\frac{\partial e_{b_{\lambda,\eta}}}{\partial T}\right|_{T_1}(T - T_1) + \cdots \tag{13-117a}$$

such that the difference in $e_{b_{\lambda,\eta}}$ in Eq. (13-116) is

$$e_{b_{\lambda,\eta}}(T_1) - e_{b_{\lambda,\eta}}(T_2) \approx \left.\frac{\partial e_{b_{\lambda,\eta}}}{\partial T}\right|_{T_1}(T_1 - T_2) + \cdots \tag{13-117b}$$

and Eq. (13-116) becomes

$$q_{r_{\lambda,\eta}} = c_{\lambda,\eta}\left[\left.\frac{\partial e_{b_{\lambda,\eta}}}{\partial T}\right|_{T_1}(T_1 - T_2)\right] \tag{13-118}$$

Thus the total, net radiative heat flux is

$$q_r = (T_1 - T_2)\int_0^\infty c_{\lambda,\eta}\left.\frac{\partial e_{b_{\lambda,\eta}}}{\partial T}\right|_{T_1} d\lambda,\eta \tag{13-119}$$

Introducing the definition of the radiative heat transfer coefficient as

$$h_r = \frac{q_r}{T_1 - T_2} \tag{13-120}$$

gives the following expression for h_r

$$h_r = \int_0^\infty c_{\lambda,\eta} \frac{\partial e_{b_{\lambda,\eta}}}{\partial T}\bigg|_{T_1} d\lambda, \eta \tag{13-121}$$

By introducing the internal fractional function, Eq. (11-79b), the expression for h_r can also be written as

$$h_r = 4\sigma T_1^3 \int_0^1 c_{\lambda,\eta}\, df_i(\lambda T_1) \tag{13-122}$$

Example 13-3

Obtain an expression for the radiative heat transfer coefficient for an isothermal, nongray gas at T_g in an isothermal enclosure at T_w.

From Eq. (10-19) the expression for the function $c_{\lambda,\eta}$ for a two-node, gas-wall configuration is

$$c_{\lambda,\eta} = \frac{1}{1/\varepsilon_{w_\eta} + 1/\bar{\varepsilon}_\eta - 1} \tag{13-123}$$

where ε_{w_η} is the spectral wall emissivity and $\bar{\varepsilon}_\eta$ is the spectral emissivity factor for the gas. The spectral emissivity factor for a nongray gas can be approximated as

$$\bar{\varepsilon}_\eta = \begin{cases} 0 & \text{outside the bands} \quad (13\text{-}124a) \\ \dfrac{\bar{A}}{\Delta\eta} & \text{inside the bands} \quad (13\text{-}124b) \end{cases}$$

For an interpretation of $\bar{A}/\Delta\eta$ see Fig. 8-15. Therefore, the function $c_{\lambda,\eta}$ becomes

$$c_{\lambda,\eta} = \begin{cases} 0 & \text{outside the bands} \\ \dfrac{1}{\dfrac{1}{\varepsilon_{w_\eta}} + \dfrac{\Delta\eta}{\bar{A}} - 1} & \text{inside the bands} \end{cases}$$

This expression for $c_{\lambda,\eta}$ is then substituted into either Eq. (13-121) or (13-122) and the integration carried out to determine h_r. In the case of vibration-rotation bands (which are usually relatively narrow compared with the spectral distribution of the Planck function), the band approximation applies and (13-121) is the form of choice. This procedure gives

$$h_r = \sum_k e'_{b_{\eta k}} \frac{\Delta\eta_k}{\dfrac{1}{\varepsilon_{w_k}} + \dfrac{\Delta\eta_k}{\bar{A}_k} - 1} \quad \text{(vibration-rotation bands)} \quad (13\text{-}125)$$

where e'_{b_η} means differentiation with respect to temperature. If the wall is black, this expression reduces to (13-112b) except for the wall layer transmissivity factor, which in this example would be 1. In the case of pure rotational bands (which are not narrow compared to the Planck function), the equation of choice to determine h_r is (13-122), which gives

$$h_r = 4\sigma T_g^3 \sum_k \frac{1}{1/\varepsilon_{w_k} + \Delta\eta_k/\bar{A}_k - 1} \Delta f_i(\lambda_k T_g) \quad \text{(pure rotation bands)} \quad (13\text{-}126)$$

SUMMARY

The problem of combined conduction, convection, and radiation heat transfer is extremely complex. All the issues of radiative transfer such as optical depth, spectral property variations, and directional intensity variations are still involved. In addition, all the complexities of convective heat transfer may be involved, such as viscous effects, inertial effects, and turbulence. When conduction, convection, and radiation occur in a participating fluid medium, the radiative transfer and energy conservation equations are, in general, coupled. Under such circumstances it is appropriate to introduce simplifications in the radiative transfer description to make it more consistent with the rest of the analysis. Such simplifications include (i) decoupling the radiative heat transfer from the other modes, (ii) linearizing the radiation formulation, and (iii) classifying the radiative transfer as either optically thin or optically thick and using concepts from Chapters 10 or 11. For cases where the radiative transfer is optically intermediate (t_{δ_T} or $t_L \approx 1$) and accuracy greater than 10 to 20% is required, a coupled, nonlinear, numerical solution of the problem is necessary.

REFERENCES

Balakrishnan, A. and Edwards, D. K. (1979), "Molecular Gas Radiation in the Thermal Entrance Region of a Duct," *J. Heat Transfer*, Vol. 101, pp. 489–495.

Brewster, M. Q. (1989), "Radiation-Stagnation Flow Model of Aluminized Solid Rocket Motor Internal Insulator Heat Transfer," *J. Thermophysics and Heat Transfer*, Vol. 3, No. 2, pp. 132–139.

Calia, V., Celentano, A., Soel, M., Konopka, W., Gutowski, R., and Ryan, R. (1989), "Measurements of UV/VIS/LWIR Optical Properties of Al_2O_3 Particles," 18th

JANNAF Exhaust Plume Technology Subcommittee Meeting, CPIA Publication 530, Nov., pp. 183–192.

Kays, W. M. and Crawford, M. E. (1980), *Convective Heat and Mass Transfer*, 2nd ed., McGraw-Hill, New York.

Ozisik, M. N. (1985), *Radiative Transfer and Interactions with Conduction and Convection*, Werbel and Peck, New York.

Parker, T. E., R. R. Foutter, J. C. Person, and W. T. Rawlins, "Experimental Measurements of the Optical Properties of Al_2O_3 and Rocket Exhaust Particles at High Temperatures," 18th JANNAF Exhaust Plume Technology Subcommittee Meeting, CPIA Publication 530, Nov. 1989, pp. 193–201.

Parry, D. L. and Brewster, M. Q. (1991), "Optical Constants of Al_2O_3 Smoke in Propellant Flames," *J. Thermophysics and Heat Transfer*, Vol. 5, No. 2, pp. 142–149.

White, F. M. (1974), *Viscous Fluid Flow*, McGraw-Hill Book Co., New York.

REFERENCES FOR FURTHER READING

Ahluwalia, R. K. and K. H. Im, "Structure of Thermal Radiation Field in an Optically Thick Limit," *AIAA J.*, Vol. 20, 1982, pp. 1766–1769.

Azad, F. H. and M. F. Modest, "Combined Radiation and Convection in Absorbing, Emitting and Anisotropically Scattering Gas-Particulate Tube Flow," *Int. J. Heat Mass Transfer*, Vol. 24, 1981, pp. 1681–1698.

Bergquam, J. B. and R. A. Seban, "Heat Transfer by Conduction and Radiation in Absorbing and Scattering Materials," *J. Heat Transfer*, Vol. 93, 1971, pp. 236–239.

Cess, R. D., "The Interaction of Thermal Radiation with Conduction and Convection Heat Transfer," *Advances in Heat Transfer*, Vol. 1, 1964, pp. 1–50.

Cess, R. D., "The Interaction of Thermal Radiation with Free Convection Heat Transfer," *Int. J. Heat Mass Transfer*, Vol. 9, 1966, pp. 1269–1277.

Chan, C. K. and C. L. Tien, "Combined Radiation and Conduction in Packed Spheres," *Proc. 5th Int. Heat Trans. Conf.*, Tokyo, Vol. 1, 1974, pp. 72–74.

Chiba, Z. and R. Greif, "Heat Transfer to Steam Flowing Turbulently in a Pipe," *Int. J. Heat Mass Transfer*, Vol. 16, 1973, p. 1645.

Echigo, R., S. Hasegawa, and S. Nakano, "Simultaneous Radiative and Convective Heat Transfer in a Packed Bed with High Porosity," *Transactions of JSME*, Vol. 40, 1974, p. 479.

Edwards, D. K., *Radiation Heat Transfer Notes*, Hemisphere Publishing Company, 1981.

Edwards, D. K. and A. Balakrishnan, "Self-Absorption of Radiation in Turbulent Molecular Gases," *Combustion and Flame*, Vol. 20, 1973, pp. 401–417.

Edwards, D. K. and A. T. Wassel, "Interpretation of a Turbulent Radiating Gas as a Low-Prandtl-Number Fluid," *Letters in Heat and Mass Transfer*, Vol. 1, 1974, pp. 19–24.

Greif, R., "Energy Transfer by Radiation and Conduction with Variable Gas Properties," *Int. J. Heat Mass Transfer*, Vol. 7, 1964, pp. 891–900.

Greif, R., "Laminar Convection with Radiation: Experimental and Theoretical Results," *Int. J. Heat Mass Transfer*, Vol. 21, 1978, p. 477.

Habib, I. S. and R. Greif, "Heat Transfer to a Flowing Non-gray Radiating Gas: An Experimental and Theoretical Study," *Int. J. Heat Mass Transfer*, Vol. 13, 1970, p. 1571.

Ho, C. H. and M. N. Ozisik, "Combined Conduction and Radiation in a Two-Dimensional Rectangular Enclosure," *Numerical Heat Transfer*, Vol. 13, No. 2, 1988, pp. 229–239.

Howell, J. R. and M. E. Goldstein, "Effective Slip Coefficients for Coupled Conduction-Radiation Problems," *J. Heat Transfer*, Vol. 91, 1969, pp. 165–166.

Kim, T. K. and T. F. Smith, "Radiative and Conductive Transfer for a Real Gas in a Cylindrical Enclosure with Gray Walls," *J. Heat Transfer*, Vol. 28, 1985, pp. 2269–2277.

Kounalakis, M. E., J. P. Gore, and G. M. Faeth, "Mean and Fluctuating Radiation Properties of Nonpremixed Turbulent Carbon Monoxide/Air Flames," *J. Heat Transfer*, Vol. 111, 1989, pp. 1021–1030.

Kubo, S., "Stagnation Point Flow of a Radiating Gas of a Large Optical Thickness," *J. Phys. Soc. of Japan*, Vol. 36, 1974, pp. 293–297.

Lee, H. and R. O. Buckius, "Combined Mode Heat Transfer Analysis Using Radiation Scaling," *J. Heat Transfer*, Vol. 108, 1986, pp. 626–632.

Martin, J. K. and C. C. Hwang, "Combined Radiant and Convective Heat Transfer to Laminar Steam Flow between Gray Parallel Plates with Uniform Heat Flux," *J. Quant. Spectr. Rad. Transfer*, Vol. 15, 1975, pp. 1071–1081.

Mattick, A. T., "Coaxial Radiative and Convective Heat Transfer in Gray and Nongray Gases," *J. Quant. Spectr. Rad. Transfer*, Vol. 24, 1980, pp. 323–334.

Nelson, D. A. "On the Uncoupled Superposition Approximation for Combined Conduction-Radiation through Infrared Radiating Gases," *Int. J. Heat Mass Transfer*, Vol. 18, 1975, pp. 711–713.

Novotny, J. L. and Kelleher, M. D., "Free Convection Stagnation Flow of an Absorbing-Emitting Gas," *Int. J. Heat Mass Transfer*, Vol. 10, 1967, pp. 1171–1178.

Novotny, J. L. and K. T. Yang, "The Interaction of Thermal Radiation in Optically Thick Boundary Layers," *J. Heat Transfer*, Vol. 89, 1967, pp. 309–312.

Pearce, B. E. and A. F. Emery, "Heat Transfer by Thermal Radiation and Laminar Forced Convection to an Absorbing Fluid in the Entry Region of a Pipe," *J. Heat Transfer*, Vol. 92, 1970, pp. 221–230.

Soufiani, A., J. M. Hartmann and J. Taine, "Validity of Band Model Calculations for CO_2 and H_2O Applied to Radiative Properties and Conductive-Radiative Transfer," *J. Quant. Spectr. Rad. Transfer*, Vol. 33, 1985, p. 243.

Soufiani, A. and J. Taine, "Application of Statistical Narrow Band Model to Coupled Radiation and Convection at High Temperature," *Int. J. Heat Mass Transfer*, Vol. 30, 1987, p. 437.

Soufiani, A. and J. Taine, "Experimental and Theoretical Studies of Combined Radiative and Convective Transfer in CO_2 and H_2O Laminar Flows," *Int. J. Heat Mass Transfer*, Vol. 32, No. 3, 1989, pp. 477–486.

Tabanfar, S. and M. F. Modest, "Combined Radiation and Convection in Absorbing, Emitting, Nongray Gas-Particulate Tube Flow," *J. Heat Transfer*, Vol. 109, 1987, pp. 478–484.

Tan, Z., "Combined Radiative and Conductive Heat Transfer in Two-Dimensional Emitting, Absorbing and Anisotropic Scattering Square Media," *Int. Comm. Heat Mass Transfer*, Vol. 16, No. 3, 1989, pp. 391–401.

Thynell, S. and C. L. Merkle, "Analysis of Volumetric Absorption of Solar Energy and its Interaction with Convection," *J. Heat Transfer*, Vol. 111, 1989, pp. 1006–1014.

Tien, C-L. and M. M. Abu-Romia, "Perturbation Solutions in the Differential Analysis of Radiation Interaction with Conduction and Convection," *AIAA J.*, Vol. 4, 1966, pp. 732–733.

Viskanta, R., "Radiative Heat Transfer: Interaction with Conduction and Convection and Approximate Methods in Radiation," *Proc. 7th Int. Heat Transfer Conf.*, Munich, Vol. 1, 1982, p. 103.

Viskanta, R., "Radiation Transfer and Interaction of Convection with Radiation Heat Transfer," *Advances in Heat Transfer*, T. F. Irvine and J. P. Hartnett, eds., Academic Press, New York, Vol. 3, 1966, pp. 176–252.

Viskanta, R., "Heat Transfer by Conduction and Radiation in Absorbing and Scattering Materials," *J. Heat Transfer*, Vol. 87, 1965, pp. 143–150.

Viskanta, R. and Grosh, R. J., "Heat Transfer by Simultaneous Conduction and Radiation in an Absorbing Medium," *J. Heat Transfer*, Vol. 84, No. 1, 1965, pp. 63–72.

Wang, L. S. and C-L. Tien, "A Study of Various Limits in Radiation Heat Transfer Problems," *Int. J. Heat Mass Transfer*, Vol. 10, 1967, pp. 1327–1338.

Wassel, A. T. and D. K. Edwards, "Molecular Gas Radiation in a Laminar or Turbulent Pipe Flow," *J. Heat Transfer*, Vol. 98, 1976, pp. 101–107.

Webb, B. W. and R. Viskanta, "Radiation-Induced Buoyancy-Driven Flow in Rectangular Enclosures: Experiment and Analysis," *J. Heat Transfer*, Vol. 109, 1987, pp. 427–433.

Yener, Y. and M. N. Ozisik, "Simultaneous Radiation and Forced Convection in Thermally Developing Turbulent Flow through a Parallel-Plate Channel," *J. Heat Transfer*, Vol. 108, 1986, pp. 985–986.

Yucel, A., H. Kehtarnavaz, and Y. Bayazitoglu, "Boundary Layer Flow of a Non-gray Radiating Fluid," *Int. Comm. Heat Mass Transfer*, Vol. 16, No. 3, 1989, pp. 415–425.

Yuen, W. W. and L. W. Wong, "Heat Transfer by Conduction and Radiation in a One-Dimensional Absorbing, Emitting, and Anistropically Scattering Medium," *J. Heat Transfer*, Vol. 102, 1980, pp. 303–307.

PROBLEMS

1. Show that the ratio of radiative to total heat flux at the boundary of an optically thick, gray, conducting medium is

$$\frac{q_{r_w}}{q_w} = \frac{1}{1 + 3N_w/4}$$

where N_w is the conduction-radiation parameter evaluated at the temperature of the medium at the wall.

2. Consider a plane parallel slab of thickness L consisting of a gray, absorbing, emitting, nonscattering, conducting medium. For the same parameters as those of Fig. 13-2 ($\varepsilon_1 = \varepsilon_2 = 1, T_2/T_1 = 0.5$), compare the exact result for dimensionless heat flux $q/\sigma T_1^4$ with (a) the diffusion approximation (13-38) for $t_L = 0.1$ and (b) the decoupled approximation (13-23) for $t_L = 10$. Comment on the accuracy of these approximations outside their ranges of applicability. Numerical values for the exact results are given in the following table. Note that $E_3(10) \sim 10^{-6}$.

N_1	$t_L = 0.1$	$t_L = 10$
0.01	1.074	0.114
0.1	2.880	0.131
1	20.88	0.315
10	200.88	2.114

3. Consider a plane parallel slab of thickness L consisting of a gray, absorbing, emitting, isotropic scattering, conducting medium that is optically thick ($t_L > 1$). The effective thermal conductivity of the medium is k and the extinction coefficient is K_e. Energy is being generated uniformly inside the medium at a local volumetric rate of q'''. The walls are diffuse and gray with emissivity ε_w and temperature T_w.

(a) Apply the diffusion approximation to the radiative transfer and obtain an equation for the temperature distribution in the medium and heat flux at the wall for arbitrary value of conduction-radiation parameter N_w.

(b) From the result of part (a) obtain the temperature for the limiting cases of conduction dominated transfer ($N_w \gg 1$) and radiation-dominated transfer ($N_w \ll 1$). Discuss the role of the radiative parameters t_L and ε_w in these two limiting cases.

4. Consider a plane parallel slab of thickness L consisting of a gray, absorbing, emitting, nonscattering, conducting medium that is optically thin ($t_L < 1$). The effective thermal conductivity of the medium is k and the absorption coefficient is K_a. Energy is being generated uniformly inside the medium at a local volumetric rate of q'''. The walls are diffuse and gray with emissivity ε_w and temperature T_w.

(a) Obtain an expression for the temperature in the medium as a function of T_w, ε_w, q''', L, k, and t_L for the limiting cases of conduction dominated transfer ($N_w \gg 1$) and radiation dominated transfer ($N_w \ll 1$). Discuss the role of the parameters k, t_L, and ε_w in these two limiting cases.

(b) What is the temperature in the medium (kelvin) for

$$T_w = 1000 \text{ K} \qquad L = 10 \text{ cm} \qquad k = 10^{-3} \text{ W/(cm K)}$$

$$K_a = 0.01 \text{ cm}^{-1} \qquad \varepsilon_w = 0.3 \qquad q''' = 0.555 \text{ W/cm}^3$$

5. Consider planar, two-dimensional stagnation flow of an optically thick, incompressible, gray, inviscid fluid (or viscous fluid with $N_0 \ll 1$ and $\delta_T \gg \delta$). The free stream fluid temperature is T_∞ and the constant wall temperature is T_0. Find a closed-form linearized solution for the nondimensional radiative and conductive heat fluxes $q^*_{r,c} = q_{r,c}/\sigma T_\infty^4$ as a function of t_δ, N_0, Pr, ε_w, and T_0/T_∞. Plot $q^*_{r,c}$ vs. t_δ for $T_0/T_\infty = 0.9$, Pr $= 0.5$, $\varepsilon_w = 1$ and $N_0 = 0.1$, 1, and 10. Compare with the viscous linearized solution of Fig. 13-7 and comment on the agreement. For convenience the numerical values of Fig. 13-7 are tabulated below.

t_δ	$q_r^*(0.01)$	$q_c^*(0.01)$	$q_r^*(0.1)$	$q_c^*(0.1)$	$q_r^*(1)$	$q_c^*(1)$
1E − 1	1.40E − 1	1.05E − 3	2.87E − 1	2.16E − 2	5.49E − 1	4.11E − 1
2E − 1	7.96E − 2	5.97E − 4	1.86E − 1	1.40E − 2	3.75E − 1	2.81E − 1
5E − 1	3.47E − 2	2.60E − 4	9.08E − 2	6.81E − 3	1.92E − 1	1.44E − 1
1E + 0	1.79E − 2	1.34E − 4	4.89E − 2	3.67E − 3	1.06E − 1	7.97E − 2
2E + 0	9.07E − 3	6.80E − 5	2.55E − 2	1.91E − 3	5.61E − 2	4.20E − 2
5E + 0	3.66E − 3	2.75E − 5	1.04E − 2	7.83E − 4	2.32E − 2	1.74E − 2
1E + 1	1.84E − 3	1.38E − 5	5.26E − 3	3.95E − 4	1.17E − 2	8.80E − 3

6. Obtain an expression for the temperature distribution $T(y)$ in a stagnant, conducting, absorbing, emitting, isotropic scattering, gray slab of thickness L with no internal generation in terms of the wall temperatures $T_{1,2}$ and emissivities $\varepsilon_{1,2}$, the thermal conductivity of the medium k, the extinction coefficient K_e, and L.

7. A stagnant, conducting, absorbing, emitting, scattering, gray medium of thickness $L = 10$ cm is heated on one side with a constant heat flux of 1.588 W/cm^2. The other side is maintained at a constant temperature of 500 K. The effective (constant) thermal conductivity of the medium is 0.02 W/cm K and the effective extinction coefficient for isotropic scattering is 1 cm^{-1}. Both walls have an emissivity of 0.8. Determine the temperature of the heated wall. Compare this result with the limiting results of pure conduction and radiation.

8. Verify that applying Eqs. (13-43) and (13-44) to the radiative terms and (13-29) and (13-30) to the conduction terms in (13-42) gives (13-37).

9. Verify that applying Eqs. (13-45) and (13-46) uniformly to the radiative and conductive terms in (13-42) gives (13-37).

10. Obtain the solution for temperature field and heat flux for an optically thick, incompressible stagnation flow of a gray fluid on an isothermal wall [Eqs. (13-86), (13-87), and (13-92)] by integrating Eq. (13-82) with (13-84) and (13-85) as boundary conditions.

11. A mixture of steam and nitrogen at 3 atm enters a black, 2-m-diameter pipe with a constant wall temperature. The steam mole fraction is 0.18. The Reynold's number is 10,250. Pr = 0.875 and $k = 0.069$ W/mK. Assuming an average bulk temperature of 1000 K, determine the average Nusselt number over the 10 m length of the pipe.

12. **(a)** Investigate the accuracy of the decoupled approximation for an optically thick plane parallel slab of thickness L consisting of a gray, absorbing, emitting, nonscattering, conducting medium. Use the same parameters as those of Fig. 13-2 ($\varepsilon_1 = \varepsilon_2 = 1, T_2/T_1 = 0.5$) and compare the dimensionless heat flux $q/\sigma T_1^4$ with the exact results given in Problem 2 for $t_L = 10$.

(b) Use the analysis of part (a) to predict $q/\sigma T_1^4$ for $t_L = 0.1$. Comment on the agreement.

13. A one-room structure with interior dimensions 20 × 20 × 3 m contains air at 1 atm and 90% relative humidity. Circulation maintains a nearly uniform temperature of 298 K. Estimate the radiative heat transfer coefficient h_r for radiative heat loss from the air in the room to the walls which are black to infrared radiation. Compared with natural convection ($h_c \sim 1$ W/m^2 K) is radiation significant? Is treating air as a nonparticipating medium a good approximation in this case?

CHAPTER 14

MONTE CARLO IN PARTICIPATING MEDIA

The Monte Carlo method was introduced in Chapter 6 as a technique for modeling radiative transfer between surfaces in an enclosure with a nonparticipating medium inside the enclosure. This technique can also be extended to include radiative transfer in a participating medium. There is no limit (except computer time) on the complexity of problems that can be modeled by this method. The effects of anisotropic scattering, nonuniform properties, nongray properties, and even polarization can all be included by this method, when it is important to do so.

The basic steps that must be followed to formulate a Monte Carlo representation of radiative transfer in a participating medium are similar to those from Chapter 6:

1. Assign discrete values of energy to photon bundles.
2. Write an algorithm to track bundles and keep score of absorption.

ASSIGNING DISCRETE VALUES OF ENERGY TO PHOTON BUNDLES

The basic equation for accounting for radiant energy in a participating medium is a balance on radiant energy for a differential volume element (see Fig. 14-1). The net radiant energy per unit volume leaving an elemental

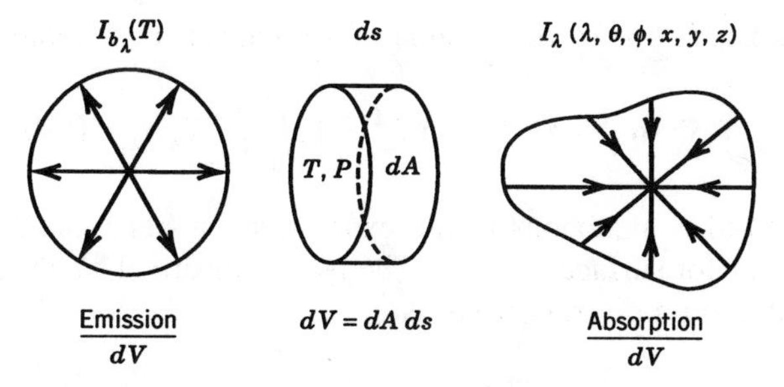

Figure 14-1 Radiant energy balance on a differential volume.

volume is the emitted energy per unit volume minus the absorbed energy per unit volume.

$$\frac{Q_{\text{net out}}}{\Delta V} = \frac{\text{emission}}{\Delta V} - \frac{\text{absorption}}{\Delta V} \tag{14-1}$$

The left-hand side is the divergence of the radiative heat flux vector, and the two terms on the right-hand side are obtained by integrating the product of blackbody intensity and actual incident intensity, respectively, times the absorption coefficient over all directions

$$\nabla \cdot \mathbf{q}_r = \int_0^\infty \int_{4\pi} K_{a_\lambda}(\lambda, T, P) I_{b_\lambda}(\lambda, T)\, d\Omega\, d\lambda - \int_0^\infty \int_{4\pi} K_{a_\lambda} I_\lambda\, d\Omega\, d\lambda \tag{14-2}$$

Since spontaneous emission* is isotropic, the emission per unit volume term can be integrated over direction to give

$$\nabla \cdot \mathbf{q}_r = 4\int_0^\infty K_{a_\lambda} e_{b_\lambda}\, d\lambda - \int_0^\infty \int_{4\pi} K_{a_\lambda} I_\lambda\, d\Omega\, d\lambda \tag{14-3}$$

Using the definition of the Planck mean emission coefficient from Eq. (7-44b)

$$K_{\text{em}_p} = \frac{1}{\sigma T^4} \int_0^\infty K_{a_\lambda} e_{b_\lambda}\, d\lambda = \int_0^1 K_{a_\lambda}\, df(\lambda T) \tag{14-4}$$

Eq. (14-3) becomes

$$\nabla \cdot \mathbf{q}_r = 4K_{\text{em}_p} \sigma T^4 - \int_0^\infty \int_{4\pi} K_{a_\lambda} I_\lambda\, d\Omega\, d\lambda \tag{14-5}$$

*Stimulated emission is treated as negative absorption.

Multiplying Eq. (14-5) by the volume of a homogeneous element V_i gives

$$V_i \nabla \cdot \mathbf{q}_{r_i} = 4V_i K_{\mathrm{em}_{p_i}} \sigma T_i^4 - V_i \int_0^\infty \int_{4\pi} K_{a_{\lambda_i}} I_{\lambda_i} \, d\Omega \, d\lambda \tag{14-6}$$

Equation (14-6) is analogous to the expression for net radiant heat transfer rate, Eq. (6-1), for surface transfer. The power absorbed by a volume element is represented on a discrete basis as

$$V_i \int_0^\infty \int_{4\pi} K_{a_{\lambda_i}} I_{\lambda_i} \, d\Omega \, d\lambda = w S_i \tag{14-7}$$

where w is the energy per bundle and S_i is the number of bundles absorbed (per unit time) by volume V_i. Thus Eq. (14-6) becomes

$$V_i \nabla \cdot \mathbf{q}_{r_i} = 4V_i K_{\mathrm{em}_{p_i}} \sigma T_i^4 - w S_i = w(N_i - S_i) \tag{14-8}$$

where the energy per bundle is defined as

$$w = \frac{4V_i K_{\mathrm{em}_{p_i}} \sigma T_i^4}{N_i} = \text{constant} \tag{14-9}$$

In Eq. (14-9) N_i is the number of bundles emitted (per unit time) by volume V_i. The energy per bundle is maintained constant for all of the bundles regardless of which volume or surface they are emitted from. The number of bundles that should be emitted from any given volume varies as the emissive power of the volume.

Assuming the temperature field is known, the solution scheme is relatively simple. The number of bundles (N_i) that should be emitted from each volume element is determined from Eq. (14-9). Each bundle is tracked through the medium until it is absorbed at some volume or surface element. A count is kept of the number of bundles absorbed at each volume element (S_i), and the source term is calculated from Eq. (14-8). If the temperature field is not known, the source term is substituted into the energy equation, which is solved for a new temperature field. Iteration of this procedure leads to a converged solution.

ALGORITHM TO TRACK BUNDLES

The algorithm to track bundles consists of the following steps:

1. Choose a location inside or on the boundaries of the medium for emission or incidence of a photon bundle.
2. Choose a wavelength for the bundle if the properties of the medium or the boundaries are nongray.

3. Choose the direction of propagation of the bundle.
4. Choose the path length to the next interaction between the bundle and the medium or the boundaries.
5. Calculate the new position of bundle (check if it left the medium).
6. Decide if the bundle is absorbed or scattered at the new location.
7. If the bundle is scattered go to step 3. If the bundle is absorbed keep score and go to step. 1.
8. Repeat steps 1 to 7 for many bundles.

Choose a Location for Emission of a Photon Bundle

The location inside a volume V where a photon bundle should be emitted can be determined from probability laws that represent the emission process. The probability of emission from a differential volume dV into a small solid angle $d\Omega$ in a small wavelength interval $d\lambda$ is the ratio of the power emitted from dV into the $d\Omega$ and $d\lambda$ intervals divided by the total power emitted from V into all solid angles at all wavelengths.

$$\begin{array}{l}\text{Probability of emission}\\ \text{from } dV \text{ into } d\Omega\\ \text{and } d\lambda \text{ intervals}\end{array} = \frac{\text{power emitted from } dV \text{ into } d\Omega\, d\lambda \text{ intervals}}{\text{total power emitted from } V}$$

This probability is given by the product of the probability distribution function $P(\lambda, \theta, \phi, \mathbf{r})$ and the small intervals $d\Omega\, d\lambda\, dA$ where $\mathbf{r}$ represents the position vector as shown in Fig. 14-2.

$$P(\lambda, \Omega, \mathbf{r})\, d\Omega\, d\lambda\, d\mathbf{r} = \frac{K_{a_\lambda} I_{b_\lambda}(T)\, d\Omega\, d\lambda\, dV}{\int_V \int_0^\infty \int_{4\pi} K_{a_\lambda} I_{b_\lambda}(T)\, d\Omega\, d\lambda\, dV} \tag{14-10}$$

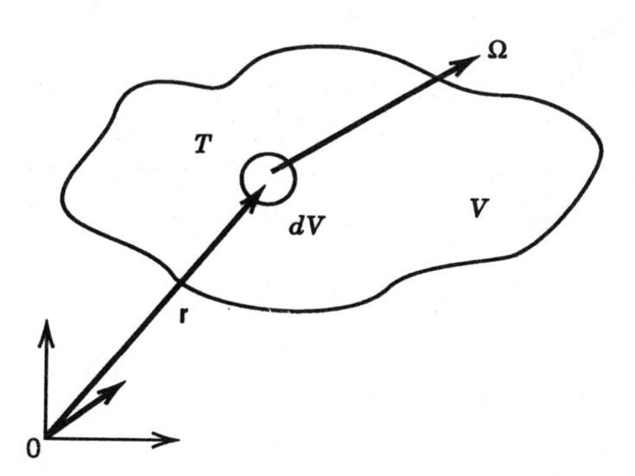

Figure 14-2 Probability of emission from a volume element dV.

Since spontaneous emission is isotropic, Eq. (14-10) gives (14-11)

$$P(\lambda, \Omega, \mathbf{r})\, d\Omega\, d\lambda\, d\mathbf{r} = \frac{K_{a_\lambda} I_{b_\lambda}(T)\, d\Omega\, d\lambda\, dV}{4\pi \int_V \int_0^\infty K_{a_\lambda} I_{b_\lambda}(T)\, d\lambda\, dV} \tag{14-11}$$

Substituting the definition of the Planck mean emission coefficient in the denominator of (14-11) gives (14-12)

$$P(\lambda, \Omega, \mathbf{r})\, d\Omega\, d\lambda\, d\mathbf{r} = \frac{K_{a_\lambda} I_{b_\lambda}(T)\, d\Omega\, d\lambda\, dV}{4\pi \int_V K_{\mathrm{em}_p} I_b(T)\, dV} \tag{14-12}$$

Assuming the temperature and absorption coefficient are uniform over V [this assumption was made with (14-6)], Eq. (14-12) can be integrated over all directions and wavelengths to give the probability of emission from dV as

$$P(\mathbf{r})\, d\mathbf{r} = \frac{dV}{V} \tag{14-13}$$

Equation (14-13) indicates that the probability of emission from a small element dV of an isothermal volume V is proportional to dV. Thus the location for emission can be determined by dividing V into equal subvolumes and emitting the same number of bundles from the center of each subvolume. If V itself is small enough relative to the overall system dimensions to be considered an elemental volume, then the emission events can take place from the center of V without further subdivision.

If it is necessary to determine the location for emission from a surface that bounds the participating medium, then the procedure outlined in Chapter 6 is followed; namely, photon bundles are uniformly distributed over surfaces that are uniform and isothermal.

Choose a Wavelength for the Bundle if Medium or Boundary Properties are Nongray

If either the participating medium or the boundaries of the medium have nongray radiative properties, the Monte Carlo analysis must be done on a spectral basis. A wavelength is assigned to the bundle using the probability law for volume emission, Eq. (14-12). Integrating (14-12) over V and over all directions gives (14-14)

$$P(\lambda)\, d\lambda = \frac{K_{a_\lambda} e_{b_\lambda}}{K_{\mathrm{em}_p} e_b}\, d\lambda \tag{14-14}$$

The probability of emission in the wavelength interval between 0 and λ (in any direction) is the integral of (14-14) between 0 and λ.

$$R(\lambda) = \int_0^{\lambda} P(\lambda')\,d\lambda' = \frac{1}{K_{\mathrm{em}_p} e_b} \int_0^{\lambda} K_{a_\lambda} e_{b_\lambda}\,d\lambda' = \frac{1}{K_{\mathrm{em}_p}} \int_0^{f(\lambda T)} K_{a_\lambda}\,df(\lambda' T) \tag{14-15}$$

$R(\lambda)$ is the cumulative probability function for emission in the wavelength band between 0 and λ. For a gray emitter (14-15) reduces to the fractional function.

$$R(\lambda) = f(\lambda T) \quad \text{(gray emitter)} \tag{14-16}$$

Choose the Direction of Propagation of the Bundle

After the initial location and wavelength for emission have been chosen (or after a photon bundle has been scattered), it is necessary to determine a direction of propagation for the bundle. For emission this step is easy since emission is isotropic. For anisotropic scattering this step is more complicated and requires further considerations.

The geometry for a single-scattering event is shown in Fig. 14-3. The azimuthal single-scattering angle is ϕ_p and the polar single-scattering angle is θ_p. The probability of scattering into a small, conical solid angle $d\theta_p$ is the ratio of the power scattered into $d\theta_p$ to the total power scattered into all directions.

$$\text{Probability of scattering into small } d\theta_p = \frac{\text{power scattered into } d\theta_p}{\text{total power scattered}}$$

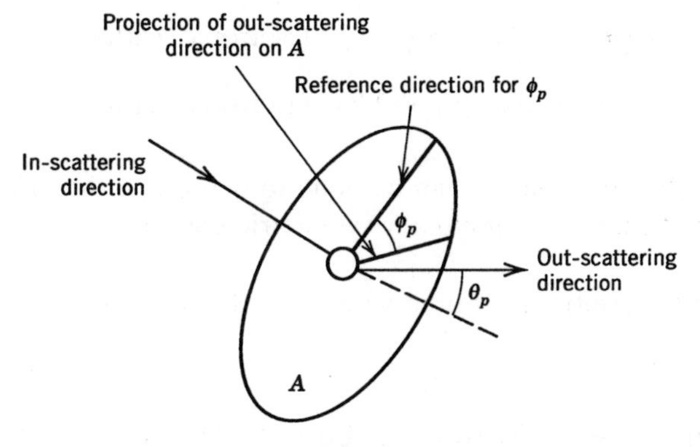

Figure 14-3 Azimuthal and polar single-particle scattering angles.

Assuming that the single scattering is azimuthally symmetric (p independent of ϕ_p), this expression becomes

$$P(\theta_p)\,d\theta_p = \frac{C_s\,\Delta\Omega' I' p(\theta_p) 2\pi \sin\theta_p\,d\theta_p}{\int_0^{\pi} C_s\,\Delta\Omega' I' p(\theta_p) 2\pi \sin\theta_p\,d\theta_p} \tag{14-17}$$

In (14-17) $\Delta\Omega'$ is the solid angle associated with the in-coming intensity I'. If p was a function of ϕ_p, then an azimuthal averaging would be necessary to obtain $p(\theta_p)$. Using the normalization condition for the phase function (14-18)

$$\frac{1}{2}\int_0^{\pi} p(\theta_p)\sin\theta_p\,d\theta_p = 1 \tag{14-18}$$

the probability distribution function (14-17) can also be written as

$$P(\theta_p)\,d\theta_p = \tfrac{1}{2}p(\theta_p)\sin\theta_p\,d\theta_p \tag{14-19}$$

The probability of scattering in the solid angle interval between 0 and θ_p (cumulative probability) can be obtained from integrating (14-19) between 0 and θ_p.

$$R(\theta_p) = \int_0^{\theta_p} P(\theta_p^*)\,d\theta_p^* \tag{14-20}$$

The polar single-scattering angle θ_p can thus be determined by generating a random number between 0 and 1 and solving (14-20) for θ_p. For isotropic scattering [$p(\theta_p) = 1$] Eq. (14-20) gives

$$R(\theta_p) = \tfrac{1}{2}(1 - \cos\theta_p) \quad \text{(isotropic scattering)} \tag{14-21}$$

$$\theta_p = \cos^{-1}(1 - 2R) \quad \text{(isotropic scattering)} \tag{14-22}$$

A similar probability analysis can be done to determine the azimuthal angle, with the result that, for azimuthally symmetric scattering,

$$\phi_p = 2\pi R \quad \text{(isotropic or otherwise azimuthally symmetric scattering)} \tag{14-23}$$

Since volume emission is isotropic, Eqs. (14-22) and (14-23) are also the appropriate expressions for determining the direction for emission by a volume element.

Figure 14-4 Fraction of power extinguished in ds.

Choose the Path Length to the Next Interaction

After a photon bundle has been either emitted or scattered by a volume element, it is necessary to determine how far it travels before it is absorbed or scattered. Probability concepts are again used to make this determination. The probability of absorption or scattering occurring in a small differential path length element ds at path length location s (see Fig. 14-4) is the ratio of power extinguished in ds to the total power extinguished along the path.

$$\text{Probability of extinction in } ds = \frac{\text{power extinguished in } ds}{\text{total power extinguished}}$$

Referring to Fig. 14-4, the intensity at $s = 0$ is I_0, and the intensity at s is $I_0 \exp(-K_e s)$ assuming that the extinction coefficient is constant along the path $[K_e \neq K_e(s)]$. From the definition of extinction coefficient [Eqs. (7-5) and (7-21)] the fraction of power either absorbed or scattered in ds is $K_e\, ds$. Thus, the probability distribution function is

$$P(s)\, ds = \frac{[I_0 \exp(-K_e s)](K_e\, ds)\Delta A\, \Delta\Omega}{\int_0^\infty I_0 \exp(-K_e s) K_e\, ds\, \Delta A\, \Delta\Omega} \tag{14-24}$$

Using the assumption that $K_e \neq K_e(s)$, (14-24) gives

$$P(s)\, ds = \exp(-K_e s) K_e\, ds \tag{14-25}$$

The cumulative probability (probability of extinction between 0 and s) is

$$R(s) = \int_0^s P(s^*)\, ds^* = 1 - \exp(-K_e s) \tag{14-26}$$

The path length to the next interaction s can thus be obtained by generating a random number and solving (14-26) for s.

$$s = -\frac{1}{K_e} \ln(1 - R) \tag{14-27}$$

In (14-27), $(1 - R)$ can be replaced by R since it is uniformly distributed between 0 and 1. The incremental optical depth to the next interaction t_s

would thus be given by

$$t_s = -\ln R \tag{14-28}$$

where

$$t_s = K_e s \qquad [K_e \neq K_e(s)] \tag{14-29}$$

Equation (14-28) is also valid in a medium with nonuniform properties, although (14-27) is not. The derivation of an expression that is equivalent to (14-27) and is valid for nonuniform properties follows from the definition of optical depth and is left as an exercise (Problem 14-6).

Decide if the Bundle is Absorbed or Scattered

The decision of whether a bundle is absorbed or scattered by a single-scattering volume element is based on comparing the single-scattering albedo ω_0 with a random number.

$$R < \omega_0 \quad \text{(scattered)} \tag{14-30}$$

$$R > \omega_0 \quad \text{(absorbed)} \tag{14-31}$$

If the bundle is scattered, a new direction is determined, and if it is absorbed, a counter is incremented and a new bundle is emitted.

These introductory comments are intended as a general outline of the Monte Carlo method in participating media and not as a precise formula for all situations. Perhaps the method is best learned by studying specific examples. Two specific problems introduced in Chapter 12, the effective absorptivity and emissivity of a semi-infinite, planar, absorbing, scattering, and emitting medium will be considered next.

Example 14-1. Effective Absorptivity of a Gray, Semi-Infinite, Isotropic Scattering Medium

Consider a cold, nonemitting, semi-infinite, planar medium subject to a diffuse hemispherical flux, q_w^+, as shown in Fig. 14-5. Apply the Monte Carlo method to determine the effective absorptivity of the medium (i.e., the fraction of the incident flux absorbed).

---

1 Assign Discrete Values of Energy to Photon Bundles

In this problem bundles will be "emitted" from the boundary at $x = 0$ and tracked until they are either scattered back into the surroundings or absorbed in the medium. The number of bundles "emitted" per unit time at $x = 0$ is N_w. The number per unit time scattered back into the surroundings $(x < 0)$ is S_w. The (constant) energy per bundle is w. The total power

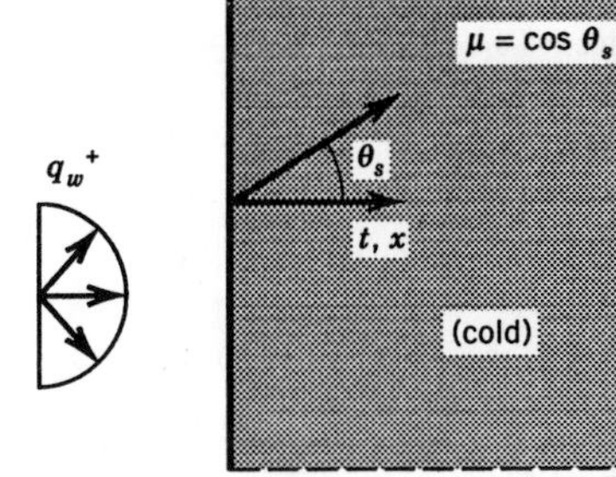

Figure 14-5 Effective absorptivity of semi-infinite slab.

emitted by the boundary is $q_w^+ A$. Setting $q_w^+ A$ equal to the product of w times N_w gives the energy per bundle as

$$w = \frac{q_w^+ A}{N_w} \tag{14-32}$$

The flux scattered back into the surroundings is

$$q_w^- = \frac{S_w w}{A} = \frac{S_w q_w^+}{N_w} \tag{14-33}$$

The effective absorptivity can then be determined as one minus the effective reflectively ($\tau_{\text{eff}} \to 0$ for semi-infinite medium) where ρ_{eff} is the ratio of the reflected to incident flux.

$$\alpha_{\text{eff}} = 1 - \rho_{\text{eff}} = 1 - \frac{q_w^-}{q_w^+} = 1 - \frac{S_w}{N_w} \tag{14-34}$$

As a check on consistency α_{eff} could also be determined by adding up all the bundles absorbed inside the medium, S_i

$$\alpha_{\text{eff}} = \frac{\sum_i S_i}{N_w} \tag{14-35}$$

Equation (14-35) is equivalent to Eq. (14-34) since all the bundles emitted by the boundary must either be absorbed at the boundary or inside the medium as shown by Eq. (14-36).

$$N_w = S_w + \sum_i S_i \tag{14-36}$$

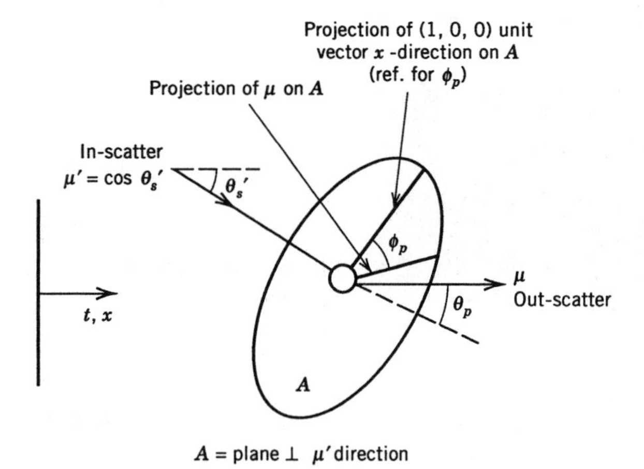

Figure 14-6 Relation between in-scattering and out-scattering angles in slab and single-particle coordinates.

In order to track bundles through the medium, it is necessary to know the relationship between the in-scattering direction ($\mu' = \cos\theta_s'$ in the slab coordinates) and the out-scattering direction ($\mu = \cos\theta_s$ in slab coordinates). Figure 14-6 shows the relationship between the in-scattering direction and out-scattering direction in both slab and single-particle coordinates. This relationship is also discussed in Chapter 12 in connection with Fig. 12-1.

The out-scattering direction in slab coordinates μ can be obtained from the in-scattering direction, μ', and the single-scattering angles θ_p and ϕ_p using the following relation, which is derived from analytic geometry.

$$\mu = \cos\phi_p\sqrt{1-\mu'^2}\sin\theta_p + \mu'\cos\theta_p \tag{14-37}$$

The slab azimuthal angle ϕ_s could be determined from Eq. (12-2) if it was needed.

2 Algorithm to Track Bundles

1. *Choose Point of Incidence.* The initial point of emission is chosen as $x = 0$ (or $t = 0$).
2. *Choose Wavelength.* In this example the properties are gray, hence there is no distinction between bundles on the basis of wavelength. All bundles are assumed to have the same spectral distribution, equal to that of q_w^+.
3. *Choose Direction*
   a. Original "emission" direction: This process is the same as diffuse emission discussed in Chapter 6 since the incident flux is diffuse. The cumulative probability function for diffuse emission from Eq.

(6-17) is

$$R(\theta_s) = \sin^2\theta_s \tag{14-38}$$

which gives the initial direction cosine as

$$\mu = \sqrt{1 - R} \tag{14-39}$$

Since R is uniformly distributed between 0 and 1, μ can also be calculated using (14-40)

$$\mu = \sqrt{R} \tag{14-40}$$

b. Scattering direction: For a given in-scattering direction cosine μ', the out-scattering direction cosine μ can be calculated from Eq. (14-37) if the single-scattering polar (θ_p) and azimuthal (ϕ_p) angles are known. The single-scattering angles can be calculated using Eqs. (14-22) and (14-23) for isotropic scattering.

4. *Choose Path Length to Next Intersection*. The path length to the next interaction (s) is determined from Eq. (14-27). In terms of optical depth the path length is given by Eq. (14-28).
5. *Calculate New Position of Bundle*. Once the distance to the next interaction (s) and the direction cosine of travel (μ) are known, the new x position can be calculated from the old position (see Fig. 14-7) using Eq. (14-41) or (14-42).

$$x_{\text{new}} = x_{\text{old}} + s\mu \tag{14-41}$$

$$t_{\text{new}} = t_{\text{old}} + t_s\mu \tag{14-42}$$

If $x_{\text{new}} < 0$, then the bundle left the medium and S_w should be incremented.

6. *Decide if Bundle is Scattered or Absorbed at New Location*. The decision of whether the bundle is absorbed or scattered is made using Eqs. (14-30) and (14-31). If the bundle is scattered, a new direction is determined (step 3b), and if it is absorbed, a counter is incremented ($S_i = S_i + 1$, optional) and a new bundle is emitted (step 1).

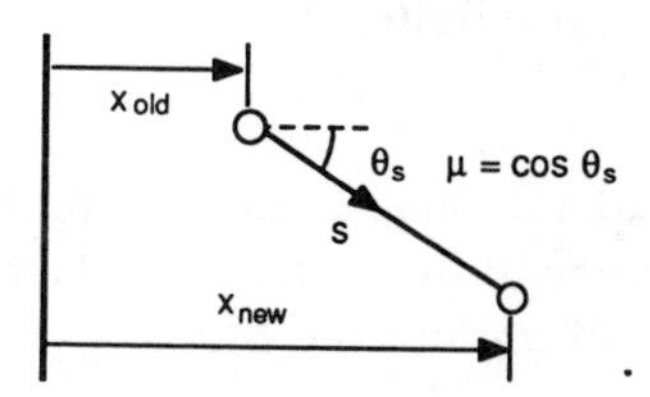

Figure 14-7 Geometry for tracking position of bundle in slab.

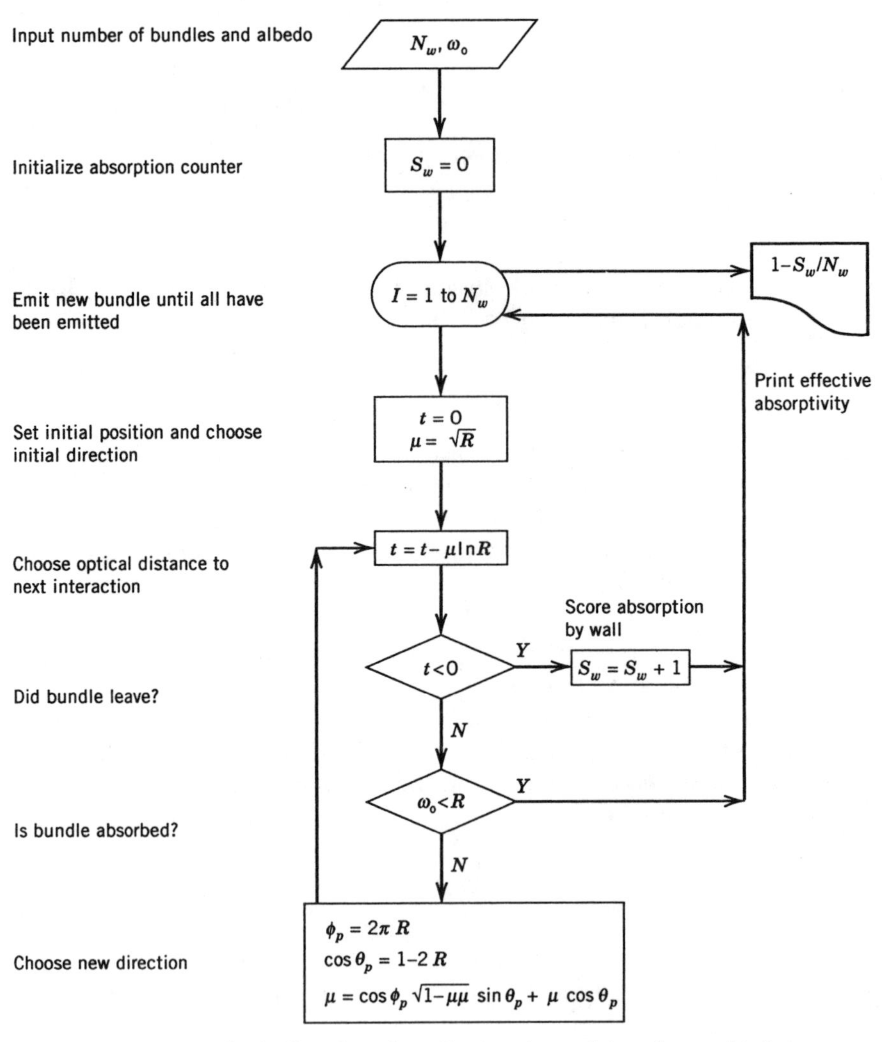

Figure 14-8 Monte Carlo flowchart for effective absorptivity of a semi-infinite, gray, isotropic scattering medium.

A flowchart for this algorithm is given in Fig. 14-8. The programming is left as an exercise (Problem 14-1).

Example 14-2. Effective Emissivity of a Gray, Semi-Infinite, Isotropic Scattering Medium

Next consider the problem of how much radiant flux is emitted by a semi-infinite, gray medium with a prescribed temperature profile into black-body surroundings at 0 K. To formulate this problem, use the Monte Carlo

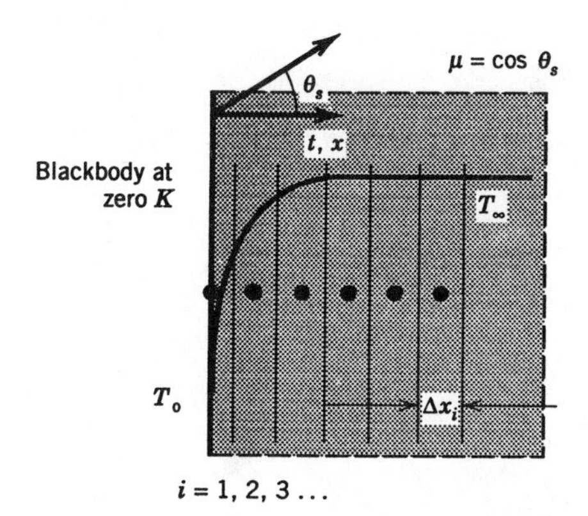

Figure 14-9 Effective emissivity of semi-infinite slab.

method and divide the medium into increments of optical depth Δt_i, where

$$\Delta t_i = K_e \, \Delta x_i \tag{14-43}$$

as shown in Fig. 14-9.

1 Assign Discrete Values of Energy to Bundles

Let N_i be the number of bundles emitted per unit time from the ith element of the medium. Similarly, let S_i be the number of bundles absorbed per unit time by the ith element. The number scattered into the blackbody surroundings S_w is therefore the total number emitted by the medium $\Sigma_i N_i$ minus the total number absorbed by the medium $\Sigma_i S_i$.

$$N_i = \text{number emitted from } i\text{th element} \tag{14-44}$$

$$S_i = \text{number absorbed by } i\text{th element} \tag{14-45}$$

$$S_w = \text{number absorbed by wall} = \sum_i (N_i - S_i) \tag{14-46}$$

Next a relationship is established between the energy balance on a differential volume, Eq. (14-5), and the number of discrete photon bundles. Applying Eq. (14-5) to a finite, one-dimensional elemental volume ($V_i = A \, \Delta x_i$) with gray properties gives

$$\frac{\Delta q_{r_i}}{\Delta x_i} = 4K_a \sigma T_i^4 - \int_{4\pi} K_a I_i \, d\Omega \tag{14-47}$$

as the net radiant power loss per unit volume (W/cm^3) from the ith element. Multiplying (14-47) by Δx_i gives the net radiant power loss per unit area

(W/cm^2) from the ith element.

$$\Delta q_{r_i} = 4(1 - \omega_0)\sigma T_i^4 \, \Delta t_i - \int_{4\pi} (1 - \omega_0) I_i \, d\Omega \, \Delta t_i \tag{14-48}$$

Introducing (14-44), (14-45), and the energy per photon bundle w, Eq. (14-48) becomes

$$\Delta q_{r_i} = \frac{w}{A}(N_i - S_i) \tag{14-49}$$

where

$$\frac{wN_i}{A} = 4(1 - \omega_0)\sigma T_i^4 \, \Delta t_i \tag{14-50}$$

$$\frac{wS_i}{A} = \int_{4\pi} (1 - \omega_0) I_i \, d\Omega \, \Delta t_i \tag{14-51}$$

The energy per bundle from (14-50) is thus

$$w = \frac{4(1 - \omega_0)\sigma T_i^4 \, \Delta t_i A}{N_i} = \text{constant} \tag{14-52}$$

Evaluating (14-52) at $i = 1$ ($T_1 = T_0$) gives

$$w = \frac{4(1 - \omega_0)\sigma T_0^4 \, \Delta t_1 A}{N_1} \tag{14-53}$$

and eliminating w between (14-52) and (14-53) gives N_i, which is the number of bundles that should be emitted from each element, relative to the reference number N_1.

$$\frac{N_i}{N_1} = \left(\frac{T_i}{T_1}\right)^4 \frac{\Delta t_i}{\Delta t_1} \tag{14-54}$$

Summing up the net radiant loss per unit area from each volume gives the net radiant flux at $x \to \infty$ (which is zero) minus the net radiant flux at $x = 0$.

$$\sum_i \Delta q_{r_i} = q_{r_\infty} - q_{r_0} = -q_{r_0} \tag{14-55}$$

The effective emissivity of the medium is the ratio of the net flux at $x = 0$ to a reference blackbody hemispherical flux, here chosen to be σT_∞^4.

$$\varepsilon_{\text{eff}} = \frac{-q_{r_0}}{\sigma T_\infty^4} \tag{14-56}$$

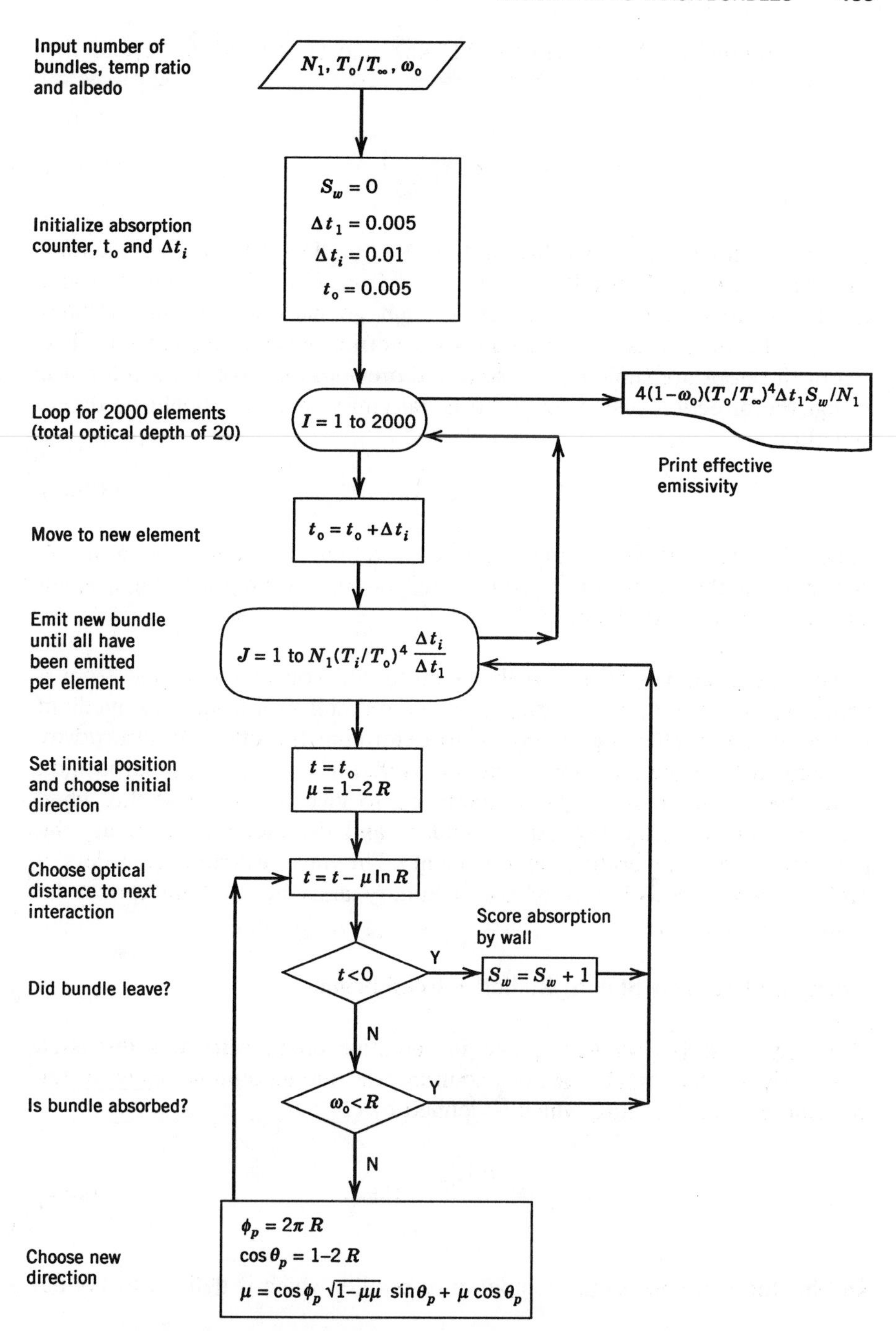

Figure 14-10 Monte Carlo flowchart for effective emissivity of a semi-infinite, gray, isotropic scattering medium.

Combining (14-56), (14-55), (14-49), (14-46), and (14-53) gives the effective emissivity as

$$\varepsilon_{\text{eff}} = 4(1 - \omega_0)\left(\frac{T_0}{T_\infty}\right)^4 \Delta t_1 \frac{S_w}{N_1} \tag{14-57}$$

Thus, ε_{eff} can be determined by emitting N_1 bundles at $t = 0$, incrementing t by Δt, emitting N_i bundles [as obtained from (14-54)] at each step and keeping score of the total number S_w that go back to the surroundings ($t < 0$). This process is continued until ε_{eff} approaches a constant value. The rest of the steps are similar to those of the previous example. One difference is that the direction cosine for the original emission event should be determined by

$$\mu = \cos\theta_s = 1 - 2R \tag{14-58}$$

rather than (14-40), since emission from a volume element is isotropic. A flowchart for this algorithm is given in Fig. 14-10. The programming is again left as an exercise (Problem 14-2).

The two examples that have been presented consider emission from a boundary into a cold, nonemitting medium and emission from a hot medium into a cold, nonreflecting boundary, to determine the effective absorptivity and emissivity of the medium. As demonstrated in Chapter 12, superposition allows these two solutions to be combined to give the net heat flux when there is emission from both the boundary and the medium, assuming the properties are temperature independent. The next example considers a problem where emission from the boundary and the medium cannot be treated independently.

Example 14-3. Gray Slab in Radiative Equilibrium

The gray, isotropic scattering slab in radiative equilibrium was discussed previously in Chapter 11. An exact solution was obtained numerically for the nondimensional heat flux, which is defined by (14-59)

$$q^* = \frac{q_r|_{\varepsilon_{1,2}=1,\, T_{w_2}=0}}{e_{b_{w_1}}} \tag{14-59}$$

and for the nondimensional temperature profile, which is defined by (14-60)

$$e_b^* = \frac{e_b|_{\varepsilon_{1,2}=1,\, T_{w_2}=0}}{e_{b_{w_1}}} \tag{14-60}$$

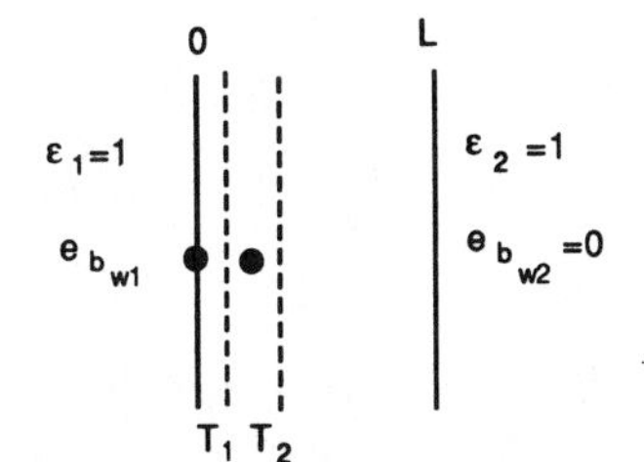

Figure 14-11 Gray slab in radiative equilibrium.

For nonblack walls ($\varepsilon_{1,2} \neq 1$) and an emitting boundary at 2 ($T_{w_2} > 0$), the actual heat flux and temperature are calculated from (11-30) and (11-29). In accordance with (14-59) and (14-60) set up the solution to this problem using the Monte Carlo technique with black walls and a nonemitting boundary at $x = L$ ($T_{w_2} = 0$) in order to determine q^* and e_b^* (Fig. 14-11).

The net flux q_r is constant through the medium and hence can be evaluated at any location. Evaluating the net flux at $x = 0$ gives the net flux as the flux emitted from the wall minus the flux absorbed by the wall.

$$q_{r_0} = e_{b_{w_1}} - q_{w_1}^- \tag{14-61}$$

Defining the number of bundles emitted from wall one per unit time as N_{w_1} and the number of bundles absorbed at wall one per unit time as S_{w_1} gives (14-62) and (14-63)

$$\frac{wN_{w_1}}{A} = e_{b_{w_1}} \tag{14-62}$$

$$\frac{wS_{w_1}}{A} = q_{w_1}^- \tag{14-63}$$

Combining (14-61), (14-62), and (14-63) gives the net flux at the wall as

$$q_{r_0} = \frac{w}{A}(N_{w_1} - S_{w_1}) \tag{14-64}$$

where the energy per bundle is

$$w = \frac{\sigma T_{w_1}^4 A}{N_{w_1}} \tag{14-65}$$

From (14-59) the nondimensional heat flux is given in terms of the photon bundle numbers as

$$q^* = \frac{N_{w_1} - S_{w_1}}{N_{w_1}} \tag{14-66}$$

Since the medium is in radiative equilibrium, the number of bundles emitted from each element must equal the number absorbed by each element

$$N_i = S_i \quad \text{(radiative equilibrium)} \tag{14-67}$$

This is evident from Eq. (14-49) upon setting $\Delta q_{r_i} = 0$. Thus every time a bundle is absorbed it should be re-emitted. If scattering is isotropic, then scattering is equivalent to absorption and re-emission, and the albedo falls out of the problem. Hence the medium can be taken as either a perfect absorber or perfect scatterer, whichever is more convenient.

Returning to (14-66), since a bundle will be emitted by the medium every time one is absorbed, the Monte Carlo formulation can be accomplished by emitting N_{w_1} bundles from wall 1 and re-emitting them whenever they are absorbed by the medium until they get absorbed at one wall or the other. Thus, a balance of photon bundles indicates that the sum of the bundles absorbed at walls 1 and 2 must equal the total number emitted from wall 1.

$$N_{w_1} = S_{w_1} + S_{w_2} \tag{14-68}$$

Substituting (14-68) into (14-66) gives the nondimensional heat flux as

$$q^* = \frac{S_{w_2}}{N_{w_1}} \tag{14-69}$$

The temperature distribution can be obtained from the energy balance on an element, Eqs. (14-49) and (14-52). Setting $N_i = S_i$ in (14-52) and eliminating w using (14-65) yields

$$e_b^* = \left.\frac{T_i^4}{T_{w_1}^4}\right|_{\varepsilon_{1,2}=1,\, T_{w_2}=0} = \frac{S_i}{4(1-\omega_0)\,\Delta t_i\, N_{w_1}} \tag{14-70}$$

It should be noted that for isotropic scattering, S_i will be proportional to $(1-\omega_0)$ and ω_0 falls out of the problem as it should. Also, the temperature is indeterminate if $\omega_0 = 1$.

A flowchart for determining q^* is given in Fig. 14-12 and the programming is again left as an exercise (Problem 14-3).

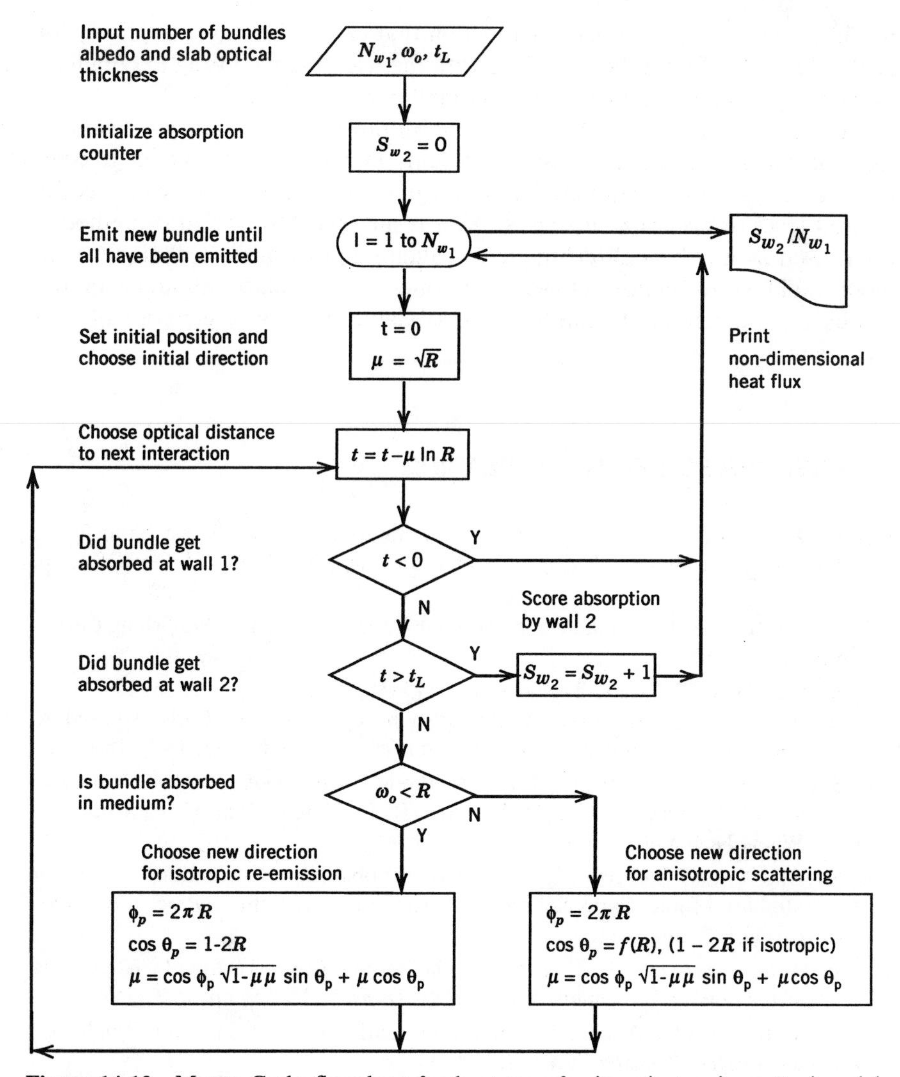

Figure 14-12 Monte Carlo flowchart for heat transfer in anisotropic scattering slab in radiative equilibrium.

SUMMARY

The Monte Carlo technique is a powerful tool for solving radiative transfer problems in participating media. Complexities such as anisotropic scattering and polarization do not involve any significant changes in the approach. Anisotropic scattering, for example, can be included by modifying the probability functions for scattering according to (14-19). While this procedure may result in a cumulative probability function that cannot be solved explicitly for angle, this difficulty can be overcome with the use of curve fitting to obtain an explicit expression or implicit solution methods can be used.

Currently the main drawback to using Monte Carlo methods as a general tool for solving radiative transfer problems in participating media is computer time. The time required for convergence increases very rapidly as the desired level of accuracy increases. As a result, the Monte Carlo method is often used as an "exact" technique to validate approximate techniques with more rapid convergence. However, if one- or two-place accuracy in the results can be tolerated, Monte Carlo is viable even as a general solution technique.

REFERENCES FOR FURTHER READING

Dunn, W. L. (1983), "Inverse Monte Carlo Solutions for Radiative Transfer in Inhomogeneous Media," *J. Quant. Spectr. and Rad. Transfer*, Vol. 29, No. 1, pp. 19–26.

Edwards, D. K. (1981), *Radiation Heat Transfer Notes*. Hemisphere Publishing Corporation, New York.

Gupta, R. P., Wall, T. F., and Truelove, J. S. (1983), "Radiative Scatter by Fly Ash in Pulverized Coal Fired Furnaces: Application of the Monte Carlo Method to Anisotropic Scatter," *Int. J. Heat Mass Transfer*, Vol. 26, No. 11, 1649–1660.

Haji-Sheikh, A. (1988), "Monte Carlo Methods," *Handbook of Numerical Heat Transfer*, W. J. Minkowycz, E. M. Sparrow, R. H. Pletcher, and G. E. Schneider, Eds., Wiley, New York.

Haji-Sheikh, A. and Sparrow, E. M. (1969), "Probability Distributions and Error Estimates for Monte Carlo Solutions of Radiation Problems," *Prog. Heat Mass Transfer*, Vol. 2, pp. 1–22.

House, L. L. and Avery, L. L. (1969), "The Monte Carlo Technique Applied to Radiative Transfer," *J. Quant. Spectr. Rad. Transfer*, Vol. 9, pp. 1579–1591.

Howell, J. R. (1966), "Application of Monte Carlo to Heat Transfer Problems," *Advances in Heat Transfer*, Vol. 5, pp. 1–54.

Howell, J. R. (1988), "Thermal Radiation in Participating Media: The Past, the Present, and Some Possible Futures," *J. Heat Transfer*, Vol. 110, Nov., pp. 1220–1229.

Howell, J. R. and Perlmutter, M. (1964a), "Monte Carlo Solution of Thermal Transfer through Radiant Media between Gray Walls," *J. Heat Transfer*, Vol. 86, No. 1, pp. 116–122.

Howell, J. R. and Perlmutter, M. (1964b), "Monte Carlo Solution of Radiant Heat Transfer in a Nongray, Nonisothermal Gas with Temperature Dependent Properties," *AIChE J.*, Vol. 10, No. 4, pp. 562–567.

Kobiyama, M. (1989), "Reduction of Computing Time and Improvement of Convergence Stability of the Monte Carlo Method Applied to Radiative Heat Transfer with Variable Properties," *J. Heat Transfer*, Vol. 111, pp. 135–140.

Siegel, R. and Howell, J. R. (1988), *Thermal Radiation Heat Transfer*, 2nd ed., Taylor, Francis, Hemisphere, New York.

Vercammen, H. A. J. and Froment, G. F. (1980), "An Improved Zone Method Using Monte Carlo Techniques for the Simulation of Radiation in Industrial Furnaces," *Int. J. Heat Mass Transfer*, Vol. 23, No. 3, pp. 329–336.

PROBLEMS

1. Formulate the radiative transfer problem for effective absorptivity of a gray, scattering, absorbing, nonemitting, semi-infinite slab with diffuse irradiation using the Monte Carlo method. Compare the Monte Carlo solution with the two discrete ordinates solution, Eq. (12-72). Use the forward scattering phase function

$$p(\theta_p) = 0.49360 \exp(2.20 \cos \theta_p) \qquad (B = 0.28,\ p^* = 0.57)$$

Do the calculations for two albedos ($\omega_0 = 0.1$ and 0.9).

2. Formulate the radiative transfer problem for effective emissivity of a gray, scattering, absorbing, emitting, semi-infinite slab with a 0 K blackbody boundary using the Monte Carlo method. Assume an exponential temperature profile

$$\frac{T(x) - T_\infty}{T_0 - T_\infty} = \exp\left[-\frac{x}{\delta_T}\right]$$

Compare the Monte Carlo solution with the two discrete ordinates solution, Eq. (12-68). Use the same phase function as Problem 1. Do the calculations for two albedos, ($\omega_0 = 0.1$ and 0.9), two optical depths, ($t_{\delta_T} = 0.15$ and 1.5) and two temperature ratios ($T_0/T_\infty = 0.7$ and 1).

3. Consider a gray, scattering, absorbing, emitting slab of thickness L in radiative equilibrium between diffuse gray walls. Scattering is anisotropic with the phase function given in Problem 1. Calculate and plot the nondimensional heat transfer

$$q^* = \frac{q}{q_1^+ - q_2^-} = \left.\frac{q}{e_{b_1} - e_{b_2}}\right|_{\varepsilon_1 = \varepsilon_2 = 1}$$

as a function of t_L for $t_L = 0.1$, 1, and 3 and for $\omega_0 = 0$ and 1. Compare with the two discrete ordinates result, Eq. (12-50). How do the results for $\omega_0 = 0$ compare with Fig. 11-3? For $\omega_0 = 1$, which asymmetry parameter works better in the discrete ordinate solution, $g = p^*$ or $1 - 2B$? Does this depend on optical thickness [see Eq. (12-36)]?

4. Consider a gray, nonscattering, absorbing, emitting isothermal slab of thickness L. Use the Monte Carlo method to calculate the hemispherical emissivity of the slab as a function of t_L for $t_L = 0.1$, 1, 3, and 10. Compare with the exact result from Chapter 10 of $1 - 2E_3(t_L)$.

5. Consider a gray, isotropic scattering, absorbing slab of thickness L. Use the Monte Carlo method to calculate the hemispherical absorptivity, reflectivity, and transmissivity of the slab with diffuse irradiation as a function of t_L and ω_0 for $t_L = 0.1$ and 1 and $\omega_0 = 0.1$ and 0.9. Compare the results with the nonscattering solution from Chapter 10 and the two-flux solution from Chapter 12 (Problem 12-3).

6. Derive an expression equivalent to Eq. (14-27) that holds for nonuniform properties, $K_e = K_e(s)$. Show that Eq. (14-28) is still valid.

APPENDIX A

PLANCK FUNCTION AND FRACTIONAL FUNCTION

$$e_{b_\lambda} = \frac{C_1}{\lambda^5(\exp[C_2/(\lambda T)] - 1)}$$

$$C_1 = 2\pi h c_0^2 = 37{,}413 \frac{\text{W}\,\mu\text{m}^4}{\text{cm}^2}$$

$$C_2 = \frac{hc_0}{k_B} = 14{,}388\ \mu\text{m K}$$

$$\sigma = \frac{C_1 \pi^4}{15 C_2^4} = 5.67 \times 10^{-12} \frac{\text{W}}{\text{cm}^2\ \text{K}^4}$$

$$f(\lambda T) = \frac{1}{\sigma T^4}\int_0^\lambda e_{b_\lambda}\, d\lambda = \frac{15}{\pi^4}\int_x^\infty \frac{x^3}{e^x - 1}\, dx, \qquad x = \frac{C_2}{\lambda T}$$

λT (μm K)	$f(\lambda T)$	$\frac{e_{b_\lambda}}{\sigma T^5}\left(\frac{1}{\text{cm K}}\right)$	λT (μm K)	$f(\lambda T)$	$\frac{e_{b_\lambda}}{\sigma T^5}\left(\frac{1}{\text{cm K}}\right)$
1000	.0003	.0372	4500	.5643	1.5238
1100	.0009	.0855	4600	.5793	1.4679
1200	.0021	.1646	4700	.5937	1.4135
1300	.0043	.2774	4800	.6075	1.3604
1400	.0078	.4222	4900	.6209	1.3089
1500	.0128	.5933	5000	.6337	1.2590
1600	.0197	.7825	5100	.6461	1.2107
1700	.0285	.9809	5200	.6579	1.1640
1800	.0393	1.1797	5300	.6694	1.1190
1900	.0521	1.3713	5400	.6803	1.0756
2000	.0667	1.5499	5500	.6909	1.0339
2100	.0830	1.7111	5600	.7010	.9938
2200	.1009	1.8521	5700	.7108	.9552
2300	.1200	1.9717	5800	.7201	.9181
2400	.1402	2.0695	5900	.7291	.8826
2500	.1613	2.1462	6000	.7378	.8485
2600	.1831	2.2028	6100	.7461	.8158
2700	.2053	2.2409	6200	.7541	.7844
2800	.2279	2.2623	6300	.7618	.7543
2900	.2505	2.2688	6400	.7692	.7255
3000	.2732	2.2624	6500	.7763	.6979
3100	.2958	2.2447	6600	.7832	.6715
3200	.3181	2.2175	6700	.7897	.6462
3300	.3401	2.1824	6800	.7961	.6220
3400	.3617	2.1408	6900	.8022	.5987
3500	.3829	2.0939	7000	.8081	.5765
3600	.4036	2.0429	7100	.8137	.5552
3700	.4238	1.9888	7200	.8192	.5348
3800	.4434	1.9324	7300	.8244	.5152
3900	.4624	1.8745	7400	.8295	.4965
4000	.4809	1.8157	7500	.8344	.4786
4100	.4987	1.7565	7600	.8391	.4614
4200	.5160	1.6974	7700	.8436	.4449
4300	.5327	1.6387	7800	.8480	.4291
4400	.5488	1.5807	7900	.8522	.4140

λT (μm K)	$f(\lambda T)$	$\frac{e_{b_\lambda}}{\sigma T^5}\left(\frac{1}{\text{cm K}}\right)$	λT (μm K)	$f(\lambda T)$	$\frac{e_{b_\lambda}}{\sigma T^5}\left(\frac{1}{\text{cm K}}\right)$
8000	.8562	.3995	20000	.9856	.0196
8100	.8602	.3856	21000	.9873	.0164
8200	.8640	.3722	22000	.9889	.0139
8300	.8676	.3594	23000	.9901	.0118
8400	.8712	.3472	24000	.9912	.0101
8500	.8746	.3354	25000	.9922	.0087
8600	.8779	.3241	26000	.9930	.0075
8700	.8810	.3132	27000	.9937	.0065
8800	.8841	.3028	28000	.9943	.0057
8900	.8871	.2928	29000	.9948	.0050
9000	.8900	.2832	30000	.9953	.0044
9100	.8928	.2739	31000	.9957	.0039
9200	.8955	.2650	32000	.9961	.0035
9300	.8981	.2565	33000	.9964	.0031
9400	.9006	.2483	34000	.9967	.0028
9500	.9030	.2404	35000	.9970	.0025
9600	.9054	.2328	36000	.9972	.0022
9700	.9077	.2255	37000	.9974	.0020
9800	.9099	.2185	38000	.9976	.0018
9900	.9121	.2117	39000	.9978	.0016
10000	.9142	.2052	40000	.9979	.0015
11000	.9318	.1518	41000	.9981	.0014
12000	.9451	.1145	42000	.9982	.0012
13000	.9551	.0878	43000	.9983	.0011
14000	.9628	.0684	44000	.9984	.0010
15000	.9689	.0540	45000	.9985	.0009
16000	.9738	.0432	46000	.9986	.0009
17000	.9777	.0349	47000	.9987	.0008
18000	.9808	.0285	48000	.9988	.0007
19000	.9834	.0235	49000	.9988	.0007

The following curve fit of $f(\lambda T)$ from Wiebelt* is useful for computer solutions:

$$x = \frac{C_2}{\lambda T} > 2: f(\lambda T) = \frac{15}{\pi^4} \sum_{m=1,2,\ldots} \frac{e^{-mx}}{m^4}\{[(mx+3)mx+6]mx+6\}$$

$$x = \frac{C_2}{\lambda T} < 2: f(\lambda T)$$

$$= 1 - \frac{15}{\pi^4}x^3\left(\frac{1}{3} - \frac{x}{8} + \frac{x^2}{60} - \frac{x^4}{5040} + \frac{x^6}{272{,}160} - \frac{x^8}{13{,}305{,}600}\right)$$

*Wiebelt, John A., *Engineering Radiation Heat Transfer*, Holt, Rinehart and Winston, Inc., New York, 1966.

APPENDIX B

SELECTED VIEW FACTORS

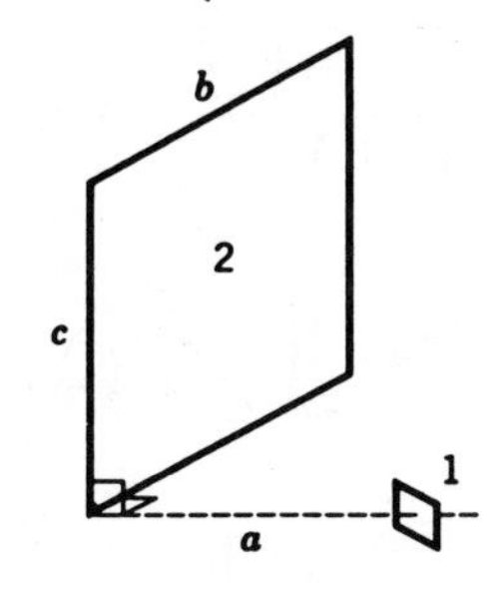

Element to Parallel Rectangle

$$X = \frac{b}{a} \qquad Y = \frac{c}{a}$$

$$F_{12} = \frac{1}{2\pi}\left[\frac{X}{\sqrt{1+X^2}}\tan^{-1}\left(\frac{Y}{\sqrt{1+X^2}}\right) + \frac{Y}{\sqrt{1+Y^2}}\tan^{-1}\left(\frac{X}{\sqrt{1+Y^2}}\right)\right]$$

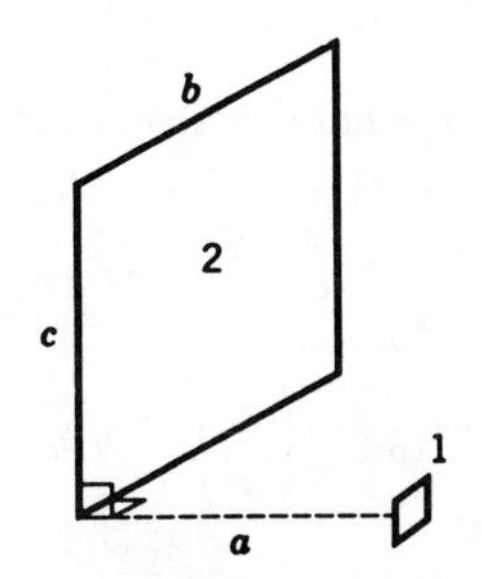

Element to Perpendicular Rectangle

$$X = \frac{b}{a} \qquad Y = \frac{c}{a}$$

$$F_{12} = \frac{1}{2\pi}\left[\tan^{-1}\left(\frac{1}{Y}\right) - \frac{Y}{\sqrt{X^2+Y^2}}\tan^{-1}\left(\frac{1}{\sqrt{X^2+Y^2}}\right)\right]$$

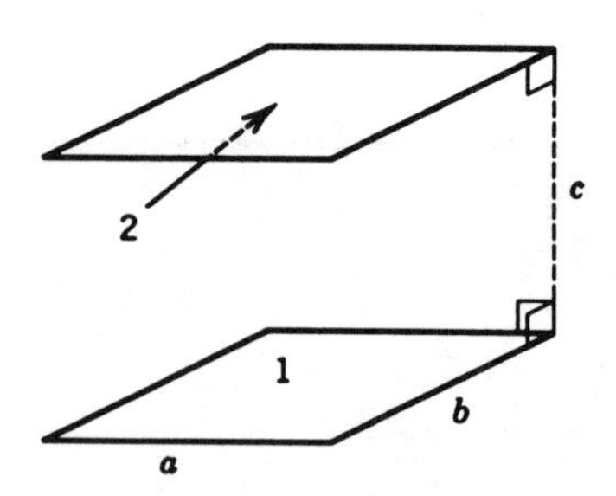

Parallel Rectangles

$$X = \frac{a}{c} \qquad Y = \frac{b}{c}$$

$$F_{12} = \frac{2}{\pi XY}\left\{\ln\sqrt{\frac{(1+X^2)(1+Y^2)}{1+X^2+Y^2}} + X\sqrt{1+Y^2}\tan^{-1}\left(\frac{X}{\sqrt{1+Y^2}}\right)\right\}$$

$$+\frac{2}{\pi XY}\left\{Y\sqrt{1+X^2}\tan^{-1}\left(\frac{Y}{\sqrt{1+X^2}}\right) - X\tan^{-1}X - Y\tan^{-1}Y\right\}$$

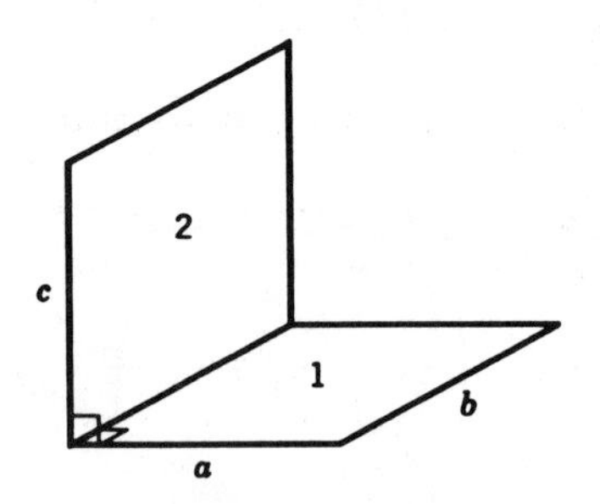

Perpendicular Rectangles

$$X = \frac{a}{c} \qquad Y = \frac{b}{c}$$

$$F_{12} = \frac{1}{\pi X}\left\{X\tan^{-1}\left(\frac{1}{X}\right) + Y\tan^{-1}\left(\frac{1}{Y}\right) - \sqrt{X^2+Y^2}\tan^{-1}\left(\frac{1}{\sqrt{X^2+Y^2}}\right)\right\}$$

$$+\frac{1}{4\pi X}\left\{\ln\left[\frac{(1+X^2)(1+Y^2)}{1+X^2+Y^2}\right] + X^2\ln\left[\frac{X^2(1+X^2+Y^2)}{(1+X^2)(X^2+Y^2)}\right]\right.$$

$$\left. + Y^2\ln\left[\frac{Y^2(1+X^2+Y^2)}{(1+Y^2)(X^2+Y^2)}\right]\right\}$$

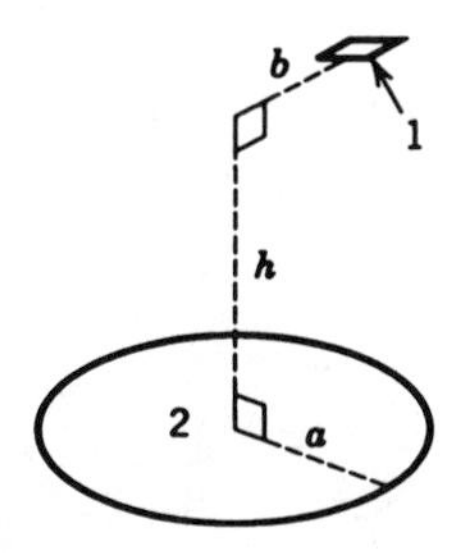

Element to Parallel, Offset Circle

$$X = \frac{h}{b} \qquad Y = \frac{a}{h} \qquad Z = 1 + (1 + Y^2)X^2$$

$$F_{12} = \frac{1}{2}\left[1 - \frac{Z - 2X^2Y^2}{\sqrt{Z^2 - 4X^2Y^2}}\right]$$

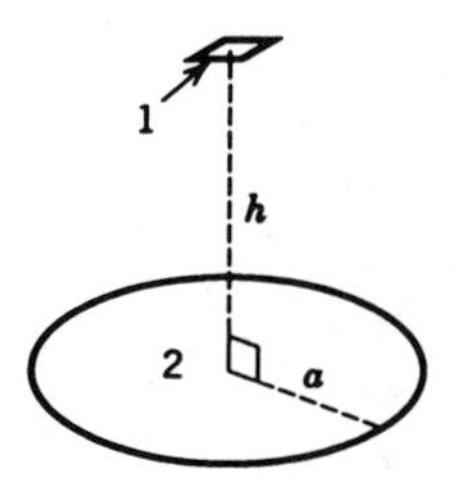

Element to Parallel, Coaxial Circle

$$X = \frac{h}{a}$$

$$F_{12} = \frac{1}{1 + X^2}$$

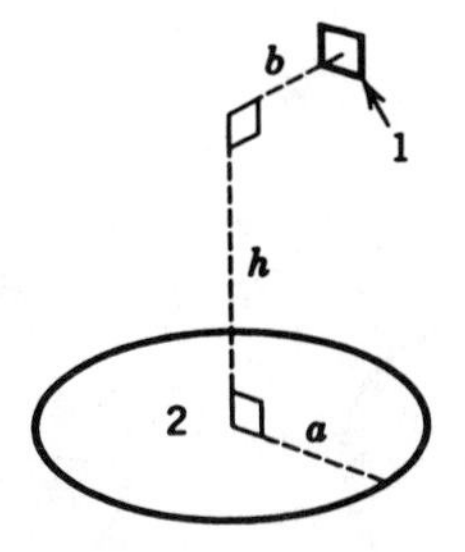

Element to Perpendicular, Offset Circle

$$X = \frac{h}{b} \qquad Y = \frac{a}{b}$$

$$F_{12} = \frac{X}{2}\left[\frac{1 + X^2 + Y^2}{\sqrt{(1 + X^2 + Y^2)^2 - 4Y^2}} - 1\right]$$

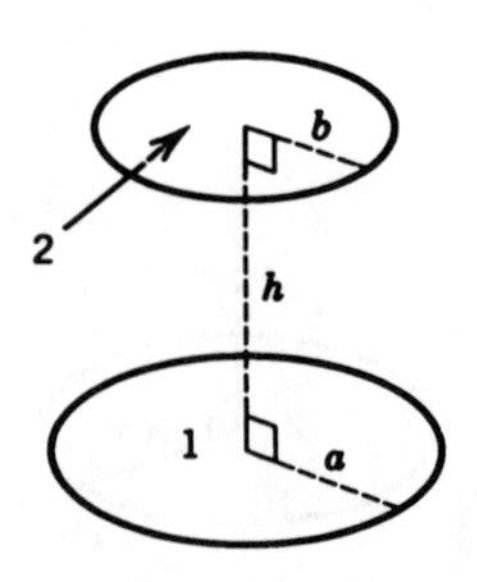

Parallel Coaxial Circles

$$X = \frac{a}{h} \qquad Y = \frac{b}{h} \qquad Z = 1 + \frac{1 + Y^2}{X^2}$$

$$F_{12} = \frac{1}{2}\left[Z - \sqrt{Z^2 - 4\left(\frac{Y}{X}\right)^2}\right]$$

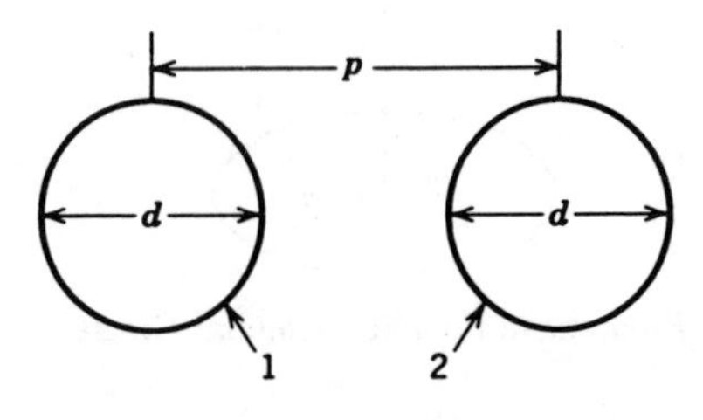

Parallel Circular Cylinders

$$X = \frac{p}{d}$$

$$F_{12} = \frac{1}{\pi}\left[\sqrt{X^2 - 1} + \sin^{-1}\left(\frac{1}{X}\right) - X\right]$$

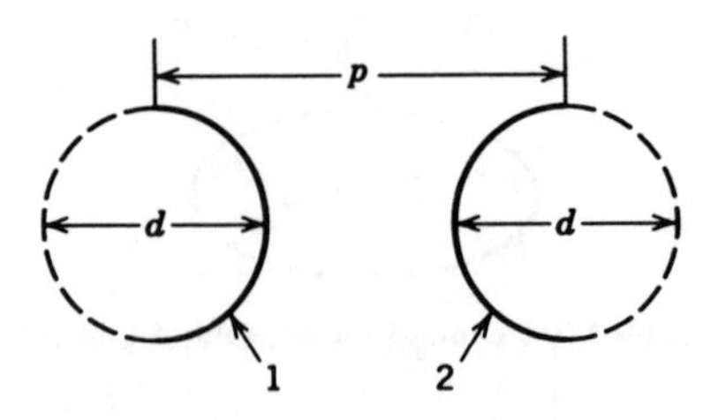

Parallel Half-Circular Cylinders

$$X = \frac{p}{d}$$

$$F_{12} = \frac{2}{\pi}\left[\sqrt{X^2 - 1} + \sin^{-1}\left(\frac{1}{X}\right) - X\right]$$

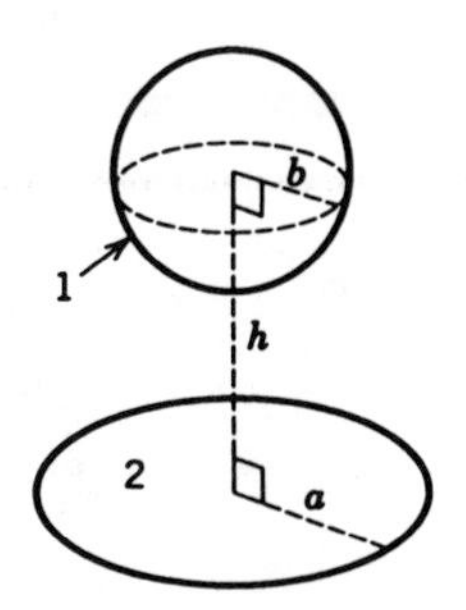

Sphere to Centered Circle

$$X = \frac{a}{h}$$

$$F_{12} = \frac{1}{2}\left[1 - \frac{1}{\sqrt{1 + X^2}}\right]$$

Infinite Strip to Parallel Circular Cylinder

$$X = \frac{2c}{d} \qquad Y = \frac{2a}{d} \qquad Z = \frac{2(a + b)}{d}$$

$$F_{12} = \frac{1}{Z - Y}\left[\tan^{-1}\left(\frac{Z}{X}\right) - \tan^{-1}\left(\frac{Y}{X}\right)\right]$$

Element to Parallel Circular Cylinder

$$X = \frac{b}{a} \qquad Y = \frac{h}{a}$$

$$A = (1 + Y^2) + X^2 \qquad B = (1 - Y^2) + X^2$$

$$F_{12} = \frac{1}{\pi Y}\tan^{-1}\left(\frac{X}{\sqrt{Y^2 - 1}}\right) + \frac{X}{\pi}\left[\frac{A - 2Y}{Y\sqrt{AB}}\tan^{-1}\sqrt{\frac{A(Y - 1)}{B(Y + 1)}} - \frac{1}{Y}\tan^{-1}\sqrt{\frac{Y - 1}{Y + 1}}\right]$$

APPENDIX C

OPTICAL CONSTANTS OF SELECTED MATERIALS

Metals: Aluminum, copper, gold
Semiconductors: Silicon, silicon carbide
Dielectrics: Aluminum oxide (sapphire), silicon dioxide (glass), water, ice

Aluminum

λ (μm)	n	k	λ (μm)	n	k
0.0165	1.01	2.40E-2	0.500	0.769	6.08
0.0207	0.987	4.41E-3	0.600	1.20	7.26
0.0310	0.940	8.16E-3	0.700	1.83	8.31
0.0400	0.885	1.16E-2	0.800	2.80	8.45
0.0517	0.789	1.90E-2	0.900	2.06	8.30
0.0620	0.668	2.68E-2	1.00	1.35	9.58
0.0708	0.520	3.81E-2	1.13	1.20	1.12E1
0.0800	0.258	7.77E-2	1.20	1.21	1.20E1
0.0918	0.0409	5.17E-1	1.38	1.26	1.40E1
0.103	0.0328	7.91E-1	1.50	1.38	1.54E1
0.207	0.130	2.39	1.77	1.77	1.83E1
0.300	0.276	3.61	2.00	2.15	2.07E1
0.400	0.490	4.86	3.00	4.24	3.06E1

Aluminum *(Continued)*

λ (μm)	n	k	λ (μm)	n	k
4.00	6.43	3.98E1	14.0	40.5	1.16E2
5.00	8.67	4.86E1	16.0	47.7	1.27E2
6.00	11.1	5.76E1	18.0	54.7	1.37E2
7.00	14.0	6.62E1	20.0	60.7	1.47E2
8.00	17.5	7.49E1	22.0	66.9	1.57E2
9.00	21.5	8.26E1	24.0	72.2	1.68E2
10.0	25.3	8.98E1	26.0	79.1	1.78E2
11.0	29.2	9.66E1	28.0	86.3	1.89E2
12.0	33.0	1.03E2	30.0	94.2	1.99E2
13.0	36.6	1.09E2			

From Shiles, E., Sasaki, T., Inokuti, M., and Smith, D. Y., *Phys. Rev. B*, Vol. 22, 1980, p. 1612.

Copper

λ (μm)	n	k	λ (μm)	n	k
0.0103	0.965	5.40E-2	0.620	0.272	3.24
0.0200	0.965	1.07E-1	0.708	0.214	3.67
0.0310	0.934	1.67E-1	0.826	0.260	5.26
0.0413	0.856	2.62E-1	1.26	0.496	6.78
0.0517	0.961	3.73E-1	2.00	0.850	1.06E1
0.0620	0.882	4.55E-1	3.10	1.59	1.65E1
0.0729	0.913	6.20E-1	4.13	2.59	2.22E1
0.0800	0.981	6.95E-1	5.17	3.81	2.75E1
0.0886	1.06	7.24E-1	6.20	5.23	3.30E1
0.103	1.09	7.31E-1	6.89	6.23	3.63E1
0.207	1.04	1.59	7.75	7.66	4.03E1
0.310	1.34	1.72	8.26	8.57	4.26E1
0.413	1.18	2.21	8.86	9.64	4.51E1
0.517	1.12	2.60	9.54	10.8	4.75E1

From Dold, B. and Mecke, R., *Optik*, Vol. 22, 1965, p. 435; Hagemann, H. J., *J. Opt. Soc. Am.*, Vol. 65, 1975, p. 742.

Gold

λ (μm)	n	k	λ (μm)	n	k
0.0103	0.929	1.78E-2	0.653	0.166	3.15
0.0200	0.859	1.56E-1	0.729	0.164	4.35
0.0302	0.799	3.17E-1	0.827	0.188	5.39
0.0400	0.851	5.96E-1	0.954	0.236	6.47
0.0500	0.850	6.45E-1	1.03	0.272	7.07
0.0600	1.11	8.13E-1	2.00	0.850	1.26E1
0.0700	1.08	6.78E-1	3.10	1.73	1.92E1
0.0800	1.03	7.13E-1	4.00	2.60	2.46E1
0.0900	1.09	7.98E-1	5.17	4.01	3.17E1
0.100	1.20	8.36E-1	6.20	5.42	3.75E1
0.200	1.43	1.22	7.09	6.94	4.20E1
0.302	1.81	1.92	8.00	8.50	4.64E1
0.400	1.66	1.96	9.18	10.8	5.16E1
0.517	0.608	2.12			

From Hagemann, H. J., Gudat, W., and Kunz, C., *J. Opt. Soc. Am.*, Vol. 65, 1975, p. 742; Canfield, L. R., Hass, G., and Hunter, W. R., *J. Physique*, Vol. 25, 1964, p. 124; Theye, M. L., *Phys. Rev. B*, Vol. 2, 1970, p. 3060; Dold, B. and Mecke, R., *Optik*, Vol. 22, 1965, p. 435.

Silicon

λ (μm)	n	k	λ (μm)	n	k
0.0102	0.998	2.12E-2	0.602	3.94	2.5E-2
0.0200	0.978	3.93E-3	0.705	3.78	1.2E-2
0.0300	0.930	1.00E-2	0.805	3.69	6E-3
0.0400	0.869	1.35E-2	0.886		7.75E-3
0.0500	0.766	2.23E-2	1.00		4.0E-4
0.0600	0.61	6.5E-2	1.107		2.5E-5
0.0708	0.426	2.55E-1	1.170		1.5E-6
0.0800	0.323	4.50E-1	2.0	3.45	
0.0902	0.269	6.77E-1	2.71	3.44	2.5E-9
0.101	0.256	9.18E-1	3.00	3.44	
0.150	0.478	2.00	3.80		1.3E-8
0.207	1.01	2.91	4.00	3.43	
0.250	1.58	3.63	5.00	3.43	2.0E-7
0.300	5.00	4.20	6.00	3.42	5E-7
0.350	5.44	2.99	6.67	3.42	2.66E-6
0.400	5.57	3.87E-1	7.00	3.42	2E-5
0.452	4.75	1.33E-1	8.00	3.42	1.53E-5
0.500	4.30	7.3E-2	9.00	3.42	8.4E-5
0.553	4.07	3.2E-2	10.00	3.42	6.76E-5

From Edwards, D. F., "Silicon (Si)," in *Handbook of Optical Constants of Solids*, ed. E. D. Palik, Academic Press, New York, 1985, pp. 547–569.

Silicon Carbide

λ (μm)	n_o	k_o	λ (μm)	n_o	k_o
0.0413	0.739	1.14E-1	6.0	2.4	6.2E-3
0.0506	0.562	2.00E-1	7.0	2.3	5.6E-3
0.0605	0.347	5.32E-1	8.0	2.0	8.3E-3
0.0708	0.424	8.73E-1	9.0	1.8	1.1E-2
0.0800	0.512	1.10	10.0	1.1	3.4E-2
0.0918	0.649	1.36	10.3	0.274	1.8E-1
0.103	0.769	1.58	10.5	0.066	9.5E-1
0.151	1.77	2.57	11.1	0.062	2.18
0.200	3.96	1.06	12.0	0.215	5.27
0.248	3.16	2.59E-1	12.5	2.22	1.28E1
0.318	2.92	7.56E-3	12.7	17.7	6.03
0.400	2.76	1.91E-4	13.0	8.00	3.64E-1
0.496	2.68	1.18E-5	14.1	4.8	4.7E-2
2.0	2.57	4.0E-4	16.7	3.8	1.1E-2
4.0	2.52	6.4E-4			

Subscript *o* in column heads = ordinary ray. From Choyke, W. J. and Edwards, D. F., "Silicon Carbide (SiC)," in *Handbook of Optical Constants of Solids*, ed. E. D. Palik, Academic Press, New York, 1985, pp. 587–595.

Aluminum Oxide (sapphire)

λ (μm)	n_o	n_e	k_o	k_e
0.20	1.91	1.91	1E-6	1E-6
0.25	1.85	1.85	8E-7	8E-7
0.30	1.81	1.81	5E-7	5E-7
0.40	1.79	1.79	3E-7	3E-7
0.50	1.77	1.77	2E-7	2E-7
1.00	1.76	1.76	6E-8	6E-8
2.00	1.73	1.73	5E-8	5E-8
3.00	1.71	1.71	2E-7	2E-7
4.00	1.68	1.68	1.5E-6	1.5E-6
4.6	1.65	1.65	1.1E-5	1.1E-5
5.0	1.62	1.62	3.1E-5	3.1E-5
5.6	1.58	1.58	1.1E-4	1.1E-4
6.0	1.56	1.56	2.4E-4	2.4E-4
7.0	1.50	1.48	1.1E-3	1.1E-3
8.0	1.38	1.36	3.3E-3	3.3E-3
9.0	1.20	1.20	8.3E-3	1.0E-2
10.0	0.92	0.96	1.8E-2	2.0E-2
11.0	0.30	0.53	8.9E-2	5.9E-2
12.0	0.05	0.07	9.9E-1	7.5E-1

Aluminum Oxide (Sapphire) *(Continued)*

λ (μm)	n_o	n_e	k_o	k_e
13.0	0.05	0.07	1.6	1.3
14.0	0.08	0.09	2.2	1.8
15.0	0.18	0.16	3.2	2.5
16.0	1.02	0.40	2.9	3.5
17.0	0.98	3.32	6.3	5.8
18.0	7.52	3.59	1.5	8.6E-1
19.0	4.23	1.74	3.4E-1	5.3E-1
20.0	2.66	0.38	2.7E-1	1.7
21.0	0.45	0.24	1.9	2.9
22.0	0.73	0.27	5.9	4.0
23.0	9.37	0.42	1.3	5.5
24.0	5.71	1.11	2.1E-1	8.4
25.0	4.57	13.2	1.8E-1	1.3E1
26.0	5.17	9.59	2.0	1.1
27.0	4.67	7.28	1.2E-1	4.3E-1
28.0	4.30	6.26	5.9E-2	2.4E-1
29.0	4.09	5.67	5E-2	1.6E-1
30.0	3.95	5.28	4E-2	1.2E-1
35.0	3.59	4.38	2E-2	5.2E-2
40.0	3.43	4.04	1.5E-2	3.3E-2
45.0	3.34	3.86	1.3E-2	2.4E-2
50.0	3.26	3.75	1.1E-2	2.0E-2

Subscripts in column heads: o = ordinary ray; e = extraordinary ray.
From Barker, A. S., Jr., *Phys, Rev.*, Vol. 132, 1963, p. 1474; Gryvnak, D. A. and Burch, D. E., *J. Opt. Soc. Am.*, Vol. 55, 1965, p. 625; Lowenstein, E. V., Smith, D. R., and Morgan, R. L., *Appl. Opt.*, Vol. 12, 1973, p. 398; Malitson, I. H., *J. Opt. Soc. Am.*, Vol. 52, 1962, p. 1377; Toon, O. B., Pollack, J. B., and Khare, B. N., *J. Geophys. Res.*, Vol. 81, 1976, p. 5733.

Silicon Dioxide (glass)

λ (μm)	n	k	λ (μm)	n	k
0.0100	0.987	9.7E-3	0.101	1.41	8.24E-1
0.0200	0.942	3.74E-2	0.207	1.543	< 1.0E-6
0.0300	0.913	9.0E-2	0.302	1.487	
0.0400	0.851	1.56E-1	0.405	1.469	
0.0500	0.803	3.00E-1	0.509	1.462	
0.0600	0.862	4.97E-1	0.589	1.458	
0.0700	0.975	7.31E-1	0.707	1.455	
0.0800	1.17	6.96E-1	0.852	1.453	
0.0902	1.32	7.95E-1	0.894	1.452	

Silicon Dioxide (glass) *(Continued)*

λ (μm)	n	k	λ (μm)	n	k
1.01	1.450		9.52	2.76	1.65
2.06	1.437		10.0	2.69	5.09E-1
3.24	1.413	< 1.0E-6	11.1	1.87	5.06E-2
3.64	1.40	2.25E-5	12.0	1.62	2.67E-1
4.00	1.39	5.79E-5	13.0	1.81	2.27E-1
4.55	1.36	2.56E-4	14.3	1.64	1.57E-1
5.00	1.34	3.98E-3	15.4	1.56	1.82E-1
5.88	1.28	5.94E-3	16.0	1.50	2.02E-1
6.90	1.12	8.51E-3	18.5	1.23	3.41E-1
7.14	1.05	1.06E-2	20.0	0.662	8.22E-1
7.69	0.772	4.0E-2	21.0	1.00	2.22
8.00	0.411	3.23E-1	22.7	2.94	1.29
8.55	0.466	9.78E-1	23.8	2.91	7.38E-1
8.85	0.356	1.53	25.0	2.74	3.97E-1
9.01	0.585	2.27	30.8	2.28	9.2E-2
9.17	1.62	2.63	40.0	2.10	4.6E-2
9.30	2.25	2.26	100	1.97	1.59E-2

From Philip, H. R., *J. Appl. Phys.*, Vol. 50, 1979, p. 1053; Brixner, B., *J. Opt. Soc. Am.*, Vol. 57, 1967, p. 674; Philip. H. R., *J. Phys. Chem. Solids*, Vol. 32, 1971, p. 1935; Weinberg, Z. A., Rubloff, G. W., and Bassous, E., *Phys. Rev. B.*, Vol. 19, 1979, p. 3107; Appleton, A., Chiranjivi, T., and Jafaripour-Ghazvini, M., in *The Physics of* SiO_2 *and Its Interfaces*, S. T. Pantelides, Ed., Pergamon, New York, 1978, p. 94; Kaminow, I. P., Bagley, B. G., and Olson, C. G., *Appl. Phys. Lett.*, Vol. 32, 1978, p. 98.

Water

λ (μm)	n	k	λ (μm)	n	k
0.20	1.424	1.3E-7	0.90	1.328	4.80E-7
0.25	1.377	6.0E-8	0.95	1.327	2.76E-6
0.30	1.359	3.6E-8	1.00	1.326	2.82E-6
0.35	1.349	8E-9	1.05	1.325	1.10E-6
0.40	1.343	3E-9	1.10	1.324	1.66E-6
0.45	1.339	7E-10	1.15	1.3235	7.32E-6
0.50	1.336	8E-10	1.20	1.323	9.74E-6
0.55	1.334	1.5E-9	1.30	1.321	1.117E-5
0.60	1.332	7E-9	1.35	1.320	2.90E-5
0.65	1.331	1.3E-8	1.40	1.320	1.448E-4
0.70	1.330	3.3E-8	1.45	1.319	3.00E-4
0.75	1.329	1.49E-7	1.50	1.318	2.065E-4
0.80	1.328	1.34E-7	1.55	1.317	1.184E-4
0.85	1.327	2.77E-7	1.60	1.316	7.89E-5

Water *(Continued)*

λ (μm)	n	k	λ (μm)	n	k
1.65	1.316	6.70E-5	5.0	1.331	1.225E-2
1.70	1.315	6.97E-5	5.26	1.318	9.59E-3
1.75	1.314	8.91E-5	5.5	1.303	1.208E-2
1.80	1.312	1.146E-4	5.8	1.266	3.226E-2
1.85	1.311	1.40E-4	6.0	1.313	1.021E-1
1.90	1.309	1.217E-3	6.05	1.324	1.121E-1
1.95	1.307	1.707E-3	6.4	1.347	4.339E-2
2.00	1.304	1.082E-3	6.5	1.338	4.107E-2
2.05	1.302	6.689E-4	7.0	1.323	3.443E-2
2.10	1.300	4.345E-4	7.5	1.303	3.456E-2
2.15	1.296	3.25E-4	8.0	1.293	3.603E-2
2.20	1.293	2.80E-4	8.5	1.286	3.768E-2
2.21	1.292	2.72E-4	9.0	1.269	4.054E-2
2.25	1.290	3.04E-4	10.0	1.214	5.316E-2
2.30	1.286	4.21E-4	10.5	1.185	6.90E-2
2.35	1.282	5.61E-4	11.0	1.151	1.021E-1
2.40	1.276	8.02E-4	11.5	1.145	1.530E-1
2.45	1.270	1.189E-3	12.0	1.160	2.092E-1
2.50	1.246	1.651E-3	12.5	1.190	2.438E-1
2.55	1.213	1.968E-3	13.0	1.220	2.918E-1
2.60	1.180	2.048E-3	13.5	1.245	3.202E-1
2.65	1.140	2.763E-3	14.0	1.275	3.582E-1
2.70	1.134	5.049E-3	15.0	1.330	4.298E-1
2.75	1.133	2.401E-2	16.0	1.400	4.088E-1
2.80	1.232	9.361E-2	17.5	1.485	3.955E-1
2.90	1.310	2.435E-1	18.0	1.535	3.939E-1
2.95	1.325	2.740E-1	20.0	1.722	3.899E-1
3.00	1.351	2.586E-1	25.0	1.690	3.734E-1
3.10	1.426	1.828E-1	30.0	1.652	3.369E-1
3.20	1.509	9.422E-2	35.0	1.615	3.373E-1
3.30	1.470	4.333E-2	40.0	1.577	3.801E-1
3.40	1.449	1.888E-2	42.0	1.567	4.048E-1
3.50	1.423	9.30E-3	50.0	1.590	5.165E-1
3.60	1.402	5.67E-3	60.0	1.670	5.945E-1
3.75	1.372	3.55E-3	75.0	1.760	5.604E-1
3.83	1.358	3.38E-3	83.0	1.801	5.284E-1
4.00	1.349	4.81E-3	100	1.878	5.292E-1
4.50	1.341	1.472E-2	117	1.942	5.025E-1
4.66	1.338	1.735E-2	150	1.995	4.918E-1
4.80	1.336	1.647E-2	200	2.025	5.189E-1

From Irvine, W. M. and Pollack, J. B., *ICARUS*, Vol. 8, 1968, p. 324.

Ice

λ (μm)	n	k	λ (μm)	n	k
0.95	1.302	8.3E-7	3.075	1.225	3.428E-1
1.00	1.302	1.99E-6	3.10	1.280	3.252E-1
1.03	1.301	2.8E-6	3.15	1.547	2.502E-1
1.05	1.301	2.5E-6	3.20	1.557	1.562E-1
1.10	1.300	1.84E-6	3.25	1.550	9.00E-2
1.15	1.299	2.93E-6	3.30	1.530	6.25E-2
1.20	1.298	9.18E-6	3.35	1.515	4.40E-2
1.25	1.297	1.41E-5	3.40	1.490	3.07E-2
1.30	1.296	1.24E-5	3.45	1.445	2.26E-2
1.40	1.295	2.04E-5	3.50	1.422	1.63E-2
1.45	1.294	1.18E-4	3.55	1.408	1.29E-2
1.50	1.294	5.58E-4	3.60	1.395	1.05E-2
1.52	1.294	6.41E-4	3.80	1.356	8.2E-3
1.55	1.294	5.83E-4	3.90	1.340	1.04E-2
1.60	1.293	3.80E-4	4.00	1.327	1.24E-2
1.65	1.293	3.03E-4	4.10	1.316	1.50E-2
1.70	1.2925	1.92E-4	4.20	1.307	1.75E-2
1.75	1.292	1.42E-4	4.30	1.299	2.18E-2
1.80	1.292	1.13E-4	4.40	1.288	2.82E-2
1.85	1.292	6.33E-5	4.50	1.280	3.30E-2
1.90	1.292	3.91E-4	4.60	1.273	2.87E-2
1.95	1.291	1.11E-3	4.70	1.266	2.15E-2
2.00	1.291	1.61E-3	4.80	1.258	1.73E-2
2.05	1.289	1.40E-3	4.90	1.252	1.47E-2
2.10	1.288	8.35E-4	5.00	1.247	1.33E-2
2.15	1.286	6.76E-4	5.10	1.241	1.30E-2
2.20	1.282	3.08E-4	5.20	1.236	1.33E-2
2.25	1.278	2.13E-4	5.30	1.231	1.52E-2
2.30	1.275	1.98E-4	5.40	1.227	1.75E-2
2.35	1.270	3.50E-4	5.50	1.226	2.10E-2
2.40	1.258	5.71E-4	5.60	1.226	2.44E-2
2.45	1.247	7.07E-4	5.70	1.226	3.13E-2
2.50	1.235	7.95E-4	5.80	1.227	4.24E-2
2.55	1.220	8.50E-4	5.90	1.232	5.26E-2
2.60	1.206	8.02E-4	6.00	1.235	6.17E-2
2.65	1.193	1.43E-3	6.05	1.235	6.46E-2
2.80	1.152	1.23E-2	6.10	1.234	6.43E-2
2.85	1.140	3.53E-2	6.20	1.232	6.16E-2
2.90	1.132	1.014E-1	6.30	1.228	5.84E-2
2.95	1.125	1.805E-1	6.40	1.226	5.55E-2
3.00	1.130	2.273E-1	6.50	1.225	5.51E-2
3.05	1.192	3.178E-1	6.60	1.223	5.59E-2

Ice *(Continued)*

λ (μm)	n	k	λ (μm)	n	k
6.65	1.222	5.64E-2	13.5	1.613	9.35E-2
6.70	1.222	5.54E-2	15.0	1.550	7.62E-2
6.80	1.221	5.35E-2	17.5	1.486	3.47E-2
6.90	1.221	5.08E-2	20.0	1.455	2.55E-2
7.00	1.221	4.91E-2	25.0	1.425	2.98E-2
7.10	1.221	4.73E-2	30.0	1.427	5.25E-2
7.20	1.221	4.53E-2	35.0	1.440	1.11E-1
7.50	1.220	3.99E-2	40.0	1.460	1.78E-1
8.00	1.219	3.69E-2	43.4	1.49	3.21E-1
8.50	1.217	3.52E-2	44.8	1.49	5.81E-1
9.00	1.210	3.65E-2	46.5	1.50	4.33E-1
9.50	1.192	3.10E-2	52.0	1.530	2.69E-1
10.0	1.152	4.13E-2	56.0	1.545	1.52E-1
10.5	1.195	6.02E-2	62.0	1.57	2.86E-1
11.0	1.290	9.54E-2	83.0	1.620	2.2E-1
11.5	1.393	1.14E-1	100	1.650	8E-2
12.0	1.480	1.20E-1	117	1.690	3E-2
12.5	1.565	1.19E-1	150	1.722	3E-2
13.0	1.612	1.08E-1	152	1.730	3E-2

From Irvine, W. M. and Pollack, J. B., *ICARUS*, Vol. 8, 1968, p. 324.

APPENDIX D

DIFFUSE GRAY SURFACE INTERCHANGE FORTRAN PROGRAM

DIFGRAY.FOR

```
C
C------ THIS IS A GENERALIZED PROGRAM FOR SOLVING DIF-
C------ FUSE-GRAY RADIANT INTERCHANGE PROBLEMS. THE
C------ BOUNDARY CONDITION FOR EACH NODE CAN BE EITHER
C------ PRESCRIBED TEMPERATURE OR HEAT TRANSFER RATE.
C------ NODAL PROPERTIES ARE SPECIFIED IN THE USER-
C------ DEFINED INPUT FILE ('DIFGRAY.INP'). ONLY THE
C------ UPPER RIGHT TRIANGLE OF THE VIEW FACTOR MATRIX
C------ NEEDS TO BE SPECIFIED. THE BOTTOM LEFT TRIANGLE
C------ WILL BE CALCULATED FROM RECIPROCITY AND THE DI-
C------ AGONAL WILL BE CALCULATED FROM SUMMATION. OUT-
C------ PUT IS GIVEN IN FILE 'DIFGRAY.OUT'. THE INPUT
C------ FILE FORMAT IS GIVEN IN TABLE 3-1.
C

      PROGRAM DIFGRAY
      IMPLICIT REAL*8 (A-H,O-Z)
      CHARACTER*50 COMMNT
      DIMENSION F(20,20),E(20),R(20),WKAREA(20),
```

```
  1B(1),Q(20),T(20),EB(20),PSI(20,20),RLAM(20,20),
  2PHI(20,20),A(20),IPT(20),PT(20)
   DATA PI/3.1415927d0/
   DATA SIGMA/5.67d-8/
   OPEN(UNIT=13,FILE='DIFGRAY.INP',STATUS='OLD')
C
C***SET OUTPUT DEVICE
C
   NOUT=11
   OPEN(UNIT=11,FILE='DIFGRAY.OUT',STATUS='NEW')
C READ COMMENT STATEMENT
   READ(13,*)
   READ(13,5)COMMNT
  5 FORMAT(1X,A50)
C READ IN # OF NODES
   READ(13,*)
   READ(13,*)N
C***SET UP PRESCRIBED TEMP OR HEAT FLUX MATRICES
C***READ IN IPT (FLAG), A (AREA), PT (PRESCRIBED T OR
C***Q), AND E (NODE EMISSIVITY)
C***A VALUE OF 0 IN IPT MEANS PRESCRIBED Q
C***A VALUE OF 1 IN IPT MEANS PRESCRIBED T
C***TO AVOID ZERO DIVIDE ERROR, E IS SET TO 0.9999 FOR
C***BLACK SURFACES AND 0.0001 FOR PERFECT REFLECTORS
C
   DO 20 K=1,3
  20 READ(13,*)
   DO 627 JJ=1,N
 626 READ (13,*) I,ANUMB,INUMB,RNUMB,ENUMB
   IPT(I)=INUMB
   A(I)=ANUMB
   PT(I)=RNUMB
   IF(ENUMB.GT.0.9999d0) THEN
      E(I)=0.9999d0
   ELSE IF(ENUMB.LT.0.0001d0) THEN
     E(I)=0.0001d0
   ELSE
     E(I)=ENUMB
   END IF
 627 CONTINUE
C
C***SET UP REFLECTIVITY VECTORS.
C
   DO 50 I=1,N
 50R(I)=1.d0-E(I)
```

```
C
C
C***SET UP VIEW FACTOR MATRIX - F(I,J)
C
   DO 25 I = 1,N
   DO 25 J = 1,N
 25 F(I,J) = 0.d0
C
   READ(13,*)
   DO 26 I = 1,N - 1
     READ(13,*)(F(I,J), J = I + 1,N)
 26 CONTINUE
C
C***CLEAN UP F(I,J) TO SATISFY RECIPROCITY EXACTLY--
C***IF TOP RIGHT TRIANGLE OF F(I,J) HAS BEEN SPECIFIED
C***AND BOTTOM LEFT SET TO ZERO, BOTTOM LEFT TRIANGLE
C***WILL BE SET ACCORDING TO RECIPROCITY
C
 DO 525 I = 1,N - 1
 IP1 = I + 1
 DO 525 J = IP1,N
 F(J,I) = A(I)*F(I,J) / A(J)
 525 CONTINUE
C
C***SET DIAGONAL OF F(I,J) TO SATISFY SUMMATION
  EXACTLY
C
 DO 540 I = 1,N
 FTEMP = 0.d0
 DO 530 J = 1,N
 IF (I.EQ.J) GOTO 530
 FTEMP = FTEMP + F(I,J)
  530 CONTINUE
 F(I,I) = 1.d0 - FTEMP
 IF(F(I,I).LT.0.d0) WRITE(NOUT,555)I,I,F(I,I)
 555 FORMAT( / 1X,'*** Warning: F(',I2,',',I2,') =
   ',F15.6)
 540 CONTINUE
C***SET UP PSI, RLAM, AND PHI MATRICES
C
 DO 110 J = 1,N
 DO 110 I = 1,N
```

```
      IF (IPT(I).EQ.1) THEN
         PSI(I,J) = (DELTA(I,J) - R(I)*F(I,J)) / E(I)
      ELSE
         PSI(I,J) = DELTA(I,J) - F(I,J)
      END IF
  110 CONTINUE
      CALL INVMAT(PSI,N)
C
      DO 101 I = 1,N
      DO 101 J = 1,N
      PHI(I,J) = DELTA(I,J)*R(I) / E(I) + PSI(I,J)
  101 RLAM(I,J) = DELTA(I,J) - PSI(I,J))*E(I) / R(I)
C
C***INPUT KNOWN T'S AND Q'S
C
      DO 105 I = 1,N
      T(I) = 0.d0
      EB(I) = 0.d0
  105 Q(I) = 0.d0
      DO 115 I = 1,N
      IF (IPT(I).EQ.0) THEN
         Q(I) = PT(I)
      ELSE
         T(I) = PT(I)
      END IF
  115 CONTINUE
C
      DO 343 I = 1,N
  343 EB(I) = SIGMA*T(I)**4
C
C***CALCULATE UNKNOWN T'S AND Q'S
C
      DO 351 I = 1,N
      DO 352 J = 1,N
      IF (IPT(I).EQ.0) THEN
         IF(IPT(J).EQ.0) THEN
            EB(I) = EB(I) + PHI(I,J)*Q(J) / A(J)
         ELSE
            EB(I) = EB(I) + PSI(I,J)*EB(J)
         END IF
      ELSE
         IF(IPT(J).EQ.0) THEN
            Q(I) = Q(I) - PSI(I,J)*Q(J) / A(J)*A(I)*E(I) / R(I)
         ELSE
            Q(I) = Q(I) + RLAM(I,J)*EB(J)*A(I)
```

```
      END IF
   END IF
 352 CONTINUE
 351 CONTINUE
C***CALCULATE TEMPERATURES FROM BLACKBODY HEMISPHERI-
CAL FLUXES
   DO 366 I=1,N
      T(I)=(EB(I)/SIGMA)**0.25
 366 CONTINUE
C
C***PRINT OUT RESULTS
C
   WRITE(NOUT,730) COMMNT
 730 FORMAT(/5X,'***',A50,/)
   WRITE(NOUT,731)
 731 FORMAT(/,1X,'View Factor Matrix:',/)
   WRITE(NOUT,733)(J,J=1,N)
 733 FORMAT(2X,'NODE',3X,20(I2,13X))
   DO 734 I=1,N
   WRITE(NOUT,732)I,(F(I,J),J=1,N)
 732 FORMAT(3X,I2,3X,20(F6.4,9X))
 734 CONTINUE
C
   WRITE(NOUT,356)
 356 FORMAT(//12X,'BB.Hem. Fluxes',3X,
   $'Temperatures',3X,'Heat Transfer Rates')
   WRITE(NOUT,360)
 360 FORMAT(16X,'(W/m**2)',11X,'(K)',15X,'(W)')
   WRITE(NOUT,361)
 361 FORMAT(12X,'------------',3X,'----------',3X,
  $'-------------------')
   WRITE(NOUT,357)(I,EB(I),T(I),Q(I),I=1,N)
 357 FORMAT(/1X,'NODE',I2,5X,E12.5,7X,F7.1,9X,E12.5)
C
   QSUM=0.
   DO 400 I=1,N
 400 QSUM=QSUM+Q(I)
   WRITE(NOUT,401)QSUM
 401 FORMAT(//1X,'QSUM=',F15.6)
C
   STOP
   END
C
C***KRONECKER DELTA FUNCTION
C
```

```
      FUNCTION DELTA(I,J)
      REAL*8 DELTA
      IF(I.EQ.J) DELTA = 1.d0
      IF(I.NE.J) DELTA = 0.d0
      RETURN
      END
C
C
C
C
C
      SUBROUTINE INVMAT(C,N)
      IMPLICIT REAL*8(A - H,O - Z)
      DIMENSION C(20,20),J(50)
C
C THIS ROUTINE TO INVERT A MATRIX BY GAUSS - JORDAN
C ELIMINATION WITH COLUMN SHIFTING TO MAXIMIZE PIVOTS
C WAS TAKEN FROM 'NUMERICAL METHODS' BY ROBERT W. HORN-
BECK.
C
      DO 125 L = 1,N
      J(L + 20) = L
  120 CONTINUE
      DO 144 L = 1,N
      CC = 0.d0
      M = L
      DO 135 K = L,N
      IF ((ABS(CC) - ABS(C(L,K))).GE.0.d0) GO TO 135
  126 M = K
      CC = C(L,K)
  135 CONTINUE
  127 IF (L.EQ.M) GO TO 138
  128 K = J(M + 20)
      J(M + 20) = J(L + 20)
      J(L + 20) = K
      DO 137 K = 1,N
      S = C(K,L)
      C(K,L) = C(K,M)
  137 C(K,M) = S
  138 C(L,L) = 1.d0
      DO 139 M = 1,N
      C(L,M) = C(L,M) / CC
  139 CONTINUE
      DO 142 M = 1,N
      IF (L.EQ.M) GO TO 142
```

```
129 CC = C(M,L)
    IF (CC.EQ.0.) GO TO 142
130 C(M,L) = 0.
    DO 141 K = 1,N
141 C(M,K) = C(M,K) - CC*C(L,K)
142 CONTINUE
144 CONTINUE
    DO 143 L = 1,N
    IF (J(L + 20).EQ.L) GO TO 143
131 M = L
132 M = M + 1
    IF (J(M + 20).EQ.L) GO TO 133
136 IF (N.GT.M) GO TO 132
133 J(M + 20) = J(L + 20)
    DO 163 K = 1,N
    CC = C(L,K)
    C(L,K) = C(M,K)
163 C(M,K) = CC
    J(L + 20) = L
143 CONTINUE
    RETURN
    END
```

APPENDIX E

DIFFUSE NONGRAY SURFACE INTERCHANGE FORTRAN PROGRAM

NONGRAY.FOR

```
C

C------ THIS IS A GENERALIZED PROGRAM FOR SOLVING DIF-
C------ FUSE / NON - GRAY RADIANT INTERCHANGE PROBLEMS.
C------ THE BOUNDARY CONDITION FOR EACH NODE MUST BE
C------ PRESCRIBED TEMPERATURE. NODAL PROPERTIES ARE
C------ SPECIFIED IN THE USER - DEFINED INPUT FILE
C------ ('NONGRAY.INP'). ONLY THE UPPER RIGHT TRIANGLE
C------ OF THE VIEW FACTOR MATRIX NEEDS TO BE SPECI-
C------ FIED. THE BOTTOM LEFT TRIANGLE WILL BE CALCU-
C------ LATED FROM RECIPROCITY AND THE DIAGONAL WILL BE
C------ CALCULATED FROM SUMMATION. OUTPUT IS GIVEN IN
C------ FILE 'NONGRAY.OUT'. INPUT FILE FORMAT IS GIVEN
C------ IN TABLE 3 - 3.
C------

C
      PROGRAM DNONGRAY
      IMPLICIT REAL*8 (A - H,O - Z)
      REAL LAMBDA(20)
      CHARACTER*50 COMMNT
      DIMENSION F(20,20),E(20,20),R(20,20),B(1),Q(20,20),
     1QTOT(20),T(20),EB(20),PSI(20,20),RLAM(20,20),
```

```
      2ENUMB(20),A(20)DATA PI/3.1415927d0/
      DATA SIGMA/5.67d-8/
      DATA C2/14388.d0/    !μmK
      OPEN(UNIT=13,FILE='NONGRAY.INP',STATUS='OLD')
C
C***SET OUTPUT DEVICE
C
      NOUT=11
      OPEN(UNIT=11,FILE='NONGRAY.OUT',STATUS='NEW')
C READ COMMENT STATEMENT
      READ(13,*)
      READ(13,5) COMMNT
    5FORMAT(1X,A50)
C READ IN # OF NODES
      READ(13,*)
      READ(13,*)N
C***SET UP PRESCRIBED TEMP MATRIX
C***READ IN A (AREA) AND T (PRESCRIBED T)
      DO 105 I=1,N
      T(I)=0.d0  !initialize all temperatures to zero
  105 CONTINUE
C
      READ(13,*)
      DO 627 JJ=1,N
  626 READ (13,*)I,ANUMB,TNUMB
      A(I)=ANUMB
      T(I)=TNUMB
  627 CONTINUE
C*** CALCULATE HEMISPHERICAL BLACK-BODY FLUX
      DO 343 I=1,N
  343 EB(I)=SIGMA*T(I)**4
C
C***READ IN NUMBER OF SPECTRAL BANDS
      READ(13,*)
      READ(13,*) NUME
      DO 11, I=1,N
      DO 11, J=1, NUME
        E(I,J)=0.d0  !initialize values of emissivity
        Q(I,J)=0.d0  !and heat transfer rate to zero
  11  CONTINUE
C***READ BOUNDARIES OF SPECTRAL REGIONS
      READ(13,*)
      READ(13,*)(LAMBDA(J),J=1,NUME-1)
C
      DO 10 K=1,3
```

```
 10 READ(13,*)
C***READ IN EMISSIVITY VALUES FOR EACH NODE
C***TO AVOID ZERO DIVIDE ERROR, E IS SET TO 0.9999 FOR
C***BLACK SURFACES AND 0.0001 FOR PERFECT REFLECTORS
   DO 15 K=1,N
    READ(13,*)I,(ENUMB(J),J=1,NUME)
    DO 16 J=1, NUME
     IF(ENUMB(J).GT.0.9999d0) THEN
         E(I,J)=0.9999d0
   ELSE IF (ENUMB(J).LT.0.0001d0) THEN
         E(I,J)=0.0001d0
     ELSE
         E(I,J)=ENUMB(J)
     END IF
 16 CONTINUE
 15 CONTINUE
C
C***SET UP REFLECTIVITY VECTORS.
C
   DO 50 I=1,N
   DO 51, J=1, NUME
 51   R(I,J)=1.d0-E(I,J)
 50 CONTINUE
C
C
C***SET UP VIEW FACTOR MATRIX-F(I,J)
C
   DO 25 I=1,N
   DO 25 J=1,N
 25F(I,J)=0.d0
C
   READ(13,*)
   DO 26 I=1,N-1
     READ(13,*)(F(I,J), J=I+1,N)
 26 CONTINUE
C
C***CLEAN UP F(I,J) TO SATISFY RECIPROCITY EXACTLY--
C***IF TOP RIGHT TRIANGLE OF F(I,J) HAS BEEN SPECIFIED
C***AND BOTTOM LEFT SET TO ZERO, BOTTOM LEFT TRIANGLE
C***WILL BE SET ACCORDING TO RECIPROCITY
C
   DO 525 I=1,N-1
   IP1=I+1
   DO 525 J=IP1,N
   F(J,I)=A(I)*F(I,J)/A(J)
   525 CONTINUE
```

```
C
C***SET DIAGONAL OF F(I,J) TO SATISFY SUMMATION
     EXACTLY
C
    DO 540 I=1,N
    FTEMP=0.d0
    DO 530 J=1,N
    IF (I.EQ.J) GOTO 530
    FTEMP=FTEMP+F(I,J)
 530 CONTINUE
    F(I,I)=1.d0-FTEMP
    IF(F(I,I).LT.0.D0)WRITE(NOUT,555)I,I,F(I,I)
 555 FORMAT(/1X,'*** Warning: F(',I2,',',I2,')=',
      F15.6)
 540 CONTINUE
C
C******** CALCULATE Q(I)'s FOR EACH SPECTRAL BAND
********
C
    DO 1000 K=1,NUME
C
C***SET UP PSI, RLAM, AND PHI MATRICES
C
    DO 110 J=1,N
    DO 110 I=1,N
      PSI(I,J)=(DELTA(I,J)-R(I,K)*F(I,J))/E(I,K)
 110 CONTINUE
    CALL INVMAT(PSI,N)
C
    DO 101 I=1,N
    DO 101 J=1,N
 101 RLAM(I,J)=(DELTA(I,J)-PSI(I,J))*E(I,K)/R(I,K)
C
C***CALCULATE UNKNOWN Q'S
C***USE FRACTIONAL FUNCTION TO WEIGHT EACH VALUE
C***OF PARTIAL HEAT TRANSFER FOR THE VARIOUS
C***SPECTRAL BANDS
C
    DO 351 I=1,N
    DO 352 J=1,N
      IF(K.EQ.1) THEN
        V=C2/LAMBDA(K)/T(J)
        Q(I,K)=Q(I,K)+RLAM(I,J)*EB(J)*A(I)*FRACT(V)
      ELSE IF(K.EQ.NUME) THEN
        V=C2/LAMBDA(K-1)/T(J)
        Q(I,K)=Q(I,K)+RLAM(I,J)*EB(J)*A(I)*
        (1.-FRACT(V))
```

```
      ELSE
        V2 = C2 / LAMBDA(K) / T(J)
        V1 = C2 / LAMBDA(K - 1) / T(J)
        Q(I,K) = Q(I,K) + RLAM(I,J)*EB(J)*A(I)*(FRACT(V2) -
FRACT(V1))
      END IF
 352 CONTINUE
 351 CONTINUE
C
 1000 CONTINUE
C
C
C***CALCULATE TOTAL NET HEAT TRANSFER FOR EACH NODE
C
   DO 1048 I = 1,N
 1048 QTOT(I) = 0.
   DO 1050 I = 1,N
C
   DO 1060 K = 1,NUME
    QTOT(I) = QTOT(I) + Q(I,K)
 1060 CONTINUE
 1050 CONTINUE
C
C
C*****************************************************
C*** PRINT OUT RESULTS
C
   WRITE(NOUT,730) COMMNT
 730 FORMAT( / 5X,'***',A50,//,60('-'))
   WRITE(NOUT,731)
 731 FORMAT(1X,'View Factor Matrix: ',/ )
   WRITE(NOUT,733) (J,J = 1,N)
 733 FORMAT(2X,'NODE',3X,20(I2,13X))
   DO 734 I = 1,N
   WRITE(NOUT,732)I,(F(I,J),J = 1,N)
 732 FORMAT(3X,I2,3X,20(F6.4,9X))
 734 CONTINUE
C
C***PRINT OUT SPECTRAL HEAT TRANSFER RATES
   WRITE(NOUT,201)
 201 FORMAT(//,60('-'),/,20X,'SPECTRAL RESULTS',
     /20X,16('-'))
   WRITE(NOUT,200)
 200 FORMAT( / 1X,'LAMBDA(min)',5X,'LAMBDA(max)',
     6X,'NODE',
```

```
      $7X,'PARTIAL Q',/,2X,
      $'(micron)',7x,'(micron)',6X,'NUMBER',9X,'(W)',/ 1X,
      $2('-----------',5X),'------',6X,'---------')
      DO 205 K=1,NUME
      IF(K.EQ.1) THEN
         RLAMO=0.
         WRITE(NOUT,210)RLAMO,LAMBDA(K),Q(1,K)
  210 FORMAT(1X,F9.4,7X,F9.4,10X,'1',6X,E12.5)
      ELSE IF(K.EQ.NUME) THEN
         WRITE(NOUT,215)LAMBDA(K-1),Q(1,K)
  215 FORMAT(1X,F9.4,8X,'infinity',10X,'1',6X,E12.5)
      ELSE
        WRITE(NOUT,210)LAMBDA(K-1),LAMBDA(K),Q(1,K)
      END IF
      WRITE(NOUT,220)(I,Q(I,K),I=2,N)
  220 FORMAT(20(35X,I2,6X,E12.5,/))
  205 CONTINUE
C
C***PRINT OUT TOTAL HEAT TRANSFER RATES
      WRITE(NOUT,300)
  300 FORMAT(/,60('-'),/,21X,'TOTAL RESULTS',/ 21X,
         13('-'))
      WRITE(NOUT,356)
  356FORMAT(/ 12X,'BB.Hem.Fluxes',3X,'Temperatures',3X,
     $'Heat Transfer Rates')
      WRITE(NOUT,360)
  360 FORMAT(16X,'(W/m**2)',11X,'(K)',15X,'(W)')
      WRITE(NOUT,361)
  361 FORMAT(12X,'---------------',3X,
     $'------------',3X,'-------------------')
      WRITE(NOUT,357)(I,EB(I),T(I),QTOT(I),I=1,N)
  357 FORMAT(/ 1X,'NODE',I2,5X,E12.5,7X,F7.1,9X,E12.5)
C
      QSUM=0.
      DO 400 I=1,N
  400 QSUM=QSUM+QTOT(I)
      WRITE(NOUT,401)QSUM
  401 FORMAT(// 1X,'Qsum=',F15.6)
C
      STOP
      END
C
C***KRONECKER DELTA FUNCTION
C
      FUNCTION DELTA(I,J)
```

```
      REAL*8 DELTA
      IF(I.EQ.J)DELTA = 1.d0
      IF(I.NE.J)DELTA = 0.d0
      RETURN
      END
C
C
      function fract(v)
      implicit real*8 (a - h,o - z)
      data pi / 3.141592654d0 /
      if(v.ge.2.0d0) then
      fract = 0.
      do 5 m = 1,100
      f1 = (dexp( - m*v) / m**4)*(((m*v + 3.0d0)*m*v +
          6.0d0)*m*v + 16.0d0)
      fract = fract + f1
      if(dabs(f1).1e.1.d - 7) goto 10
    5 continue
   10fract = fract*(15.d0 / pi**4)
      else
      fract = 1.0d0 - (15.d0 / pi**4)*v**3*(1.d0 / 3.d0 -
      v / 8.d0 + v**2 / 6.d1 - v**4 / 5040.d0 + v**6 /
      272160.d0 - v**8 / 13305600.d0)
      endif
      return
      end
C
      SUBROUTINE INVMAT(C,N)
      IMPLICIT REAL*8 (A - H,O - Z)
      DIMENSION C(20,20),J(50)
C
C THIS ROUTINE TO INVERT A MATRIX BY GAUSS - JORDAN
C ELIMINATION WITH COLUMN SHIFTING TO MAXIMIZE PIVOTS
C WAS TAKEN FROM 'NUMERICAL METHODS' BY
C ROBERT W. HORNBECK.
C
      DO 125 L = 1,N
      J(L + 20) = L
  125 CONTINUE
      DO 144 L = 1,N
      CC = 0.D0
      M = L
      DO 135 K = L,N
      IF ((ABS(CC) - ABS(C(L,K))).GE.0.D0) GO TO 135
  126 M = K
      CC = C(L,K)
```

```
135 CONTINUE
127 IF (L.EQ.M) GO TO 138
128 K = J(M + 20)
    J(M + 20) = J(L + 20)
    J(L + 20) = K
    DO 137 K = 1,N
    S = C(K,L)
    C(K,L) = C(K,M)
137 C(K,M) = S
138 C(L,L) = 1.d0
    DO 139 M = 1,N
    C(L,M) = C(L,M) / CC
139 CONTINUE
    DO 142 M = 1,N
    IF (L.EQ.M) GO TO 142
129 CC = C(M,L)
    IF (CC.EQ.0.d0) GO TO 142
130 C(M,L) = 0.
    DO 141 K = 1,N
141 C(M,K) = C(M,K) - CC*C(L,K)
142 CONTINUE
144 CONTINUE
    DO 143 L = 1,N
    IF (J(L + 20).EQ.L) GO TO 143
131 M = L
132 M = M + 1
    IF (J(M + 20).EQ.L) GO TO 133
136 IF (N.GT.M) GO TO 132
133 J(M + 20) = J(L + 20)
    DO 163 K = 1,N
    CC = C(L,K)
    C(L,K) = C(M,K)
163 C(M,K) = CC
    J(L + 20) = L
143 CONTINUE
    RETURN
    END
```

APPENDIX F

EXPONENTIAL WIDE BAND MODEL (H_2O 2.7 μm)

```
c program for calculating h2o 2.7 micron bandwidth us-
c  ing exp. w.b. model
c
     real nsum
     dimension eta(3),delta(3,3),g(3),vok(3),
     1 alpha0(3),alpha(3),gamma(3),delt(3)
c*******************************************
     2 write(9,3)
     3 format ( / 1x,'input T(K),Pa(atm),P(atm),Path(m)')
      read(9,*) t,pa,p,path
      if (t.lt.0.) stop
c*******************************************
     m = 3
     data (delta(1,i),i = 1,3) / 0.,2.,0./
     data (delta(2,i),i = 1,3) / 1.,0.,0./
     data (delta(3,i),i = 1,3) / 0.,0.,1./
     data (alpha0(i),i = 1,3) / 0.19,2.3,22.4 /!m**2 / (g cm)
     data (eta(i),i = 1,3) / 3652.,1595.,3756./
     data (g(i),i = 1,3) / 1.,1.,1./
     t0 = 100    !K
     eta0 = 3760.
     b = 8.6*(t0 / t)**.5 + .5
     rn = 1.
```

```
      gamma0 = 0.13219
      w0 = 60.    !1 / cm
c*******************************************
      hck = 1.4388       !cm K
      inf = 10
      wmol = 18.
      p0 = 1.    !atm
      x = path*pa*101.*1000.*wmol / 8.314 / t      !g / m**2
      pe = (p / p0 + (pa / p0)*(b - 1.))**rn
      w = w0*sqrt(t / t0)
       write(9,7)x,pe,w
      7 format( / 1x,'X = ',f10.4,3x,'(g / m**2)',2x,
        1'pe = ',f8.4,3x,'w = ',f8.4,3x,'(1 / cm)')
      do 444 i = 1,3
c
c determine psi and phi functions at t and t0
c
      do 222 j = 1,3
      delt(j) = delta(i,j)
      if (delt(j).ge.0.) then
      vok(j) = 0.
      else
      vok(j) = -delt(j)
      endif
 222 continue
      call psi(t,psit,hck,m,inf,eta,delt,g,vok)
      call psi(t0,psit0,hck,m,inf,eta,delt,g,vok)
      call phi(t,phit,hck,m,inf,eta,delt,g,vok)
      call phi(t0,phit0,hck,m,inf,eta,delt,g,vok)
c
c calculate alpha(t) and gamma(t)
c
      nsum = 0.
      dsum = 0.
      do 10 k = 1,m
      uk = hck*eta(k) / t
      uok = hck*eta(k) / t0
      nsum = nsum + uk*delt(k)
      dsum = dsum + uok*delt(k)
  10 continue
      nsum = (1.- exp( - nsum))*psit
      dsum = (1.- exp( - dsum))*psit0
      alpha(i) = alpha0(i)*nsum / dsum
      gamma(i) = gamma0*(phit / phit0)*(t0 / t)**0.5
 444 continue
```

```
    alph = alpha(1) + alpha(2) + alpha(3)
    gamm = 0.
    do 555 i = 1,3
 555 gamm = gamm + sqrt(gamma(i)*alpha(i))
    gamm = gamm*gamm / alph
    write (9,12) alpha,gamm
  12 format( / 1x,'alpha = ',e9.3,3x,'(m**2 / g
      cm)',3x,'gamma = ',e12.4)
c
c calculate u and beta
c
    u = x*alph / w
    beta = gamm*pe
    write(9,14) u,beta
      14 format ( / 1x,'u = ',f8.3,3x,'beta = ',f8.3)
    if (beta.lt.1.) then
     if (u.le.beta) then
      astar = u
         em = 1.- u / astar
    elseif (u.le.1./ beta) then
      astar = (2.*sqrt(beta*u) - beta)
         em = 1.- sqrt(u) / astar
    else
      astar = (log(beta*u) + 2.- beta)
         em = 1.- 1./ astar
    endif
    else
     if (u.le.1.) then
       astar = u
         em = 1.- u / astar
    else
      astar = (log(u) + 1.)
         em = 1.- 1./ astar
    endif
    endif
    if (em.lt.0.1) em = 0.1
    abar = astar*w
    deleta = abar / em
    etaup = eta0 + deleta*0.5
    etalo = eta0 - deleta*0.5
    write (9,33) astar,abar,deleta,etaup,etalo,em
      33 format( / 1x,'a* = ',f6.4,2x,'a = ',f8.3,2x,'(1 /
      cm)',2x,'deleta = ',f5.0,2x,'etaup = ',f5.0,2x,
      'etal0 = ',f5.0,2x'em = ',f4.3)
    pause
```

```
      goto 2
      end
c
      subroutine psi(t,psit,hck,m,inf,eta,delta,g,vok)
      external fact
      real nsum,num
      dimension eta(3),delta(3),g(3),vok(3)
      psin = 1.
      psid = 1.
      do 30 k = 1,m
      uk = hck*eta(k) / t
      nsum = 0.
      dsum = 0.
      vk = vok(k)
      do 10 i = 1,inf
      num = fact(vk + g(k) + abs(delta(k)) - 1.) / (fact(g(k) -
     1.)*fact(vk))
      num = num*exp( - uk*vk)
      nsum = nsum + num
      vk = vk + 1.
   10 continue
      vk = 0.
      do 20 i = 1,inf
      den = fact(vk + g(k) - 1.) / (fact(g(k) - 1.)*fact(vk))
      den = den*exp( - uk*vk)
      dsum = dsum + den
      vk = vk + 1.
   20 continue
c
      psin = psin*nsum
      psid = psid*dsum
   30 continue
c
      psit = psin / psid
      return
      end
c
      subroutine phi(t,phit,hck,m,inf,eta,delta,g,vok)
      external fact
      real nsum,num
      dimension eta(3),delta(3),g(3),vok(3)
      phin = 1.
      phid = 1.
      do 30 k = 1,m
      uk = hck*eta(k) / t
```

```
      nsum = 0.
      dsum = 0.
      vk = vok(k)
      do 10 i = 1,inf
      num = fact(vk + g(k) + abs(delta(k)) - 1.) / (fact(g(k) -
        1.)*fact(vk))
      num = (num*exp( - uk*vk))**0.5
      nsum = nsum + num
      vk = vk + 1.
   10 continue
c
      vk = vok(k)
      do 20 i = 1,inf
      den = fact(vk + g(k) + abs(delta(k)) - 1.) / (fact(g(k) -
        1.)*fact(vk))
      den = den*exp( - uk*vk)
      dsum = dsum + den
      vk = vk + 1.
   20 continue
c
      phin = phin*nsum
      phid = phid*dsum
   30 continue
c
      phit = phin*phin / phid
      return
      end
c
      function fact(y)
      fact = 1.
      n = ifix(y)
      do 10 i = 1,n
      fact = fact*float(i)
   10 continue
return
end
```

APPENDIX G

EXPONENTIAL INTEGRAL FUNCTION

$$E_n(t) = \int_0^1 \mu^{n-2} \exp\left(\frac{-t}{\mu}\right) d\mu, \qquad t > 0$$

$$\frac{dE_n(t)}{dt} = -E_{n-1}(t), \qquad n > 2$$

$$\frac{dE_1(t)}{dt} = -\frac{1}{t}\exp(-t)$$

t	$E_1(t)$	$E_2(t)$	$E_3(t)$
0.	∞	1.000	0.5000
0.01	4.0379	0.9497	0.4903
0.02	3.3547	0.9131	0.4810
0.03	2.9591	0.8817	0.4720
0.04	2.6813	0.8535	0.4633
0.05	2.4679	0.8278	0.4549
0.06	2.2953	0.8040	0.4468
0.07	2.1508	0.7818	0.4388
0.08	2.0269	0.7610	0.4311
0.09	1.9187	0.7412	0.4236
0.10	1.8229	0.7225	0.4163
0.20	1.2227	0.5742	0.3519
0.30	0.9057	0.4691	0.3000
0.40	0.7024	0.3894	0.2573
0.50	0.5598	0.3266	0.2216

t	$E_1(t)$	$E_2(t)$	$E_3(t)$
0.60	0.4544	0.2762	0.1916
0.70	0.3738	0.2349	0.1661
0.80	0.3106	0.2009	0.1443
0.90	0.2602	0.1724	0.1257
1.00	0.2194	0.1485	0.1097
1.25	0.1464	0.1035	0.0786
1.50	0.1000	0.0731	0.0567
1.75	0.0695	0.0522	0.0412
2.00	0.0489	0.0375	0.0301
2.25	0.0348	0.0272	0.0221
2.50	0.0249	0.0198	0.0163
2.75	0.0180	0.0145	0.0120
3.00	0.0130	0.0106	0.0089
3.25	0.0095	0.0078	0.0066
3.50	0.0070	0.0058	0.0049

APPENDIX H

INTERNAL FRACTIONAL FUNCTION

$$f_i(\lambda T) = \frac{1}{4\sigma T^3}\int_0^\lambda \frac{\partial e_{b_\lambda}}{\partial T}\,d\lambda = \frac{15}{4\pi^4}\int_x^\infty \frac{x^4 e^x}{(e^x - 1)^2}\,dx, \qquad x = \frac{C_2}{\lambda T}$$

$\lambda T(\mu\text{m K})$	f_i	$\lambda T(\mu\text{m K})$	f_i
1000	.0012	2800	.3862
1100	.0033	2900	.4150
1200	.0071	3000	.4429
1300	.0133	3100	.4697
1400	.0225	3200	.4955
1500	.0351	3300	.5201
1600	.0510	3400	.5437
1700	.0702	3500	.5661
1800	.0924	3600	.5875
1900	.1172	3700	.6077
2000	.1442	3800	.6270
2100	.1729	3900	.6452
2200	.2027	4000	.6624
2300	.2334	4100	.6788
2400	.2644	4200	.6942
2500	.2955	4300	.7088
2600	.3263	4400	.7226
2700	.3566		

$\lambda T(\mu m\ K)$	f_i	$\lambda T(\mu m\ K)$	f_i
4500	.7357	8500	.9458
4600	.7481	8600	.9475
4700	.7597	8700	.9492
4800	.7708	8800	.9507
4900	.7812	8900	.9522
5000	.7911	9000	.9537
5100	.8004	9100	.9551
5200	.8093	9200	.9564
5300	.8176	9300	.9577
5400	.8255	9400	.9589
5500	.8330	9500	.9601
5600	.8401	9600	.9613
5700	.8469	9700	.9624
5800	.8533	9800	.9634
5900	.8593	9900	.9645
6000	.8650	10000	.9654
6100	.8705	11000	.9736
6200	.8757	12000	.9794
6300	.8806	13000	.9836
6400	.8853	14000	.9868
6500	.8897	15000	.9892
6600	.8939	16000	.9910
6700	.8980	17000	.9925
6800	.9018	18000	.9936
6900	.9055	19000	.9946
7000	.9089	20000	.9953
7100	.9123	21000	.9959
7200	.9154	22000	.9965
7300	.9184	23000	.9969
7400	.9213	24000	.9973
7500	.9241	25000	.9976
7600	.9267	26000	.9978
7700	.9292	27000	.9981
7800	.9316	28000	.9983
7900	.9339	29000	.9984
8000	.9361	30000	.9986
8100	.9382	31000	.9987
8200	.9403	32000	.9988
8300	.9422	33000	.9989
8400	.9440	34000	.9990

λT(μm K)	f_i	λT(μm K)	f_i
35000	.9991	43000	.9995
36000	.9992	44000	.9995
37000	.9992		
38000	.9993		
39000	.9993	45000	.9996
		46000	.9996
40000	.9994	47000	.9996
41000	.9994	48000	.9996
42000	.9995	49000	.9997

INDEX